Methods in Enzymology

Volume 68

RECOMBINANT DNA

METHODS IN ENZYMOLOGY

EDITORS-IN-CHIEF

Sidney P. Colowick   Nathan O. Kaplan

Methods in Enzymology

Volume 68

Recombinant DNA

EDITED BY

Ray Wu

DIVISION OF BIOLOGICAL SCIENCES
SECTION OF BIOCHEMISTRY, MOLECULAR AND CELL BIOLOGY
CORNELL UNIVERSITY
ITHACA, NEW YORK

1979

ACADEMIC PRESS

A Subsidiary of Harcourt Brace Jovanovich, Publishers

New York London Toronto Sydney San Francisco

ACADEMIC PRESS, INC.
111 Fifth Avenue, New York, New York 10003

United Kingdom Edition published by
ACADEMIC PRESS, INC. (LONDON) LTD.
24/28 Oval Road, London NW1 7DX

Library of Congress Cataloging in Publication Data

Main entry under title:

Recombinant DNA.

(Methods in enzymology ; v. 68)
Bibliography: p.
Includes index.
1. Recombinant DNA. I. Wu, Ray.
II. Series.
QP601.M49 vol. 68 [QH442] 574.1'925'08s
ISBN 0-12-181968-X [574.8'732] 79-26584

PRINTED IN THE UNITED STATES OF AMERICA

81 82 9 8 7 6 5 4 3 2

Table of Contents

Section I. Introduction

Section II. Enzymes Used in Recombinant DNA Research

Section III. Synthesis, Isolation, and Purification of DNA

Section IV. Vehicles and Hosts for the Cloning of Recombinant DNA

Section V. Screening and Selection of Cloned DNA

Section VI. Detection and Analysis of Expression of Cloned Genes

Contributors to Volume 68

Article numbers are in parentheses following the names of contributors.
Affiliations listed are current.

DAVID ANDERSON (30), *Genex Corporation, Rockville, Maryland 20852*

JAMES C. ALWINE (15), *Laboratory of Molecular Virology, National Cancer Institute, National Institutes of Health, Bethesda, Maryland 20014*

S. L. AUCKERMAN (38), *Department of Biology, The Johns Hopkins University, Baltimore, Maryland 21218*

KEITH BACKMAN (16), *Department of Biology, Massachusetts Institute of Technology, Cambridge, Massachusetts 02139*

C. P. BAHL (7), *Cetus Corporation, Berkeley, California 94710*

RAMAMOORTHY BELAGAJE (8), *Lilly Research Laboratories, Indianapolis, Indiana 46206*

HANS-ULRICH BERNARD (35), *Department of Biology, University of California, San Diego, La Jolla, California 92093*

DALE BLANK (33), *Rosenstiel Basic Medical Sciences Research Center and Department of Biology, Brandeis University, Waltham, Massachusetts 02154*

FRANCISCO BOLIVAR (16), *Deparamento Biologia Molecular, Instituto de Investigaciones Biomedicas, Universidad Nacional Autonoma de Mexico, Mexico 20, D.F., Mexico Apdo Postal 70228*

ROLAND BROUSSEAU (6), *Division of Biological Sciences, National Research Council of Canada, Ottawa K1A OR6, Canada*

EUGENE L. BROWN (8), *Synthex Research, Palo Alto, California 94304*

DOUGLAS BRUTLAG (3), *Department of Biochemistry, Stanford University School of Medicine, Stanford, California 94305*

JOHN CARBON (27, 31), *Department of Biological Sciences, University of California, Santa Barbara, Santa Barbara, California 93106*

P. CHIU (29), *Section of Biochemistry, Molecular and Cell Biology, Cornell University, Ithaca, New York 14853*

LOUISE CLARKE (27, 31), *Department of Biological Sciences, University of California, Santa Barbara, Santa Barbara, California 93106*

STANLEY N. COHEN (32), *Departments of Genetics and Medicine, Stanford University School of Medicine, Stanford, California 94305*

JOHN COLLINS (2), *Gesellschaft für Biotechnologische Forschung mbH. Mascheroder Weg 1, D-3300 Braunschweig-Stöckheim, West Germany*

NICHOLAS R. COZZARELLI (4), *Departments of Biochemistry and Biophysics and Theoretical Biology, University of Chicago, Chicago, Illinois 60637*

J. L. CULLETON (38), *Department of Biology, The Johns Hopkins University, Baltimore, Maryland 21218*

R. P. DOTTIN (38), *Department of Biology, The Johns Hopkins University, Baltimore, Maryland 21218*

L. ENQUIST (18), *Laboratory of Molecular Virology, National Cancer Institute, National Institutes of Health, Bethesda, Maryland 20014*

HENRY A. ERLICH (32), *Department of Medicine, Stanford University School of Medicine, Stanford, California 94305*

KAREN FAIIRNER (33), *Rosenstiel Basic Medical Sciences Research Center and Department of Biology, Brandeis University, Waltham, Massachusetts 02154*

G. C. FAREED (24), *Department of Microbiology and Immunology, Molecular Biology Institute, University of California, Los Angeles, Los Angeles, California 90024*

DAVID FIGURSKI (17), *Department of Microbiology, College of Physicians and Surgeons, Columbia University, New York, New York 10032*

S. G. FISCHER (11), *Department of Biological Sciences, State University of New York at Albany, Albany, New York 12222*

B. R. FISHEL (38), *Department of Biology, The Johns Hopkins University, Baltimore, Maryland 21218*

MICHAEL L. GOLDBERG (14), *Abteilung Zelliologie, Biozentrium Der Universität Basel, CH-4056 Basel, Switzerland*

HOWARD M. GOODMAN (5), *Howard Hughes Medical Institute Laboratory and the Department of Biochemistry and Biophysics, University of California, San Francisco, California, 94143*

MICHAEL GRUNSTEIN (25), *Department of Biology, University of California, Los Angeles, Los Angeles, California 90024*

DONALD R. HELINSKI (17, 35), *Department of Biology, University of California, San Diego, La Jolla, California 92093*

LYNNA HEREFORD (33), *Rosenstiel Basic Medical Sciences Research Center and Department of Biology, Brandeis University, Waltham, Massachusetts 02154*

N. PATRICK HIGGINS (4), *Department of Biochemistry, University of Wyoming, Laramie, Wyoming*

RONALD HITZEMAN (31), *Department of Biological Sciences, University of California, Santa Barbara, Santa Barbara, California 93106*

BARBARA HOHN (19), *Friedrich Miescher Institut, CH-4002 Basel, Switzerland*

JANICE P. HOLLAND (28), *Department of Biochemistry, University of Connecticut Health Center, Farmington, Connecticut 06032*

MICHAEL J. HOLLAND (28), *Department of Biochemistry, University of Connecticut Health Center, Farmington, Connecticut 06032*

HANSEN M. HSIUNG (6), *Division of Biological Sciences, National Research Council of Canada, Ottawa K1A OR6, Canada*

KIMBERLY A. JACKSON (28), *Department of Biochemistry, University of Connecticut Health Center, Farmington, Connecticut 06032*

MICHAEL KAHN (17), *Department of Bacteriology and Public Health, Washington State University, Pullman, Washington 99164*

KATHLEEN M. KEGGINS (23), *Department of Biological Sciences, University of Maryland, Baltimore County, Catonsville, Maryland 21228*

DAVID J. KEMP (15), *The Walker and Eliza Hall Institute of Medical Research, Post Office, Royal Melbowne Hospital, Victoria 3050, Australia*

H. GOBIND KHORANA (8), *Departments of Biology and Chemistry, Massachusetts Institute of Technology, Cambridge, Massachusetts 02139*

ROBERTO KOLTER (17), *Department of Biology, University of California, San Diego, La Jolla, California 92093*

L. F. LAU (7), *Section of Biochemistry, Molecular and Cell Biology, Cornell University, Ithaca, New York 14853*

GAIL D. LAUER (34), *The Biological Laboratories, Harvard University, Cambridge, Massachusetts 02138*

LEONARD S. LERMAN (11), *Department of Biological Sciences, State University of New York at Albany, Albany, New York 12222*

RICHARD P. LIFTON (14), *Department of Biochemistry, Stanford University School of Medicine, Stanford, California 94305*

JOHN LIS (10), *Section of Biochemistry, Molecular and Cell Biology, Cornell University, Ithaca, New York 14853*

SHIRLEY LONGACRE (12), *Parasitologie Experimentale, Institut Pasteur, 75724 Paris, Cedex 15, France*

Paul S. Lovett (23), *Department of Biological Sciences, University of Maryland, Baltimore County, Catonsville, Maryland 21228*

Hugh O. McDevitt (32), *Departments of Medicine and Medical Microbiology, Stanford University School of Medicine, Stanford, California 94305*

Raymond J. MacDonald (5), *Howard Hughes Medical Institute Laboratory and the Department of Biochemistry and Biophysics, University of California, San Francisco, San Francisco, California 94143*

Bernard Mach (12), *Department of Microbiology, University of Geneva, CH 1205 Geneva, Switzerland*

R. E. Manrow (38), *Department of Biology, The Johns Hopkins University, Baltimore, Maryland 21218*

Richard Meyer (17), *Department of Microbiology, University of Texas, Austin, Texas 78712*

D. A. Morrison (21), *Department of Biological Sciences, University of Illinois, Chicago Circle, Chicago, Illinois 60680*

John F. Morrow (1), *Department of Microbiology, The Johns Hopkins School of Medicine, Baltimore, Maryland 21205*

S. A. Narang (6, 7), *Division of Biological Sciences, National Research Council of Canada, Ottawa K1A OR6, Canada*

Timothy Nelson (3), *Department of Biochemistry, Stanford University School of Medicine, Stanford, California 94305*

Barbara A. Parker (15), *Department of Biochemistry, Stanford University School of Medicine, Stanford, California 94305*

Barry Polisky (37), *Department of Biology, Indiana University, Bloomington, Indiana 47401*

A. F. Purchio (24), *Department of Microbiology and Immunology, Molecular Biology Institute, University of California, Los Angeles, Los Angeles, California 90024*

John Reeve (36), *Department of Microbiology, Ohio State University, Columbus, Ohio 43210*

Jakob Reiser (15), *Department of Biochemistry, Stanford University School of Medicine, Stanford, California 94305*

Eric Remaut (17), *Laboratorium voor Moleculaire Biologie, Bijksuniversiteit Gent, B-9000 Gent, Belgium*

Jaime Renart (15), *Instituto de Enzimologia del C.S.I.C., Facultad de Medicina de la Universidad Autonoma, Arzobispo Morcillo s/n. Madrid-34, Spain*

Robert Ricciardi (33), *Department of Biological Chemistry, Harvard Medical School, Boston, Massachusetts 02115*

Richard J. Roberts (2), *Cold Spring Harbor Laboratory, Cold Spring Harbor, New York 11724*

Bryan Roberts (33), *Rosentiel Basic Medical Sciences Research Center and Department of Biology, Brandeis University, Waltham, Massachusetts 02154*

Thomas M. Roberts (34), *Department of Biochemistry and Molecular Biology, Harvard University, Cambridge, Massachusetts 02138*

Michael Rosbash (33), *Rosenstiel Basic Medical Sciences Research Center and Department of Biology, Brandeis University, Waltham, Massachusetts 02154*

R. J. Rothstein (7), *Department of Microbiology, New Jersey School of Medicine, Newark, New Jersey 07103*

Stephanie Ruby (33), *Department of Biological Chemistry, Harvard Medical School, Boston, Massachusetts 02115*

Michael J. Ryan (8), *Microbiological Sciences, Schering Corporation, Bloomfield, New Jersey 07003*

A. Sen (13), *Meloy Laboratories Inc., Springfield, Virginia 22151*

LUCILLE SHAPIRO (30), *Department of Molecular Biology, Albert Einstein College of Medicine, Bronx, New York*

H. MICHAEL SHEPARD (37), *Department of Biology, Indiana University, Bloomington, Indiana 47401*

F. SHERMAN (29), *Department of Radiation Biology and Biophysics, University of Rochester, School of Medicine, Rochester, New York 14642*

M. SHOYAB (13), *Laboratory of Viral Carcinogenesis, National Cancer Institute, National Institutes of Health, Bethesda, Maryland 20014*

A. M. SKALKA (30), *Department of Cell Biology, Roche Institute of Molecular Biology, Nutley, New Jersey 07110*

EDWIN SOUTHERN (9), *M.R.C. Mammalian Genome Unit, King's Building, Edinburgh EH9 3JT, Scotland*

GEORGE R. STARK (14, 15), *Department of Biochemistry, Stanford University School of Medicine, Stanford, California 94305*

N. STERNBERG (18), *Cancer Biology Program, Frederick Cancer Research Center, Frederick, Maryland 21701*

J. I. STILES (29), *Department of Radiation Biology and Biophysics, University of Rochester, School of Medicine, Rochester, New York 14642*

M. SUZUKI (22), *Boyce Thompson Institute, Cornell University, Ithaca, New York 14853*

A. A. SZALAY (22), *Boyce Thompson Institute, Cornell University, Ithaca, New York 14853*

J. W. SZOSTAK (29), *Sidney Faber Cancer Institute, Boston, Massachusetts 02115*

CHRISTOPHER THOMAS (17), *Department of Biology, University of California, San Diego, La Jolla, California 92093*

B.-K. TYE (29), *Section of Biochemistry, Molecular and Cell Biology, Cornell University, Ithaca, New York 14853*

GEOFFREY M. WAHL (15), *Department of Biochemistry, Stanford University School of Medicine, Stanford, California 94305*

JOHN WALLIS (25), *Department of Microbiology and Immunology, Molecular Biology Institute, University of California, Los Angeles, Los Angeles, California 90024*

JEFFREY G. WILLIAMS (14), *Imperial Cancer Research Fund, Mill Hill, London NW7, England*

SAVIO L. C. WOO (26), *Howard Hughes Medical Institute Laboratory and Department of Cell Biology, Baylor College of Medicine, Texas Medical Center, Houston, Texas 77030*

JOHN WOOLFORD (33), *Rosenstiel Basic Medical Sciences Research Center and Department of Biology, Brandeis University, Waltham, Massachusetts 02154*

RAY WU (7, 10, 29), *Section of Biochemistry, Molecular and Cell Biology, Cornell University, Ithaca, New York 14853*

ROBERT C.-A. YANG (10), *Section of Biochemistry, Molecular and Cell Biology, Cornell University, Ithaca, New York 14853*

Preface

DNA is the genetic material of virtually all living organisms. The physical mapping of genes, the sequence analysis of DNA, and the identification of regulatory elements for DNA replication and transcription depend on the availability of pure specific DNA segments. The DNA of higher organisms is so complex that it is often impossible to isolate DNA molecules corresponding to a single gene in sufficient amounts for analysis at the molecular level. However, exciting new developments in recombinant DNA research make possible the isolation and amplification of specific DNA segments from almost any organism. These new developments have revolutionized our approaches in solving complex biological problems.

Recombinant DNA technology also opens up new possibilities in medicine and industry. It allows the manipulation of genes from different organisms or genes made synthetically for the large-scale production of medically and agriculturally useful products.

This volume includes a number of the specific methods employed in recombinant DNA research. Other related methods can be found in "Nucleic Acids," Volume 65, Part I, of this series.

I wish to thank the numerous authors who have contributed to this volume, as well as the very capable staff of Academic Press, for their assistance and cooperation. I also wish to extend my appreciation to Stanley Cohen and Lawrence Grossman for their advice in planning the contents of this volume.

RAY WU

METHODS IN ENZYMOLOGY

EDITED BY

Sidney P. Colowick and Nathan O. Kaplan

VANDERBILT UNIVERSITY
SCHOOL OF MEDICINE
NASHVILLE, TENNESSEE

DEPARTMENT OF CHEMISTRY
UNIVERSITY OF CALIFORNIA
AT SAN DIEGO
LA JOLLA, CALIFORNIA

METHODS IN ENZYMOLOGY

EDITORS-IN-CHIEF

Sidney P. Colowick   Nathan O. Kaplan

VOLUME VIII. Complex Carbohydrates
Edited by ELIZABETH F. NEUFELD AND VICTOR GINSBURG

VOLUME IX. Carbohydrate Metabolism
Edited by WILLIS A. WOOD

VOLUME X. Oxidation and Phosphorylation
Edited by RONALD W. ESTABROOK AND MAYNARD E. PULLMAN

VOLUME XI. Enzyme Structure
Edited by C. H. W. HIRS

VOLUME XII. Nucleic Acids (Parts A and B)
Edited by LAWRENCE GROSSMAN AND KIVIE MOLDAVE

VOLUME XIII. Citric Acid Cycle
Edited by J. M. LOWENSTEIN

VOLUME XIV. Lipids
Edited by J. M. LOWENSTEIN

VOLUME XV. Steroids and Terpenoids
Edited by RAYMOND B. CLAYTON

VOLUME XVI. Fast Reactions
Edited by KENNETH KUSTIN

VOLUME XVII. Metabolism of Amino Acids and Amines (Parts A and B)
Edited by HERBERT TABOR AND CELIA WHITE TABOR

VOLUME XVIII. Vitamins and Coenzymes (Parts A, B, and C)
Edited by DONALD B. MCCORMICK AND LEMUEL D. WRIGHT

VOLUME XIX. Proteolytic Enzymes
Edited by GERTRUDE E. PERLMANN AND LASZLO LORAND

VOLUME XXXIV. Affinity Techniques (Enzyme Purification: Part B)
Edited by WILLIAM B. JAKOBY AND MEIR WILCHEK

VOLUME XXXV. Lipids (Part B)
Edited by JOHN M. LOWENSTEIN

VOLUME XXXVI. Hormone Action (Part A: Steroid Hormones)
Edited by BERT W. O'MALLEY AND JOEL G. HARDMAN

VOLUME XXXVII. Hormone Action (Part B: Peptide Hormones)
Edited by BERT W. O'MALLEY AND JOEL G. HARDMAN

VOLUME XXXVIII. Hormone Action (Part C: Cyclic Nucleotides)
Edited by JOEL G. HARDMAN AND BERT W. O'MALLEY

VOLUME XXXIX. Hormone Action (Part D: Isolated Cells, Tissues, and Organ Systems)
Edited by JOEL G. HARDMAN AND BERT W. O'MALLEY

VOLUME XL. Hormone Action (Part E: Nuclear Structure and Function)
Edited by BERT W. O'MALLEY AND JOEL G. HARDMAN

VOLUME XLI. Carbohydrate Metabolism (Part B)
Edited by W. A. WOOD

VOLUME XLII. Carbohydrate Metabolism (Part C)
Edited by W. A. WOOD

VOLUME XLIII. Antibiotics
Edited by JOHN H. HASH

VOLUME XLIV. Immobilized Enzymes
Edited by KLAUS MOSBACH

VOLUME XLV. Proteolytic Enzymes (Part B)
Edited by LASZLO LORAND

VOLUME XLVI. Affinity Labeling
Edited by WILLIAM B. JAKOBY AND MEIR WILCHEK

VOLUME XLVII. Enzyme Structure (Part E)
Edited by C. H. W. HIRS AND SERGE N. TIMASHEFF

VOLUME XLVIII. Enzyme Structure (Part F)
Edited by C. H. W. HIRS AND SERGE N. TIMASHEFF

VOLUME XLIX. Enzyme Structure (Part G)
Edited by C. H. W. HIRS AND SERGE N. TIMASHEFF

VOLUME L. Complex Carbohydrates (Part C)
Edited by VICTOR GINSBURG

VOLUME LI. Purine and Pyrimidine Nucleotide Metabolism
Edited by PATRICIA A. HOFFEE AND MARY ELLEN JONES

VOLUME LII. Biomembranes (Part C: Biological Oxidations)
Edited by SIDNEY FLEISCHER AND LESTER PACKER

VOLUME LIII. Biomembranes (Part D: Biological Oxidations)
Edited by SIDNEY FLEISCHER AND LESTER PACKER

VOLUME LIV. Biomembranes (Part E: Biological Oxidations)
Edited by SIDNEY FLEISCHER AND LESTER PACKER

VOLUME LV. Biomembranes (Part F: Bioenergetics)
Edited by SIDNEY FLEISCHER AND LESTER PACKER

VOLUME LVI. Biomembranes (Part G: Bioenergetics)
Edited by SIDNEY FLEISCHER AND LESTER PACKER

VOLUME LVII. Bioluminescence and Chemiluminescence
Edited by MARLENE A. DELUCA

VOLUME LVIII. Cell Culture
Edited by WILLIAM B. JAKOBY AND IRA H. PASTAN

VOLUME LIX. Nucelic Acids and Protein Synthesis (Part G)
Edited by KIVIE MOLDAVE AND LAWRENCE GROSSMAN

VOLUME LX. Nucleic Acids and Protein Synthesis (Part H)
Edited by KIVIE MOLDAVE AND LAWRENCE GROSSMAN

VOLUME 61. Enzyme Structure (Part H)
Edited by C. H. W. HIRS AND SERGE N. TIMASHEFF

VOLUME 62. Vitamins and Coenzymes (Part D)
Edited by DONALD B. MCCORMICK AND LEMUEL D. WRIGHT

VOLUME 63. Enzyme Kinetics and Mechanisms (Part A)
Edited by DANIEL L. PURICH

VOLUME 64. Enzyme Kinetics and Mechanisms (Part B)
Edited by DANIEL L. PURICH

VOLUME 65. Nucleic Acids (Part I)
Edited by LAWRENCE GROSSMAN AND KIVIE MOLDAVE

VOLUME 66. Vitamins and Coenzymes (Part E)
Edited by DONALD B. MCCORMICK AND LEMEUL D. WRIGHT

VOLUME 67. Vitamins and Coenzymes (Part F) (in preparation)
Edited by DONALD B. MCCORMICK AND LEMUEL D. WRIGHT

VOLUME 68. Recombinant DNA
Edited by RAY WU

VOLUME 69. Photosynthesis and Nitrogen Fixation (Part C) (in preparation)
Edited by ANTHONY SAN PIETRO

Section I

Introduction

[1] Recombinant DNA Techniques

By JOHN F. MORROW

The recombinant DNA method consists of joining DNA molecules *in vitro* and introducing them into living cells where they replicate. Research using this method is relatively new and fast-moving. In only 6 years of recombinant DNA research, a number of significant accomplishments have been made. Two mammalian hormones have been produced in bacteria by means of synthetic DNA.[1,2] Polypeptides similar or identical to several found in eukaryotes have been synthesized in *Escherichia coli*.[3-10] These achievements promise a new, inexpensive means of large-scale production of selected peptides or proteins. Furthermore, using recombinant DNA, somatic recombination of immunoglobulin genes has been established,[11] and a large number of variable-region genes have been found.[12] Intervening sequences (introns) have been found in the DNA of eukaryotic cells.[13-16]

I would like to mention the origins of this versatile new technology before describing recent advances. The isolation of mutant *E. coli* strains unable to restrict foreign DNA (cleave it specifically and degrade it) laid

[1] K. Itakura, T. Hirose, R. Crea, A. D. Riggs, H. L. Heyneker, F. Bolivar, and H. W. Boyer, *Science* **198,** 1056 (1977).

[2] D. V. Goeddel, D. G. Kleid, F. Bolivar, H. L. Heyneker, D. G. Yansura, R. Crea, T. Hirose, A. Kraszewski, K. Itakura, and A. D. Riggs, *Proc. Natl. Acad. Sci. U.S.A.* **76,** 106 (1979).

[3] K. Struhl, J. R. Cameron, and R. W. Davis, *Proc. Natl. Acad. Sci. U.S.A.* **73,** 1471 (1976).

[4] B. Ratzkin and J. Carbon, *Proc. Natl. Acad. Sci. U.S.A.* **74,** 487 (1977).

[5] D. Vapnek, J. A. Hautala, J. W. Jacobson, N. H. Giles, and S. R. Kushner, *Proc. Natl. Acad. Sci. U.S.A.* **74,** 3508 (1977).

[6] R. C. Dickson and J. S. Markin, *Cell* **15,** 123 (1978).

[7] L. Villa-Komaroff, A. Efstratiadis, S. Broome, P. Lomedico, R. Tizard, S. P. Naber, W. L. Chick, and W. Gilbert, *Proc. Natl. Acad. Sci. U.S.A.* **75,** 3727 (1978).

[8] A. C. Y. Chang, J. H. Nunberg, R. J. Kaufman, H. A. Erlich, R. T. Schimke, and S. N. Cohen, *Nature (London)* **275,** 617 (1978).

[9] O. Mercereau-Puijalon, A. Royal, B. Cami, A. Garapin, A. Krust, F. Gannon, and P. Kourilsky, *Nature (London)* **275,** 505 (1978).

[10] T. H. Fraser and B. J. Bruce, *Proc. Natl. Acad. Sci. U.S.A.* **75,** 5936 (1978).

[11] C. Brack, M. Hirama, R. Lenhard-Schuller, and S. Tonegawa, *Cell* **15,** 1 (1978).

[12] J. G. Seidman, A. Leder, M. Nau, B. Norman, and P. Leder, *Science* **202,** 11 (1978).

[13] D. M. Glover and D. S. Hogness, *Cell* **10,** 167 (1977).

[14] R. L. White and D. S. Hogness, *Cell* **10,** 177 (1977).

[15] P. K. Wellauer and I. B. Dawid, *Cell* **10,** 193 (1977).

[16] S. M. Tilghman, D. C. Tiemeier, J. G. Seidman, B. M. Peterlin, M. Sullivan, J. V. Maizel, and P. Leder, *Proc. Natl. Acad. Sci. U.S.A.* **75,** 725 (1978).

ISBN 0-12-181968-X

part of the foundation.[17] The discovery of site-specific restriction endonucleases[18,19] also contributed (see Nathans and Smith,[20] Roberts,[21] and this volume [2], for review). Two general methods for joining DNA molecules from different sources were found.[22–24] Particularly useful was the first enzyme found to create self-complementary, cohesive termini on DNA molecules by specific cleavage at staggered sites in the two DNA strands, the *Eco*RI restriction endonuclease.[25–27] It was used in the first *in vitro* construction of recombinant molecules that subsequently replicated *in vivo*.[28]

What can be done by the recombinant DNA method? Principally three sorts of things:

1. Isolation of a desired sequence from a complex mixture of DNA molecules, such as a eukaryotic genome, and replication of it to provide milligram quantities for biochemical study.
2. Alteration of a DNA molecule. One can insert restriction endonuclease recognition sites, or other DNA segments, at random or predetermined locations. One can also delete restriction sites, or DNA segments between such sites, by techniques that permit joining any two DNA termini after their appropriate modification. Such an alteration can be helpful in determining the functions performed by various parts of a DNA sequence. This is attractive where efficient means of fine-structure genetic analysis of random mutations are lacking, as in animals and plants.
3. Synthesis in bacteria of large amounts of peptides or proteins that are of interest to science, medicine, or commerce.

Before indicating specifically the most useful methods for obtaining each of the above goals, we look at recent advances in the basic techniques. The essential ingredients of a recombinant DNA experiment are:

[17] W. B. Wood, *J. Mol. Biol.* **16,** 118 (1966).
[18] H. O. Smith and K. W. Wilcox, *J. Mol. Biol.* **51,** 379 (1970).
[19] T. J. Kelly, Jr. and H. O. Smith, *J. Mol. Biol.* **51,** 393 (1970).
[20] D. Nathans and H. O. Smith, *Annu. Rev. Biochem.* **44,** 273 (1975).
[21] R. J. Roberts, *Gene* **4,** 183 (1978).
[22] P. E. Lobban and A. D. Kaiser, *J. Mol. Biol.* **78,** 453 (1973).
[23] D. A. Jackson, R. H. Symons, and P. Berg, *Proc. Natl. Acad. Sci. U.S.A.* **69,** 2904 (1972).
[24] V. Sgaramella, J. H. van de Sande, and H. G. Khorana, *Proc. Natl. Acad. Sci. U.S.A.* **67,** 1468 (1970).
[25] J. E. Mertz and R. W. Davis, *Proc. Natl. Acad. Sci. U.S.A.* **69,** 3370 (1972).
[26] J. Hedgpeth, H. M. Goodman, and H. W. Boyer, *Proc. Natl. Acad. Sci. U.S.A.* **69,** 3448 (1972).
[27] V. Sgaramella, *Proc. Natl. Acad. Sci. U.S.A.* **69,** 3389 (1972).
[28] S. N. Cohen, A. C. Y. Chang, H. W. Boyer, and R. B. Helling, *Proc. Natl. Acad. Sci. U.S.A.* **70,** 3240 (1973).

1. A DNA vehicle (vector, replicon) which can replicate in living cells after foreign DNA is inserted into it.

2. A DNA molecule to be replicated (passenger), or a collection of them.

3. A method of joining the passenger to the vehicle.

4. A means of introducing the joined DNA molecule into a host organism in which it can replicate (DNA transformation or transfection).

5. A means of screening or genetic selection for those cells that have replicated the desired recombinant molecule. This is necessary since transformation and transfection methods are inefficient, so that most members of the host cell population have no recombinant DNA replicating in them. This selection or screening for desired recombinants provides a route to recovery of the recombinant DNA of interest in pure form.

Since a thorough review of recombinant DNA was completed in 1976,[29] I will concentrate on progress since then.

Cloning Vehicles

Plasmids

Many bacterial plasmids have been used as cloning vehicles. Currently, *E. coli* and its plasmids constitute the most versatile type of host–vector system for DNA cloning.

A number of derivatives of natural plasmids have been developed for cloning. Most of these new plasmid vehicles were made by combining DNA segments, and desirable qualities, of older vehicles (Table I). All those listed have a "relaxed" mode of replication, such that plasmid DNA accumulates to make up about one-third of the total cellular DNA when protein synthesis is inhibited by chloramphenicol or spectinomycin.[30]

pBR322 is now the most widely used plasmid for cloning of DNA. One of its virtues is that it has six different types of restriction cleavage termini at which foreign DNA can be inserted. A very detailed restriction enzyme cleavage map and DNA sequence information are also important.[31,32] The *Pst*I site in the Ap (penicillinase) gene has further advantages. If dG homopolymer tails are added to *Pst*-cleaved pBR322 DNA, and dC homopolymer tails to the DNA to be inserted, the *Pst*I sites are reconstituted in

[29] R. L. Sinsheimer, *Annu. Rev. Biochem.* **46,** 415 (1977).

[30] A. C. Y. Chang and S. N. Cohen, *J. Bacteriol.* **134,** 1141 (1978).

[31] J. G. Sutcliffe, *Proc. Natl. Acad. Sci. U.S.A.* **75,** 3737 (1978).

[32] J. G. Sutcliffe, *Nucleic Acids Res.* **5,** 2721 (1978).

TABLE I
NEW PLASMID CLONING VEHICLES[a]

Plasmid	Cleavage sites at which foreign DNA may be inserted	Markers of transformed cells	Sites at which insertion inactivates a marker	Molecular weight	Other characteristics	Reference
pACYC184	*Bam*HI, *Eco*RI, *Hin*cII (two sites), *Hin*dIII, *Sal*I	Cm^R, Tc^R	All except *Hin*cII	4.0 kb	Compatible with ColEl-derived plasmids	[b]
pBR322	*Bam*HI, *Eco*RI, *Hin*cII (two sites), *Hin*dIII, *Pst*I, *Sal*I	Ap^R, Tc^R	All except *Eco*RI	4.362 kb	Derived from pMB9 and RSF2124	[c]
pBR324	*Bam*HI, *Eco*RI, *Hin*cII (two sites), *Hin*dIII, *Sal*I, *Sma*I (*Xma*I)	Ap^R, Tc^R, *Imm*El	All	8.3 kb	Derived from pBR322 and pMB1	[d]
pBR325	*Bam*HI, *Eco*RI, *Hin*cII (two sites), *Hin*dIII, *Pst*I, *Sal*I	Cm^R, Ap^R, Tc^R	All	5.4 kb	Derived from pBR322 and PlCm^R	[d]
pJC74	*Bgl*II, *Eco*RI	Ap^R	—	16 kb	Derived from ColEl, λdrif D18, and Rldrd19; λ *in vitro* packaging can be used	[e]
pJC75-58	*Bam*HI, *Bgl*II, *Eco*RI	Ap^R	—	12 kb		[e]
pJC720	*Hin*dIII	Rif^R	—	24 kb		[e, f]
pKB158	*Bam*HI, *Bgl*II, *Eco*RI, *Hpa*I, *Pst*I, *Sal*I	Tc^R, *Imm*λ, *Imm*El	*Bam*HI, *Sal*I	6.5 kb	Derived from pMB9 and λ	[g]
pKB166	*Bgl*II, *Eco*RI, *Hae*II (two sites), *Hin*dIII	*Imm*λ, *Imm*El	*Hin*dIII	2.4 kb	Derived from pKB158	[g]

pKC16	*Bam*HI, *Eco*RI, *Hin*dIII	Ap^R, *Imm*λ	*Hin*dIII	11.4 kb	Plasmid copy number increases after temperature shift	[h]
pOPl	*Eco*RI	Ap^R	—	17.4 kb	Inserted DNA is transcribed under control of *lac* repressor plus CRP	[i]
pPCϕ1, pPCϕ2, pPCϕ3	*Eco*RI	Ap^R	—	4.4 kb	DNA is inserted into *lacZ* gene in any of the three translation frames	[j]
RSF1030	*Bam*HI	Ap^R	—	8.3 kb	Clinical isolate; it does not cross-anneal with pMB9 or ColEl	[k]

[a] Key to abbreviations: Ap^R, Cm^R, Rif^R, Tc^R, symbolize resistance to ampicillin, chloramphenicol, rifampicin, and tetracycline, respectively; *Imm*λ, immunity to coliphage λ; *Imm*El, immunity to colicin El; kb, 1000 base pairs of DNA.

[b] A. C. Y. Chang and S. N. Cohen, *J. Bacteriol.* **134,** 1141 (1978).

[c] F. Bolivar, R. L. Rodriguez, P. J. Greene, M. C. Betlach, H. L. Heyneker, H. W. Boyer, J. H. Crosa, and S. Falkow, *Gene* **2,** 95 (1977).

[d] F. Bolivar, *Gene* **4,** 121 (1978).

[e] J. Collins and H. J. Brüning, *Gene* **4,** 85 (1978).

[f] J. Collins and B. Hohn, *Proc. Natl. Acad. Sci. U.S.A.* **75,** 4242 (1978).

[g] K. Backman, D. Hawley, and M. J. Ross, *Science* **196,** 182 (1977).

[h] R. N. Rao and S. G. Rogers, *Gene* **3,** 247 (1978).

[i] D. H. Gelfand, H. M. Shepard, P. H. O'Farrell, and B. Polisky, *Proc. Natl. Acad. Sci. U.S.A.* **75,** 5869 (1978).

[j] P. Charnay, M. Perricaudet, F. Galibert, and P. Tiollais, *Nucleic Acids Res.* **5,** 4479 (1978).

[k] D. M. Coen, J. R. Bedbrook, L. Bogorad, and A. Rich, *Proc. Natl. Acad. Sci. U.S.A.* **74,** 5487 (1977).

many of the resulting recombinant plasmids.[7,33,34] The inserted DNA segment and the short dG:dC homopolymer segments flanking it may be cleaved from the vehicle by *Pst*I digestion. Furthermore, the segment of DNA inserted into the Ap gene is transcribed from the penicillinase promoter. This has permitted transcription and translation of mouse dihydrofolate reductase[8] and rat proinsulin, evidently fused to the N-terminus of penicillinase,[7] in *E. coli*.

Several of the vehicles in Table I have the advantage of inactivation of a genetic marker by insertion of DNA at a particular restriction site. For instance, the *Bam*HI and *Sal*I sites of pBR322 are within the Tc gene. Insertion of DNA at either of these sites generates an Ap^R Tc^S plasmid.[33] Success or failure of the procedure to form recombinants may then be checked by replica-plating Ap^R transformants on tetracycline–agar plates to test for the Tc^S phenotype. Furthermore, the recombinants may be selected by culture in the presence of tetracycline and D-cycloserine, which kills exponentially growing cells. Bacteria containing the Ap^R Tc^R vehicle plasmids are killed, while those carrying the AP^R Tc^S recombinant plasmids survive.[29,33]

Similarly, the *Pst*I site of pBR322 is within the Ap gene. Insertion of DNA there creates an Ap^S Tc^R plasmid.[33] However, most rat insulin cDNA recombinant plasmids gave rise to colonies on ampicillin plates.[7] These appeared after a longer incubation period than that needed for colonies containing pBR322 alone. There is evidence that they resulted from infrequent deletion of the cDNA passenger segment which was inserted at the *Pst*I site of pBR322 by means of dG and dC homopolymer tails.

Unfortunately, insertion of DNA at the *Eco*RI site of pBR322 does not result in any known marker inactivation. However, several other plasmids do have this advantage. For example, RSF2124 fails to produce colicin E1 if DNA is inserted at its single *Eco*RI site.[35] Also, pACYC184 and pBR325 were constructed to provide useful cloning vehicles with inactivation of a drug resistance marker, chloramphenicol resistance, by DNA insertion at their single *Eco*RI sites (Table I).

Insertional marker inactivation is a useful feature in a cloning vehicle. However, it is not essential, for other methods can ensure that virtually all transformed *E. coli* contain a recombinant plasmid rather than the unaltered vehicle plasmid. If restriction enzyme-cleaved cohesive DNA termini are used for joining, treatment of the vehicle DNA with phosphatase removes terminal phosphoryl groups and prevents rejoining without an

[33] F. Bolivar, R. L. Rodriguez, P. J. Greene, M. C. Betlach, H. L. Heyneker, H. W. Boyer, J. H. Crosa, and S. Falkow, *Gene* **2,** 95 (1977).

[34] M. B. Mann, R. N. Rao, and H. O. Smith, *Gene* **3,** 97 (1978).

[35] M. So, R. Gill, and S. Falkow, *Mol. Gen. Genet.* **142,** 238 (1975).

insert.[36] DNA joining by means of homopolymer extensions added by terminal deoxynucleotidyltransferase[22,23] also can ensure that most plasmids have inserted DNA. Of course, both of these methods require that nicked circular vehicle plasmid DNA be removed from the cleaved linear vehicle DNA before use, either by exhaustive restriction endonuclease digestion or by gel electrophoresis. Since nicked circular plasmid DNA transforms bacteria with the same high efficiency as supercoiled plasmid DNA, it can give rise to clones containing vehicle plasmid rather than a recombinant.

Inactivation of a genetic marker is not an infallible indication of DNA insertion. For instance, when pBR324 is digested by *Bam*HI, *Hin*dIII, or *Sal*I restriction endonuclease, ligated with other similarly digested DNA, and used to transform bacteria to ampicillin resistance, 5–10% of the Tc^S transformants do not carry recombinant DNA.[37] Such Tc^S transformants, possibly resulting from deletion of a portion of the tetracycline resistance gene, have also been observed for pBR322. Corresponding Cm^S Ap^R Tc^R transformants, carrying no inserted DNA, represent about 10% of the Cm^S transformants in recombinant DNA experiments using *Eco*RI-cleaved pBR325 DNA.[37]

The pJC plasmids in Table I are representatives of a class of plasmids called cosmids because they have the λ phage *cos* DNA site required for packaging into λ phage particles, and they replicate as plasmids because of their ColE1 DNA segments. Their chief advantage is the fact that pJC DNA ligated to foreign DNA can be packaged *in vitro* and used to infect *E. coli* efficiently, yielding as many as 500,000 recombinant clones per microgram of inserted DNA. Using the $CaCl_2$ transformation procedure[38] instead, or minor modifications of it, one usually obtains 1000–50,000 transformants per microgram inserted DNA after cohesive-end ligation[39] or homopolymer tail annealing.[7,40] However, the efficiency of transformation can be improved by modifications such as the method of Suzuki and Szalay (this volume [22]), so that it rivals the efficiency of λ phage *in vitro* packaging and infection. The *in vitro* packaging also exerts a size selection on DNA to be packaged, so that all surviving pJC cosmids have inserted DNA.

Several plasmids contain the lactose operon promoter and operator and a single *Eco*RI site for inserting DNA within the *lacZ* gene. They can

[36] A. Ullrich, J. Shine, J. Chirgwin, R. Pictet, E. Tischer, W. J. Rutter, and H. M. Goodman, *Science* **196,** 1313 (1977).

[37] F. Bolivar, *Gene* **4,** 121 (1978).

[38] S. N. Cohen, A. C. Y. Chang, and L. Hsu, *Proc. Natl. Acad. Sci. U.S.A.* **69,** 2110 (1972).

[39] J. F. Morrow, S. N. Cohen, A. C. Y. Chang, H. W. Boyer, H. M. Goodman, and R. B. Helling, *Proc. Natl. Acad. Sci. U.S.A.* **71,** 1743 (1974).

[40] P. C. Wensink, D. J. Finnegan, J. E. Donelson, and D. S. Hogness, *Cell* **3,** 315 (1974).

be useful for synthesizing a polypeptide encoded by the cloned DNA segment. Besides the *lacPO* region, pOP1 contains most of the *lacZ* gene. It is a mutant plasmid with an unusually high copy number (74 per bacterial chromosome). When it is maintained in an *E. coli cya*$^-$ strain, induction results from the addition of IPTG and cyclic AMP. When fully induced, more than half of the synthesized protein is the β-galactosidase fragment, to which a foreign polypeptide will presumably be fused in a recombinant strain.

pPCϕ1, pPCϕ2, and pPCϕ3 contain *lacPO* and a much smaller portion of the *Z* gene than pOP1 does. They were constructed (from pBR322) in order to overcome the translation reading frame problem for inserted DNA, which is transcribed from the *lac* promoter in these plasmids. One end of the DNA inserted at the *Eco*RI site of these plasmids is 22, 24, or 26 base pairs, respectively, downstream from the initial methionine codon of the β-galactosidase gene fragment. The inserted DNA sequence will be translated in one of the three frames unless a noncoding sequence within it has termination codons in all three frames. This array of three plasmids is not needed if the inserted DNA is joined to the vehicle by homopolymer tails of various lengths, since one-third of the junctions should give the right frame.[7] They should be useful for synthesis of proteins encoded by restriction cleavage fragments of DNA or by synthetic DNA, however.

A plasmid similar to pPCϕ1 is pBH20.[1] It was derived from pBR322 and a 203-base-pair *Hae*III fragment of λ*plac5* DNA. It has a single *Eco*RI site 22 base pairs from the initial methionine codon of the *lacZ* gene.

pKC16 (Table I) is derived from pBR322. The plasmid copy number increases after thermal induction, since the plasmid contains the λ*cI857* temperature-sensitive repressor gene and the λ *O* and *P* genes. It reaches more than 100 copies per bacterial chromosome.

Several other plasmids are useful for special purposes. pBR324 and pKB158 have restriction cleavage sites creating blunt, base-paired DNA termini (the *Sma*I and *Hpa*I sites, respectively). Blunt-ended foreign DNA fragments can be joined to these vehicles and replicated (see DNA joining methods below). RSF1030 is noteworthy because its DNA does not cross-hybridize with ColE1 or pMB9 DNA (pMB9 is a Tc^R plasmid related to ColE1, widely used for DNA cloning[29]). Consequently, a DNA sequence cloned in pMB9, e.g., a reverse transcript (cDNA) plasmid, can easily be used to screen RSF1030 recombinants by hybridization,[41] e.g., for the corresponding gene from a eukaryotic genome, for complementary repeated sequences, etc.

Only pBR322, pMB9, pBR313, and pSC101 are certified now as EK2 plasmid vectors for cloning DNA from warm-blooded vertebrates and

[41] M. Grunstein and D. S. Hogness, *Proc. Natl. Acad. Sci. U.S. A.* **72,** 3961 (1975).

other sources.[29,33,42] The other plasmids described (Table I) are EK1 vectors.

Bacteriophage

Derivatives of λ phage were developed as cloning vehicles early,[29] and they are probably the best vehicles for the cloning and isolation of particular genes from eukaryotic genome DNA. λ derivatives have three main advantages over plasmids for this purpose. First, thousands of recombinant phage plaques on a single 88-mm petri dish can easily be screened for a given DNA sequence by nucleic acid hybridization.[43] Second, *in vitro* packaging of recombinant DNA molecules provides a very efficient means of infecting bacteria with them.[44,45] Finally, millions of independently packaged recombinant phage can be replicated conveniently and stored in a single solution as a "library" in which all sequences of a large genome (e.g., rabbit) are likely to be represented.[46]

The Charon phages are λ derivatives, some of which approach the maximum possible capacity for inserted DNA in a nondefective λ phage vector (Table II). Charons 4 and 8–11 are believed to be able to replicate more than 22 kb of inserted DNA. A number of Charon phages contain a *lacZ* (β-galactosidase) gene. Substitution of foreign DNA in place of the *lacPOZ* DNA segment makes the phage $lacO^-$. Growth on a lac^+ bacterial strain on plates containing a chromogenic noninducing β-galactosidase substrate then provides a quick indication of insertion of foreign DNA or failure to do so. Charon 4 has this feature and has been used extensively, as has its EK2 derivative, Charon 4A.[46]

λgt4 · λB was constructed to provide a phage vector which has the λ attachment site and is able to form temperature-inducible lysogens, since it has the *cI857* allele. This has been helpful in overproduction of DNA ligase encoded by an inserted *E. coli* DNA segment.

λgtWES · λB and λgtvirJZ · λB are EK2 vectors, useful for replicating DNA of warm-blooded vertebrates, etc.[42]

λΔz1, λΔz2, λΔz3, and their corresponding derivatives with amber mutations provide phage vectors with the same advantages for protein production as pPCϕ1, pPCϕ2, and pPCϕ3 plasmids (see above).

M13mp2 (Table II) is a derivative of M13, a single-stranded DNA

[42] "Guidelines for Research Involving Recombinant DNA Molecules" (rev.), *Fed. Regist.* **43,** 60108 (1978).

[43] W. D. Benton and R. W. Davis, *Science* **196,** 180 (1977).

[44] N. Sternberg, D. Tiemeier, and L. Enquist, *Gene* **1,** 255 (1977).

[45] B. Hohn and K. Murray, *Proc. Natl. Acad. Sci. U.S.A.* **74,** 3259 (1977).

[46] T. Maniatis, R. C. Hardison, E. Lacy, J. Lauer, C. O'Connell, D. Quon, G. K. Sim, and A. Efstratiadis, *Cell* **15,** 687 (1978).

TABLE II
NEW VIRAL CLONING VEHICLES[a]

Vehicle	Cleavage sites at which foreign DNA may be inserted	Sites at which insertion inactivates a marker	Molecular weight (kb)	Other characteristics	Reference
Bacteriophage					
Charons 1–16, 3A, 4A, 16A	*Eco*RI, *Hin*dIII, *Sst*I	*Eco*RI (*lacZ*)	43–50	From 0 to 24 kb of DNA can be inserted	[b]
λgt4 · λB	*Eco*RI	—	44	Recombinant phage can lysogenize	[c]
λgtWES · λB	*Eco*RI, *Sst*I	—	41	—	[d]
λgtvirJZ · λB	*Eco*RI, *Sst*I	—	41	—	[e]
λΔZ1, λΔZ2, λΔZ3	*Eco*RI	—	41	DNA is inserted into *lacZ* gene in any of three translation frames	[f]

M13mp2	*Eco*RI	*Eco*RI (*lacZ* α-peptide)	7.2	Single-stranded DNA is produced	[g]
Eukaryotic vector					
SV40	*Eco*RI and *Hpa*II termini	—	3.8	—	[h,i]
SVGT5	*Bam*HI and *Hin*dIII termini	—	4.18	Inserted DNA was transcribed and translated	[j]

[a] Key to abbreviations: *Imm*λ, λ phage immunity; kb, 1000 base pairs of DNA.

[b] F. R. Blattner, B. G. Williams, A. E. Blechl, K. Denniston-Thompson, H. E. Faber, L.-A. Furlong, D. J. Grunwald, D. O. Kiefer, D. D. Moore, J. W. Schumm, E. L. Sheldon, and O. Smithies, *Science* **196,** 161 (1977).

[c] S. M. Panasenko, J. R. Cameron, R. W. Davis, and I. R. Lehman, *Science* **196,** 188 (1977).

[d] D. Tiemeier, L. Enquist, and P. Leder, *Nature (London)* **263,** 526 (1976).

[e] D. J. Donoghue and P. A. Sharp, *Gene* **1,** 209 (1977).

[f] P. Charnay, M. Perricaudet, F. Galibert, and P. Tiollais, *Nucleic Acids Res.* **5,** 4479 (1978).

[g] B. Gronenborn and J. Messing, *Nature (London)* **272;** 375 (1978).

[h] D. H. Hamer, D. Davoli, C. A. Thomas, Jr., and G. C. Fareed, *J. Mol. Biol.* **112,** 155 (1977).

[i] N. Muzyczka, *Gene* **6,** 107 (1979).

[j] R. C. Mulligan, B. H. Howard, and P. Berg, *Nature (London)* **277,** 108 (1979).

phage, which contains the *lac* promoter and operator and the proximal portion of the *lacZ* gene. Mutagenesis generated an *Eco*RI site at the codon for the fifth amino acid from the N-terminus of β-galactosidase. Insertion of foreign DNA at this *Eco*RI site inactivates *lacZ* α-complementation, producing colorless or light-blue plaques on appropriate host bacteria with a chromogenic substrate. The significant aspect of M13mp2 is that the phage particles provide only one strand of the cloned DNA, i.e., they strand-separate the DNA for the investigator. This is very useful for the DNA-sequencing method of Sanger *et al.*,[47] for nucleic acid hybridization, etc. If two recombinants are isolated with the inserted DNA in opposite orientations each will serve as a source for a different strand.

Unfortunately, M13mp2 needs an F^+ *E. coli* host to make plaques, and conjugative plasmids are not acceptable in EK1 host–vector systems.[42] A strain carrying a conjugation-deficient derivative of F may be approved soon for the EK1 level.

Eukaryotic Vectors

Only SV40 virus has been used as a cloning vehicle to replicate introduced DNA independently of the chromosomes in eukaryotic cells. The SV40 vehicle with the largest capacity for inserted DNA permits encapsidation of about 4.3 kb of added DNA, a small capacity compared to plasmid or phage vectors.[29]

An *Eco*RI–*Hpa*II fragment of SV40 DNA has been used to replicate the *E. coli* Su^+III tRNA gene (Table II). This recombinant DNA has been used to transform rat cells.[48] The *Eco*RI–*Hpa*II SV40 fragment has also been joined at its *Eco*RI terminus to a larger fragment of λ phage DNA, and the linear molecule has been used to transform mouse cells (Table II).

Another fragment of SV40 DNA called SVGT5 was chosen to permit transcription and translation of inserted DNA sequences. The "body" (main exon) of the VP1 gene was excised, leaving the gene's 5′-end "leaders," its intervening sequence, and its 3′-end. A recombinant derived from it formed rabbit β-globin in monkey kidney cells (Table II).

An *in situ* nucleic acid hybridization method for SV40 recombinants has been described.[49]

In other experiments involving introduction of DNA into eukaryotic

[47] F. Sanger, S. Nicklen, and A. R. Coulson, *Proc. Natl. Acad. Sci. U.S.A.* **74,** 5463 (1977).

[48] P. Upcroft, H. Skolnik, J. A. Upcroft, D. Solomon, G. Khoury, D. H. Hamer, and G. C. Fareed, *Proc. Natl. Acad. Sci. U.S.A.* **75,** 2117 (1978).

[49] L. P. Villarreal and P. Berg, *Science* **196,** 183 (1977).

cells, yeast has been transformed with yeast DNA sequences of recombinants grown in *E. coli*.[50] It appears likely that the 2-μm circular DNA of many yeast strains can be developed as a cloning vehicle.[29]

Transformation of cultured mouse cells with DNA from several vertebrate species has also been demonstrated.[51] This transformation method seems rather general, enabling one to introduce DNA from many sources into an integrated state in the DNA of mouse cells. It provides a means of replacing a eukaryotic gene, cloned in bacteria, into a eukaryotic cell for studies on its function in an environment closer to its natural one.

DNA To Be Replicated

DNA extracted from an organism can be prepared for cloning in a variety of ways after its purification, notably by restriction endonuclease digestion or by shearing to a selected length.[40,52] Unfractionated DNA representing the entire genome can be utilized.[3–6,46] However, if a particular gene is desired, purification before cloning reduces the number of recombinants that must be screened. General methods for this are RPC-5 chromatography[53] and agarose gel electrophoresis[11,54] after restriction endonuclease cleavage.

Cloning of DNA synthesized by reverse transcription of polyadenylated RNA has been applied widely.[29] Recent work has defined the best experimental conditions.[55–57] DNA synthesized chemically, up to 207 base pairs in length, has also been employed.[1,2,58–61]

[50] A. Hinnen, J. B. Hicks, and G. R. Fink, *Proc. Natl. Acad. Sci. U.S.A.* **75,** 1929 (1978).

[51] M. Wigler, A. Pellicer, S. Silverstein, and R. Axel, *Cell* **14,** 725 (1978).

[52] L. Clarke and J. Carbon, *Cell* **9,** 91 (1976).

[53] S. C. Hardies and R. D. Wells, *Proc. Natl. Acad. Sci. U.S.A.* **73,** 3117 (1976).

[54] S. M. Tilghman, D. C. Tiemeier, F. Polsky, M. H. Edgell, J. G. Seidman, A. Leder, L. W. Enquist, B. Norman, and P. Leder, *Proc. Natl. Acad. Sci. U.S.A.* **74,** 4406 (1977).

[55] E. Y. Friedman and M. Rosbash, *Nucleic Acids Res.* **4,** 3455 (1977).

[56] G. N. Buell, M. P. Wickens, F. Payvar, and R. T. Schimke, *J. Biol. Chem.* **253,** 2471 (1978).

[57] M. P. Wickens, G. N. Buell, and R. T. Schimke, *J. Biol. Chem.* **253,** 2483 (1978).

[58] K. J. Marians, R. Wu, J. Stawinski, T. Hozumi, and S. A. Narang, *Nature (London)* **263,** 744 (1976).

[59] H. L. Heyneker, J. Shine, H. M. Goodman, H. W. Boyer, J. Rosenberg, R. E. Dickerson, S. A. Narang, K. Itakura, S. Lin, and A. D. Riggs, *Nature (London)* **263,** 748 (1976).

[60] J. R. Sadler, J. L. Betz, M. Tecklenburg, D. V. Goeddel, D. G. Yansura, and M. H. Caruthers, *Gene* **3,** 211 (1978).

[61] H. G. Khorana, *Science* **203,** 614 (1979).

DNA Joining Methods

Cohesive Ends

A number of restriction enzymes make staggered cuts in the two DNA strands so that single-stranded termini are produced.[20,21] These can be annealed with DNA from another source and ligated to form recombinant DNA.[28,39]

The cloning vehicle DNA can also recircularize by itself, with no inserted DNA. As a result, 75–90% of the transformants usually contain vehicle alone instead of recombinant DNA.[36,39] A method to prevent resealing of the vehicle DNA is removal of its terminal phosphate groups by incubation with nuclease-free alkaline phosphatase.[36,62] The vehicle DNA can then be ligated into a recombinant circle with the passenger DNA, or the passenger DNA can be cyclized to yield molecules which generally cannot replicate, but the vehicle cannot be ligated alone. This phosphatase method has been used very effectively on *Hin*dIII and *Eco*RI termini; all clones examined had inserted foreign DNA.[36,63,64] It has also been used on *Bam*HI termini with success.[1] Note that a higher concentration of DNA ligase is needed for complete joining than if DNA were not incubated with phosphatase.

Cohesive ends can be added to blunt-ended DNA molecules by ligation with synthetic DNA linkers.[60,65,66] These are duplex, blunt-ended DNA molecules, from 8 to 14 base pairs in length, containing the recognition site for a restriction endonuclease that produces cohesive termini. Linkers with an *Eco*RI, a *Bam*HI, or a *Hin*dIII site are available commercially. Linkers are joined to blunt-ended passenger DNA molecules by T4 ligase. After digestion with the relevant restriction enzyme and removal of excess linker, the passenger DNA is ligated to vehicle DNA via the complementary termini and cloned. *Eco*RI[10,36,64,67–69] and *Hin*dIII[36,63,70]

[62] B. Weiss, T. R. Live, and C. C. Richardson, *J. Biol. Chem.* **243,** 4530 (1968).

[63] P. H. Seeburg, J. Shine, J. A. Martial, J. D. Baxter, and H. M. Goodman, *Nature (London)* **270,** 486 (1977).

[64] J. Shine, P. H. Seeburg, J. A. Martial, J. D. Baxter, and H. M. Goodman, *Nature (London)* **270,** 494 (1977).

[65] C. P. Bahl, K. J. Marians, R. Wu, J. Stawinsky, and S. A. Narang, *Gene* **1,** 81 (1976).

[66] R. H. Scheller, R. E. Dickerson, H. W. Boyer, A. D. Riggs, and K. Itakura, *Science* **196,** 177 (1977).

[67] R. H. Scheller, T. L. Thomas, A. S. Lee, W. H. Klein, W. D. Niles, R. J. Britten, and E. H. Davidson, *Science* **196,** 197 (1977).

[68] F. Heffron, M. So, and B. J. McCarthy, *Proc. Natl. Acad. Sci. U.S.A.* **75,** 6012 (1978).

[69] P. Charnay, M. Perricaudet, F. Galibert, and P. Tiollais, *Nucleic Acids Res.* **5,** 4479 (1978).

[70] H. Lehrach, A. M. Frischauf, D. Hanahan, J. Wozney, F. Fuller, R. Crkvenjakov, H. Boedtker, and P. Doty, *Proc. Natl. Acad. Sci. U.S.A.* **75,** 5417 (1978).

linkers have been widely used. The optimal conditions for joining the *Eco*RI decamer linker have been determined.[71] A variation of the linker method utilizes the appropriate modification methylase to protect internal sites in the passenger DNA from cleavage by the restriction endonuclease.[46]

The conditions affecting cohesive end ligation have been explored.[72] A useful variation is to ligate termini made by cleavage with one restriction endonuclease to those made by another. For instance, *Dpn*II and *Mbo*I leave the single-stranded 5′-terminus GATC (the recognition site for both is GATC). These DNA ends can be joined to GATC 5′-termini generated by *Bam*HI, whose complete recognition sequence is GGATCC. *Bgl*II and *Bcl*I make ends that can be joined to the preceding ones, too; their recognition sequences are AGATCT and TGATCT, respectively. Similarly, *Sal*I and *Xho*I both leave TCGA single-stranded termini, though they recognize GTCGAC and CTCGAG, respectively.[21]

Homopolymer Tails

Terminal deoxynucleotidyltransferase can be used to add a homopolymer extension, e.g., polydeoxyadenylate, to each 3′-end of the vehicle DNA, and a complementary extension to each 3′-end of the passenger DNA (see this volume [3]). An attractive application is to add dG tails to *Pst*I-cleaved pBR322 DNA and dC tails to the DNA to be inserted, anneal, and transform *E. coli* with the DNA. *Pst*I sites are reconstituted on both sides of the inserted DNA in many of the resulting recombinant plasmids.[7,33,34] Ligation of the vehicle with the insert before transformation is unnecessary if they have complementary homopolymer tails. Indeed, it has not been possible to ligate them efficiently *in vitro*.[22,23]

When homopolymer tails were first used for joining DNA molecules, an exonuclease was employed to render the vehicle and passenger DNA termini single-stranded before incubation with terminal transferase.[22,23] The exonuclease is not necessary.[73] However, a technical precaution which is useful in cloning large eukaryotic DNA fragments is to eliminate small polynucleotides (≤ 1 kb) before annealing vehicle and passenger DNAs. Preparations of eukaryotic DNA with *weight-average* molecular weights of 10 kb or more often contain a significant *number* of much smaller DNA molecules. Also, terminal transferase can initiate homo-

[71] A. Sugino, H. M. Goodman, H. L. Heyneker, J. Shine, H. W. Boyer, and N. R. Cozzarelli, *J. Biol. Chem.* **252,** 3987 (1977).

[72] A. Dugaiczyk, H. W. Boyer, and H. M. Goodman, *J. Mol. Biol.* **96,** 171 (1975).

[73] R. Roychoudhury, E. Jay, and R. Wu, *Nucleic Acids Res.* **3,** 863 (1976).

polymer chains *de novo*.[74] Sucrose gradient centrifugation is useful for removing these small molecules, which can interfere with cloning of larger DNA segments.

Under partially denaturing conditions, dA:dT homopolymer joints can be digested by S1 nuclease to permit separation of the passenger DNA from the vehicle.[29]

Blunt-End Joining

T4 DNA ligase joins DNA molecules with duplex, base-paired termini.[24] The apparent K_m is about 50 μM DNA 5′-ends, which corresponds to 80 mg/ml of DNA molecules 5000 base pairs long.[71] Nevertheless, a reasonable concentration of T4 ligase joins enough DNA termini to permit construction of new plasmids.[1,75] The concentration of DNA fragments employed, between 200 and 5000 base pairs long, has been 100–300 μg/ml. Blunt-end joining has also been used extensively to attach synthetic restriction site linkers to DNA molecules (see above). T4 RNA ligase stimulates the reaction.[71]

Blunt ends can be produced on a DNA fragment by cleavage with any of a number of restriction endonucleases. Alternatively, random shear breakage or a restriction enzyme making staggered cuts may be used, but the DNA termini must then be made blunt by biochemical methods. This has been done by removal of single-stranded termini by incubation with single-strand-specific nuclease S1.[36,46,67,69] Alternatively, T4 DNA polymerase,[1,10] *E. coli* DNA polymerase I,[63,68,75] or reverse transcriptase,[36] with added deoxynucleoside triphosphates, has been used. Sometimes combined treatment with nuclease S1 followed by DNA polymerase I and dNTPs has been employed.[63,70]

Introducing Recombinant DNA into a Host

Almost all recombinant DNA research has used, as host cells, *E. coli* K12 mutants lacking restriction of foreign DNA.[29] Increased biological containment is provided by the approved EK2 host for plasmids, χ1776.[42,76] Low efficiency of transformation has been obtained with this strain in some studies (e.g., 20,000 Ap^R transformants per microgram of supercoiled pBR322 DNA,[63] compared to the usual 10^6 per microgram

[74] K. Kato, J. M. Goncalves, G. E. Houts, and F. J. Bollum, *J. Biol. Chem.* **242,** 2780 (1967).

[75] K. Backman, M. Ptashne, and W. Gilbert, *Proc. Natl. Acad. Sci. U.S.A.* **73,** 4174 (1976).

[76] R. Curtiss, III, M. Inoue, D. Pereira, J. C. Hsu, L. Alexander, and L. Rock, *in* "Molecular Cloning of Recombinant DNA" (W. A. Scott and R. Werner, eds.), p. 99. Academic Press, New York, 1977.

with other *E. coli* strains). Improved transformation procedures now yield as much as 10^6–10^7 transformants per microgram of supercoiled pBR322 DNA, and a reported 10^4 recombinants per microgram when made by joining with homopolymer tails.[7,76,77]

In vitro packaging provides an efficient means of introducing recombinant phage DNA or cosmid DNA into *E. coli*.[44–46] Up to 700,000 plaques per microgram of inserted DNA have been obtained.[78]

Several other species of microorganisms may prove useful as hosts for replication of recombinant DNA. A number of plasmid vectors have been found for *Bacillus subtilis*.[79–81] The U.S. National Institutes of Health has only permitted cloning of DNA from *Bacillus* species with these, to date. Genetic transformation of *Saccharomyces cerevisiae* has been demonstrated,[50] but the NIH currently only permits cloning of yeast DNA, or certain prokaryotic DNAs, in yeast. A vehicle for recombinant DNA and a transformation procedure are also available for *Staphylococcus aureus*.[82]

Many recombinants containing known *E. coli* genes have been isolated by genetic selection of transformed cells expressing the desired function (reviewed in Sinsheimer[29]). A number of genes of fungi have been expressed in *E. coli* and have complemented bacterial mutations when replicated as part of a recombinant DNA molecule.[3–6] Four out of 15 genetic complementations tested by J. Carbon *et al.*, using yeast DNA in *E. coli*, were successful.[83] These studies used DNA extracted from fungi, not reverse transcripts of fungal mRNA.

It is unlikely that most genes of animals will encode functional proteins in *E. coli*, because it appears that most such genes contain intervening sequences. At least one of these is usually located at a site on the DNA near that corresponding to the 5′-end of the mRNA sequence. However, this problem can be bypassed by using reverse transcripts of mRNAs. Several of these have been expressed (transcribed and translated into protein) after introduction into *E. coli* as part of recombinant plasmids.[7–10] One of these conferred a phenotype, trimethoprim resistance, which could be selected as a result of synthesis of mouse dihydro-

[77] M. V. Norgard, K. Keem, and J. J. Monahan, *Gene* **3**, 279 (1978).

[78] R. Lenhard-Schuller, B. Hohn, C. Brack, M. Hirama, and S. Tonegawa, *Proc. Natl. Acad. Sci. U.S.A.* **75**, 4709 (1978).

[79] K. M. Keggins, P. S. Lovett, and E. J. Duvall, *Proc. Natl. Acad. Sci. U.S.A.* **75**, 1423 (1978).

[80] T. J. Gryczan and D. Dubnau, *Proc. Natl. Acad. Sci. U.S.A.* **75**, 1428 (1978).

[81] S. D. Ehrlich, *Proc. Natl. Acad. Sci. U.S.A.* **75**, 1433 (1978).

[82] S. Löfdahl, J. Sjöström, and L. Philipson, *Gene* **3**, 161 (1978).

[83] J. Carbon, B. Ratzkin, L. Clarke, and D. Richardson, *in* "Molecular Cloning of Recombinant DNA" (W. A. Scott and R. Werner, eds.), p. 59. Academic Press, New York, 1977.

folate reductase in *E. coli*.[8] The eukaryotic proteins made in bacteria were identified by immunological methods. Several such methods have been described which are capable of screening large number of recombinant clones for a protein of interest.[84-86]

Methods of screening that are independent of protein synthesis in bacteria are commonly used, however. Clones of interest can often be identified by screening methods utilizing hybridization with a pure radioactively labeled nucleic acid probe.[41,43,49] If a pure nucleic acid probe is not available, one can still identify a cloned protein structural gene if an RNA preparation containing some of the mRNA of interest is available. The desired clone inhibits the *in vitro* translation of a particular mRNA by forming a DNA–RNA hybrid with it (hybrid-arrested translation).[87] A more sensitive type of method utilizes the cloned DNA to purify the particular mRNA by DNA–RNA hybridization; the mRNA is then identified by *in vitro* translation. Either gel filtration (this volume [33]) or cloned DNA linked to cellulose[88] can be used to purify the DNA–RNA hybrids, which are then dissociated by heating before *in vitro* translation.

Fidelity of Recombinant DNA Cloning

The passenger segment of a recombinant DNA molecule is usually under no selection pressure for genetic function. Thus its nucleotide sequence is subject to evolutionary drift. Nevertheless, consideration of the low spontaneous mutation frequency in wild-type bacterial strains suggests that growth for a few hundred generations should not alter the DNA sequence of *most* individual molecules in a recombinant DNA preparation. This has been confirmed most clearly for rabbit globin cDNA clones pβG1[89] (β-globin) and pHB72[90] (α-globin). pβG1 contains the entire coding region, and its nucleotide sequence agrees with partial mRNA sequence data and the primary structure of the protein. pHB72 similarly agrees with mRNA and protein data and represents 361 of the 423 base pairs of the α-globin coding region. The nucleotide sequence of an oval-

[84] H. A. Erlich, S. N. Cohen, and H. O. McDevitt, *Cell* **13,** 681 (1978).

[85] S. Broome and W. A. Gilbert, *Proc. Natl. Acad. Sci. U.S.A.* **75,** 2746 (1978).

[86] A. Skalka and L. Shapiro, *Gene* **1,** 65 (1976).

[87] B. M. Paterson, B. E. Roberts, and E. L. Kuff, *Proc. Natl. Acad. Sci. U.S.A.* **74,** 4370 (1977).

[88] M. E. Sobel, T. Yamamoto, S. L. Adams, R. DiLauro, V. E. Avvedimento, B. deCrombrugghe, and I. Pastan, *Proc. Natl. Acad. Sci. U.S.A.* **75,** 5846 (1978).

[89] A. Efstratiadis, F. C. Kafatos, and T. Maniatis, *Cell* **10,** 571 (1977).

[90] H. C. Heindell, A. Liu, G. V. Paddock, G. M. Studnicka, and W. A. Salser, *Cell* **15,** 43 (1978).

bumin cDNA plasmid also indicates faithful cloning of the mRNA's sequence.[91]

The faithfully replicated sequences above are relatively short and free of internal repetition. In contrast, long DNA segments with tandem sequence repetition have occasionally been partly deleted from recombinant DNA cloned in *E. coli*. The rRNA gene unit of *Xenopus* contains, in its nontranscribed spacer, up to 5000 base pairs of tandem repetition of a short sequence, each repeat of which is probably less than 50 base pairs.[92] Nevertheless, this DNA has been cloned with a high degree of stability of the *Xenopus* DNA sequence.[39] After hundreds of generations, 98% of the molecules had not lost or gained repeated sequence elements. Furthermore, they were stable in *recA*$^-$ or *rec*$^+$ *E. coli*.[92] Five copies of the *Xenopus* 5 S rRNA genes, containing over 100 repeats of a 15-nucleotide sequence, were also cloned stably.[93]

On the other hand, deletions have occurred in clones of other repeating sequences. Plasmids containing *Drosophila* satellite DNAs with tandem repeats of 5- or 7- base-pair sequences lost part of the inserted DNA unless the insert size was 1 kb or less.[94] Since this occurred even in *recA*$^-$ bacteria, unequal intramolecular recombination of replicating DNA molecules was proposed as a mechanism. In the case of silk fibroin gene plasmids, 90% of subclones of a plasmid with 1.3 kb composed of 18-nucleotide tandem repeats were unaltered.[95] On the other hand, 15 kilobase pairs of these tandem repeats were cloned in a pMB9 recombinant plasmid, but progressive loss of the fibroin repeated sequences occurred, down to 4–6 kb, at which point they were stable.[96,97] Deletions also occurred at a low frequency in plasmids containing a tandem duplication of length 4.7 or 8.6 kb including yeast rRNA genes.[98] This was found in both *rec*$^+$ and *recA*$^-$ bacteria. Repeated DNA sequences were lost from a λ phage recombinant carrying three copies of a 2.8-kb fragment of adenovirus-2 DNA. One copy of the inserted fragment remained.[99]

[91] L. McReynolds, B. W. O'Malley, A. D. Nisbet, J. E. Fothergill, D. Givol, S. Fields, M. Robertson, and G. G. Brownlee, *Nature (London)* **273,** 723 (1978).

[92] P. K. Wellauer, I. B. Dawid, D. D. Brown, and R. H. Reeder, *J. Mol. Biol.* **105,** 461 (1976).

[93] D. Carroll and D. D. Brown, *Cell* **7,** 477 (1976).

[94] D. Brutlag, K. Fry, T. Nelson, and P. Hung, *Cell* **10,** 509 (1977).

[95] J. F. Morrow, N. T. Chang, J. M. Wozney, A. C. Richards, and A. Efstratiadis, *in* "Molecular Cloning of Recombinant DNA" (W. A. Scott and R. Werner, eds.), p. 161. Academic Press, New York, 1977.

[96] Y. Ohshima and Y. Suzuki, *Proc. Natl. Acad. Sci. U.S.A.* **74,** 5363 (1977).

[97] T. Mukai and J. F. Morrow, in preparation.

[98] A. Cohen, D. Ram, H. O. Halvorson, and P. C. Wensink, *Gene* **3,** 135 (1978).

[99] M. Perricaudet, A. Fritsch, U. Pettersson, L. Philipson, and P. Tiollais, *Science* **196,** 208 (1977).

In summary, DNA sequences that are not internally repetitious can be cloned faithfully. Internally repeated DNAs can be cloned, but partial loss of the passenger DNA may result from a homologous recombination process. This is independent of the *recA* gene product.

Alteration of DNA Molecules

This methodology, often called site-directed mutagenesis, does not depend entirely on recombinant DNA. It has been applied intensively to SV40 DNA. It does depend on a DNA transformation method by which mutagenized DNA can be inserted into cells, replicated, and cloned, so that a homogeneous preparation can subsequently be analyzed. Recombinant DNA makes this possible not only for viruses and naturally occurring plasmids, but for any DNA segment. One can create DNA sequence deletions at restriction endonuclease sites, between two such sites, or randomly. A restriction endonuclease site, within a synthetic DNA linker, can also be inserted either at a cleavage site for another restriction enzyme or at random. Point mutations can also be induced efficiently at selected sites.

In vitro DNA alteration has been used to map genes of SV40[100–104] and to map functions of plasmid DNAs, including replication.[105,106] Related recombinant DNA methods have been used to demonstrate that the intervening sequences within the β-globin gene are not essential for its transcription and translation.[107]

Two ways of creating a small deletion at DNA termini produced by restriction enzyme cleavage have been described. In the first, one digests with an exonuclease until about 30 nucleotides have been removed from each DNA strand (λ 5′-exonuclease has been used). Cells are then infected with the linear DNA molecules. At a low efficiency, the DNA ends are joined *in vivo*, creating deletion mutations lacking 15–50 base pairs.[101] The joining presumably depends on partial homology between different sequences in the single-stranded ends. *Hpa*II and *Eco*RI sites have been

[100] C. Lai and D. Nathans, *J. Mol. Biol.* **89,** 179 (1974).
[101] J. Carbon, T. E. Shenk, and P. Berg, *Proc. Natl. Acad. Sci. U.S.A.* **72,** 1392 (1975).
[102] T. E. Shenk, J. Carbon, and P. Berg, *J. Virol.* **18,** 664 (1976).
[103] C. Lai and D. Nathans, *Virology* **75,** 335 (1976).
[104] D. Shortle and D. Nathans, *Proc. Natl. Acad. Sci. U.S.A.* **75,** 2170 (1978).
[105] F. Bolivar, M. C. Betlach, H. L. Heyneker, J. Shine, R. L. Rodriguez, and H. W. Boyer, *Proc. Natl. Acad. Sci. U.S.A.* **74,** 5265 (1977).
[106] F. Heffron, M. So, and B. J. McCarthy, *Proc. Natl. Acad. Sci. U.S.A.* **75,** 6012 (1978).
[107] R. C. Mulligan, B. H. Howard, and P. Berg, *Nature (London)* **277,** 108 (1979).

deleted in this way.[108,109] The partial exonuclease digestion is probably not necessary for creating a deletion upon infection of mammalian cells, even if the restriction endonuclease employed generates cohesive termini.[100,102,103]

The second method requires a restriction enzyme cleavage that makes single-stranded termini. It involves removal of those with S1 single-strand-specific nuclease followed by blunt-end joining by T4 DNA ligase. *Xma*I, *Hae*II, and *Eco*RI sites have been deleted by this method.[37]

A fragment of DNA between two restriction cleavage sites can be deleted. SV40 DNA was partially digested by either *Hin*dIII or *Hin*dII, each of which can cleave it at several sites. DNA molecules shorter than the genome's length were then used to infect cells. Some of the mutants recovered were joined *in vivo* cleanly at a restriction site. Others acquired deletions extending a few hundred base pairs beyond the relevant restriction site.[100,103] Alternatively, the partially digested DNA fragments can be joined *in vitro* before transformation of cells, as was done with *Eco*RI*-cleaved DNA.[33] Deletions can be made by joining dissimilar restriction termini *in vitro* if both ends are made blunt before incubation with T4 ligase.[75]

Deletions can also be made at random locations by cleaving both strands of DNA with DNase I in the presence of Mn^{2+}. Cells are then infected with the linear molecules, and the progeny resulting from *in vivo* DNA joining are examined.[102]

Insertion mutations can be equally useful, especially if the inserted DNA contains a restriction endonuclease site that permits easy detection and mapping. Inserts consisting of large pieces of naturally occurring DNA may contain unidentified control sequences which produce artifacts in subsequent experiments. Short synthetic linkers containing restriction sites are better-defined inserts. An *Eco*RI linker has been used to alter the translation reading frame in cloning vehicles.[69] Clearly, a linker can eliminate a restriction site, even one that leaves single-stranded termini, if the termini are made blunt before ligation to the linker.[63] *Eco*RI linkers have also been inserted at random sites of cleavage by DNase I.[106]

Three methods for induction of point mutations at selected sites have been described. One involves conversion of cytosine residues to uracil at a localized single-stranded gap, produced at the site of a nick made by a restriction endonuclease.[104] Another utilizes nick translation in the presence of N^4-hydroxy dCTP. This induces T-to-C transition mutations.[110] A

[108] C. Covey, D. Richardson, and J. Carbon, *Mol. Gen. Gent.* **145,** 155 (1976).

[109] B. Polisky, R. J. Bishop, and D. H. Gelfand, *Proc. Natl. Acad. Sci. U.S.A.* **73,** 3900 (1976).

[110] W. Müller, H. Weber, F. Meyer, and C. Weissmann, *J. Mol. Biol.* **124,** 343 (1978).

more difficult, but very specific, mutagenesis method utilizes a synthetic oligodeoxynucleotide, 12 residues or longer, differing from the "wild-type" DNA sequence. This is annealed to a single-stranded circular DNA molecule and extended with a DNA polymerase. As many as one-third of the progeny of infection of cells with these molecules have the desired sequence change.[111,112]

These methods of *in vitro* DNA alteration are more complex than the time-honored investigation of mutations that occur *in vivo*. They represent a new way of doing genetic research. They have the advantage that a large proportion of the progeny carry the desired mutation (sometimes all do). Also, some of the DNA sequence changes can be very easily and precisely mapped, e.g., an inserted restriction site linker. Site-directed mutagenesis is complementary to study of *in vivo* mutations, in a sense. The former focuses on a portion of a DNA molecule and can show what physiological processes are affected by a change in it. The latter begins with a phenotype resulting from altered physiological processes and can discover what portions of the DNA are most closely related to that phenotype.

Conclusion

Recombinant DNA methods have become a mainstay of molecular genetics. They are also contributing to the solution of practical problems, for example, supplying hormones. The techniques described in this volume will surely play an important part in advancing our knowledge and command of heredity.

Acknowledgments

I am grateful to H. O. Smith and B. Weiss for comments on the manuscript for this article and to M. A. Kahler and J. Olsen for help in its preparation.

Work on this review was aided by Grants GM 26557 and CA 16519 from the National Institutes of Health.

[111] C. A. Hutchison, III, S. Phillips, M. H. Edgell, S. Gillam, P. Jahnke, and M. Smith, *J. Biol. Chem.* **253,** 6551 (1978).

[112] A. Razin, T. Hirose, K. Itakura, and A. D. Riggs, *Proc. Natl. Acad. Sci. U.S.A.* **75,** 4268 (1978).

Section II

Enzymes Used in Recombinant DNA Research

[2] Directory of Restriction Endonucleases

By RICHARD J. ROBERTS

Table I is intended to serve as a directory to the restriction endonucleases that have now been characterized. In forming the list, all endonucleases that cleave DNA at a specific sequence have been considered restriction enzymes, although in most cases there is no direct genetic evidence for the presence of a host-controlled restriction–modification system.

Certain strains have been omitted from this list to save space. Thus the many different *Staphylococcus aureus* isolates containing an isoschizomer of *Sau*3A[1] are not listed individually. Similarly the numerous strains of gliding bacteria (orders *Myxobacterales* and *Cytophagales*) that showed evidence of specific endonucleases during a large-scale screening[2] are still rather poorly characterized.

Within Table I the source of each microorganism is given either as an individual or a national culture collection. The enzymes are named in accordance with the proposal of Smith and Nathans.[3] When two enzymes recognize the same sequence (i.e., are isoschizomers), the prototype (i.e., the first example isolated) is indicated in parentheses in column 3. The recognition sequences (column 4) are abbreviated so that only one strand, reading $5' \rightarrow 3'$, is indicated and the point of cleavage, when known, is indicated by an arrow ($\downarrow$). When two bases appear in parentheses, either one may appear at that position within the recognition sequence. Where known, the base modified by the corresponding methylase is indicated by an asterisk. $\overset{*}{\mathrm{A}}$ is N^6-methyladenosine; $\overset{*}{\mathrm{C}}$ is 5-methylcytosine. The frequency of cleavage (columns 5–8) has been experimentally determined for bacteriophage lambda (λ) and adenovirus-2 (Ad2) DNAs, but represents the computer-derived values from the published sequences of SV40[4] and ϕX174[5] DNAs. When more than one reference appears (column 9), the first contains the purification procedure for the restriction enzyme, the second concerns its recognition sequence, the third contains the purification procedure for the methylase,

1. E. E. Stobberingh, R. Schiphof, and J. S. Sussenbach, *J. Bacteriol.* **131**, 645 (1977).
2. H. Mayer and H. Reichenbach, *J. Bacteriol.* **136**, 708 (1978).
3. H. O. Smith and D. Nathans, *J. Mol. Biol.* **81**, 419 (1973).
4. V. B. Reddy, B. Thimmappaya, R. Dhar, K. N. Subramanian, B. S. Zain, J. Pan, P. K. Ghosh, M. L. Celma, and S. M. Weissman, *Science* **200**, 494 (1978).
5. F. Sanger, G. M. Air, B. G. Barrell, N. L. Brown, A. R. Coulson, J. C. Fiddes, C. A. Hutchison, III, P. M. Slocombe, and M. Smith, *Nature* (*London*) **265**, 687 (1977).

METHODS IN ENZYMOLOGY, VOL. 68

ISBN 0-12-181968-X

TABLE I
RESTRICTION ENDONUCLEASES

Microorganism	Source	Enzyme	Sequence	Number of cleavage sites λ	Ad2	SV40	ϕX174	References[e]
Achromobacter immobilis	ATCC 15934	*Aim*I	?	?	?	?	?	6
Acinetobacter calcoaceticus	R. J. Roberts	*Acc*I	GT↓$\binom{A}{C}\binom{G}{T}$AC	7	8	1	3	7
	R. J. Roberts	*Acc*II (*Fnu*DII)	CGCG	>50	>50	0	14	7
Agrobacterium tumefaciens	ATCC 15955	*Atu*AI	?	>30	>30	?	?	8
Agrobacterium tumefaciens B6806	E. Nester	*Atu*BI (*Eco*RII)	CC$\binom{A}{T}$GG	>35	>35	16	2	9
Agrobacterium tumefaciens ID 135	C. Kado	*Atu*II (*Eco*RII)	CC$\binom{A}{T}$GG	>35	>35	16	2	10
Agrobacterium tumefaciens C58	E. Nester	*Atu*CI (*Bcl*I)	TGATCA	7	5	1	0	8
Anabaena catanula	CCAP 1403/1	*Aca*I	?	?	?	?	?	11
Anabaena cylindrica	A. deWaard	*Acy*I	GPu↓CGPyC	>14	>14	0	7	12
Anabaena subcylindrica	K. Murray	*Asu*I	G↓GNCC	>30	>30	11	2	11
Anabaena variabilis	K. Murray	*Ava*I	C↓PyCGPuG	8	?	0	1	13
	K. Murray	*Ava*II	G↓G$\binom{A}{T}$CC	>17	>30	6	1	13, 14 and 15
	K. Murray	*Ava*III	ATGCAT	?	?	3	0	16, 17 and 18
Anabaena variabilis [uw]	E. C. Rosenvold	*Avr*I (*Ava*I)	CPyCGPuG	8	?	0	1	19
	E. C. Rosenvold	*Avr*II	CCTAGG	1	2	2	0	19
Arthrobacter luteus	ATCC 21606	*Alu*I	AG↓CT	>50	>50	35	24	20
Arthrobacter pyridinolis	R. DiLauro	*Apy*I	CC$\binom{A}{T}$GG	>35	>35	16	2	21
Bacillus amyloliquefaciens F	ATCC 23350	*Bam*FI (*Bam*HI)	GGATCC	5	3	1	0	22
Bacillus amyloliquefaciens H	F. E. Young	*Bam*HI	G↓GATCC	5	3	1	0	23, 24

Bacillus amyloliquefaciens K	T. Kaneko	*Bam*KI (*Bam*HI)	GGATCC	5	3	1	0	22
Bacillus amyloliquefaciens N	T. Ando	*Bam*NI (*Bam*HI)	GGATCC	5	3	1	0	25
	T. Ando	*Bam*N_x	?	?	?	?	?	25 and 26
Bacillus brevis S	A. P. Zarubina	*Bbv*SI	GC($\overset{*T}{\underset{A}{}}$)GC	Specific methylase				27
Bacillus brevis	ATCC 9999	*Bbv*I	GC($\overset{T}{\underset{A}{}}$)GC	>30	>30	23	14	28
Bacillus caldolyticus	A. Atkinson	*Bcl*I	T↓GATCA	7	5	1	0	29
Bacillus cereus	ATCC 14579	*Bce*14579	?	>10	?	?	?	22
Bacillus cereus	1AM 1229	*Bce*1229	?	>10	?	?	?	22
Bacillus cereus	T. Ando	*Bce*170 (*Pst*I)	CTGCAG	18	25	2	1	22
Bacillus cereus Rf sm st	T. Ando	*Bce*R (*Fnu*DII)	CGCG	>50	>50	0	14	22
Bacillus globigii	G. A. Wilson	*Bgl*I	GCCNNNN↓NGGC	22	12	1	0	30 and 31, 32
	G. A. Wilson	*Bgl*II	A↓GATCT	>6	12	0	0	30 and 31, 33
Bacillus megaterium 899	B899	*Bme*899	?	>5	?	?	?	22
Bacillus megaterium B205-3	T. Kaneko	*Bme*205	?	>10	?	?	?	22
Bacillus megaterium	J. Upcroft	*Bme*I	?	>10	>20	4	?	34
Bacillus pumilus AHU1387	T. Ando	*Bpu*I	?	6	>30	2	?	35
Bacillus sphaericus	1AM 1286	*Bsp*1286	?	?	?	?	?	22
Bacillus sphaericus R	P. Venetianer	*Bsp*RI (*Hae*III)	GGCC	>50	>50	19	11	36
Bacillus stearothermophilus 1503-4R	N. Welker	*Bst*I (*Bam*HI)	GGATCC	5	3	1	0	37
Bacillus stearothermophilus 240	A. Atkinson	*Bst*AI	?	?	?	?	?	38
Bacillus stearothermophilus ET	N. Welker	*Bst*EI	?	?	?	?	?	39
	N. Welker	*Bst*EII	?	11	8	0	0	39
	N. Welker	*Bst*EIII	?	>7	?	?	?	39
Bacillus subtilis strain X5	T. Trautner	*Bsu*RI (*Hae*III)	GG↓$\overset{*}{C}$C	>50	>50	19	11	40, 41, 42
Bacillus subtilis Marburg 168	T. Ando	*Bsu*M	?	>10	?	?	?	22
Bacillus subtilis	ATCC 6633	*Bsu*6663	?	>20	?	?	?	22
Bacillus subtilis	1AM 1076	*Bsu*1076 (*Hae*III)	GGCC	>50	>50	19	11	22

(Continued)

TABLE I—*Continued*

Microorganism	Source	Enzyme	Sequence	Number of cleavage sites				References[e]
				λ	Ad2	SV40	φX174	
Bacillus subtilis	1AM 1114	*Bsu*1114 (*Hae*III)	GGCC	>50	>50	19	11	22
Bacillus subtilis	1AM 1247	*Bsu*1247 (*Pst*I)	CTGCAG	18	25	2	1	22, 43
Bacillus subtilis	ATCC 14593	*Bsu*1145	?	>20	?	?	?	22
Bacillus subtilis	1AM 1192	*Bsu*1192	?	>10	?	?	?	22
Bacillus subtilis	1AM 1193	*Bsu*1193	?	>30	?	?	?	22
Bacillus subtilis	1AM 1231	*Bsu*1231	?	>20	?	?	?	22
Bacillus subtilis	1AM 1259	*Bsu*1259	?	>8	?	?	?	22
Bordetella bronchiseptica	ATCC 19395	*Bbr*I (*Hin*dIII)	AAGCTT	6	11	6	0	44
Brevibacterium albidum	ATCC 15831	*Bal*I	TGG ↓ CCA	15	17	[illegible]	0	45
Brevibacterium luteum	ATCC 15830	*Blu*I (*Xho*I)	C ↓ TCGAG	1	5	[illegible]	1	46
	ATCC 15830	*Blu*II (*Hae*III)	GGCC	>50	>50	19	11	47
Caryophanon latum L.	H. Mayer	*Cla*I	AT ↓ CGAT	12	?	0	0	48
Chloroflexus aurantiacus	A. Bingham	*Cau*I (*Ava*II)	GG(A/T)CC	>30	>30	6	1	49
		*Cau*II	?	>30	>30	0	?	49
Chromobacterium violaceum	ATCC 12472	*Cvi*I	?	?	?	?	?	6
Corynebacterium humiferum	ATCC 21108	*Chu*I (*Hin*dIII)	AAGCTT	6	11	6	0	6
	ATCC 21108	*Chu*II (*Hin*dII)	GTPyPuAC	34	>20	7	13	6
Corynebacterium petrophilum	ATCC 19080	*Cpe*I (*Bcl*I)	TGATCA	7	5	1	0	50
Diplococcus pneumoniae	S. Lacks	*Dpn*I	GA* ↓ TC	?	?	?	0	51, 52 and 53
Diplococcus pneumoniae	S. Lacks	*Dpn*II (*Mbo*I)	GATC	>50	>50	7	0	51, 52
Enterobacter cloacae	H. Hartmann	*Ecl*I	?	15	?	?	?	54
	H. Hartmann	*Ecl*II (*Eco*RII)	CC(A/T)GG	>35	>35	16	2	54
Enterobacter cloacae	DSM 30056	*Eca*I	G ↓ GTNACC	12	?	0	0	55

Escherichia coli RY13	R. N. Yoshimori	*Eco*RI	G ↓ A$\overset{*}{\text{A}}$TTC	5	5	1	0	56, 57, 56, 58
	R. N. Yoshimori	*Eco*RI′	PuPuA ↓ TPyPy	>10	>10	24	16	59
Escherichia coli R245	R. N. Yoshimori	*Eco*RII	↓ $\overset{*}{\text{C}}$C($^{A}_{T}$)GG	>35	>35	16	2	60, 61 and 62, 60
Escherichia coli B	W. Arber	*Eco*B	TG$\overset{*}{\text{A}}$(N)$_8$TGCT	?	?	?	?	63, 64 and 65, 66
Escherichia coli K	M. Meselson	*Eco*K	AAC(N)$_6$GTGC	?	?	?	?	67, 68, 69
Escherichia coli (PI)	K. Murray	*Eco*PI	AGACC	?	?	?	?	70, 71, 72 and 73, 74
Escherichia coli P15	W. Arber	*Eco*P15	?	?	?	?	?	75
Fusobacterium nucleatum A	M. Smith	*Fnu*AI (*Hin*fI)	G ↓ ANTC	>50	>50	10	21	76
	M. Smith	*Fnu*AII (*Mbo*I)	GATC	>50	>50	7	0	44
Fusobacterium nucleatum C	M. Smith	*Fnu*CI (*Mbo*I)	↓ GATC	>50	>50	7	0	76
Fusobacterium nucleatum D	M. Smith	*Fnu*DI (*Hae*III)	GG ↓ CC	>50	>50	19	11	76
	M. Smith	*Fnu*DII	CG ↓ CG	>50	>50	0	14	76
	M. Smith	*Fnu*DIII (*Hha*I)	GCG ↓ C	>50	>50	2	18	76
Fusobacterium nucleatum E	M. Smith	*Fnu*EI (*Sau*3A)	↓ GATC	>50	>50	7	0	76
Fusobacterium nucleatum 48	M. Smith	*Fnu*48 I	?	>50	?	?	?	76
Haemophilus aegyptius	ATCC 11116	*Hae*I	($^{A}_{T}$)GG ↓ CC($^{T}_{A}$)	?	?	11	6	77
	ATCC 11116	*Hae*II	PuGCGC ↓ Py	>30	>30	1	8	78, 79
	ATCC 11116	*Hae*III	GG ↓ $\overset{*}{\text{C}}$C	>50	>50	19	11	80, 41, 81
Haemophilus aphrophilus	ATCC 19415	*Hap*I	?	>30	?	?	?	44
	ATCC 19415	*Hap*II (*Hpa*II)	C ↓ CGG	>50	>50	1	5	82, 83
Haemophilus gallinarum	ATCC 14385	*Hha*I[a]	GACGC	>50	>50	0	14	82, 84 and 85
Haemophilus haemo-globinophilus	ATCC 19416	*Hhg*I (*Hae*III)	GGCC	>50	>50	19	11	44
Haemophilus haemolyticus	ATCC 10014	*Hgc*I	G$\overset{*}{\text{C}}$G ↓ C	>50	>50	2	18	86, 86, 87
	ATCC 10014	*Hha*II (*Hin*fI)	GANTC	>50	>50	10	21	88
Haemophilus influenzae 1056	J. Stuy	*Hin*1056I (*Fnu*DII)	CGCG	>50	>50	0	14	89
	J. Stuy	*Hin*1056II	?	>30	>30	0	4	89

(*Continued*)

TABLE I—*Continued*

Microorganism	Source	Enzyme	Sequence	Number of cleavage sites				References[e]
				λ	Ad2	SV40	φX174	
Haemophilus influenzae serotype b, 1076	J. Stuy	*Hin*bIII (*Hin*dIII)	AAGCTT	6	11	6	0	89
Haemophilus influenzae R_b	C. A. Hutchison	*Hin*bIII (*Hin*dIII)	AAGCTT	6	11	6	0	90 and 42
Haemophilus influenzae serotype c, 1160	J. Stuy	*Hin*cII (*Hin*dII)	GTPyPuAC	34	>20	7	13	89
Haemophilus influenzae serotype c, 1161	J. Stuy	*Hin*cII (*Hin*dII)	GTPyPuAC	34	>20	7	13	89
Haemophilus influenzae R_c	A. Landy, G. Leidy	*Hin*cII (*Hin*dII)	GTPyPuAC	34	>20	7	13	91
Haemophilus influenzae R_d (exo mutant)	S. H. Goodgal	*Hin*dI	C$\overset{*}{A}$C	Specific methylase				92, 93
	S. H. Goodgal	*Hin*dII	GTPy ↓ Pu$\overset{*}{A}$C	34	>20	7	13	94, 95, 92, 93
	S. H. Goodgal	*Hin*dIII	$\overset{*}{A}$ ↓ AGCTT	6	11	6	0	96, 96, 92, 93
	S. H. Goodgal	*Hin*dIV	G$\overset{*}{A}$C	Specific methylase				92, 93
Haemophilus influenzae R_d 123	V. Tanyashin	*Hin*dGLU	?	?	?	?	?	97
Haemophilus influenzae R_f	C. A. Hutchison	*Hin*fI	G ↓ ANTC	>50	>50	10	21	90, 98 and 99
	C. A. Hutchison	*Hin*fII (*Hin*dIII)	AAGCTT	6	11	6	0	87
Haemophilus influenzae H-1	M. Takanami	*Hin*HI (*Hae*II)	PuGCGCPy	>30	>30	1	8	82
Haemophilus parahaemolyticus	C. A. Hutchison	*Hph*I[b]	GGTGA	>50	>50	4	9	90, 100
Haemophilus parainfluenzae	J. Setlow	*Hpa*I	GTT ↓ AAC	11	6	4	3	101, 102
	J. Setlow	*Hpa*II	C ↓ $\overset{*}{C}$GG	>50	>50	1	5	101, 102, 81
Haemophilus suis	ATCC 19417	*Hsu*I (*Hin*dIII)	A ↓ AGCTT	6	11	6	0	44
Herpetosiphon giganteus HP1023	J. H. Parish	*Hgi*AI	G$\binom{T}{A}$GC$\binom{T}{A}$ ↓ C	20	?	0	3	103
Klebsiella pneumoniae OK8	J. Davies	*Kpn*I	GGTAC ↓ C	2	8	1	0	104, 105
Microcoleus species	D. Comb	*Mst*I	TGCGCA	>10	>15	0	1	106, 106a

Moraxella bovis	ATCC 10900	*Mbo*I	↓GATC	>50	>50	7	0	107
	ATCC 10900	*Mbo*II[c]	GAAGA	>50	>50	15	11	107, 108 and 109
Moraxella glueidi LG1	J. Davies	*Mgl*I	?	?	?	?	?	104
Moraxella glueidi LG2	J. Davies	*Mgl*II	?	?	?	?	?	104
Moraxella nonliquefaciens	ATCC 19975	*Mno*I (*Hpa*II)	C↓CGG	>50	>50	1	5	44, 110
	ATCC 19975	*Mno*II	?	>10	>6	2	?	44
Moraxella nonliquefaciens	ATCC 17953	*Mnl*I[d]	CCTC	>100	>100	52	35	111
Moraxella nonliquefaciens	ATCC 17954	*Mnn*I (*Hind*II)	GTPyPuAC	34	>20	7	13	112
	ATCC 17954	*Mnn*II (*Hae*III)	GGCC	>50	>50	19	11	112
	ATCC 17954	*Mnn*III	?	>50	>50	?	?	112
	ATCC 17954	*Mnn*IV (*Hha*I)	GCGC	>50	>50	2	18	112
Moraxella osloensis	ATCC 19976	*Mos*I (*Mbo*I)	GATC	>50	>50	7	0	107
Moraxella species	R. J. Roberts	*Msp*I (*Hpa*II)	CCGG	>50	>50	1	5	113
Myxococcus virescens	H. Reichenbach	*Mvi*I	?	1	?	?	?	114
	H. Reichenbach	*Mvi*II	?	?	?	?	?	114
Neisseria gonorrhoea	G. Wilson	*Ngo*I (*Hae*II)	PuGCGCPy	>30	>30	1	8	115
Neisseria gonorrhoea	CDC 66	*Ngo*II (*Hae*III)	GGCC	>50	>50	19	11	116
Oerskovia xanthineolytica	R. Shekman	*Oxa*I (*Alu*I)	AGCT	>50	>50	35	24	117
	R. Shekman	*Oxa*II	?	?	?	?	?	117
Proteus vulgaris	ATCC 13315	*Pvu*I	CGATCG	4	7	0	0	28
	ATCC 13315	*Pvu*II	CAG↓CTG	15	22	3	0	28
Providencia alcalifaciens	ATCC 9886	*Pal*I (*Hae*III)	GGCC	>50	>50	19	11	34
Providencia stuartii 164	J. Davies	*Pst*I	CTGCA↓G	18	25	2	1	104, 118
Pseudomonas facilis	M. VanMontagu	*Pfa*I	?	>30	>30	?	?	47, 89
Rhodopseudomonas sphaeroides	R. Lascelles	*Rsp*I	?	3	12	0	?	119
Rhodopseudomonas sphaeroides	S. Kaplan	*Rsh*I (*Pvu*I)	CGATCG	4	7	0	0	120
Serratia marcescens S_b	C. Mulder	*Sma*I	CCC↓GGG	3	12	0	0	121, 122
Serratia species SAI	B. Torheim	*Ssp*I	?	?	?	?	?	123
Staphylococcus aureus 3A	E. E. Stobberingh	*Sau*3A (*Mbo*I)	GATC	>50	>50	7	0	124
Staphylococcus aureus PS96	E. E. Stobberingh	*Sau*96I (*Asu*I)	G↓GNCC	>30	>30	11	2	125

(Continued)

TABLE I—*Continued*

Microorganism	Source	Enzyme	Sequence	Number of cleavage sites				References[e]
				λ	Ad2	SV40	φX174	
Streptococcus faecalis subsp. *zymogenes*	R. Wu	*Sfa*I (*Hae*III)	GG ↓ CC	>50	>50	19	11	126
Streptococcus faecalis ND547	D. Clewell	*Sfa*NI	GATGC	>50	>30	6	12	8
Streptomyces achromogenes	ATCC 12767	*Sac*I	GAGCT ↓ C	2	7	0	0	127
	ATCC 12767	*Sac*II	CCGC ↓ GG	3	>25	0	1	127
	ATCC 12767	*Sac*III	?	>30	>30	?	?	127
Streptomyces albus	CM1 52766	*Sal*PI (*Pst*I)	CTGCAG	18	25	2	1	128
Streptomyces albus subsp. *pathociclicus*	KCC S0166	*Spa*I (*Xho*I)	CTCGAG	1	5	0	1	129
Streptomyces albus G	J. M. Ghuysen	*Sal*I	G ↓ TCGAC	2	3	0	0	130
	J. M. Ghuysen	*Sal*II	?	>30	?	?	?	130
Streptomyces bobiliae	ATCC 3310	*Sbo*I	?	?	?	?	?	131
Streptomyces bradiae	ATCC 3535	*Sbr*I	?	?	?	?	?	131
Streptomyces cupidosporus	KCC S0316	*Scu*I (*Xho*I)	CTCGAG	1	5	0	1	131
Streptomyces exfoliatus	H. Takahashi	*Sex*I (*Xho*I)	CTCGAG	1	6	0	1	129
Streptomyces goshikiensis	H. Takahashi	*Sgo*I (*Xho*I)	CTCGAG	1	6	0	1	129
Streptomyces griseus	ATCC 23345	*Sgr*I	?	0	7	0	?	127
Streptomyces hygroscopicus	?	*Shy*I	?	2	?	?	?	132
Streptomyces lavendulae	ATCC 8644	*Sla*I (*Xho*I)	C ↓ TCGAG	1	6	0	1	131
Streptomyces luteoreticuli	H. Takahashi	*Slu*I (*Xho*I)	CTCGAG	1	6	0	1	129
Streptomyces stanford	S. Goff, A. Rambach	*Sst*I (*Sac*I)	GAGCT ↓ C	2	7	0	0	133, 134
	S. Goff, A. Rambach	*Sst*II (*Sac*II)	CCGC ↓ GG	3	>25	0	1	133
	S. Goff, A. Rambach	*Sst*III (*Sac*III)	?	>30	>30	?	?	133

Thermoplasma acidophilum	D. Searcy	*Tha*I (*Fnu*DII)	CG ↓ CG	>50	>50	0	14	135
Thermopolyspora glauca	ATCC 15345	*Tgl*I (*Sac*II)	CCGCGG	3	>25	0	1	28
Thermus aquaticus YTI	J. I. Harris	*Taq*I	T ↓ CGA	>50	>50	1	10	136
	J. I. Harris	*Taq*II	?	>30	>30	4	6	44
Xanthomonas amaranthicola	ATCC 11645	*Xam*I (*Sal*I)	GTCGAC	2	3	0	0	130
Xanthomonas badrii	ATCC 11672	*Xba*I	T ↓ CTAGA	1	4	0	0	137
Xanthomonas holcicola	ATCC 13461	*Xho*I	C ↓ TCGAG	1	6	0	1	46
	ATCC 13461	*Xho*II	Pu ↓ GATCPy	>20	>20	3	0	89, 28
Xanthomonas malvacearum	ATCC 9924	*Xma*I	C ↓ CCGGG	3	12	0	0	122
	ATCC 9924	*Xma*II (*Pst*I)	CTGCAG	18	25	2	1	122
Xanthomonas nigromaculans	ATCC 23390	*Xni*I (*Pvu*I)	CGATCG	4	7	0	0	112
Xanthomonas oryzae	M. Ehrlich	*Xor*I (*Pst*I)	CTGCAG	18	25	2	1	138
	M. Ehrlich	*Xor*II (*Pvu*I)	CGATCG	4	7	0	0	138
Xanthomonas papavericola	ATCC 14180	*Xpa*I (*Xho*I)	C ↓ TCGAG	1	6	0	1	138

[a] *Hga*I cleaves as indicated: 5′ GACGCNNNNN ↓ 3′
3′ CTGCGNNNNNNNNNN ↓ 5′.

[b] *Hph*I cleaves as indicated: 5′ GGTGANNNNNNNN ↓ 3′
3′ CCACTNNNNNNN ↑ 5′.

[c] *Mbo*II cleaves as indicated: 5′ GAAGANNNNNNNN ↓ 3′
3′ CTTCTNNNNNNN ↑ 5′.

[d] *Mnl*I cleaves 5 to 10 bases from the recognition sequence.

[e] Key to references:

6. S. A. Endow and R. J. Roberts, unpublished observations.
7. M. Zabeau and R. J. Roberts, unpublished observations.
8. D. Sciaky and R. J. Roberts, unpublished observations.
9. G. Roizes, M. Patillon, and A. Kovoor, *FEBS Lett*. **82,** 69 (1977).
10. J. M. LeBon, C. Kado, L. J. Rosenthal, and J. Chirikjian, *Proc. Natl. Acad. Sci. U.S.A.* **75,** 4097 (1978).
11. S. G. Hughes, T. Bruce, and K. Murray, unpublished observations.
12. A. DeWaard, J. Korsuize, C. P. van Beveren, and J. Maat, *FEBS Lett*. **96,** 106 (1978).

TABLE I—*Continued*

13. K. Murray, S. G. Hughes, J. S. Brown, and S. Bruce, *Biochem. J.* **159,** 317 (1976).
14. J. G. Sutcliffe and G. Church, unpublished observations.
15. C. Fuchs, E. C. Rosenvold, A. Honigman, and W. Szybalski, *Gene* **4,** 1 (1978).
16. G. Roizes, P.-C. Nardeux, and R. Monier, *FEBS Lett.* **104,** 39 (1979).
17. H. Shimatake and M. Rosenberg, unpublished observations.
18. K. Denniston-Thompson, D. D. Moore, K. E. Kruger, M. E. Furth, and F. R. Blattner, *Science* **198,** 1051 (1978).
19. E. C. Rosenvold and W. Szybalski, unpublished observations.
20. R. J. Roberts, P. A. Myers, A. Morrison, and K. Murray, *J. Mol. Biol.* **102,** 157 (1976).
21. R. DiLauro, unpublished observations.
22. T. Shibata, S. Ikawa, C. Kim, and T. Ando, *J. Bacteriol.* **128,** 473 (1976).
23. G. A. Wilson and F. E. Young, *J. Mol. Biol.* **97,** 123 (1975).
24. R. J. Roberts, G. A. Wilson, and F. E. Young, *Nature (London)* **265,** 82 (1977).
25. T. Shibata and T. Ando, *Biochim. Biophys. Acta* **442,** 184 (1976).
26. T. Shibata and T. Ando, *Mol. Gen. Genet.* **138,** 269 (1975).
27. B. F. Vanyushin and A. P. Dobritsa, *Biochim. Biophys. Acta* **407,** 61 (1975).
28. T. R. Gingeras and R. J. Roberts, unpublished observations.
29. A. H. A. Bingham, T. Atkinson, D. Sciaky, and R. J. Roberts, *Nucleic Acids Res.* **5,** 3457 (1978).
30. G. A. Wilson and F. E. Young, *in* "Microbiology 1976" (D. Schlessinger, ed.), p. 350. Am. Soc. Microbiol., Washington, D.C., 1976.
31. C. H. Duncan, G. A. Wilson, and F. E. Young, *J. Bacteriol.* **134,** 338 (1978).
32. T. Bickle, unpublished observations.
33. V. Pirrotta, *Nucleic Acids Res.* **3,** 1747 (1976).
34. R. E. Gelinas, P. A. Myers, and R. J. Roberts, unpublished observations.
35. S. Ikawa, T. Shibata, and T. Ando, *J. Biochem. (Tokyo)* **80,** 1457 (1976).
36. A. Kiss, B. Sain, E. Csordas-Toth, and P. Venetianer, *Gene* **1,** 323 (1977).
37. J. Catterall and N. Welker, *J. Bacteriol.* **129,** 1110 (1977).
38. A. H. A. Bingham, R. J. Sharp, and A. Atkinson, unpublished observations.
39. R. B. Meagher, unpublished observations.
40. S. Bron, K. Murray, and T. A. Trautner, *Mol. Gen. Genet.* **143,** 13 (1975).
41. S. Bron and K. Murray, *Mol. Gen. Genet.* **143,** 25–33 (1975).
42. U. Gunthert, M. Freund, and T. A. Trautner, *FEBS Symp.* **12,** (abstr.) (1978).
43. T. Hoshino, T. Uozumi, S. Horinouchi, A. Ozaki, T. Beppu, and K. Arima, *Biochim. Biophys. Acta* **479,** 367 (1977).
44. R. J. Roberts and P. A. Myers, unpublished observations.

45. R. E. Gelinas, P. A. Myers, G. A. Weiss, R. J. Roberts, and K. Murray, *J. Mol. Biol.* **114,** 433 (1977).
46. T. R. Gingeras, P. A. Myers, J. A. Olson, F. A. Hanberg, and R. J. Roberts, *J. Mol. Biol.* **118,** 113 (1978).
47. M. Van Montagu, unpublished observations.
48. H. Mayer, R. Grosschedl, H. Schutte, and G. Hobom, unpublished observations.
49. A. H. A. Bingham and J. Darbyshire, unpublished observations.
50. J. Fisherman, T. R. Gingeras, and R. J. Roberts, unpublished observations.
51. S. Lacks and B. Greenberg, *J. Biol. Chem.* **250,** 4060 (1975).
52. S. Lacks and B. Greenberg, *J. Mol. Biol.* **114,** 153 (1977).
53. G. E. Geier and P. Modrich, *J. Biol. Chem.* **254,** 1408 (1979).
54. H. Hartmann and W. Goebel, *FEBS Lett.* **80,** 285 (1977).
55. H. Mayer, E. Schwarz, M. Melzer, and G. Hobom, unpublished observations.
56. P. J. Greene, M. C. Betlach, H. M. Goodman, and H. W. Boyer, *Methods Mol. Biol.* **7,** 87 (1974).
57. J. Hedgpeth, H. M. Goodman, and H. W. Boyer, *Proc. Natl. Acad. Sci. U.S.A.* **69,** 3448 (1972).
58. A. Dugaiczyk, J. Hedgpeth, H. W. Boyer, and H. M. Goodman, *Biochemistry* **13,** 503 (1974).
59. K. Murray, J. S. Brown, and S. A. Bruce, unpublished observations.
60. R. N. Yoshimori, Ph.D. Thesis, University of California, San Francisco (1971).
61. C. H. Bigger, K. Murray, and N. E. Murray, *Nature (London), New Biol.* **244,** 7 (1973).
62. H. W. Boyer, L. T. Chow, A. Dugaiczyk, J. Hedgpeth, and H. M. Goodman, *Nature (London), New Biol.* **244,** 40 (1973).
63. B. Eskin and S. Linn, *J. Biol. Chem.* **247,** 6183 (1972).
64. J. A. Lautenberger, N. C. Kan, D. Lackey, S. Linn, M. H. Edgell, and C. A. Hutchison, III, *Proc. Natl. Acad. Sci. U.S.A.* **75,** 2271 (1978).
65. J. V. Ravetch, K. Horiuchi, and N. D. Zinder, *Proc. Natl. Acad. Sci. U.S.A.* **75,** 2266 (1978).
66. J. A. Lautenberger and S. Linn, *J. Biol. Chem.* **247,** 6176 (1972).
67. M. Meselson and R. Yuan, *Nature (London)* **217,** 1110 (1968).
68. N. C. Kan, J. A. Lautenberger, M. H. Edgell, and C. A. Hutchison, III, *Fed. Proc., Fed. Am. Soc. Exp. Biol.* **37,** 1499 (1978); also unpublished observations.
69. A. Haberman, J. Heywood, and M. Meselson, *Proc. Natl. Acad. Sci. U.S.A.* **69,** 3138 (1972).
70. A. Haberman, *J. Mol. Biol.* **89,** 545 (1974).
71. B. Bachi and V. Pirrotta, unpublished observations.
72. J. P. Brockes, *Biochem. J.* **133,** 629 (1973).
73. J. P. Brockes, P. R. Brown, and K. Murray, *Biochem. J.* **127,** 1 (1972).
74. J. P. Brockes, P. R. Brown, and K. Murray, *J. Mol. Biol.* **88,** 437 (1974).
75. J. Reiser and R. Yuan, *J. Biol. Chem.* **252,** 451 (1977).
76. A. Lui, B. C. McBride, and M. Smith, unpublished observations.
77. K. Murray, A. Morrison, H. W. Cooke, and R. J. Roberts, unpublished observations.

TABLE I—*Continued*

78. R. J. Roberts, J. B. Breitmeyer, N. F. Tabachnik, and P. A. Myers, *J. Mol. Biol.* **91,** 121 (1975).
79. C.-P.D. Tu, R. Roychoudhury, and R. Wu, *Biochem. Biophys. Res. Commun.* **72,** 355 (1976).
80. J. H. Middleton, M. H. Edgell, and C. A. Hutchison, III, *J. Virol.* **10,** 42 (1972).
81. M. B. Mann and H. O. Smith, *Nucleic Acids Res.* **4,** 4211 (1977).
82. M. Takanami, *Methods Mol. Biol.* **7,** 113 (1974).
83. H. Sugisaki and K. Takanami, *Nature (London), New Biol.* **246,** 138 (1973).
84. N. L. Brown and M. Smith, *Proc. Natl. Acad. Sci. U.S.A.* **74,** 3213 (1977).
85. H. Sugisaki, *Gene* **3,** 17 (1978).
86. R. J. Roberts, P. A. Myers, A. Morrison, and K. Murray, *J. Mol. Biol.* **103,** 199 (1976).
87. M. B. Mann and H. O. Smith, unpublished observations.
88. M. B. Mann, R. N. Rao, and H. O. Smith, *Gene* **3,** 97 (1978).
89. J. A. Olson, P. A. Myers, and R. J. Roberts, unpublished observations.
90. J. H. Middleton, P. V. Stankus, M. H. Edgell, and C. A. Hutchison, III, unpublished observations.
91. A. Landy, E. Ruedisueli, L. Robinson, C. Foeller, and W. Ross, *Biochemistry* **13,** 2134 (1974).
92. P. H. Roy and H. O. Smith, *J. Mol. Biol.* **81,** 427 (1973).
93. P. H. Roy and H. O. Smith, *J. Mol. Biol.* **81,** 445 (1973).
94. H. O. Smith and K. W. Wilcox, *J. Mol. Biol.* **51,** 379 (1970).
95. T. J. Kelly, Jr. and H. O. Smith, *J. Mol. Biol.* **51,** 393 (1970).
96. R. Old, K. Murray, and G. Roizes, *J. Mol. Biol.* **92,** 331 (1975).
97. V. I. Tanyashin, L. I. Li, I. O. Muizhnieks, and A. A. Baev, *Dokl. Akad. Nauk SSSR* **231,** 226 (1976).
98. C. A. Hutchison, III and B. G. Barrell, unpublished observations.
99. K. Murray and A. Morrison, unpublished observations.
100. D. Kleid, Z. Humayun, A. Jeffrey, and A. Ptashne, *Proc. Natl. Acad. Sci. U.S.A.* **73,** 293 (1976).
101. P. A. Sharp, B. Sugden, and J. Sambrook, *Biochemistry* **12,** 3055 (1973).
102. D. E. Garfin and H. M. Goodman, *Biochem. Biophys. Res. Commun.* **59,** 108 (1974).
103. N. L. Brown, M. McClelland, and P. R. Whitehead, unpublished observations.
104. D. L. Smith, F. R. Blattner, and J. Davies, *Nucleic Acids Res.* **3,** 343 (1976).
105. J. Tomassini, R. Roychoudhury, R. Wu, and R. J. Roberts, *Nucleic Acids Res.* **5,** 4055 (1978).
106. D. Comb, I. Schildkraut, and R. J. Roberts, unpublished observations.
106a. T. R. Gingeras, J. P. Milazzo, and R. J. Roberts, *Nucleic Acids Res.* **5,** 4105 (1978).

107. R. E. Gelinas, P. A. Myers, and R. J. Roberts, *J. Mol. Biol.* **114,** 169 (1977).
108. N. L. Brown, C. A. Hutchison, III, and M. Smith, *J. Mol. Biol.* (in press).
109. S. A. Endow, *J. Mol. Biol.* **114,** 441 (1977).
110. U. L. RajBhandary and B. Baumstark, unpublished observations.
111. M. Zabeau, R. Greene, P. A. Myers, and R. J. Roberts, unpublished observations.
112. F. Hanberg, P. A. Myers, and R. J. Roberts, unpublished observations.
113. M. Van Montagu, P. A. Myers, and R. J. Roberts, unpublished observations.
114. D. W. Morris and J. H. Parish, *Arch. Microbiol.* **108,** 227 (1976).
115. G. A. Wilson and F. E. Young, unpublished observations.
116. D. J. Clanton, J. M. Woodward, and R. V. Miller, *J. Bacteriol.* **135,** 270 (1978).
117. A. Stotz and P. Philippson, unpublished observations.
118. N. L. Brown and M. Smith, *FEBS Lett.* **65,** 284 (1976).
119. A. H. A. Bingham, A. Atkinson, and J. Darbyshire, unpublished observations.
120. J. Gardner and S. Kaplan, unpublished observations.
121. R. Greene and C. Mulder, unpublished observations.
122. S. A. Endow and R. J. Roberts, *J. Mol. Biol.* **112,** 521 (1977).
123. B. Torheim, personal communication.
124. J. S. Sussenbach, C. H. Monfoort, R. Schiphof, and E. E. Stobberingh, *Nucleic Acids Res.* **3,** 3193 (1976).
125. J. S. Sussenbach, P. H. Steenbergh, J. A. Rost, W. J. Van Leeuwen, and J. D. A. van Embden, *Nucleic Acids Res.* **5,** 1153 (1978).
126. R. Wu, C. T. King, and E. Jay, *Gene* **4,** 329 (1978).
127. J. R. Arrand, P. A. Myers, and R. J. Roberts, unpublished observations.
128. K. Chater, *Nucleic Acids Res.* **4,** 1989 (1977).
129. H. Shinotsu, H. Takahashi, and H. Saito, unpublished observations.
130. J. R. Arrand, P. A. Myers, and R. J. Roberts, *J. Mol. Biol.* **118,** 127 (1978).
131. H. Takahashi, M. Shimizu, H. Saito, Y. Ikeda, and H. Sugisaki, *Gene* **5,** 9 (1979).
132. F. Walter, M. Hartmann, and M. Roth, *FEBS Symp.* **12,** (abstr.) (1978).
133. S. Goff and A. Rambach, *Gene* **3,** 347 (1978); also unpublished observations.
134. F. Muller, S. Stoffel, and S. G. Clarkson, unpublished observations.
135. D. McConnell, D. Searcy, and J. G. Sutcliffe, *Nucleic Acids Res.* **5,** 1729 (1978).
136. S. Sato, C. A. Hutchison, III, and J. I. Harris, *Proc. Natl. Acad. Sci. U.S.A.* **74,** 542 (1977).
137. B. S. Zain and R. J. Roberts, *J. Mol. Biol.* **115,** 249 (1977).
138. J. Shedlarski, M. Farber, and M. Ehrlich, unpublished observations.

TABLE II
LIST OF ENZYMES WITH KNOWN RECOGNITION SEQUENCES

Terminal extension	Restriction enzyme	Recognition sequence
Blunt ends	*Dpn*I	G$\overset{*}{\mathrm{A}}$↓TC
	*Eco*RI′	PuPuA↓TPyPy
	*Sma*I	CCC↓GGG
	*Alu*I	AG↓CT
	*Pvu*II	CAG↓CTG
	*Fnu*DII	CG↓CG
	*Hae*I	$\binom{A}{T}$GG↓CC$\binom{A}{T}$
	*Hpa*I	GTT↓AAC
5′↓GATC	*Mbo*I	↓GATC
	*Bgl*II	A↓GATCT
	*Bam*HI	G↓GATCC
	*Bcl*I	T↓GATCA
	*Xho*II	Pu↓GATCPy
5′↓CG	*Hpa*II	C↓CGG
	*Taq*I	T↓CGA
	*Cla*I	AT↓CGAT
	*Acy*I	GPu↓CGPyC
5′↓TCGA	*Xho*I	C↓TCGAG
	*Sal*I	G↓TCGAC
5′↓AATT	*Eco*RI	G↓AATTC
5′↓AGCT	*Hin*dIII	A↓AGCTT
5′↓CCGG	*Xma*I	C↓CCGGG
5′↓CTAG	*Xba*I	T↓CTAGA
3′ TGCA↓	*Pst*I	CTGCA↓G
3′ GTAC↓	*Kpn*I	GGTAC↓C
3′ GC↓	*Sac*II	CCGC↓GG
3′ GCGC↓	*Hae*II	PuGCGC↓Py
3′ CG↓	*Hha*I	GCG↓C
3′ AGCT↓	*Sac*I	GAGCT↓C
5′ ↓CC$\binom{A}{T}$GG	*Eco*RII	↓CC$\binom{A}{T}$GG
5′ ↓GTNAC	*Eca*I	G↓GTNACC
5′ ↓NNNNN	*Hga*I	5′ ↓NNNNNNNNNNGCGTC 3′ 3′ ↑NNNNNCGCAG 5′
5′ ↓PyCGPu	*Ava*I	C↓PyCGPuG
5′ ↓ANT	*Hin*fI	G↓ANTC
5′ ↓GNC	*Asu*I	G↓GNCC
5′ ↓G$\binom{A}{T}$C	*Ava*II	G↓G$\binom{A}{T}$CC
5′ ↓$\binom{A}{C}\binom{G}{T}$	*Acc*I	GT↓$\binom{A}{C}\binom{G}{T}$AC
3′ N↓	*Hph*I	5′ GGTGANNNNNNNN↓ 3′ 3′ CCACTNNNNNNN↑ 5′
	*Mbo*II	5′ GAAGANNNNNNNN↓ 3′ 3′ CTTCTNNNNNNN↑ 5′

and the fourth describes its recognition sequence. In some cases two references appear in one of these categories when two independent groups have reached similar conclusions.

Table II contains a listing of enzymes for which the recognition sequence is known and which might be useful for preparing recombinant DNAs. They are grouped according to the nature of the fragment ends produced. Thus, fragments generated by all enzymes within any group can be joined to one another.

[3] Addition of Homopolymers to the 3′-Ends of Duplex DNA with Terminal Transferase[1]

By TIMOTHY NELSON and DOUGLAS BRUTLAG

The linkage of two DNAs *in vitro* to form recombinant molecules first became possible with the discovery of DNA ligases.[2] These enzymes, which seal nicks in DNA, can covalently join two DNAs that have complementary sticky ends such as the short, staggered ends generated by many restriction endonucleases.[3] Lobban and Kaiser[4] and Jackson *et al.*[5] showed that complementary ends could be added to DNA molecules *in vitro* with terminal transferase, thus allowing any two DNAs to be linked. These workers added complementary single-stranded homopolymers to two DNA molecules, annealed the homopolymer regions, and covalently closed the resulting hybrid *in vitro* with DNA polymerase I and DNA ligase from *Escherichia coli.* The DNA polymerase was necessary to trim any excess unpaired nucleotides at the 3′-ends or to fill in gaps generated by unequal lengths of the complementary homopolymer regions. Wensink *et al.*[6] simplified this procedure by showing that the annealed recombinant molecules were infectious and that they would be covalently closed *in vivo* during transfection.

Lobban and Kaiser[4] originally found that completely duplex molecules were inefficient primers for the terminal transferase reaction and that pretreatment of the DNA with lambda exonuclease to expose single-stranded

[1] This work was supported by a Basil O'Connor starter grant from the National Foundation-March of Dimes.

[2] I. R. Lehman, *Science* **186,** 790 (1974).

[3] J. E. Mertz and R. W. Davis, *Proc. Natl. Acad. Sci. U.S.A.* **69,** 3370 (1972).

[4] P. E. Lobban and A. D. Kaiser, *J. Mol. Biol.* **78,** 453 (1973).

[5] D. A. Jackson, R. H. Symons, and P. Berg, *Proc. Natl. Acad. Sci. U.S.A.* **69,** 2904 (1972).

[6] P. C. Wensink, D. J. Finegan, J. E. Donelson, and D. S. Hogness, *Cell* **3,** 315 (1974).

ISBN 0-12-181968-X

3′-termini remarkably improved their priming ability. This degradation of the 5′-ends of duplex termini allowed terminal transferase to extend every 3′-end quantitatively, a prerequisite for the formation of biologically active recombinant molecules. Subsequently, Roychoudhury *et al.*[7,8] and Brutlag *et al.*[9] found experimental conditions that allowed terminal transferase to utilize duplex termini without prior exonuclease treatment. Both studies showed that conditions that destabilize DNA duplexes, lower ionic strength or the substitution of Co^{2+} for Mg^{2+} or both, permitted terminal transferase to extend duplex termini. Similar conditions had previously been found by Kato *et al.*[10] to be optimal for terminal transferase action upon single-stranded primers, presumably because they disrupted secondary structure in the primer.

The conditions for homopolymer addition described in this chapter are designed to unravel duplex DNA at the ends. Consequently they are of general application for addition to 3′-primer termini that are recessed, flush, or extended with respect to the 5′-end. Under these conditions addition to randomly sheared DNA is also highly efficient. Unfortunately, addition to nicks is also encouraged,[11] and thus intact DNA and terminal transferase free of endonuclease are required if homopolymers are to be added only at the ends of molecules.

The primary advantage of using homopolymer termini for forming recombinant DNA is that randomly sheared DNA segments can be cloned, eliminating the need for DNA containing specific restriction sites. A second advantage is that all successful infection events result from hybrid molecules, since the vector cannot anneal with itself, and hence is noninfectious. The major disadvantage of this method is that the infectivity of annealed molecules is usually much lower than that of covalently closed forms.

Principle

The principle of this method is extension of the 3′-end of duplex DNA with a single-stranded homopolymer by treatment with terminal transferase in the presence of a single deoxynucleoside triphosphate. The reaction conditions encourage fraying of the duplex ends and thus permit terminal transferase to act upon a duplex primer. Conditions for adding each of the four deoxynucleotide triphosphates are described to allow the use of either pair of complementary homopolymers for recombinant DNA formation. The average length of the homopolymer can be controlled by

[7] R. Roychoudhury, E. Jay, and R. Wu, *Nucleic Acids Res.* **3,** 863 (1976).

[8] R. Roychoudhury and R. Wu, this series, Vol. 65, p. 43.

[9] D. Brutlag, K. Fry, T. Nelson, and P. Hung, *Cell* **10,** 509 (1977).

[10] K. Kato, J. M. Gonçalves, G. E. Houts, and F. J. Bollum, *J. Biol. Chem.* **242,** 2780 (1967).

[11] W. Bender and N. Davidson, *Cell* **7,** 595 (1976).

varying the time or rate of the reaction, or by limiting the level of deoxynucleotide substrate. Since terminal transferase is a nonprocessive enzyme,[12,13] the reaction rate can be decreased by lowering the level of enzyme below its dissociation constant for the homopolymer being synthesized, or by decreasing the reaction temperature.

Materials and Reagents

Terminal transferase (terminal deoxynucleotidyl transferase, specific activity 34,000 units/mg) was purified according to Chang and Bollum[14] and was the gift of R. L. Ratliff. The commerical preparations of P-L Biochemicals and Bethesda Research Laboratories have also proved satisfactory. Labeled deoxynucleoside triphosphates from New England Nuclear and unlabeled triphosphates from P-L Biochemicals were analyzed for their content of nucleoside di- and monophosphates as described below. *Eco*RI endonuclease was purified according to Modrich and Zabel[15] and was the gift of Paul Modrich.

Methods

Preparation of Substrates

Linear duplex DNA samples are extracted with phenol, passed over a BioRad A-15m agarose column (in 10 m*M* Tris, pH 7.4, 1 m*M* EDTA) to remove the phenol, ethanol-precipitated twice, and resuspended in 10 m*M* Tris-HCl, 1 m*M* EDTA, pH 7.4. The efficiency of homopolymer addition can be seriously reduced by failure to extract with phenol, particularly following treatment with restriction enzymes which may remain associated with the termini of DNA. For example, terminating an *Eco*RI endonuclease digestion by heating at 65° inactivates the endonuclease and results in DNA that is an efficient primer for DNA polymerase I. However, this DNA is not an effective primer for terminal transferase prior to phenol extraction.[16] Further, it is important that the DNA samples be free of nicks and short oligonucleotides, both of which can serve as efficient primers in the reactions to follow. Utilization of such primers will severely reduce the efficiency of subsequent annealing steps in the formation of hybrid molecules. The A-15m agarose chromatography helps eliminate very short primers. The number-average molecular weight of

[12] L. M. S. Chang and F. J. Bollum, *Biochemistry* **10,** 536 (1971).
[13] D. Brutlag and A. Kornberg, *J. Biol. Chem.* **247,** 241 (1972).
[14] L. M. S. Chang and F. J. Bollum, *J. Biol. Chem.* **246,** 909 (1971).
[15] P. Modrich and B. Zabel, *J. Biol. Chem.* **251,** 5866 (1976).
[16] T. Nelson, unpublished observations.

sheared preparations needed for the calculation of end concentration is most readily determined by electron microscopy.[17]

Unlabeled nucleoside triphosphates are dissolved in water and adjusted to pH 7.0. The relative amounts of nucleoside mono- and diphosphate contaminants are determined by polyethyleneimine chromatography[18] with 0.4 M NH_4HCO_3 as solvent. The total nucleotide concentration is determined from the extinction coefficient and then corrected for the actual amount of nucleoside triphosphate determined above. [^{3}H]Nucleoside triphosphates are similarly chromatographed to determine radiochemical purity, and the specific activity is diluted at least 10-fold with the unlabeled form.

Synthesis of Polypyrimidine Homopolymers on Duplex DNA

Optimal conditions for the synthesis of polypurines and polypyrimidines by terminal transferase are different[9,10] and will be treated separately. Polypyrimidines [poly(dT) and poly(dC)] are added to 10–20 pmol/ml of DNA termini with 40 μg/ml terminal transferase in 100 mM potassium cacodylate, pH 7.0, 1 mM $CoCl_2$, 0.1 mM dithiothreitol, and 100 μM ^{3}H-dTTP (0.25 mCi/μmol). The cobalt salt must be added to the reaction last. Since EDTA is usually present in the DNA samples, it is essential that the level of cobalt be 1 mM in excess of the final EDTA concentration. Also, other weaker chelators of Co^{2+}, such as phosphate, which may be present in the DNA or in preparations of terminal transferase, should be avoided. Higher amounts of $CoCl_2$ (5–10 mM) are inhibitory to terminal transferase.The entire reaction mixture is usually warmed to 37° prior to addition of the enzyme that starts the reaction. Roychoudhury[7,8] has described similar conditions for polypyrimidine addition that use lower amounts of terminal transferase. Lower levels of terminal transferase should be used with caution when employing sheared DNA primers. Since terminal transferase is a nonprocessive enzyme,[12,13] its ability to use a primer will depend on its affinity for that terminus. Sheared DNA will contain both recessed and protruding 3′-ends, and efficient utilization of all ends is assured only if the concentration of enzyme is in excess of its dissociation constant for all types of primers. As a result of the nonprocessive action of terminal transferase, different enzyme concentrations are necessary for maximal rates of polypyrimidine and polypurine synthesis (Table I). Once the maximal rate is attained, no further increase in rate with increasing enzyme is observed. In all the reactions reported in Table I the molar concentration of enzyme is in vast excess of the molar concentration of primer termini. Higher terminal transferase

[17] R. Davis, M. Simon, and N. Davidson, this series, Vol. 12, Part D, p. 412.

[18] K. Randerath and E. Randerath, *J. Chromatogr.* **16,** 111 (1964).

Addition of Homopolymers to Termini Generated by *Eco*RI[a]

Terminal transferase (μM)	Length of poly (dT) (dTMP/end/5 min)	Length of poly (dA) (dAMP/end/5 min)
0.42	95	19
1.05	291	17
2.11	300	42
4.22	333	130
10.5	424	140
21.1	—	124

[a] *Eco*RI-cleaved pSC101 DNA (0.03 μM in termini) was incubated with the indicated concentration of terminal transferase (13.6–683 μg/ml or 462–23,100 units/ml) under the conditions described in the methods for each reaction.

concentrations in general favor both more efficient utilization of all available primers and more uniform extension of each.

In preparative reactions the extent of addition is monitored by transferring the reaction to ice while aliquots are removed for acid precipitation. The reaction may then be returned to 37° if further chain extension is desired. The actual rate of the reaction with a particular enzyme preparation and primer sample is usually determined beforehand with a small-scale reaction at the desired concentrations of enzyme and substrate. The reactions are terminated by the addition of EDTA to 10 m*M*, followed by phenol extraction and A-15m chromatography in 10 m*M* Tris, pH 7.4, 1 m*M* EDTA. Since DNA containing single-stranded regions often adheres to glass surfaces, we routinely work with plastic columns and tubes. DNA in the void volume of the A-15m column is usually at a concentration in excess of that necessary for the annealing reaction and transfection without further concentration by ethanol precipitation.

Synthesis of Poly(dA) and Poly(dI) Homopolymers

Although polypurine homopolymers can be synthesized under the same conditions as described for polypyrimidine synthesis,[7,8] the use of conditions optimized for polypurine extension makes the addition reaction more rapid, more uniform, and less sensitive to contaminating enzymatic activities. Polypurine synthesis carried out in 20 m*M* potassium phosphate, pH 7.0, 4 m*M* $MgCl_2$, 0.1 m*M* dithiothreitol, and 100 μM deoxynucleoside triphosphate is at least 10 times faster than polypurine synthesis in the cacodylate–cobalt buffer described above.[10] Maximal rates of polypurine synthesis require the use of higher concentrations of primer termini (20–100 pmol/ml) and terminal transferase (400 μg/ml) than for polypyrimidine synthesis (Table I). This makes small reaction volumes essential. The use of phosphate buffer inhibits *de novo* poly(dA) synthesis frequently encountered during primed synthesis of poly(dA).[10]

Figure 1 compares the rate of addition of poly(dA) to a duplex DNA primer in the high-ionic-strength buffers normally used to assay the enzyme, compared with the lower ionic strengths described here. At low

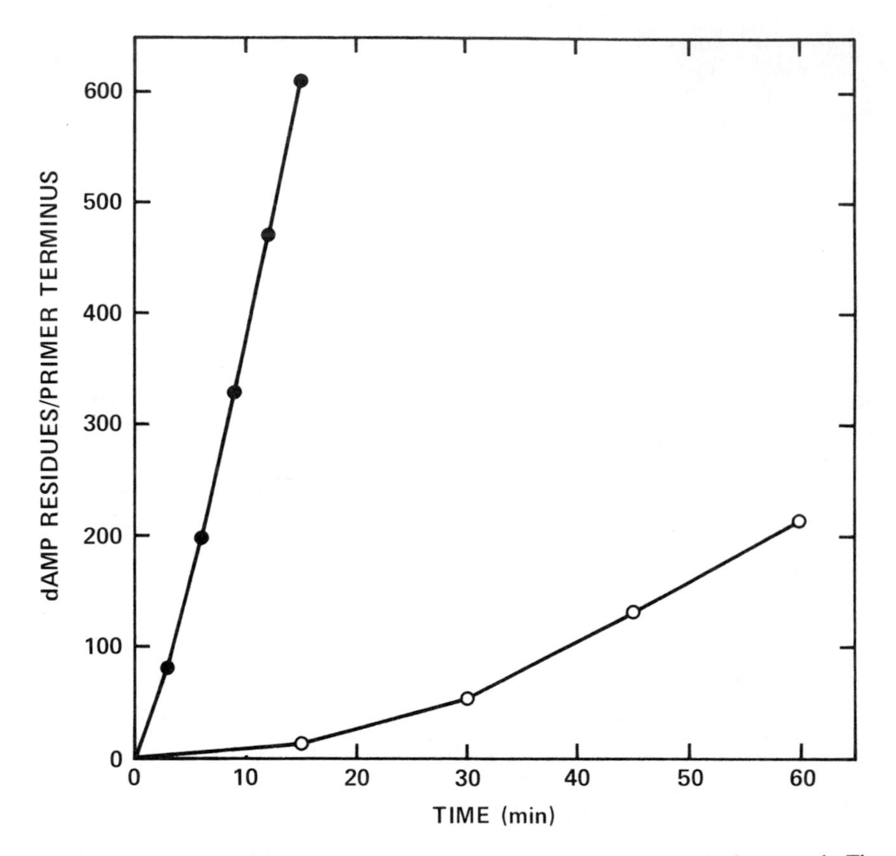

FIG. 1. Addition of poly(dA) to sheared duplex DNA at low and high ionic strength. The solid circles show the average length of poly(dA) added to hydrodynamically sheared satellite DNA (1.705 g/cm^3 DNA from *Drosophila melanogaster*,[9] 0.006 μ*M* termini) which was incubated with terminal transferase (4 μ*M*) under the low-ionic-strength conditions (20 m*M* potassium phosphate, pH 7.0, 4 m*M* $MgCl_2$) described in the text. The open circles show an identical reaction performed in 100 m*M* potassium cacodylate, 15 m*M* potassium phosphate, pH 7.0, 8 m*M* $MgCl_2$, 0.1 m*M* dithiothreitol. These higher ionic strength conditions are those originally used by Lobban and Kaiser[4] for terminal addition to duplex DNA. Kinetics with exclusively phosphate buffers give similar results.

ionic strength, the addition of poly(dA) is nearly linear for the first 600 nucleotides, while at higher ionic strength the rate progressively increases as more and more primers become available, either through extension of less favorable recessed 3′-ends or by endonuclease cleavage of growing chains to generate new primer termini. These autocatalytic reaction kinetics were first observed with duplex primers by Lobban and Kaiser[4] who obtained uniform primer extension only after digestion of the 5′-ends with lambda exonuclease.

The product of the polypurine synthesis is purified as described for the polypurine products. The A-15m chromatography is particularly important here to eliminate the phosphate buffer which strongly inhibits Ca^{2+}-mediated transfection.

Synthesis of Poly(dG) Homopolymers

The joining of vectors and inserts with poly(dG·dC) homopolymers can permit the reconstruction of certain restriction sites (such as *Pst*I, *Hae*III, *Kpn*I, etc.), resulting in two such sites on either side of the insert in the hybrid plasmid.[19] This restriction site reconstruction permits easy removal of the inserted DNA from the vector after cloning (Scheme 1): Direct synthesis of long poly(dG) homopolymers by the polypurine procedure described above is not possible because of the secondary structure formed by poly(dG).[20] While it is possible to circumvent this problem,[21] Röwekamp and Firtel[22] have found that very short poly(dG) and poly(dC) homopolymers (8 to 10 bp average length) can be used for the formation of highly infective hybrid DNA molecules. These workers use the cacodylate–$CoCl_2$ reaction conditions described by Roychoudhury *et al.*[7] (140 m*M* potassium cacodylate, 30 m*M* Tris-OH, 1 m*M* $CoCl_2$, 0.1 m*M* dithiothreitol, pH 6.9) but with a limiting level of dGTP or dCTP (only a 50- to 100-fold molar excess of nucleoside triphosphate over the available termini) and high concentrations of terminal transferase (1000 units/ml) and ends (140 pmol/ml). Under these conditions incubation for only 45 sec at 37° is reported to result in homopolymers of 10 nucleotides average length. Again we have found that it is wise to chill the reaction to 0° and check the extent of reaction before termination. Since the nucleoside triphosphates are well below the K_m for terminal transferase, the rate of the reaction is very sensitive to the absolute amount of nucleoside triphos-

[19] A. Otsuka, personal communication.

[20] F. J. Bollum, *in* "The Enzymes" (P. D. Boyer, ed.), 3rd ed., Vol. 10, p. 145. Academic Press, New York, 1974.

[21] C. F. Lefler and F. J. Bollum, *J. Biol. Chem.* **244**, 594 (1969).

[22] W. Röwekamp and R. Firtel, unpublished results.

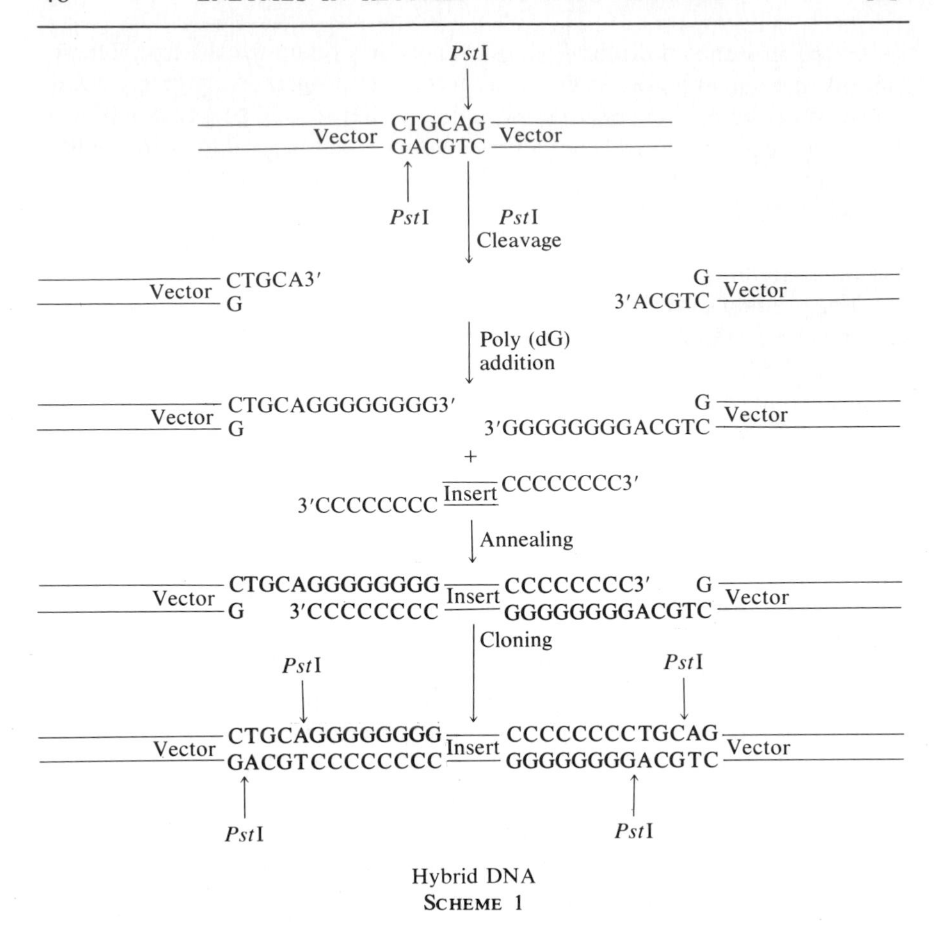

Hybrid DNA
SCHEME 1

phate and to the presence of other contaminating nucleotide species. Röwekamp and Firtel[22] report very high infectivities (1000–10,000 colonies/μg of pBR322 vector DNA) even with such short homopolymers. We have confirmed this and find the infectivity with short poly(dG)-poly(dC) homopolymers to be reproducibly 10–100 times higher than with the longer poly(dA)-poly(dT) homopolymers. The success of this method for regenerating restriction sites on either site of the inserted DNA depends strongly on the elimination of all 3′-exonuclease from the restriction endonuclease. The removal of even a single nucleotide from the 3′-end of the vector prior to the terminal transferase reaction will prevent resynthesis of the site. Commercial preparations of *Pst*I from Bethesda Research Laboratories have been satisfactory in this respect. Our current results indicate that 80% of all hybrids have both sites reconstructed, indicating that at least 90% of all the 3′-ends are intact after *Pst*I cleavage.

This method could also be used to clone *Pst*I-generated DNA fragments at any site in a vector by adding poly(dG) to the *Pst*I fragment to be inserted and poly(dC) to any linear vector DNA.

Comments

The efficiency of primer extension by these procedures has been measured in several ways including adsorption of DNA with homopolymer termini to poly(dT) or poly (dA) cellulose,[5] the ability of two DNAs with complementary homopolymers to form circles after annealing,[9] and the nearest-neighbor transfer of ^{32}P from the homopolymer to the terminal nucleotide of the DNA segment.[7] Although these methods are no longer employed routinely, they are useful when one first attempts to add homopolymers with a new preparation of terminal transferase or when a very low infectivity results from transfection. Similarly, new preparations of terminal transferase should be checked for endonuclease activity under the conditions used for homopolymer synthesis. The simplest tests include the conversion of supercoiled DNA to nicked form, monitored either by electron microscopy or by agarose gel electrophoresis. If endonuclease is detected, it may be advisable to try the Co^{2+} conditions rather than the Mg^{2+} conditions described as optimal for polypurine synthesis, since many endonucleases are less active in the presence of Co^{2+}.

One of the major disadvantages of using homopolymers to link hybrid DNAs had been the inability to excise them readily from the recombinant molecule. This disadvantage is largely overcome by the ability to resynthesize restriction sites as described above. Even if reconstruction of a restriction site is not feasible, at least two methods are available for excising fragments joined by homopolymer linkers. Hofstetter *et al.*[23] showed that the low thermal stability of poly(dA·dT) regions made possible their specific cleavage by S1 nuclease under conditions of partial denaturation (45–50% formamide, w/v). This technique has been used analytically to estimate the length of the linker regions and preparatively to isolate the cloned segment free of vector DNA. Goff and Berg[24] have described a technique for excision of cloned segments, which takes advantage of the opposite polarity of the homopolymers at either end of the insert. This procedure can be used to isolate segments flanked by either poly(dA·dT) or by poly(dG·dC).

It should also be cautioned that heterogeneity in the lengths of homopolymer joins has been observed both in cloned and subcloned recom-

[23] H. Hofstetter, A. Schambock, J. van den Berg, and C. Weissmann, *Biochim. Biophys. Acta* **454,** 587 (1976).

[24] S. Goff and P. Berg, *Proc. Natl. Acad. Sci. U.S.A.* **75,** 1763 (1978).

binant plasmids.[9,25] This length heterogeneity may be related to the instability observed for other simple repeating sequences within cloned DNA.[9] Such length variation should be considered in any sequencing strategy that involves sequencing through the homopolymer joins or that relies on a restriction site near such a join. The use of very short homopolymers as described above should largely eliminate this problem.

[25] T. Maniatis, G. K. Sim, A. Efstratiadis, and F. C. Kafatos, *Cell* **8,** 163 (1976).

[4] DNA-Joining Enzymes: A Review

By N. Patrick Higgins and Nicholas R. Cozzarelli

The first DNA-joining enzymes identified were the DNA ligases. They join together DNA chains by transmuting the high-energy pyrophosphate linkage of a nucleotide cofactor into a phosphoester bond between 5′-phosphoryl and 3′-hydroxyl termini. The impetus for the nearly simultaneous discovery of DNA ligase in bacteria[1-3] and in bacteriophage-infected cells[4,5] was the recognition that DNA fragments were joined together during genetic recombination.[6,7] Shortly thereafter, Okazaki and co-workers[8] showed that ligase played the key role in DNA replication of joining the nascent small pieces of DNA at the replication fork. Ligase has since been found to participate in the synthesis and repair of DNA in a variety of organisms.[9-11]

The termini of DNA strands to be joined by ligase must in general be abutted by base pairing to a complementary strand. This ensures proper alignment and preservation of the nucleotide sequence of DNA; thus, the reaction can be considered *conservative*. Template-dependent synthesis and faithful conservation of DNA coding information is also the hallmark

[1] M. Gellert, *Proc. Natl. Acad. Sci. U.S.A.* **57,** 148 (1967).
[2] B. M. Olivera, and I. R. Lehman, *Proc. Natl. Acad. Sci. U.S.A.* **57,** 1426 (1967).
[3] M. L. Gefter, A. Becker, and J. Hurwitz, *Proc. Natl. Acad. Sci. U.S.A.* **58,** 240 (1967).
[4] B. Weiss, and C. C. Richardson, *Proc. Natl. Acad. Sci. U.S.A.* **57,** 1021 (1967).
[5] N. R. Cozzarelli, N. E. Melechen, T. M. Jovin, and A. Kornberg, *Biochem. Biophys. Res. Commun.* **28,** 578 (1967).
[6] M. Meselson, and J. J. Weigle, *Proc. Natl. Acad. Sci. U.S.A.* **47,** 869 (1961).
[7] G. M. Kellenberger, M. L. Zichichi, and J. J. Weigle, *Proc. Natl. Acad. Sci. U.S.A.* **47,** 869 (1964).
[8] K. Sugimoto, T. Okazaki, and R. Okazaki, *Proc. Natl. Acad. Sci. U.S.A.* **60,** 1356 (1968).
[9] I. R. Lehman, *Science* **186,** 790 (1974).
[10] K. A. Nasmyth, *Cell* **12,** 1109 (1977).
[11] L. H. Johnston, and K. A. Nasmyth, *Nature (London)* **274,** 891 (1978).

ISBN 0-12-181968-X

of most DNA polymerases. There is, however, increasing interest in *radical* or template-independent reactions. Since they are not disciplined by a complementary strand, they generally change the nucleotide sequence at the site of reaction. Polynucleotide phosphorylase and terminal deoxynucleotidyltransferase are examples of radical polymerases that generate new sequences at the 3'-hydroxyl terminus of DNA molecules.[12-14] T4-induced RNA ligase is a radical ligase that catalyzes the uninstructed joining of both RNA and DNA chains. Although a role for T4 RNA ligase in the joining of nucleic acids *in vivo* has not been demonstrated, some presumably radical enzymes must circularize the single-stranded RNA viroids in plant tissues[15,16] and splice precursor RNA chains to form functional messages in eukaryotes.[17] T4 DNA ligase at high concentrations also carries out a template-independent joining of molecules with base-paired ends, so that it can operate in either a conservative or radical mode.

The term "ligase" will be reserved for proteins that join the preexisting termini of polynucleotide chains. This is the basis for the common assay for RNA and DNA ligases—conversion of a terminal 5'-^{32}P-labeled phosphoryl to a form resistant to digestion with bacterial alkaline phosphatase. However, several enzymes have been discovered that join polynucleotides in a fundamentally different manner. The substrate for these enzymes is not the ends but continuous stretches of nucleic acids which are broken and rejoined in a concerted fashion. These breakage–reunion (B-R) enzymes[98a] require no cofactor to supply energy for reforming the DNA backbone bond. Presumably, the energy released by breakage is stored in a covalent enzyme–DNA intermediate and utilized for joining. A major class of B-R enzymes catalyze the interconversion of topological isomers of DNA and are designated topoisomerases.[18] The first enzyme isolated in this class of proteins was *Escherichia coli* ω protein, and other examples are the bacterial DNA gyrases and the DNA untwisting enzymes from eukaryotic cells. The integration of bacteriophage λ DNA involves breaking at least four strands of duplex DNA and resealing the ends to different partners without alteration in the winding number. The

[12] S. Gillam, K. Waterman, and M. Smith, *Nucleic Acids Res.* **2,** 613 (1975).

[13] K. Kato, J. M. Gonçalves, G. E. Houts, and F. J. Bollum, *J. Biol. Chem.* **242,** 2780 (1967).

[14] R. Roychoudhury, E. Jay, and R. Wu, *Nucleic Acids Res.* **3,** 863 (1976).

[15] T. O. Diener, *Annu. Rev. Microbiol.* **28,** 23 (1974).

[16] H. L. Sänger, G. Klotz, D. Riesner, H. J. Gross, and A. K. Kleinschmidt, *Proc. Natl. Acad. Sci. U.S.A.* **73,** 3852 (1976).

[17] F. Crick, *Science* **204,** 264 (1979).

[18] J. C. Wang, and L. F. Liu, *In* "Molecular Genetics" (J. H. Taylor, ed.), Part III, pp. 65–88. Academic Press, New York, 1979.

reaction does not require any energy cofactor. This enzyme and the topoisomerases isolated to date are conservative B-R enzymes. However, the ϕX174 *cisA*-coded protein is a site-specific B-R enzyme which apparently is not template-instructed, and there may be additional radical B-R enzymes. Several models for movement of insertion elements have proposed radical joining reactions at the terminus of the transposing region[19,20]; these reactions might be carried out by ligases, B-R enzymes, or a combination of both. This article is a brief review of the four classes of enzymes that ligate polynucleotide chains *in vitro,* the radical and conservative ligases and B-R enzymes. Since the conservative enzymes are more often summarized, emphasis will be placed on the enzymes that carry out template-independent reactions.

DNA Ligase

The *E. coli* and bacteriophage T4-induced enzymes are the most thoroughly investigated DNA ligases, and the work up to 1974 has been summarized in the review by Lehman.[9] DNA ligase from *E. coli* is a single polypeptide chain of molecular weight 74,000, and bacteriophage T4 induces a DNA ligase having a single chain with a molecular weight of 68,000. Phosphodiester bond synthesis is coupled to the cleavage of a pyrophosphate bond in nicotinamide adenine dinucleotide (NAD) for the bacterial enzyme and in ATP for the phage enzyme (Fig. 1). Two different covalent intermediates have been identified as partial reaction products for each enzyme. The first step of the reaction is the adenylylation of free enzyme with the release of nicotinamide mononucleotide (NMN) for the *E. coli* enzyme and of pyrophosphate for the T4 enzyme. For both ligases, the epsilon amino group of a lysine residue forms a phosphoamide linkage with AMP. This reaction is noteworthy because the high-energy bond necessary to drive phosphodiester bond synthesis is actually stored in the enzyme before it encounters a polynucleotide chain. From this step, the reaction paths of both ligases are identical. The second step is transfer of the AMP moiety from the enzyme to the 5′-phosphoryl of the DNA substrate, recreating a pyrophosphate linkage and thereby preserving the high-energy bond. The third step of the reaction is the nucleophilic attack of the adjacent 3′-hydroxyl group to form a phosphodiester bond and eliminate AMP.

Purification of *E. coli* DNA ligase has recently been aided by the construction of overproducing strains of bacteria that harbor the ligase gene

[19] N. D. F. Grindley, and D. J. Sherratt, *Cole Spring Harbor Symp. Quant. Biol.* **43,** 1257 (1978).

[20] J. A. Shapiro, *Proc. Natl. Acad. Sci. U.S.A.* **76,** 1933 (1979).

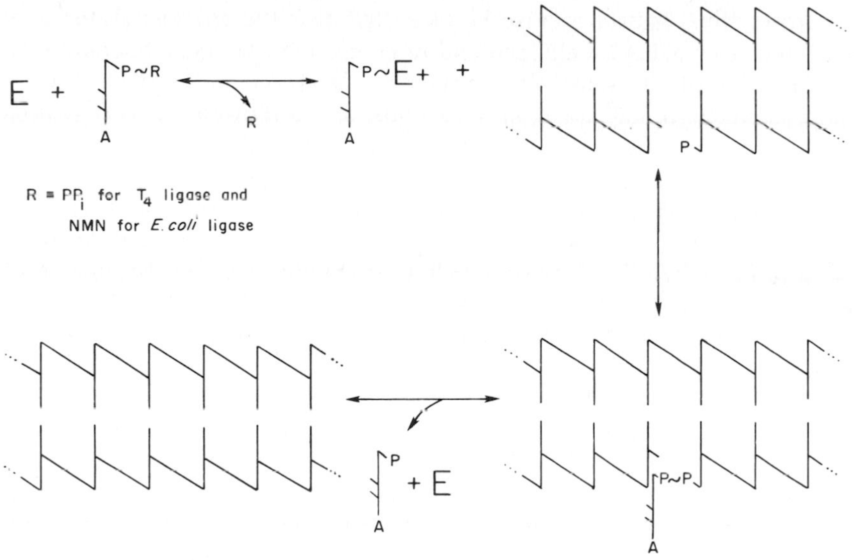

R = PP_i for T_4 ligase and
NMN for *E. coli* ligase

FIG. 1. Reaction mechanism of DNA ligase (E).

on a plasmid or episome.[21,22] Panasenko *et al.* constructed a bacterial strain lysogenic for a transducing λ phage that carries the ligase gene next to an "up" promoter mutation, *lopll*.[21] After induction, phage replication results in many ligase gene copies, and the amount of enzyme reaches 2–3% of the total cellular protein. With a simple procedure involving a single chromatographic step, 30 mg of homogeneous enzyme was prepared from 120 g of these cells.

Reliable schemes for the preparation of homogeneous T4 DNA ligase are available.[23,24] Recently a simpler protocol has been reported to yield nearly pure enzyme after only two chromatographic steps.[25] A significant advantage of preparing the T4 ligase is that at least three other valuable T4-induced enzymes can be concurrently purified from the same batch of infected cells. Panet *et al.*[24] devised a purification procedure for the simultaneous isolation of polynucleotide kinase, DNA ligase, and DNA po-

[21] S. M. Panasenko, R. J. Alazard, and I. R. Lehman, *J. Biol. Chem.* **253,** 4590 (1978).

[22] K. Borck, J. D. Beggs, W. J. Brammar, A. S. Hopkins, and N. E. Murray, *Mol. Gen. Genet.* **146,** 199 (1976).

[23] B. Weiss, A. Jacquemin-Sablon, T. R. Live, G. C. Fareed, and C. C. Richardson, *J. Biol. Chem.* **243,** 4543 (1968).

[24] A. Panet, J. H. van de Sande, P. C. Loewen, H. G. Khorana, A. J. Raae, J. R. Lillehaug, and K. Kleppe, *Biochemistry* **12,** 5045 (1973).

[25] K.-W. Knopf, *Eur. J. Biochem.* **73,** 33 (1977).

lymerase. RNA ligase can also be interdigitated; the enzyme elutes after DNA ligase on DEAE-cellulose and does not bind to phosphocellulose.

A number of different assays have been used to monitor DNA ligases. These include joining the cohesive ends of bacteriophage λ,[1,3] sealing nicks in natural DNA molecules,[4] intermolecular joining of short homopolymers annealed to a template strand,[2] linking polynucleotides to the ends of oligonucleotides immobilized on cellulose under the instruction of a complementary strand,[5] circularization of the self-complementary polymer poly[d(A-T)],[26] pyrophosphate exchange,[27,28] and formation of an acid-precipitable enzyme–adenylate complex.[23,25,29]

DNA ligases catalyze a variety of inter- and intramolecular joining reactions, and some examples are given in Table I. The reactions catalyzed by the *E. coli* enzyme are asterisked. *Escherichia coli* ligase joins oligo(dT) that is base-paired to poly(dA); it also joins oligo(dA) base-paired to poly(dT), but this is a much less favorable reaction. The self-complementary poly[d(A-T)] can fold back upon itself so as to oppose the ends and ligase seals these into closed loop molecules (circles). Joining of the 5′-phosphoryl of a DNA chain to the 3′-hydroxyl of an RNA molecule has been shown in two ways.[30] First, oligo(A) was joined to the 5′-phosphoryl of oligo(dA) on a poly(dT) template strand. Second, ligase can circularize poly[d(A-T)]pU in which UMP occupies the 3′-hydroxyl terminal position.

T4 DNA ligase is a much more permissive enzyme than *E. coli* ligase. In addition to the reactions mentioned above, it catalyzes a number of ligations that cannot as yet be performed with the *E. coli* enzyme. Kleppe *et al.*[31] and Fareed *et al.*[32] found that ligation of oligo(dT) annealed to poly(A) took place at a few percent of the rate occurring with oligo(dT)·poly(dA). The phage enzyme will also act as an RNA ligase in the strict sense and join RNA to RNA, but this is a very unfavorable reaction. The joining of oligo(I), oligo(C), or oligo(U) in the presence of their complementary polyribonucleotide partners occurs at rates at least two orders of magnitude lower than with DNA substrates.[28] Joining of the 5′-phosphoryl end of an RNA molecule to the 3′-hydroxyl of a DNA chain was first noted by Westegard *et al.*[33] who found that T4 ligase would seal

[26] P. Modrich and I. R. Lehman, *J. Biol. Chem.* **245,** 3626 (1970).

[27] V. Sgaramella and S. D. Ehrlich, *Eur. J. Biochem.* **86,** 531 (1978).

[28] H. Sano and G. Feix, *Biochemistry* **13,** 5110 (1974).

[29] S. B. Zimmerman and C. K. Oshinsky, *J. Biol. Chem.* **244,** 4689 (1969).

[30] K. Nath and J. Hurwitz, *J. Biol. Chem.* **249,** 3680 (1974).

[31] K. Kleppe, J. H. van de Sande, and H. G. Khorana, *Proc. Natl. Acad. Sci. U.S.A.* **67,** 68 (1970).

[32] G. C. Fareed, E. M. Wilt, and C. C. Richardson, *J. Biol. Chem.* **246,** 925 (1971).

[33] O. Westegaard, D. Brutlag, and A. Kornberg, *J. Biol. Chem.* **248,** 1361 (1973).

TABLE I
EXAMPLES OF DNA LIGASE REACTIONS[a]

Type of reaction	Substrate	Template	Linkages formed	References
DNA–DNA	Oligo(dT)*	Poly(dA)	dTpdT	2, 30–32, 35
DNA–DNA	Oligo(dA)*	Poly(dT)	dApdA	32
DNA–DNA	Oligo[d(A-T)]*	Self-complementary	dTpdA and dApdT	27, 30, 32
DNA–DNA	Oligo(dT)	Poly(A)	dTpdT	28, 32, 35
DNA–DNA	$(pdT)_{11}pdC$	Poly(dA)	dCpdT	36
DNA–DNA	Termini of exonuclease III-digested DNA	?	Complementary strand, terminal cross-links	37
DNA–DNA	Blunt-ended duplex DNA	None	Phosphodiester bonds between duplex ends	40
RNA–DNA	Oligo(dA) plus oligo(A)*	Poly(dT)	ApdA	30
RNA–DNA	Oligo[d(A-T)]pU*	Self-complementary	UpdA	30
RNA–DNA	Oligo(dT) plus oligo(U)	Poly(dA)	UpdT	30
RNA–RNA	Oligo(A)	Poly(dT)	ApA	28, 31, 32
RNA–RNA	Oligo(I)	Poly(C)	IpI	28
RNA–RNA	Oligo(C)	Poly(I)	CpC	28
RNA–RNA	Oligo(U)	Poly(A)	UpU	28
DNA–RNA	$(pA)_3(pdA)_n$	Poly(dT)	dApA	30

[a] All reactions listed are catalyzed by T4 ligase, but *E. coli* ligase carries out only the asterisked reactions. The convention for "Type of reaction" and "Linkages formed" is that the molecule contributing the 3′-hydroxyl group is on the left and the moiety supplying the 5′-phosphoryl is on the right.

in an RNA fragment used to prime DNA synthesis on the ϕX174 viral strand. Similarly, Nath and Hurwitz[30] showed that T4 ligase joined chains of $(pA)_3(pdA)_n$ when hybridized to poly(dT), albeit at a low rate. The closed circular form of ColE1 DNA purified from chloramphenicol-treated *E. coli* can contain ribonucleotide segments.[34] If this RNA is a remnant of the primer for DNA synthesis, then some *E. coli* ligase must be able to join RNA to DNA even though the known *E. coli* ligase has not yet demonstrated this capacity *in vitro*.

Extensive base pairing is not required for T4 DNA ligase to join nucleotide chains. Harvey and Wright[35] showed that T4 ligase joined oligonucleotides in the presence of their complement at temperatures above their T_m. For example, the rate of joining of $(pdT)_9$ of the presence of poly(dA) was optimal at 25°, and ligation was found at temperatures as high as 37°. Yet, in the absence of ligase at least, at 25° no complex between nonathymidylate and poly(dA) was observed by gel chromatography. It is quite possible that in these studies the enzyme stabilized a transient duplex since the complementary strand is required for the ligation. However, a clear example of a radical ligation was the early observation of Tsiapolis and Narang[36] that T4 ligase joined molecules with a terminal mismatch. The polymer $(dpT)_{11}$pdC hybridized to poly(dA) was joined to form products 400 nucleotides long. This type of reaction could provide an important method for creating site-specific mutations, and further investigations should be made. Another reaction of T4 ligase proceeding with an imperfectly paired substrate was reported by Weiss.[37] A mixture of T4 DNA ligase and *E. coli* exonuclease III cross-links the termini of duplex DNA producing molecules that rapidly renatured after denaturation. A mixture of T7-induced enzymes including ligase also carries out terminal cross-linking.[38] The mechanism suggested for these reactions was exonuclease digestion of the 3′-end to expose a sequence of two to three nucleotides complementary to the 5′-end of the sister strand, base pairing which created a terminal loop and apposed the 5′- and 3′-termini, and finally ligation.

The most emphatically radical of all the T4 ligase reactions was discovered by Sgaramella *et al.*[39,40] who found that it would carry out the in-

[34] D. G. Blair, D. B. Clewell, D. J. Sherratt, and D. R. Helinski, *Proc. Natl. Acad. Sci. U.S.A.* **69,** 2518 (1972).

[35] C. L. Harvey and R. Wright, *Biochemistry* **11,** 2667 (1972).

[36] C. M. Tsiapolis and S. A. Narang, *Biochem. Biophys. Res. Commun.* **39,** 631 (1970).

[37] B. Weiss, *J. Mol. Biol.* **103,** 669 (1976).

[38] P. Sadowski, A. McGreer, and A. Becker, *Can. J. Biochem.* **52,** 525 (1974).

[39] V. Sgaramella, J. H. van de Sande, and H. G. Khorana, *Proc. Natl. Acad. Sci. U.S.A.* **67,** 1468 (1970).

[40] V. Sgaramella and H. G. Khorana, *J. Mol. Biol.* **72,** 493 (1972).

termolecular joining of DNA substrates at completely base-paired ends. This reaction, called blunt-end joining, is a powerful method for structuring DNA molecules *in vitro* (see below) but proceeds much less readily than nick sealing. Blunt-end joining is not linearly dependent on enzyme concentration but increases greatly at higher enzyme levels. Although T4 RNA ligase does not catalyze blunt-end joining, it markedly stimulates the reaction, particularly at low DNA ligase concentrations.[11] As much as a 20-fold increase in the rate was observed at low DNA ligase concentrations, and the stimulation was specific for blunt-end joining because RNA ligase had only a marginal influence on the joining of cohesive ends. In the presence of RNA ligase, T4 DNA ligase had about the same turnover rate (1 mol/min per mol of ligase) for blunt-end and cohesive-end joining. The apparent K_m for blunt ends in the presence or absence of RNA ligase was 50 μM (in terms of 5′-termini) which was two orders of magnitude higher than for a nicked substrate. Joining of the base-paired ends of *Hae*III restriction enzyme fragments of ColE1 DNA was examined in detail.[41] The reaction was highly efficient with all sizes of molecules participating in the reaction; linear molecules longer than ColE1 DNA were generated, and 10% of the molecules were circular. The circles were on average smaller than the linear products. However, in a more dilute reaction, 40% of the product molecules were circular.[42] Therefore, reactions in which an intermolecular product is desired should be carried out at high substrate concentrations, whereas circularization is favored by low substrate concentrations and short polymers. The precision of the blunt-end joining reaction was demonstrated by the ability of the joined products to be redigested with *Hae*III restriction enzyme.[41,42]

Poor blunt-end joining has been reported with preparations of T4 DNA ligase active in sealing nicks, even though it has been proved that the identical enzyme is operative in both reactions.[41] There are several possible reasons for this observation. First, the degree of contamination with T4 RNA ligase sharply influences the blunt-end joining reaction. Second, the reaction exhibits bimolecular reaction kinetics at low substrate concentrations[43] and a high K_m for the substrate. Therefore, dilute reactions may result in disappointing yields. Third, in the absence of RNA ligase, the blunt-end joining reaction is not linear with respect to the DNA ligase concentration and requires much larger amounts of enzyme (10- to 30-fold) than sealing nicks.[26,41]

[41] A. Sugino, H. M. Goodman, H. L. Heyneker, J. Shine, H. W. Boyer, and N. R. Cozzarelli, *J. Biol. Chem.* **252,** 3987 (1977).

[42] M. Mottes, C. Morandi, S. Cremaschi, and V. Sgaramella, *Nucleic Acids Res.* **4,** 2467 (1977).

[43] K. V. Deugau and J. H. van de Sande, *Biochemistry* **17,** 723 (1978).

The use of DNA ligase in *de novo* construction of DNA of defined sequence has been pioneered by Khorana and his co-workers.[44] An example of the strategy employed was synthesis of the duplex DNA corresponding to yeast alanine tRNA. Overlapping polydeoxyribonucleotide segments representing the entire two strands of the DNA were synthesized by chemical methods. Their lengths ranged from 8 to 12 nucleotides, and the overlapping regions were at least 4 nucleotides long. The last step employed DNA ligase to stitch the base-paired fragments together. T4 DNA ligase has the advantage of requiring a smaller overlapping sequence than the *E. coli* enzyme.

DNA ligase is routinely used in the cloning of DNA to join large DNA fragments containing complementary and antiparallel single-strand extensions commonly called cohesive or sticky ends; the techniques have recently been reviewed by Vosberg.[45] There are three general approaches to creating cohesive ends. The simplest method is to use a type II restriction endonuclease that introduces staggered scissions at unique sequences. A second method uses terminal deoxynucleotidyltransferase to generate single-stranded tails of either poly(dA) and poly(dT) or poly(dG) and poly(dC) at the 3′-hydroxyl terminus of different populations of DNA molecules. In the third method, short duplexes containing a restriction nuclease cleavage site called "linkers" are joined to DNA by T4 ligase-catalyzed blunt-end joining. Linkers have been constructed for the *Eco*RI, *Hin*dIII, *Bam*I, *Hpa*II, *Mbo*I, and *Pst*I restriction enzymes.[46–49] The decanucleotide duplex

pCCGGATCCGG
 GGCCTAGGCCp

has sites for the *Mbo*I (↓GATC), *Hpa*II (C↓CGG), and *Bam*I (G↓GATCC) nucleases (arrows indicate the site of cleavage).[48] After digesting the product with one of the restriction enzymes that cleaves the linkers, the product can be bound by cohesive-end joining to any cloning vehicle that has one of these restriction sites. Examples of genetic se-

[44] H. G. Khorana, K. L. Agarwal, H. Büchi, M. G. Caruthers, N. K. Gupta, K. Kleppe, A. Kumar, E. Ohtsuka, U. L. RajBhandary, J. H. van de Sande, V. Sgaramella, T. Terao, H. Weber, and T. Yamada, *J. Mol. Biol.* **72,** 209 (1972).

[45] H.-P. Vosberg, *Hum. Genet.* **40,** 1 (1977).

[46] K. J. Marians, R. Wu, J. Stawinski, T. Hozumi, and S. A. Narang, *Nature (London)* **263,** 744 (1976).

[47] H. L. Heynecker, J. Shine, H. M. Goodman, H. W. Boyer, J. Rosenberg, R. E. Dickerson, S. A. Narang, K. Itakura, S. Liu, and A. D. Riggs, *Nature (London)* **263,** 748 (1976).

[48] C. P. Bahl, K. J. Marians, R. Wu, J. Stawinski, and S. A. Narang, *Gene* **1,** 81 (1976).

[49] R. H. Scheller, R. E. Dickerson, H. W. Boyer, A. D. Riggs, and K. Itakura, *Science* **196,** 177 (1977).

quences cloned using linkers are the *lac* operator region,[46–48,50] rat insulin,[51] human chorionic somatomammotropin,[52] and rat growth hormone.[53] The *c*I gene of bacteriophage λ was cloned by direct insertion of the sequence into the plasmid pCR11 with blunt-end joining.[54]

The *E. coli* and T4-induced DNA ligases have been paradigms for similar enzymes in other cells. *Bacillus subtilis* contains a DNA ligase requiring NAD,[55] like the *E. coli* enzyme, and bacteriophage T7 induces an ATP-dependent ligase.[56] DNA ligases have also been found in a large number of animal and plant cell types, and these have recently been reviewed by Soderhall and Lindahl.[57] Mammalian cells contain two different enzymes designated I and II.[58–63] DNA ligase I accounts for most of the ligase activity in dividing but not resting cells, and it fluctuates according to growth phase. The enzyme has a native molecular weight of approximately 200,000 based on sedimentation and gel filtration data.[58,64,65] DNA ligase I has an apparent K_m for ATP of 0.2–1.5 μM.[65–67] Like the *E. coli* and T4 enzymes, it forms an adenylylated enzyme intermediate[58] and joins the 5′-phosphoryl terminus of DNA to the 3′-hydroxyl end of a DNA (or less efficiently RNA) chain in a duplex.[68] DNA ligase II is not inhibited by antibodies directed against DNA ligase I.[58] DNA ligase II is also smaller, having a molecular weight of about

[50] C. P. Bahl, R. Wu, and S. A. Narang, *Gene* **3,** 123 (1978).

[51] A. Ullrich, J. Shine, J. Chirgwin, R. Pictet, E. Tischer, W. J. Rutter, and H. M. Goodman, *Science* **196,** 1313 (1977).

[52] J. Shine, P. H. Seeburg, J. A. Martial, J. D. Baxter, and H. M. Goodman, *Nature (London)* **270,** 494 (1977).

[53] P. H. Seeburg, J. Shine, J. A. Martial, J. D. Baxter, and H. M. Goodman, *Nature (London)* **270,** 486 (1977).

[54] K. Backman, M. Ptashne, and W. Gilbert, *Proc. Natl. Acad. Sci. U.S.A.* **73,** 4174 (1976).

[55] P. J. Laipis, B. M. Olivera, and A. T. Ganesan, *Proc. Natl. Acad. Sci. U.S.A.* **62,** 289 (1969).

[56] A. Becker, G. Lyn, M. Gefter, and J. Hurwitz, *Proc. Natl. Acad. Sci. U.S.A.* **58,** 1996 (1967).

[57] S. Soderhall and T. Lindahl, *FEBS Lett.* **67,** 8 (1976).

[58] S. Soderhall and T. Lindahl, *J. Biol. Chem.* **250,** 8438 (1975).

[59] S. Soderhall and T. Lindahl, *Biochem. Biophys. Res. Commun.* **53,** 910 (1973).

[60] H. Teraoka, M. Shimoyachi, and K. Tsukada, *FEBS Lett.* **54,** 217 (1975).

[61] D. G. Evans, S. H. Ton, and H. M. Kier, *Biochem. Soc. Trans.* **3,** 1131 (1975).

[62] S. B. Zimmerman and C. J. Levin, *J. Biol. Chem.* **250,** 149 (1975).

[63] S. Soderhall, *Nature (London)* **260,** 640 (1976).

[64] P. Beard, *Biochim. Biophys. Acta* **269,** 385 (1972).

[65] G. C. F. Pedrali Noy, S. Spadari, G. Ciarrocchi, A. M. Pedrini, and A. Falaschi, *Eur. J. Biochem.* **39,** 343 (1973).

[66] U. Bertazzoni, M. Mathelet, and F. Campagnari, *Biochim. Biophys. Acta* **287,** 404 (1972).

[67] H. Young, S. H. Ton, L. A. F. Morrice, R. S. Feldberg, and H. M. Keir, *Biochem. Soc. Trans.* **1,** 520 (1973).

[68] E. Bedows, J. T. Wachsman, and R. I. Gumport, *Biochemistry* **16,** 2231 (1977).

100,000,[58,60,62] and its apparent K_m for ATP of 40–100 μM is two orders of magnitude higher than that of DNA ligase I. Mitochondria contain a DNA ligase which is very similar to DNA ligase II in molecular size and in K_m for ATP.[69]

RNA Ligase

RNA ligase consists of a single 41,000-dalton polypeptide chain[70,71] encoded by bacteriophage T4 gene *63*.[72] It was discovered in T4-infected cells by Silber *et al.*[73] as an enzyme that catalyzed the circularization of single-stranded poly(A). RNA ligase has since been shown to carry out inter- as well as intramolecular joining of RNA and DNA chains. In the terminology widely employed for RNA ligase, the 5′-phosphoryl-terminated moiety is called the donor and the 3′-hydroxyl-terminated portion is termed the acceptor. The mechanism of RNA ligase-catalyzed reactions is analogous to that of DNA ligase (Fig. 1), except for the important distinction that a complementary strand is not required[74]; the abutment of ends is carried out at acceptor and donor binding sites on the enzyme. Like T4 DNA ligase, the energy cofactor is ATP. The first step in the reaction is formation of an adenylylated enzyme. This is a stable intermediate that can be easily identified by polyacrylamide gel electrophoresis in the presence of sodium dodecyl sulfate (SDS).[75] The effect of adenylylation is striking, because the decrease in electrophoretic mobility of the protein is that expected of an increase in molecular weight of 4000, although the actual increase is only 350. Adenylylation is reversible, so the enzyme catalyzes pyrophosphate exchange with ATP.[74] Transfer of AMP from the enzyme to the 5′-phosphoryl of the donor terminus requires the presence of an acceptor end.[76] The 3′-hydroxyl terminus of the acceptor may affect a conformational change necessary for adenyl transfer—the enzyme does not cock the trigger unless a target is in sight! The activated donor molecule does not necessarily join in the acceptor which stimulated its formation. The term *acceptor exchange* was coined for cases where the activated intermediate was formed in the presence of one acceptor but ligated to a different acceptor.[76] The adenylylated donor

[69] C. J. Levin and S. B. Zimmerman, *Biochem. Biophys. Res. Commun.* **69,** 514 (1976).

[70] T. J. Snopek, A. Sugino, K. Agarwal, and N. R. Cozzarelli, *Biochem. Biophys. Res. Commun.* **68,** 417 (1976).

[71] J. A. Last and W. F. Anderson, *Arch. Biochem. Biophys.* **174,** 167 (1976).

[72] T. J. Snopek, W. B. Wood, M. P. Conley, P. Chen, and N. R. Cozzarelli, *Proc. Natl. Acad. Sci. U.S.A.* **74,** 3355 (1977).

[73] R. Silber, V. G. Malathi, and J. Hurwitz, *Proc. Natl. Acad. Sci. U.S.A.* **69,** 3009 (1972).

[74] J. W. Cranston, R. Silber, V. G. Malathi, and J. Hurwitz, *J. Biol. Chem.* **249,** 7447 (1974).

[75] N. P. Higgins, A. P. Geballe, T. J. Snopek, A. Sugino, and N. R. Cozzarelli, *Nucleic Acids Res.* **4,** 3175 (1977).

[76] A. Sugino, T. J. Snopek, and N. R. Cozzarelli, *J. Biol. Chem.* **252,** 1732 (1977).

intermediate is difficult to detect with RNA substrates but accumulates in reactions with a DNA acceptor.[76] Evidently a DNA acceptor is better at evoking donor adenylylation than ligation. Proof that the adenylylated donor is a true reaction intermediate was obtained by purifying the compound and showing that it could be joined to an acceptor in the absence of ATP with the stoichiometric release of AMP.[76,77] In addition, many adenylylated synthetic compounds are donor substrates in the absence of ATP (see below).[78]

RNA ligase can be easily purified in large amounts. The most specific enzyme assay is the one originally used by Silber *et al.*[73]—cyclization of [5′-^{32}P]poly(A) which renders the ^{32}P resistant to digestion with alkaline phosphatase. At later stages in the purification, the enzyme can also be detected by exchange of labeled pyrophosphate with ATP[71] and by formation of an acid-precipitable enzyme–adenylate complex with [^{3}H]ATP.[25] The kinetics of the appearance of RNA ligase after T4 infection is highly unusual in that the enzyme is synthesized throughout the latent period. For maximum enzyme yield, harvest of infected cells should be delayed as long as possible. Moreover, a seven-fold increase in the amount of RNA ligase results from infection with T4 double-mutant strains carrying one mutation in the *regA* gene and another mutation in a gene required for DNA synthesis.[75] Using this rich source of enzyme, Higgins *et al.*[75] prepared physically homogeneous, nuclease-free enzyme in high yield. There are several other purification procedures yielding RNase-free enzyme of good purity,[70,71,73] but these preparations contain a contaminating DNA exonuclease that removes 5′-mononucleotides from the 3′-terminus of single-stranded substrates. This nuclease level, although small compared to the ligase activity, represents a major obstacle for DNA-joining reactions that often require high levels of enzyme. Chromatography on DNA–agarose removes this exonuclease.[75]

RNA ligase is a versatile reagent for the radical construction of nucleic acids. Intermolecular reactions leading to high-yield synthesis of oligoribonucleotides with a defined sequence have been described by several groups.[77,79–83] Self-polymerization of acceptors can be prevented by using

[77] E. Ohtsuka, S. Nishikawa, M. Sugiura, and M. Ikehara, *Nucleic Acids Res.* **3**, 1613 (1976).

[78] T. E. England, R. I. Gumport, and O. C. Uhlenbeck, *Proc. Natl. Acad. Sci. U.S.A.* **74**, 4839 (1977).

[79] G. C. Walker, O. C. Uhlenbeck, E. Bedows, and R. I. Gumport, *Proc. Natl. Acad. Sci. U.S.A.* **72**, 122 (1975).

[80] J. J. Sninsky, J. A. Last, and P. T. Gilham, *Nucleic Acids Res.* **3**, 3157 (1976).

[81] O. C. Uhlenbeck and V. Cameron, *Nucleic Acids Res.* **4**, 85 (1977).

[82] E. Ohtsuka, S. Nishikawa, R. Fukumoto, S. Tanaka, A. F. Markham, M. Ikehara, and M. Sugiura, *Eur. J. Biochem.* **81**, 285 (1977).

[83] Y. Kikuchi, F. Hishinuma, and K. Sakaguchi, *Proc. Natl. Acad. Sci. U.S.A.* **75**, 1270 (1978).

polynucleotides with hydroxyls at both the 5′- and 3′-ends, and several strategies have been employed to limit self-reaction of the donor. One method is to use a large excess of acceptor.[77,79] However, this may require unacceptably high acceptor concentrations, since intramolecular joining is generally a favored reaction for donors long enough to circularize (six to eight nucleotides). Donors with a phosphoryl at both the 3′- and 5′-termini are used efficiently with equimolar acceptor concentrations,[81] and Kikuchi et al. recently synthesized on a milligram scale an octanucleotide segment of bacteriophage Qβ coat protein gene using this method.[84] Periodate oxidation[85] and addition of methoxyethyl[80] and ethoxymethylidene[86] groups have been used to block the 3′-hydroxyl of donors. The latter methods are easily reversible, and Ohtsuka *et al.*[86] used this approach to synthesize a heptadecanucleotide corresponding to bases 61–77 of *E. coli* $tRNA^{fMet}$. The smallest monoaddition donors are the mononucleoside 3′,5′-bisphosphates pCp, pAp, pUp, pGp, pdCp, pdAp, pdTp, pdUp, and pdGp. They add to the 3′-terminus of a number of different acceptors including tRNA, mRNA, and double- and single-stranded viral RNA.[87,88] These studies are examples of the important application of RNA ligase technology to the precise engineering of naturally derived RNA molecules. In addition, "mutant" yeast $tRNA^{Phe}$ was synthesized by Kaufmann and Littauer[89] by joining two half-molecules in which the Y base at position 37 had been removed. More recently, Meyhack *et al.*[90] constructed six model substrates of the *B. subtilis* 5 S rRNA precursor to test the substrate recognition requirements of the M5 processing nuclease.

DNA is about as good a donor as RNA for RNA ligase-catalyzed reactions but is a poorer acceptor. Snopek *et al.*[70] showed that a number of short synthetic oligodeoxyribonucleotides participated in intra- and intermolecular reactions. Cyclization of $(pdT)_n$ from 6 to 30 nucleotides long was observed, and at the optimum chain length of 20 the rate was $\frac{1}{10}$ that of cyclization of poly(A).[76] Sugino *et al.*[76] showed that 2″,3′-deoxynucleotides such as dideoxythymidylate (d_2T) were effective 3′-blocking groups to prevent self-joining of DNA donors. Intermolecular joining of the donor $(pdC)_8pd_2T$ with $dA(pdA)_5$ was observed. Facilitation

[84] Y. Kikuchi and K. Sakaguchi, *Nucleic Acids Res.* **5,** 591 (1978).

[85] G. Kaufmann and N. R. Kallenbach, *Nature (London)* **254,** 452 (1975).

[86] E. Ohtsuka, S. Nishikawa, A. F. Markham, S. Tanaka, T. Miyake, T. Wakabayashi, M. Ikehara, and M. Sugiura, *Biochemistry* **17,** 4894 (1978).

[87] G. A. Bruce and O. C. Uhlenbeck, *Nucleic Acids Res.* **5,** 3665 (1978).

[88] T. E. England and O. C. Uhlenbeck, *Nature (London)* **275,** 560 (1978).

[89] G. Kaufmann and U. Z. Littauer, *Proc. Natl. Acad. Sci. U.S.A.* **71,** 3741 (1974).

[90] B. Meyhack, B. Pace, O. C. Uhlenbeck, and N. R. Pace, *Proc. Natl. Acad. Sci. U.S.A.* **75,** 3045 (1978).

of donor activation by acceptor exchange improved the reaction-addition of ApApA resulted in a 10-fold stimulation of DNA-to-DNA joining, and the final yield represented an acceptable 13%. As a donor for intermolecular joining with the ribonucleotide acceptor $A(pA)_{\overline{20}}$, $(dpT)_{12-18}pd_2T$ was even better than periodate-oxidized $(pA)_{\overline{20}}$. Completing the set of possible combinations, $dA(pdA)_5$ served as an acceptor for oxidized $(pA)_{\overline{20}}$, although it was roughly $\frac{1}{10}$ as effective as ApApA. Finally, Hinton *et al.*[91] have reported that oligodeoxyribonucleotide acceptors can be joined to 2′-deoxyribonucleoside 3′,5′-bisphosphates.

RNA ligase also joins 3′,5′-hydroxyl ribohomopolymers to large natural DNA substrates, generating single-strand extensions at the 5′-termini.[92] This reaction, called 5′-tailing, is surprisingly insensitive to the size and structure of the DNA donor. Double- and single-stranded DNA fragments from 10 to 6000 nucleotides long can be joined quantitatively to a number of oligoribonucleotide acceptors. The DNAs tested thus far include ColE1 DNA digested with *Eco*R1 or *Hae*III restriction endonucleases which produce cohesive and base-paired ends, respectively, heat-denatured *Hae*III-generated ColE1 DNA fragments, full-length linear ϕX174 viral DNA, and a self-complementary synthetic decanucleotide. The acceptor requirements are much more fastidious. The most efficient acceptor is $A(pA)_5$, but $I(pI)_5$ and $C(pC)_5$ are also utilized well, whereas $U(pU)_5$ is inert. The shortest acceptor joined is a trinucleoside diphosphate, and the optimum length is six nucleotides, but polymers between 10 and 20 nucleotides long are ligated at half the maximal rate. The 5′-tailing reaction provides an efficient radical method for generating extensions of predetermined length and base composition at the 5′-ends; it complements the terminal deoxynucleotidyltransferase-mediated addition of either ribo- or deoxyribonucleotides to the 3′-hydroxyl terminus.[12,14] On duplex DNA substrates the 5′-tailing product provides a natural template for reverse transcriptase, and therefore it would be possible to build duplex structures at the 5′-end of long molecules.

RNA ligase is capable of performing an amazing variety of reactions, but it is not totally permissive. As donors, it accepts molecules as diverse as a 3′,5′-mononucleoside bisphosphate and the 5′-base-paired end of duplex DNA thousands of base pairs long. There is even more latitude for the already activated donors of the general formula A-5′PP-X, since X does not even need to be a purine or pyrimidine. England *et al.*[78] have found that ADP ribose, diphospho-CoA, FAD, and even ADP cyanoethanol, along with AppA, AppC, AppG, AppU, and AppT, can be joined to

[91] D. M. Hinton, J. A. Baez, and R. I. Gumport, *Biochemistry* **17,** 5091 (1978).

[92] N. P. Higgins, A. P. Geballe, and N. R. Cozzarelli, *Nucleic Acids Res.* **6,** 1013 (1979).

oligoribonucleotides. The acceptor requirements are much stricter. The base, sugar, and chain length have all been shown to influence both the rate and the final yield of joining in a number of systems. The smallest acceptor is a trinucleoside diphosphate. A ribose moiety at the 3′-terminus is preferred by usually a factor of 10 or more to a deoxyribose. Oligo(A) at the acceptor terminus provides the best substrates, followed by oligo(I) and oligo(C); oligo(U) is generally very poor. However, this may be an oversimplification. Kikuchi *et al.*[83] found that $U(pU)_3$ would join efficiently to pCp but not at all to pAp, and there are other examples where it is not simply the acceptor that is important but the particular combination of acceptor and donor.[86]

The enzyme has thus far proved to have an unmeasurably high K_m for both the donor and acceptor termini. Therefore, the reaction increases proportionately with substrate concentration. This is sometimes an important factor in deciding whether or not a reaction with a poor substrate like a DNA acceptor can profitably be attempted. With a high concentration of molecules, there is a good chance for success; but if the molecule is extremely large, then the prospects are diminished. Fortunately large amounts of pure enzyme are easily obtainable and can be used to drive some sluggish reactions.

It is ironic that with the wealth of RNA ligase reactions demonstrated *in vitro* there is no evidence for an involvement in nucleic acid metabolism *in vivo* other than the weak indication provided by the initiation of its synthesis early after infection. However, unlike the enzymes known to be required for nucleic acid metabolism, it continues to be produced throughout the latent period. Mutants that have no detectable RNA ligase activity seem to have no abnormality in RNA or DNA metabolism but are blocked in the essential step of attachment of tail fibers to phage heads, notably the last step in virus development.[72] This morphogenetic step is efficiently catalyzed by RNA ligase *in vitro*.[72] Here is an intriguing formal analogy to the RNA ligase facilitation of blunt-end joining by DNA ligase where its proposed role is also the apposition of participating macromolecular entities. The unmeasurably high K_m for all nucleic acids tested suggests that its natural nucleic acid substrate, if it exists at all, may yet remain to be identified.

In eukaryotes, joining of RNA is clearly of widespread importance. The direct transcript is generally longer than the functional RNA.[93,94] An activity has been identified that cleaves a precursor at specific sites and joins the ends to splice the molecules into a functional unit. O'Farrell *et*

[93] D. F. Klessig, *Cell* **12,** 9 (1977).

[94] S. M. Berget, C. Moore, and P. A. Sharp, *Proc. Natl. Acad. Sci. U.S.A.* **74,** 3171 (1977).

al.[95] and Knapp *et al.*[96] have found that a yeast extract carries out site-specific cleavage and rejoining of precursors of several tRNAs. Blanchard *et al.*[97] have also reported evidence for *in vitro* splicing of adenovirus-2 mRNA in HeLa cell extracts. The mechanism of this reaction is as yet undetermined, but the requirement for ATP in the yeast system may signal that a ligase analogous to RNA ligase is needed for the joining process. The enzymology of circular RNA production in plants is unknown.

B-R Enzymes

In 1971 Wang[98] purified an enzyme from *E. coli*, the ω protein, that removed superhelical turns from covalently closed DNA molecules. The reaction must involve breakage of backbone bonds, rotation about the helix axis opposite the sense of the supertwists, and reunion of the backbone. Nonetheless, ω had no demonstrable nuclease or ligase activity and required no energy cofactor. Moreover, since relaxation could occur gradually, ω must prevent free rotation of the transiently broken ends of DNA. To account for these results Wang[98] proposed that the phosphodiester bond energy released by DNA strand rupture was stored in a high-energy complex of enzyme and DNA, and that the dissociation of this complex was concomitant with reformation of the DNA backbone. This model accounted for the lack of ligase activity in ω, because the substrate for the joining component is not nicked DNA but instead a transient nick is formed in continuous regions of DNA by the enzymes. The ends of the nick do not rotate freely because one end is covalently bound to the enzyme and the adjacent other end could easily be impeded by the large enzyme. The requirement for an exogneous energy source is circumvented by the high-energy intermediate. This should be compared to the DNA ligase reaction where the energy extracted from the cofactor in the first step is stored in subsequent reaction intermediates.

The breakage-reunion (B-R) model is probably correct and has heralded new classes of proteins that play important roles in the replica-

[95] P. Z. O'Farrell, B. Cordel, P. Valenzuela, W. J. Rutter, and H. M. Goodman , *Nature (London)* **274**, 438 (1978).

[96] G. Knapp, J. S. Beckmann, P. F. Johnson, S. A. Fuhrman, and J. Abelson, *Cell* **14**, 221 (1978).

[97] J. M. Blanchard, J. Weber, W. Jelinek, and J. E. Darnell, *Proc. Natl. Acad. Sci. U.S.A.* **75**, 5344 (1978).

[98] J. Wang, *J. Mol. Biol.* **55**, 523 (1971).

[98a] We prefer the term *breakage-reunion* to *nicking-closing* because the interruption in the DNA need not be a nick but could be across both strands.

tion, recombination, repair, and transcription of DNA. One class is the topoisomerases that carry out the interconversion of topological isomers of DNA. Examples include ω proteins, bacterial DNA gyrases, and eukaryotic untwisting enzymes.[18] B-R proteins whose primary activity is not the interconversion of topological isomers of DNA include the φX174 *cisA* protein and the λ *int* protein. The relaxation complexes of small plasmids studied by Helinski and co-workers[99] may be intermediate in B-R reactions, since denaturation of relaxation complexes and topoisomerases under defined conditions leads to DNA strand rupture with nucleic acid covalently linked to protein.[100–104]

ω proteins from *E. coli* and *Micrococcus luteus* have been purified to homogeneity and consist of single 110,000- and 120,000-dalton polypeptides, respectively.[101,105,106] Both enzymes require Mg^{2+} and act efficiently on highly negatively twisted (underwound) DNA substrates. Covalently closed DNAs containing few negative or positive superhelical turns are relaxed very poorly. With single-stranded circular DNA substrates these proteins catalyze the formation of topologically knotted rings.[105,107] In this reaction a region of the circle must be passed between the ends of the interrupted DNA backbone prior to ligation. Another reaction demonstrating the same principle is the reassociation of single-stranded complementary circular DNA molecules into a covalently closed duplex structure.[108] The high-energy reaction intermediate poised for resealing contains ω covalently linked to the 5′-phosphoryl of the transiently broken DNA chain.[101]

The rat liver DNA untwisting enzyme has also been extensively purified and consists of a single polypeptide of about 65,000 daltons.[109,110] This protein and other similar eukaryotic activities are set apart from the prokaryotic relaxing enzymes by the lack of a divalent metal ion requirement, the efficient removal of positive as well as negative superhelical coils, and, in the key reaction intermediate, the covalent attachment of pro-

[99] D. G. Blair and D. R. Helinski, *J. Biol. Chem.* **250,** 8785 (1975).

[100] D. G. Guiney and D. R. Helinski, *J. Biol. Chem.* **250,** 8796 (1975).

[101] R. E. Depew, L. F. Liu, and J. C. Wang, *J. Biol. Chem.* **253,** 511 (1978).

[102] J. J. Champoux, *Proc. Natl. Acad. Sci. U.S.A.* **74,** 3800 (1977).

[103] A. Sugino, C. L. Peebles, K. N. Kreuzer, and N. R. Cozzarelli, *Proc. Natl. Acad. Sci. U.S.A.* **74,** 4767 (1977).

[104] M. Gellert, K. Mizuuchi, M. H. O'Dea, T. Itoh, and J. Tomizawa, *Proc. Natl. Acad. Sci. U.S.A.* **74,** 4772 (1977).

[105] V. T. Kung and J. C. Wang, *J. Biol. Chem.* **252,** 5398 (1978).

[106] R. Hecht and H. W. Thielman, *Nucleic Acids Res.* **4,** 4235 (1978).

[107] L. F. Liu, R. E. Depew, and J. C. Wang, *J. Mol. Biol.* **106,** 439 (1976).

[108] K. Kirkegaard and J. C. Wang, *Nucleic Acids Res.* **5,** 3811 (1978).

[109] W. Keller, *Proc. Natl. Acad. Sci. U.S.A.* **72,** 4876 (1975).

[110] J. J. Champoux and B. L. McConaughy, *Biochemistry* **15,** 4638 (1976).

tein to a 3′-phosphoryl DNA terminus.[102] Like ω, the rat liver enzyme can intertwine single-stranded covalently closed circular molecules to form a closed duplex molecule.[111] This unusual activity may signal an involvement of untwisting enzymes and ω in the homologous recombination of topologically constrained DNA molecules.[108,111] The positive coil-relaxing activity of untwisting enzymes, histones, and an assembly factor are required for the efficient *in vitro* synthesis of chromatin-like material,[112] and this has been suggested as the mechanism for generation of negative superhelical turns in eukaryotic DNA molecules.[113]

DNA gyrase, which was discovered by Gellert *et al.*[114] as a host activity required for the *in vitro* integrative recombination of relaxed bacteriophage λ DNA molecules, introduces negative superhelical coils into relaxed DNA. The enzyme requires ATP and Mg^{2+} and is stimulated by spermidine. Temperature-sensitive mutants have been isolated that have temperature-sensitive gyrase activity.[115,116] Shift of these mutants to the nonpermissive temperature arrests DNA synthesis, as does addition of the gyrase inhibitor nalidixate or novobiocin. The enzyme is also important for transcription, recombination, and repair of cellular DNA.[114–118] For *E. coli* gyrase, the subunits designated A and B contain the 105,000-dalton product of the *nalA* gene and the 95,000-dalton product of the *cou* gene, respectively.[103,104,119,120] The *M. luteus* enzyme is composed of two polypeptides, α and β, with molecular weights of 115,000 and 97,000, respectively.[121] Gyrase can be purified as a complex of A and B proteins or in the form of subunits which can be mixed to reconstitute activity.[119] In the absence of ATP the enzyme relaxes negative superhelical turns in DNA. Thus, like other topoisomerases, ATP is not required to break and rejoin

[111] J. J. Champoux, *Proc. Natl. Acad. Sci. U.S.A.* **74,** 5328 (1977).
[112] R. A. Laskey, B. M. Honda, A. D. Mills, and J. T. Finch, *Nature* (*London*) **275,** 416 (1978).
[113] J. E. Germond, B. Hirt, P. Oudet, M. Gross-Bellard, and P. Chambon, *Proc. Natl. Acad. Sci. U.S.A.* **72,** 1843 (1975).
[114] M. Gellert, K. Mizuuchi, M. H. O'Dea, and H. A. Nash, *Proc. Natl. Acad. Sci. U.S.A.* **73,** 3872 (1976).
[115] K. N. Kreuzer, K. McEntee, A. P. Geballe, and N. R. Cozzarelli, *Mol. Gen. Genet.* **167,** 129 (1978).
[116] C. L. Peebles, N. P. Higgins, K. N. Kreuzer, A. Morrison, P. O. Brown, A. Sugino, and N. R. Cozzarelli, *Cold Spring Harbor Symp. Quant. Biol.* **43,** 41 (1978).
[117] C. L. Smith, K. Kubo, and F. Imamoto, *Nature* (*London*) **275,** 424 (1978).
[118] J. B. Hays and S. Boehmer, *Proc. Natl. Acad. Sci. U.S.A.* **75,** 4125 (1978).
[119] N. P. Higgins, C. L. Peebles, A. Sugino, and N. R. Cozzarelli, *Proc. Natl. Acad. Sci. U.S.A.* **75,** 1773 (1978).
[120] M. Gellert, M. H. O'Dea, T. Itoh, and J. Tomizawa, *Proc. Natl. Acad. Sci. U.S.A.* **73,** 4474 (1976).
[121] L. F. Liu, and J. C. Wang, *Proc. Natl. Acad. Sci. U.S.A.* **75,** 2098 (1978).

the DNA backbone bonds. Hydrolysis of ATP is not even required for introduction of negative supercoils, since the nonhydrolyzable ATP analog, adenyl-5′-yl-imidodiphosphate, promotes limited supercoiling.[122] The covalent intermediate of DNA gyrase differs from those of ω and rat liver untwisting enzymes in two important and interesting ways. First, denaturation of gyrase leads to site-specific breaks[103,104]; breakage by the relaxing enzyme is not random but occurs at many more sites. Second, gyrase breakage is across both strands of DNA, and protein becomes covalently attached to the two resulting 5′-phosphoryl groups which are separated by a four-base stagger.[123] The enzyme thus generates a four-base cohesive end which can be repaired *in vitro* with *E. coli* DNA polymerase I.[123]

The *cisA* protein coded by φX174 is a polypeptide of 60,000 daltons, which is required for virus replication.[124,125] The proposed role for this enzyme in the replication of closed duplex (RF) DNA is as follows. Upon binding to the supercoiled RF molecule it produces a site-specific break at approximately residue 4300 of the positive strand by forming a covalent bond with the 5′-phosphoryl.[126] This is like the intermediate proposed by Wang for ω, but in this case the demonstrable covalent bond and rupture of the strands is a normal event and not tied to enzyme denaturation. The *cisA* protein interacts with *rep* unwinding protein, helix-destabilizing protein, and DNA polymerase in the unwinding of the duplex concomitant with DNA replication.[127] Chain growth proceeds with the *cisA*-bound terminus locked at the replication fork.[128] Upon complete traverse of the circle, *cisA* protein cleaves the new origin and rejoins the two original 5′- and 3′-ends, releasing a single-strand circle. Thus, like topoisomerases, *cisA* protein is a B-R enzyme but one with a protracted coupling period. It may be a radical enzyme, since a template may not be required; but if so, unlike the situation in the radical ligases we have considered, the nucleotide sequence of the substrate is conserved.

The site-specific integration of bacteriophage λ DNA into host chromosomal sequences requires the *int* gene product. Recently this protein

[122] A. Sugino, N. P. Higgins, P. O. Brown, C. L. Peebles, and N. R. Cozzarelli, *Proc. Natl. Acad. Sci. U.S.A.* **75,** 4838 (1978).

[123] A. Morrison and N. R. Cozzarelli, *Cell* **17,** 175 (1979).

[124] S. Eisenberg, J. F. Scott, and A. Kornberg, *Proc. Natl. Acad. Sci. U.S.A.* **73,** 1594 (1976).

[125] J.-E. Ikeda, A. Yudelevich, and J. Hurwitz, *Proc. Natl. Acad. Sci. U.S.A.* **73,** 2269 (1976).

[126] S. Eisenberg and A. Kornberg, *J. Biol. Chem.* **254,** 5328 (1979).

[127] J. R. Scott, S. Eisenberg, L. L. Bertsch, and A. Kornberg, *Proc. Natl. Acad. Sci. U.S.A.* **74,** 193 (1977).

[128] S. Eisenberg, J. Griffith, and A. Kornberg, *Proc. Natl. Acad. Sci. U.S.A.* **74,** 3198 (1977).

was purified and shown to have a molecular weight of 40,000.[129] The purified *int* gene product is a site-specific DNA-binding protein and must be supplemented with host proteins to carry out recombination. The reaction proceeds in the absence of Mg^{2+} or an energy cofactor, but spermidine is required, as in the presence of negative supercoils in the DNA substrate under some reaction conditions. Interestingly, the superhelical turns present in the substrate are not lost during the reaction, which means that the four ends that are broken and joined to separate partners are physically restrained during the complete reaction sequence.[130] Although there is no evidence as yet for a covalent protein–DNA intermediate in the integration reaction, the lack of an energy cofactor and the conservation of the supercoiling of the DNA substrate presupposes a coupled B-R mechanism.

Perspectives

Enzymes that carry out joining of DNA chains can be grouped into two distinct categories—ligases and B-R enzymes (Table II). The former join the ends of polynucleotides and require an energy source. Both the *E. coli* and the phage T4-coded DNA ligases are primarily conservative, i.e., template-directed enzymes, and play essential roles in DNA metabolism. However, T4 DNA ligase can also act radically and alter the nucleotide sequence of the substrate, as in blunt-end joining and ligation of mismatched termini. T4 RNA ligase is strictly a radical enzyme. Unfortunately, it is unknown whether the radical properties of ligases are biologically important.

B-R enzymes, on the other hand, carry out a coupled breakage and reunion of nucleic acids at contiguous regions via covalent protein–DNA intermediates and do not require energy input. However, a cofactor may be used to drive an endergonic reaction, as in the generation of negative supercoils by DNA gyrase. The role of B-R enzymes in the replication and recombination of DNA is becoming increasingly clear. DNA gyrase maintains the negative superhelical tension in DNA for replication and transcription, and for integration of bacteriophage λ. The *cisA* protein initiates the replication of ϕX174 closed duplex DNA. Although not yet proved to be a B-R reaction, λ integration has the properties anticipated—no energy requirement for joining, and conservation of topological structure of the DNA throughout the reaction. The integration of phage mu, a transposon, may share common host factors with the λ-

[129] Y. Kikuchi and H. A. Nash, *J. Biol. Chem.* **253,** 7149 (1978).
[130] K. Mizuuchi, M. Gellert and H. A. Nash, *J. Mol. Biol.* **121,** 375 (1978).

TABLE II
JOINING ENZYMES[a]

Enzyme	Classification	Function	Genetic loci	Protomer molecular weight	Type of reaction intermediate	Energy cofactor
E. coli DNA ligase	Ligase, conservative	Replication and repair of DNA	*lig*	74,000	Protein–5′-pA	NAD
T4 DNA ligase	Ligase, conservative and radical	Replication and repair of viral DNA	T4 gene *30*	68,000	Protein–5′-pA	ATP
T4 RNA ligase	Ligase, radical	Tail fiber attachment	T4 gene *63*	41,000	Protein–5′-pA	ATP
E. coli ω	B-R, conservative	?	?	110,000	Protein–5′-pDNA	None
DNA untwisting enzyme	B-R, conservative	Chromatin assembly	?	65,000	Protein–3′-pDNA	None
E. coli DNA gyrase	B-R, conservative	Replication, transcription, repair, and recombination of DNA	*nalA* and *cou*	105,000 and 95,000	Protein–5′-pDNA	ATP
φX174 *cisA* protein	B-R, radical?	Initiation of closed duplex φX174 DNA replication	*cisA*	60,000	Protein–5′-pDNA	None
λ *int* protein	B-R, conservative	Integration and excision of λ DNA	λ *int* and host factors	40,000 and unknown	?	None

[a] Conventions are defined in the footnote to Table I.

integration system (H. Nash, personal communication). However, a short sequence of the host DNA is duplicated in the same orientation immediately adjacent to the two ends of some inserted transposable elements.[19] Therefore, if a B-R enzyme is responsible for insertion, it is likely to be radical.

The enzymes with the widest use for joining nucleic acids are the two T4 ligases. These have complementary substrate specificity. With T4 DNA and RNA ligases available, virtually all possible combinations of inter- and intramolecular ligation of single- and double-stranded RNA and DNA substrates can be performed. Their utility in constructing recombinant DNA is manifest.

Section III

Synthesis, Isolation, and Purification of DNA

[5] Cloning of Hormone Genes from a Mixture of cDNA Molecules[1]

By HOWARD M. GOODMAN AND RAYMOND J. MACDONALD

The molecular cloning of complementary DNA (cDNA) synthesized from a purified mRNA is a well-established method for obtaining purified DNA for sequence analysis and use as a hybridization probe. However, only a limited number of purified mRNAs are currently available, namely, those from specialized cells producing predominantly a single protein such as globin, immunoglubin, or ovalbumin. We have recently described the isolation and amplification in bacteria of the DNA sequences complementary to mRNA for several polypeptide hormones: rat insulin,[2] rat growth hormone,[3] human chorionic somatomammotropin,[4] and human growth hormone.[5] All these clones were isolated after transformation of bacteria with an enriched but impure population of cDNA molecules ligated to a plasmid vector. This chapter describes the procedures we have used to isolate and clone specific hormone cDNAs from an impure mRNA population.

Principle of the Method

A novel purification procedure for obtaining cDNA clones containing nucleotide sequences complementary to an individual mRNA species has been developed.[6] The method employs restriction endonuclease cleavage of double-stranded cDNA transcribed from a complex mixture of mRNA. The method does not require any extensive purification of RNA but instead makes use of transcription of RNA into cDNA, sequence-specific fragmentation of this cDNA with one or two restriction endonucleases, and fractionation of the cDNA restriction fragments on the basis of their

[1] Support for writing this chapter came from grants from the United States Public Health Service: Grant Nos. CA 14026 and AM 19997. H.M.G. is an investigator at the Howard Hughes Medical Institute. R. J. M. was supported by Grant AM21344 to Wiliam J. Rutter.

[2] A. Ullrich, J. Shine, J. Chirgwin, R. Pictet, E. Tischer, W. J. Rutter, and H. M. Goodman, *Science* **196,** 1313 (1977).

[3] P. H. Seeburg, J. Shine, J. A. Martial, J. D. Baxter, and H. M. Goodman, *Nature (London)* **270,** 486 (1977).

[4] J. Shine, P. H. Seeburg, J. A. Martial, J. D. Baxter, and H. M. Goodman, *Nature (London)* **270,** 494 (1977).

[5] H. M. Goodman, P. H. Seeburg, J. Shine, J. A. Martial, and J. D. Baxter, *Specific Eukaryotic Genes: Struct. Organ. Funct., Proc. Alfred Benzon Symp., 13th, 1978,* p. 179 (1979).

[6] P. H. Seeburg, J. Shine, J. A. Martial, A. Ullrich, J. D. Baxter, and H. M. Goodman, *Cell* **12,** 157 (1977).

ISBN 0-12-181968-X

lengths. The use of restriction endonucleases eliminates size heterogeneity and produces homogeneous-length DNA fragments from any cDNA species that contains at least two restriction sites. From the initially heterogeneous population of cDNA transcripts, uniform-sized fragments of desired sequence are produced. The fragments may be several hundred nucleotides in length and may in some instances include part or all of the structural gene for a particular protein. The length of the fragments depends on the number of nucleotides separating the restriction sites and is usually different for different regions of DNA. Fractionation by length allows purification of an enriched population of fragments containing the desired sequence. Current separation and analysis methods permit the isolation of such fragments from a RNA preparation in which a specific mRNA species represents at least 2% of the mass of the RNA transcribed. The use of standard RNA fractionation methods to prepurify the mRNA before transcription will result in lowering the actual lower limit of detection to less than 2% of the total mRNA isolated from the organism.

Specific sequences purified by the procedure outlined above may be further purified by a second specific cleavage with a restriction endonuclease capable of cleaving the desired sequence at an internal site. This cleavage results in formation of two subfragments of the desired sequence, separable on the basis of their lengths. The subfragments are separated from uncleaved and specifically cleaved contaminating sequences which had substantially the same original size prior to the second cleavage. The method is founded upon the rarity and randomness of placement of restriction endonuclease recognition sites, which results in an extremely low probability that a contaminant having the same original length as the desired molecule will be cleaved by the same second enzyme to yield fragments having the same length as those yielded by the desired sequence. After separation from the contaminants, the subfragments of the desired sequence may be rejoined using DNA ligase to reconstitute the original sequence.[4] The two subfragments must be prevented from joining together in the reverse order relative to the original sequence. This is accomplished by treatment with alkaline phosphatase prior to the second restriction endonuclease digestion and separation and rejoining of the subfragments. The purified fragments can then be recombined with a cloning vector and transformed into a suitable host strain.

Materials and Reagents

Reagents

1. Oligo(dT) (dT_{12-18}), P-L Biochemicals
2. AMV reverse transcriptase, Office of Program Resources and Logistics, Life Sciences, Inc.

3. Restriction site oligonucleotide linkers, e.g., d(CCAAGCTTGG), Collaborative Research, Inc.
4. T4 DNA ligase, Bethesda Research Laboratories, Inc. (5224) or PL Biochemicals (0870)
5. Restriction endonucleases (e.g., *Hin*dIII), Biolabs, Inc.
6. Radioactive nucleoside triphosphates, New England Nuclear or Amersham/Searle
7. S1 nuclease, Miles Laboratories
8. Nonradioactive nucleoside triphosphates, P-L Biochemicals
9. T4 polynucleotide kinase, Boehringer Mannheim Biochemicals (174 645)
10. *Escherichia coli* DNA polymerase I, Boehringer Mannheim Biochemicals (104 485)
11. Calf intestinal alkaline phosphatase (CIAP), Sigma Type VII (P-4502)

Bacterial Strains and Plasmids[7–9]

1. χ1776, EK2 host[7] constructed by Roy Curtiss, III, Department of Microbiology, University of Alabama Medical Center, Birmingham, Alabama 35294
2. Plasmids pMB9, pBR313, and pBR322[8,9] obtained from Herbert W. Boyer, University of California, San Francisco, San Francisco, California 94143

Materials

1. TEN 9 buffer: 100 m*M* NaCl, 20 m*M* Tris-HCl, pH 9.0, 1 m*M* Na_2EDTA
2. S1 buffer: 30 m*M* NaOAc, pH 4.5, 0.3 *M* NaCl, 3 m*M* $ZnCl_2$, 10 μg/ml denatured calf thymus DNA
3. S1 stock buffer: 30 m*M* NaOAc, pH 4.5, 0.3 *M* NaCl, 3 m*M* $ZnCl_2$, 50% glycerol
4. TBE buffer: 10.8 g Tris base, 5.5 g boric acid, 0.93 g Na_2EDTA per liter of H_2O
5. TBE_{ss} (sample solution): 0.2 ml 10× TBE buffer, 0.3 ml 0.2 *M* Na_2EDTA, pH 8, 0.5 ml 4% sarkosyl, 1.0 ml glycerol, 0.5 ml 0.5% bromphenol blue
6. 10× *Pol*I buffer: 0.6 *M* Tris-HCl, pH 7.5, 80 m*M* $MgCl_2$
7. 10 × *Hin*dIII buffer: 60 m*M* Tris-HCl, pH 7.5, 60 m*M* $MgCl_2$, 0.6 *M* NaCl

[7] R. Curtiss, III, D. A. Pereira, J. C. Hsu, S. C. Hull, J. E. Clark, L. F. Maturin, Sr., R. Goldschmidt, R. Moody, M. Inoue, and L. Alexander, *Proc. Miles Int. Symp., 10th,* p. 45 (1977).

[8] F. Bolivar, R. L. Rodriguez, M. C. Betlach, and H. W. Boyer, *Gene* **2,** 75 (1977).

[9] F. Bolivar, R. L. Rodriguez, P. J. Greene, M. C. Betlach, H. L. Heyneker, H. W. Boyer, J. H. Crosa, and S. Falkow, *Gene* **2,** 95 (1977).

8. 10× Kinase buffer: 0.7 *M* Tris-HCl, pH 7.6, 0.1 *M* $MgCl_2$, 50 m*M* DTT
9. 10 × Ligase buffer: 0.66*M* Tris-HCl, pH 7.6, 66 m*M* $MgCl_2$, 0.1 *M* DTT
10. χ medium: 10 g Bactotryptone, 5 g yeast extract, 10 g NaCl, 100 mg diaminopimelic acid, 40 mg thymine, 1 liter tap-distilled H_2O, 0.33 ml 6 *N* NaOH. Autoclave 20 min. Add 10 ml sterile 50% glucose and 10 ml filter-sterilized nalidixic acid (7.5 mg/ml) in 50 m*M* NaOH.
11. Mχ wash buffer: 0.1 *M* NaCl, 5 m*M* $MgCl_2$, 5 m*M* Tris-HCl, pH 7.6
12. Mχ Ca^{2+} buffer: 75 m*M* $CaCl_2$, 250 m*M* KCl, 5 m*M* $MgCl_2$, 5 m*M* Tris-HCl, pH 7.6
13. χ plates: χ medium containing 15 g Bactoagar per liter
14. Triton lysis mix: 1 ml 10% Triton, 75 ml 0.25 *M* EDTA, pH 8, 15 ml 1 *M* Tris, pH 8, 9 ml H_2O

Methods

The complete scheme for cDNA cloning is illustrated in Fig. 1.

Step 1. mRNA Purification.[6] The source of the RNA is obviously of critical importance in isolating the particular hormone cDNA clone of interest. As examples, we have used human term placentas to isolate human chorionic somatomammotropin clones,[4] rat islets of Langerhans to isolate rat insulin clones,[2] a rat pituitary tumor cell line (GC cells) to isolate rate growth hormone,[3] and growth hormone-producing tumors of pituitary origin from patients with acromegaly as a source for human growth hormone sequences.[5] In all cases the tissue is quick-frozen in liquid nitrogen and stored at −70°. For extraction of total RNA, 40 g of the frozen tissue was broken into small pieces and dissolved with the aid of a blender in 140 ml of a freshly prepared solution of 7 *M* guanidinium-HCl,[10] 20 m*M* Tris-HCl, pH 7.5, 1 m*M* EDTA, and 1% sarcosyl at 0°. After adding 0.5 g CsCl to each milliliter, the dark-brown solution was heated at 65° for 5 min, quick-cooled in ice, layered on top of a 5-ml cushion of 5.7 *M* CsCl, 10 m*M* Tris-HCl, pH 7.5, 1 m*M* EDTA in 1 × 3½ in. nitrocellulose tubes and centrifuged[11] in an SW27 rotor at 27,000 rpm for 16 hr at 15°. After centrifugation, the tube contents were decanted, the tubes were drained, and the bottom ½ cm containing the clear RNA pellet was cut off with a razor blade. Pellets were transferred to a sterile Erlenmeyer flask and dissolved in 20 ml 10 m*M* Tris-HCl, pH 7.5, 1 m*M* EDTA, 5% sarcosyl, and 5% phenol. The solution was then made 0.1 *M* in NaCl and

[10] R. A. Cox, this series, Vol. 12, p. 120.

[11] V. Glisin, R. Crkvenjakov, and C. Byus, *Biochemistry* **13,** 2633 (1974).

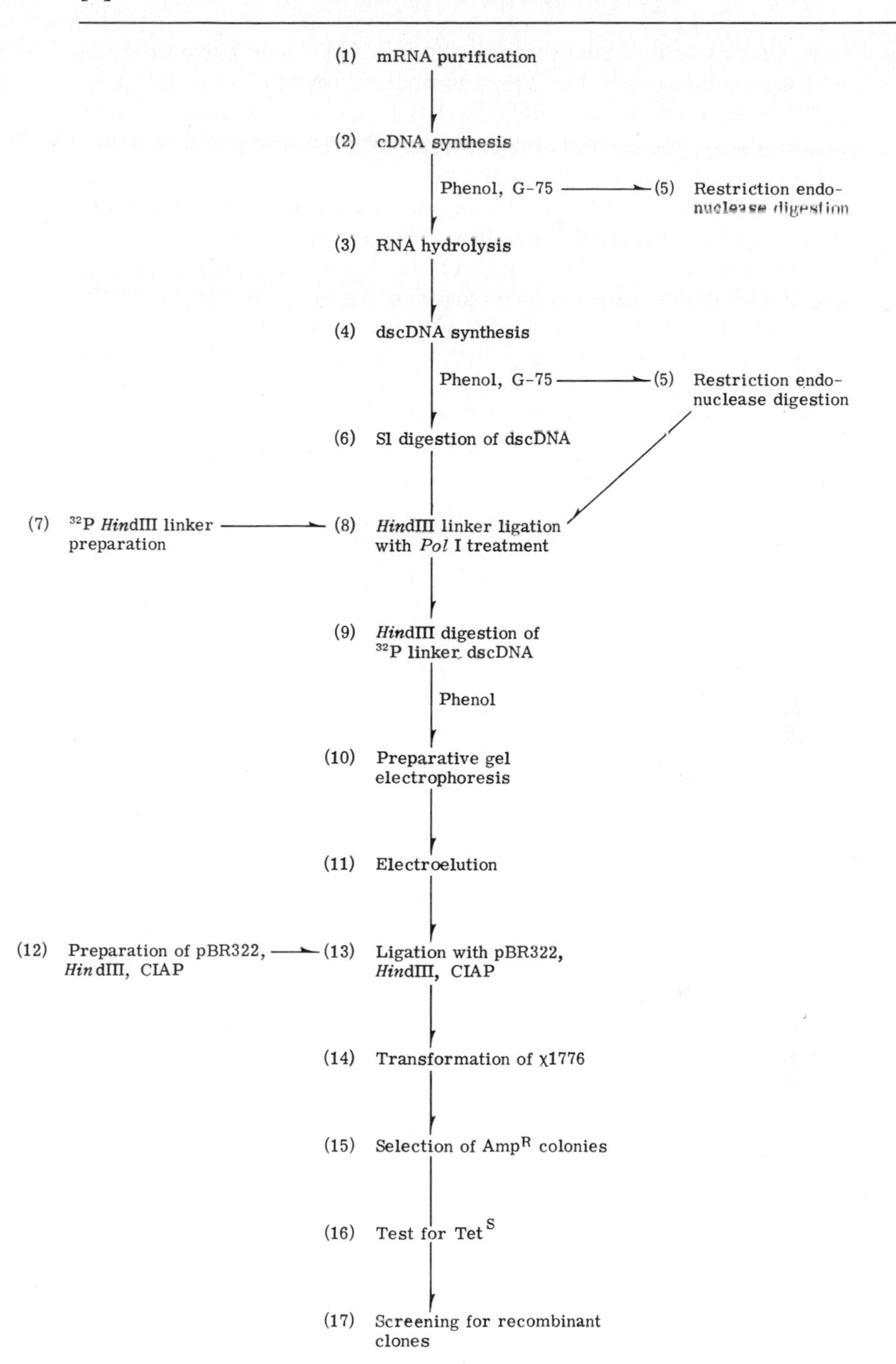

FIG. 1. Scheme for cDNA cloning.

vigorously shaken with 40 ml of a 50% phenol–50% chloroform mixture. RNA was precipitated from the aqueous phase with ethanol in the presence of 0.2 *M* sodium acetate, pH 5.5. RNA pellets were washed with 95% ethanol, dried, and dissolved in sterile H_2O. For example, 40 g of placental tissue usually yielded about 30 mg of RNA from which approximately 300 μg of polyadenylated RNA was obtained after being chromatographed twice on oligo(dT) cellulose.[12]

Step 2. cDNA Synthesis. Dry 1 mCi of ^{3}H-dCTP (~35 Ci/mmol) in an appropriate tube under a stream of nitrogen. Add 250 μl 0.2 *M* Tris-HCl, pH 8.3, 10 μl 1 *M* $MgCl_2$, 20 μl 1 *M* dithioerythritol (DTT), 10 μl of 2.5 mg/ml oligo(dT_{12-18}), 60 μl of 5 m*M* dCTP, 200 μl of 5 m*M* each dATP, dGTP, dTTP, ~50 μg poly(A^+) RNA, and H_2O to 1 ml. If a smaller amount of poly(A^+) RNA is used, scale down the reaction volumes proportionally while keeping the enzyme/RNA ratio constant. The final concentrations of dNTPs are dCTP, 0.3 m*M*, and dATP, dTTP, and dGTP, 1.0 m*M* each. The reaction is started with the addition of 40 μl of AMV reverse transcriptase (or the equivalent of 250 units). After incubation for 15 min at 45° the reaction is stopped by the addition of $\frac{1}{10}$ volume (104) μl of 0.2 *M* EDTA, pH 8.5.

The time course of the reaction is assayed by taking 2-μl samples at 0 time and 15 min and adding them to 0.5 ml 10 m*M* EDTA, pH 7.0, containing 200 μg tRNA carrier. After adding 1 ml of 10% trichloroacetic acid (TCA) and leaving on ice for 30 min, the precipitates are filtered through GFC filters, washed with 5% TCA, and counted in a liquid scintillation counter. The reaction should yield about 1–10 μg of cDNA with a specific activity of $\sim 6 \times 10^6$ dpm/μg.

The entire reaction mixture is mixed with $\frac{1}{2}$ volume (0.5 ml) of phenol and then with an equal volume (0.5 ml) of $CHCl_3$–isoamyl alcohol (24:1). After centrifugation to separate the phases [an Eppendorf microcentrifuge (Brinkman) is very convenient for all centrifugation steps] the aqueous layer is removed and the organic phase and interface reextracted with a small volume of TEN 9 buffer (100 m*M* NaCl, 20 m*M* Tris-HCl, pH 9.0, 1 m*M* Na_2EDTA).

The combined aqueous extracts are mixed with glycerol (10% final concentration) and chromatographed on a Sephadex G-75 column in TEN 9 buffer. A 5-ml plastic disposable pipette is suitable as a column for small loading volumes (<200 μl) and a 10-ml pipette for larger volumes. Collect 0.2-ml fractions for small columns and 0.4-ml fractions for large columns. After counting appropriate size aliquots, pool the excluded peak fractions containing the cDNA.

Step 3. RNA Hydrolysis

[12] H. Aviv and P. Leder, *Proc. Natl. Acad. Sci. U.S.A.* **69,** 1408 (1972).

STEP a. Ethanol precipitation of pooled G-75 fractions: Add $\frac{1}{10}$ volume of 2 *M* NaCl and 2.5 volumes 95% ethanol (EtOH). After 30 min in a dry ice–EtOH bath, or overnight at −20°, centrifuge for 20 min in the Brinkman microfuge or 60 min at 10,000 rpm in the HB4 rotor of a Sorvall centrifuge. Check the supernatant for counts before discarding and dry the pellets under vacuum briefly.

STEP b. RNA hydrolysis[13]: Dissolve the dried pellet in fresh 0.1 *M* NaOH (0.5–5 times the original reaction volume) and incubate at 70° for 20 min. Neutralize with $\frac{1}{20}$ volume 1 *M* Tris-HCl, pH 7.8, and enough 1 *M* HCl to bring to pH 8 (check with pH paper).

STEP c. Second ethanol precipitation: Precipitate hydrolyzed sample as in step a. After removing the supernatant, add 70% EtOH at −20° to the pellet, mix, and place at −70° for 30 min; centrifuge 10 min in the microfuge, remove the supernatant, and dry the pellet under vacuum. (Check the supernatant for counts before discarding.)

Step 4. Double-Stranded cDNA Synthesis. To 57 μl of single-stranded [^{3}H] cDNA (1 μg) from step 3 (add H_2O to 57 μl) add 25 μl 0.2 *M* Tris-HCl, pH 8.3, 2 μl 1 *M* DTT, 1 μl 1 *M* $MgCl_2$, and 10 μl of 5 m*M* each dNTP (0.5 m*M* each final). Heat the mixture for 3 min at 68° and quick-cool on ice. Add 5 μl (15–30 units) AMV reverse transcriptase and incubate 1.5 hr at 42°.[14,15] Take 2-μl time points at 0, 45, and 90 min and assay TCA-precipitable counts. This test measures the stability of the tritiated cDNA first strand. The double-stranded cDNA (dscDNA) is isolated by Sephadex G-75 chromatography as described for single-stranded cDNA in step 1.

The formation of a double-stranded hairpin structure can be assayed using S1 nuclease as follows.

STEP a. Add about 5000 cpm of dscDNA directly to 1 ml of S1 buffer (30 m*M* NaOAc, pH 4.5, 0.3 *M* NaCl, 3 m*M* $ZnCl_2$, 10 μg/ml denatured calf thymus DNA) at 0°. Divide into two 0.5-ml aliquots and add 5 μl S1 nuclease to one of the two tubes.[16] Stock S1 nuclease (Miles 37-637-1) is stored concentrated in S1 stock buffer (S1 buffer without DNA and containing 50% glycerol) and should be diluted before use about 10-fold in S1

[13] Hydrolysis of RNA with 0.3 *N* NaOH at room temperature overnight results in less single-strand scissions in the cDNA.

[14] An alternate protocol can be found in W. Salser, *in* "Genetic Engineering" (A. M. Chakrabarty, ed.), p. 53. CRC Press, Cleveland, Ohio, 1979.

[15] We usually obtain yields of 3–20% [micrograms of cDNA per microgram of poly(A^+) RNA] for the first-strand synthesis and 30–100% for the second strand. In preliminary experiments using a simpler method for synthesizing dscDNA as described by M. P. Wickens, G. N. Buell, and R. T. Schimke [*J. Biol. Chem.* *253*, 2483 (1978)], we obtained better yields than using the method described above.

[16] Use Siliclad-treated glass tubes.

stock buffer so that 5 μl contains 300 units. After incubation of both tubes at 37° for 60 min, TCA-precipitate and count.[17] About 70% of the counts should be S1-resistant.

STEP b. Add about 5000 cpm of dscDNA to 1 ml of S1 buffer, heat at 100° for 3 min, and quick-cool on ice. Divide in half and add 5 μl of S1 nuclease to one tube. Process as in step a. If a large fraction of the dscDNA is the form of a hairpin, about 50–60% of the counts should still be S1-resistant.

STEP c. After following the procedure in step a, heat both tubes at 100° for 3 min, quick-cool on ice, and add 5 μl S1 nuclease to the same tube that had S1 nuclease in step a. Incubate 60 min at 37° and TCA-precipitate. If the first S1 nuclease treatment digested the hairpin, only about 10% of the counts will be TCA-precipitable after the second S1 reaction.

Step 5. Restriction Endonuclease Digestion. When cDNA is transcribed from a heterogeneous mRNA preparation, it shows pronounced size heterogeneity not only because of the sequence complexity of the RNA template but most probably also as a result of "slippage" of the oligo(dT) primer on the much longer poly(A) stretch of varying length and the difficulty of obtaining full-length cDNA transcripts from long RNA molecules. To determine whether cDNA sequences from a particular mRNA are present in such a heterogeneous cDNA preparation, the cDNA is fragmented with restriction endonucleases.[6] By this approach the effect of size heterogeneity is reduced and it is possible to detect discrete restriction fragments generated from any predominant cDNA species. To facilitate radioautography in subsequent steps of this analysis [α-^{32}P]dCTP (~25 Ci/mmol) is substituted for [^{3}H]dCTP in the synthesis of the first and/or second cDNA strand (steps 1–4).

For restriction endonuclease digestions of single-stranded cDNA 5-μl analytical reactions are terminated after incubation by the addition of 20 μl of ice-cold water. They are then boiled for 2 min, quick-cooled on ice, and made 7 m*M* in $MgCl_2$. Small aliquots (~2×10^5 cpm) are digested with an excess of one of the restriction endonucleases (*Hae*III, *Hha*I, or *Hin*fI) capable of cleaving single-stranded DNA[18,19] or with any appropriate restriction enzyme when dscDNA is used as the substrate.[20] The amount of each enzyme used is empirically determined to be in excess of the amount needed to digest completely an equivalent amount of

[17] Add 2 ml of 10% TCA to each tube and retain on ice for 30 min. Collect the precipitates on GFC filters, dry them under a heat lamp, and count in a toluene-based fluor in a scintillation counter.

[18] K. Horiuchi and N. D. Zinder, *Proc. Natl. Acad. Sci. U.S.A.* **72,** 2555 (1975).

[19] R. W. Blakesley and R. D. Wells, *Nature (London)* **257,** 421 (1975).

restriction-sensitive DNA under identical reaction conditions. (Restriction endonuclease fragments of single-stranded bacteriophage DNAs such as ϕX174 and M13 are convenient markers, as they can also be isolated as double-stranded replicative forms.) Reactions are stopped with 5 μl of 20 m*M* EDTA, 20% sucrose, and 0.05% bromphenol blue, heated to 100° for 1 min, and then analyzed by polyacrylamide gel electrophoresis. For example, a convenient system is a composite 4.5–10% polyacrylamide slab gel[6] run for 2.5 hr at 150 V in Tris-borate–EDTA buffer (10.8 g Tris base, 5.5 g boric acid, 0.93 g Na_2EDTA per liter) and visualized by autoradiography of the dried gel.

Discrete bands on the polyacrylamide gel should be readily detectable on top of the heterodisperse cDNA "background" provided that the corresponding mRNA species represents at least 2% of the mass of the RNA template and that the cDNA species contains at least two cleavage sites for the enzyme(s) used. These conditions can usually be met, if necessary, by prior purification of the poly(A^+) RNA (e.g., on a sucrose gradient) and/or careful selection of the restriction enzyme(s) employed. These bands can be identified as resulting from a specific hormone mRNA by (1) direct sequence analysis of the cDNA band after excision from gel and comparison with the known amino acid sequence of the hormone,[3-6] (2) a similar analysis after cloning of the band (see below),[2-4] or (3) correlation of the intensity of a particular band with the translational activity of the mRNA after sucrose gradient fractionation.

Step 6. S1 Nuclease Digestion of Double-Stranded cDNA. To ~1 μg of full-length dscDNA in 170 μl of H_2O add 20 μl of 10× S1 buffer (0.3 *M* NaOAc, pH 4.5, 3 *M* NaCl, 45 m*M* $ZnCl_2$) and 10 μl of S1 nuclease (240 units of enzyme, final concentration 1200 units/ml). Incubate the reaction at 22° for 30 min and then at 3–4° for 5 min. Terminate the reaction by addition of $\frac{1}{10}$ volume 1 *M* Tris-HCl, pH 9, and $\frac{1}{8}$ volume 0.2 *M* Na_2EDTA, pH 8. After addition of tRNA carrier to 50 μg/ml extract once with an equal volume of phenol and then with an equal volume of $CHCl_3$–isoamyl alcohol (24:1). Separate and remove the aqueous layer and reextract the organic phase and interface with 50 μl TEN 9 buffer. After combining the aqueous phases and adding glycerol (10% final) chromatograph the dscDNA on a 4-ml column of Sephadex G-75 using TEN 9 buffer. Seven-drop fractions are collected and to the excluded peak containing the dscDNA are added NaCl to 0.2 *M* and 2.5 volumes of 95% EtOH. Pellet the cDNA in the microfuge and dissolve at about 10 ng/μl in 6 m*M* Tris-HCl, pH 7.5, and 0.8 m*M* $MgCl_2$. This mild S1 nuclease treatment digests the hairpin structure of the dscDNA and yields a final product of duplex dscDNA.

[20] J. L. Roberts, P. H. Seeburg, J. Shine, E. Herbert, J. D. Baxter, and H. M. Goodman, *Proc. Natl. Acad. Sci. U.S.A.* **76**, 2153 (1979).

Step 7. ^{32}P-Restriction Site Oligonucleotide Linker Preparation

STEP a. END-LABELING OF THE LINKER. Dry 50 μCi of [γ-^{32}P]ATP (usually ~1000 Ci/mmol) under a stream of nitrogen in a microfuge tube. Add 7 μl of *Hin*dIII (dCCAAGCTTGG) or other suitable restriction site oligonucleotide linker (e.g., *Bam*H1, *Sal*I) at 75 ng/μl in H_2O, 1 μl 10× kinase buffer (0.7 *M* Tris-HCl, pH 7.6, 0.1 *M* $MgCl_2$, 50 m*M* DTT), and 2 μl T4 polynucleotide kinase (5 units, Boehringer 174 645) and incubate 30 min at 37°. Then add 1 μl 10× kinase buffer, 1 μl 10 m*M* ATP, 2 μl T4 polynucleotide kinase, and 6 μl H_2O. Incubate for an additional 30 min at 37°. Terminate the reaction by extracting once with 20 μl phenol–$CHCl_3$–isoamyl alcohol (25:24:1) and twice with 20 μl $CHCl_3$. Remove residual $CHCl_3$ in the aqueous layer by gently blowing a stream of nitogen on it for a few minutes.

STEP b. TESTING THE ^{32}P LINKERS. Mix the following components: 1 μl ^{32}P linkers, 1 μl 10× T4 DNA ligase buffer (0.66 *M* Tris-HCl, pH 7.6, 66 m*M* $MgCl_2$, 0.1 *M* DDT), 1 μl 10 m*M* ATP, 6 μl H_2O, and 1 μl T4 DNA ligase (1 unit/μl, Bethesda Research Laboratories 5224). Incubate for 2 hr at 14°C. Remove 5 μl of the reaction mixture and combine with 5 μl TBE_{SS} (sample solution: 0.2 ml 10× TBE buffer, 0.3 ml 0.2 *M* Na_2EDTA, pH 8, 0.5 ml 4% sarkosyl, 1.0 ml glycerol, 0.5 ml 0.5% bromphenol blue). Analyze on a prerun 7% polyacrylamide gel (20:1 acrylamide/bis) and run at ~150 V for about 1 hr or until the bromphenol blue is about halfway down the gel. Wrap the gel in Saran Wrap and autoradiograph with Kodak No-Screen film for 2 hr. about ≥50% of the label should be in oligomers rather than monomers.[21]

Step 8. ^{32}P-Linker Addition to Double-Stranded cDNA. Treatment of S1 nuclease-digested dscDNA with *E. coli* DNA polymerase I increases the efficiency of blunt-end linker addition about 10-fold. Treat the dscDNA from step 6 with *E. coli* DNA polymerase I as follows. To 26.5 μl containing ~200 ng of ^{3}H- and/or ^{32}P-labeled S1 nuclease-treated dscDNA (step 6), or a restriction fragment isolated by preparative gel electrophoresis,[4] add 3.9 μl of 10× *Pol*I buffer (0.6 *M* Tris-HCl, pH 7.5, 80 m*M* $MgCl_2$), 2 μl 0.2 *M* β-mercaptoethanol, 4 μl 10 m*M* ATP, 1.6 μl of a mixture of the four dNTPs at 5 m*M* each, and 2 μl (4 units) of *E. coli* DNA polymerase I[22] (Boehringer, 104485). Incubate for 10 min at 10°. This step ensures that the dscDNA is a complete duplex structure without protruding single-strand tails.[3,4] Now add 1.5 μl 10× *Pol*I buffer, 0.8 μl

[21] Poor polymerization can be caused by a low concentration of ^{32}P-labeled 5′-ends, bad [γ-^{32}P]ATP, bad kinase, or a ligase preparation which works poorly for blunt-end ligation. See A. Sugino, H. M. Goodman, H. L. Heyneker, J. Shine, H. W. Boyer, and N. R. Cozzarelli [*J. Biol. Chem.* **252,** 3987 (1977)] for a more detailed discussion.

[22] The *E. coli Pol*I should be free of contaminating DNase activity.

0.2 M β-mercaptoethanol, 1.5 μl 10 mM ATP, 9.2 μl ^{32}P linkers, and 2 μl T4 DNA ligase (1 unit/μl). Incubate at 14° for 3–4 hr. Remove 2 μl of the reaction into 5 μl TBE_{ss} and freeze the remainder of the reaction. Analyze the 2-μl sample on a 7% polyacrylamide gel as in step 7. Autoradiograph for 4 hr. The linker polymerization should be nearly as efficient as the linker test ligation in step 7.

Step 9. Restriction Endonuclease Digestion of the ^{32}P-Linker–Double-Stranded cDNA

STEP a. RESTRICTION ENDONUCLEASE DIGESTION. When *Hin*dIII linkers have been used, dilute the ligation reaction from step 8 (~53 μl remaining) with 250 μl 7 mM $MgCl_2$ and 60 mM NaCl. Heat at 65° for 5 min. Add 7 μl of *Hin*dIII endonuclease (70 units, Biolabs, Inc.) and additional 7-μl aliquots at 1.5 hr and 3 hr of incubation at 37°. Incubate for a total time of 4–5 hr at 37°. Remove 15 μl from the reaction (freeze the remainder at −20°) and mix with 10 μl TBE_{ss}. Analyze on a 7% polyacrylamide gel. Radioautography should show that no linker polymers remain and that a band (or smear) is present at the length expected for the input dscDNA.[2–4] If the gel is satisfactory, thaw the reaction mix and extract once with phenol–$CHCl_3$–isoamyl alcohol (25:24:1). Remove the aqueous layer and add $\frac{1}{10}$ volume 2 M NaCl and 2.5 volumes 95% EtOH. Leave overnight at −20°.

STEP b. PREPARATIVE ELECTROPHORESIS. Centrifuge the sample for 15 min at 4° in the Brinkman microfuge. Dissolve the pellet in 30 μl of 0.25× TBE and add 5 μl TBE_{SS} and 5 μl glycerol. Separate on a 5% polyacrylamide gel as described in step 10.[23]

Step 10. Preparative Polyacrylamide Gel Electrophoresis.[24,25] The following stock solutions are required: (a) 10× TBE buffer (108 g Tris base, 55 g boric acid, 9.3 g Na_2EDTA, H_2O to 1 liter); (b) 20:1 AcBis (20 g acrylamide, 1 g bisacrylamide, H_2O to 100 ml); (c) 2% ammonium persulfate (store for less than 1 week at 4°); and (d) TBE_{ss} (see step 7). For a gel 15 × 13 cm and 1.5 mm thick use the amounts shown in Table I (for a longer gel, 15 cm × 1.5 mm, double the quantities). Prerun the short gel (13 cm) at 150 V for 30 min and run at 150 V. Prerun the long gel (30 cm) at 250 V for 30 min and run at 250 V.

Continue the electrophoresis for an appropriate time (depending on

[23] An alternate method for separating the ^{32}P linkers from the linkers attached to the cDNA is to chromatograph the mixture on a 10-ml column of BioGel A-5m (200–400 mesh, Bio-Rad Laboratories) in 10 mMM Tris-HCl pH 7.5, 100 mM NaCl, and 1 mM EDTA as described by A. E. Sippel, H. Land, W. Lindenmaier, M. C. Nguyen-Huu, T. Wurtz, K. N. Timmis, K. Giesecke, and G. Schutz [*Nucleic Acids Res.* **5**, 3275 (1978)]. This procedure eliminates steps 10 and 11; i.e., after step 9 and chromatography proceed to step 12.

[24] C. W. Dingman and A. L. Peacock, *Biochemistry* **7**, 659 (1968).

[25] T. Maniatis, A. Jeffrey, and H. van de Sande, *Biochemistry* **14**, 3787 (1975).

TABLE I
POLYACRYLAMIDE GEL COMPOSITION

Component	5% gel	7% gel
AcBis	10 ml	14 ml
10X TBE buffer	4 ml	4 ml
2% Ammonium persulfate	2.5 ml	2.5 ml
TEMED	13 μl	13 μl
H_2O	23.5 ml	19.5 ml
	40 ml	40 ml

the fragment size) and radioautograph the wet gel after covering it with Saran Wrap.

Step 11. Electroelution. Using the autoradiograph of the gel as a guide excise the ^{32}P-labeled "high"-molecular weight dscDNA with a scapel. Place the gel slice and 3.5 ml of 0.25× TBE buffer together with 10 μg/ml tRNA in medium-sized ($\frac{5}{8}$-in-diameter) dialysis tubing. Place in the electroelution apparatus described by McDonell *et al.*[26] Orient the bag parallel to the electrodes and fill the apparatus with about 1.5 liters of 0.5× TBE buffer. Electroelute at 100 mA constant current (~35 V) for at least 8 hr. Check the efficiency of elution by Cerenkov counting of the eluate and the gel piece plus bag. Precipitate the eluted DNA by the addition of $\frac{1}{10}$ volume of 2 *M* NaCl and 2.5 volumes of 95% EtOH. Store overnight at −20° and centrifuge. Dissolve the pellet in 0.3 ml 10 m*M* Tris-HCl, pH 8, containing 1 m*M* EDTA. Centrifuge out the debris for 1 min in the microfuge. Ethanol-precipitate the DNA again in a 1.5-ml microfuge tube.

Step 12. Preparation of Phosphatase-Treated Linear pBR322. Prepare supercoiled pBR322 plasmid DNA as described in references 8 and 9 and digest with the appropriate restriction endonuclease (i.e., the same enzyme as used for the linkers, e.g., *Hin*dIII endonuclease for *Hin*dIII linker, *Bam*H1 endonuclease for *Bam*H1 linkers). In the case of *Hin*dIII add 5 μl of 10× *Hin*dIII buffer (60 m*M* Tris-HCl, pH 7.5, 60 m*M* $MgCl_2$, 0.6 *M* NaCl) to 15 μg of pBR322 DNA and bring to a final volume of 46 μl with H_2O. Add 2 μl (20 units) of *Hin*dIII endonuclease at 0, 40, and 80 min of incubation at 37°. Incubate a total of 2 hr. Check the digest by analysis on a 1% agarose gel and save a 5-μl sample of the reaction for a control in the transformation step 14. Dilute the reaction with 450 μl H_2O and heat at 65° for 10 min to inactivate the enzyme. Add 5 μl 1 *M* Tris-HCl, pH 9.0, and 1.5 μl (3 units) of CIAP. Incubate 30 min at 65°, add another 1.5 μl CIAP, and incubate an additional 30 min at 65°. Extract the reaction three times with an equal volume of phenol–$CHCl_3$–isoamyl alcohol (25:24:1)

[26] M. W. McDonell, M. N. Simon, and F. W. Studier, *J. Mol. Biol.* **110,** 119 (1977).

and then three times with $CHCl_3$–isoamyl alcohol (24:1). Add NaCl to 0.2 *M* to the aqueous phase and precipitate with 2.5 volumes of EtOH. Centrifuge, resuspend, and reprecipitate with EtOH. Dissolve the DNA in 0.2 ml 10 m*M* Tris-HCl, pH 7.5, and 1 m*M* EDTA and determine the concentration spectrophotometrically (1 mg/ml DNA = 20 A_{260}).

Step 13. Ligation of ^{32}P-Linker–Double-Stranded cDNA to Phosphatase-Treated HindIII-Digested pBR322. Dissolve the isolated ^{32}P-linker–dscDNA after electroelution or chromatography in sufficient 0.2× ligase buffer[27] to give 50–100 ng in 8 μl (based on count recovery). To the 8 μl of ^{32}P-linker–dscDNA (~50 ng) add 2 μl *Hin*dIII-digested, phosphatase-treated pBR322 from step 12 (150 ng), 1.5 μl 10× ligase buffer, and 1.5 μl 10 m*M* ATP. Heat 2 min at 37°, add 2 μl (2 units) T4 DNA ligase, and incubate at 14° for ≥4 hr.

It is possible to assay for ligation by removing 1-μl samples at 0 time and at 4 hr and analyzing them on a 5% polyacrylamide gel. Dry the gel and expose it at −70° using a Lightning Plus intensifying screen with Kodak XR-2 film. Successful ligation is indicated by ^{32}P incorporation into the region of the gel where linear pBR322 migrates.

To the remainder of the reaction add NaCl to 0.2 *M*, 2.5 volumes of 95% EtOH, and precipitate overnight at −20°. Centrifuge for 15 min in the microfuge and dry the pellet briefly under vacuum. Dissolve the pellet in 50 μl 10 m*M* Tris-HCl, pH 7.6.

Step 14. Transformation of χ1776.[28]

STEP a. Grow a 10-ml culture of *E. coli* strain χ1776 at 37° overnight in χ medium (19 g Bactotryptone, 5 g yeast extract, 10 g NaCl, 100 mg diaminopimelic acid, 40 mg thymine, 1 liter tap-distilled H_2O, 0.33 ml 6 *N* NaOH. Autoclave 20 min and then add 10 ml sterile 50% glucose and 10 ml of 7.5 mg/ml filter-sterilized nalidixate in 50 m*M* NaOH).

STEP b. Dilute the overnight culture to 0.05 A_{600} in 100 ml χ medium. Grow at 37° with shaking to 0.4 A_{600}.

STEP c. Centrifuge the culture in the Sorvall SS34 rotor (6000 rpm, 5 min, 5°) in screw-cap polycarbonate tubes. Resuspend the cell pellets in 60 ml *cold* (4°) Mχ wash buffer (0.1 *M* NaCl, 5 m*M* $MgCl_2$, 5 m*M* Tris-HCl, pH 7.6). Centrifuge again. Resuspend the cell pellets in 30 ml of Mχ wash buffer, combine into one tube, and centrifuge (6000 rpm, 5 min, 5°).

STEP d. Resuspend the cell pellet in 30 ml *cold* Mχ Ca^{2+} buffer (75 m*M* $CaCl_2$, 250 m*M* KCl, 5 m*M* $MgCl_2$, 5 m*M* Tris-HCl, pH 7.6). Allow to stand on ice for 20 min.

STEP e. Centrifuge again (6000 rpm, 5 min, 5°). Resuspend in 0.5 ml

[27] Ligase buffer (10×): 0.66 *M* Tris–HCl, pH 7.6, 66 m*M* $MgCl_2$, 0.1 *M* DTT.

[28] See also M. V. Norgard, K. Keem, and J. J. Monahan [*Gene* **3**, 279 (1978)] for another high-efficiency χ1776 transformation procedure.

Mχ Ca^{2+} buffer. Do all further manipulations in the appropriate physical containment facility as specified by the NIH guidelines on recombinant DNA research.

STEP f. Add 100 μl of the cell suspension (from step e) to 50 μl of the ligated DNA solution (in 10 m*M* Tris, pH 7.6) in a 0.5-ml polypropylene microfuge tube. Mix well. Hold on ice for 60 min.

STEP g. Heat-shock the mixture for 90 sec at 42° and cool briefly to room temperature. Transfer the solution to a sterile 10-ml culture tube and add 1.35 ml χ medium at room temperature.

STEP h. Incubate the culture for 1 hr at 37° with shaking.

Step 15. Selection of Ampicillin Resistant (Amp^R) *Colonies.* Plate all of the 1.5 ml culture in 0.2-ml aliquots on χ-ampicillin plates. (Add 15 g Bactoagar before autoclaving and 5 ml of 10 mg/ml filter-sterilized ampicillin after autoclaving to the χ medium recipe. Glucose and nalidixate are also added after autoclaving, as in the χ medium recipe.) Incubate plates at 37° overnight or longer.

We obtain several million transformants per microgram of supercoiled pBR322 DNA. The following tests are useful in accessing the quality of the plasmid vector used in the transformation:

TEST a. EFFICIENCY OF *Hin*dIII DIGESTION. The starting plasmid, pBR322, should migrate as a supercoiled DNA molecule on a 1% agarose gel and give a high transformation frequency ($\sim 4 \times 10^6$ Amp^R transformants per microgram of DNA). After *Hin*dIII digestion the supercoils and nicked circles should disappear from the gel profile, and a low background of Amp^R transformants should be obtained after transformation ($\leq 10^4/\mu g$).

TEST b. EFFICIENCY OF CIAP TREATMENT. There should be little or no increase in Amp^R transformants after self-ligation of *Hin*dIII-CIAP-treated pBR322 as compared to *Hin*dIII-cut pBR322 without ligation. Also there should be little or no reformation of circles of the self-ligated *Hin*dIII-CIAP-treated pBR322 as analyzed on a agarose gel.

TEST c. PRESENCE OF COHESIVE, LIGATABLE ENDS. *Method a:* Ligate the treated vector with *Hin*dIII-digested yeast or *E. coli* DNA and transform a suitable host (using suitable containment conditions). There should be a large increase in the number of Amp^R transformants as compared to the self-ligated vector alone. The linear band of *Hin*dIII-CIAP-treated pBR322 should disappear from an agarose gel after ligation to the yeast or *E. coli* DNA. *Method b:* Self-ligate *Hin*dIII-CIAP-treated pBR322. To another aliquot add T4 polynucleotide kinase during the ligation. Compare these two samples to *Hin*dIII-CIAP-treated pBR322 by transformation and agarose gel electrophoresis. The ligated and untreated samples should show only monomeric linear DNA on the agarose gel and give very few

Amp^R transformants. The kinase-treated sample should show multimeric linear and closed circular DNA forms on the gel and give high transformation efficiencies (i.e., a large number of Amp^R transformants). Kinase treatment restores the 5′-terminal phosphate removed by prior CIAP digestion.

Step 16. Test for Tetracycline Sensitivity (*Tet*^{S}). Because of phosphatase treatment of the pBR322-*Hin*dIII vector most of the Amp^R transformants obtained in step 15 should be recombinants, i.e., have an inserted piece of DNA at the *Hin*dIII site. Pick and stab the Amp^R colonies onto χ–Tet plates (χ plates containing 25 μg/ml tetracycline). Recombinants should be Amp^R Tet^S,[29] while colonies transformed with the parental pBR322 will be Amp^R Tet^R. Deletions at the *Hin*dIII site of pBR322 are sometimes generated during *Hin*dIII digestion and phosphatase treatment—colonies transformed with these molecules are also Amp^R Tet^S but do not contain recombinant DNA. Further direct testing of the DNA is therefore necessary.

Step 17. Screening for Recombinant Clones. Grow each of the Amp^R Tet^S colonies from step 16 in 2 ml χ medium overnight at 37° in 10-ml sterile tubes. Centrifuge 2 min in the microfuge and after decanting the supernatant freeze the pellet at −70° for 30 min. Thaw the pellet and resuspend it in 100 μl 25% sucrose and 50 m*M* Tris, pH 8. Add 20 μl freshly prepared egg white lysozyme (10 mg/ml in 0.25 *M* Tris, pH 8). Leave on ice for 5 min and then add 5 μl of 0.25 *M* EDTA, pH 8. Leave on ice for 5–15 min, add 150 μl of Triton lysis mix (1 ml 10% Triton, 75 ml 0.25 *M* EDTA, pH 8, 15 ml 1 *M* Tris, pH 8, 9 ml H_2O) and put back on ice for 5–15 min longer. Note: Invert tube gently for mixing after adding Triton lysis mix. Centrifuge in the microfuge for 25 min (5°). Carefully remove the supernatant and after adding 100 μl H_2O treat it with 5 μl of 10 mg/ml RNase A (boiled for 10 min) for 1 hr at 37°. Extract with an equal volume of phenol–$CHCl_3$ (1:1) and add NaCl to 0.3 *M* and 2 volumes of 95% EtOH. Leave at −70° for 30 min, centrifuge, and wash the DNA pellet with ~0.3 ml 70% EtOH (cold). Finally, dissolve the DNA in 10 μl of 10 m*M* Tris, pH 7.5.

Analyze the DNA before and after appropriate restriction enzyme cleavage on an agarose or polyacrylamide gel. Supercoiled DNA which migrates slower than the pBR322 control, or release of a DNA fragment

[29] Insertion of DNA fragments into the *Hin*dIII site of pBR322 reduces the level of tetracycline resistance of cells carrying such recombinant plasmids to varying degrees dependent on the sequences cloned into this site (see footnote 2). In particular, we have observed that the insertion of DNA molecules containing poly(dA·dT) regions allows the expression of tetracycline resistance at reduced levels (5–10 μg/ml). Insertion of sequences into the *Bam*HI or *Sal*I of pBR322 does not show this phenomenon.

(other than linear pBR322) after *Hin*dIII digestion indicates that a recombinant DNA molecule has been isolated. Its restriction pattern can be compared to that obtained with the cDNA in step 5. Further identification of the clones can be made by DNA sequence analysis, hybridization analysis, or using them as reagents in the hybrid arrest translation procedure.[30]

[30] B. M. Paterson, B. E. Roberts, and E. L. Kuff, *Proc. Natl. Acad. Sci. U.S.A.* **74,** 4370 (1977).

[6] Improved Phosphotriester Method for the Synthesis of Gene Fragments[1]

By S. A. NARANG, HANSEN M. HSIUNG, and ROLAND BROUSSEAU

One of the most challenging tasks in the synthesis of DNA with defined sequences has been the unambiguous chemical synthesis of short deoxyribooligonucleotides. During the past several years, the triester method[2] of synthesis has been extensively investigated as an alternative to the diester approach.[3] This has resulted in the development of various new and novel features in the triester approach such as (1) a modified triester method,[4] (2) an improved phosphorylating reagent,[5] (3) coupling reagents,[6] and (4) a deblocking reagent.[7] These developments have played a major role in establishing the modified triester approach as an efficient method of polynucleotide synthesis in terms of simplicity, speed, and yield. As a result, the syntheses of various biologically important sequences, such as *lac* operator DNA,[8] adaptor molecules containing

[1] N.R.C.C. No. 17290.

[2] R. L. Letsinger and K. K. Ogilvie, *J. Am. Chem. Soc.* **91,** 3350 (1969); R. Arentzen and C. B. Reese, *J. Chem. Soc., Perkin Trans.* p. 445 (1977); F. Eckstein and I. Rizk, *Chem. Ber.* **102,** 2362 (1969); T. Neilson and E. S. Werstink, *J. Am. Chem. Soc.* **96,** 2362 (1974).

[3] H. G. Khorana, K. L. Agarwal, P. Bresmer, H. Buchi,, M. H. Caruthers, P. J. Cashion, M. Fridkin, E. Jay, K. Kleppe, R. Kleppe, A. Kumar, R. C. Loewen, R. C. Miller, K. Minamoto, A. Panet, U. L. RajBhandary, B. Ramamoorthy, T. Sekiya, T. Takeya, and J. H. van de Sande, *J. Biol. Chem.* **251,** 565 (1976).

[4] K. Itakura, C. P. Bahl, N. Katagiri, J. Michniewicz, S. A. Narang, and R. H. Wightman, *Can. J. Chem.* **51,** 3 (1973); T. C. Catlin and F. Cramer, *J. Org. Chem.* **38,** 245 (1973).

[5] N. Katagiri, K. Itakura, and S. A. Narang, *J. Am. Chem. Soc.* **97,** 7332 (1975).

[6] J. Stawinski, T. Hozumi, and S. A. Narang, *Can. J. Chem.* **54,** 670 (1976).

[7] J. Stawinski, T. Hozumi, S. A. Narang, C. P. Bahl, and R. Wu, *Nucleic Acids Res.* **4,** 353 (1977).

[8] C. P. Bahl, R. Wu, K. Itakura, N. Katagiri, and S. A. Narang, *Proc. Natl. Acad. Sci. U.S.A.* **73,** 91 (1976).

ISBN 0-12-181968-X

restriction sites of the endonucleases[9] used for cloning, the somatostatin[10] gene, and a tridecadeoxyoligonucleotide inhibiting the growth of Rous sarcoma virus,[11] were achieved in the last few years. In spite of these achievements, an efficient method of purifying triester oligonucleotides was still lacking. Invariably, it had been found to be quite difficult to resolve the triester product from starting material containing purine bases (especially guanine) by conventional silica gel chromatography. In this article, we update the synthetic methodology reported earlier[12] by introducing preparative reverse-phase chromatography to resolve most deoxyribooligonucleotides containing guanine bases.

Reverse-Phase Chromatography of Protected Oligonucleotides Containing Triester Groups

During our studies with the modified triester approach, we generally observed that the pure, fully protected product was difficult to separate quantitatively by conventional silica gel chromatography from the crude reaction mixture, especially from components containing several guanine bases. Previously,[7] we had attempted to remove the hydroxyl component by converting it to a phosphodiester group by treatment with β-cyanoethyl phosphate in the presence of benzenesulfonyl triazole. In the present studies, we found that thin-layer chromatography (TLC) on silanized silica gel (RP-2) and KC_{18} (RP-18) plates in an acetone–water solvent gave excellent separation of components containing trityl and hydroxyl groups and also of those differing in size. The polar component (containing 3′-phosphodiester) generally moved to the solvent front, and the fully protected component (containing the trityl group) was the slowest. The mobility of the component containing a 5′-hydroxyl group was in between that of the polar and the nonpolar compounds. A typical example is fractionation on a RP-2 TLC plate of a complex mixture formed during the preparation of protected AC as compared to fractionation of silica gel, as shown in Fig. 1A and B. Similarly, purification of GGCA could not be achieved on silica gel as determined after complete deblocking and TLC on a polyethyleneimine (PEI) plate (Fig. 1D), whereas purification by TLC on reverse-phase (RP-2) plates gave pure GGCA as analyzed on a PEI plate (Fig. 1C). These results indicated that reverse-phase chroma-

[9] C. P. Bahl, K. J. Marians, R. Wu, J. Stawinski, and S. A. Narang, *Gene* **1,** 81 (1976).
[10] K. Itakura, T. Hirose, R. Crea, A. D. Riggs, H. L. Heyneker, F. Bolivar, and H. W. Boyer, *Science* **198,** 1056 (1977).
[11] P. C. Zamecnik and M. L. Stephenson, *Proc. Natl. Acad. Sci. U.S.A.* **75,** 280 (1978).
[12] S. A. Narang, R. Brousseau, H. M. Hsiung, and J. J. Michniewicz, this series (in press).

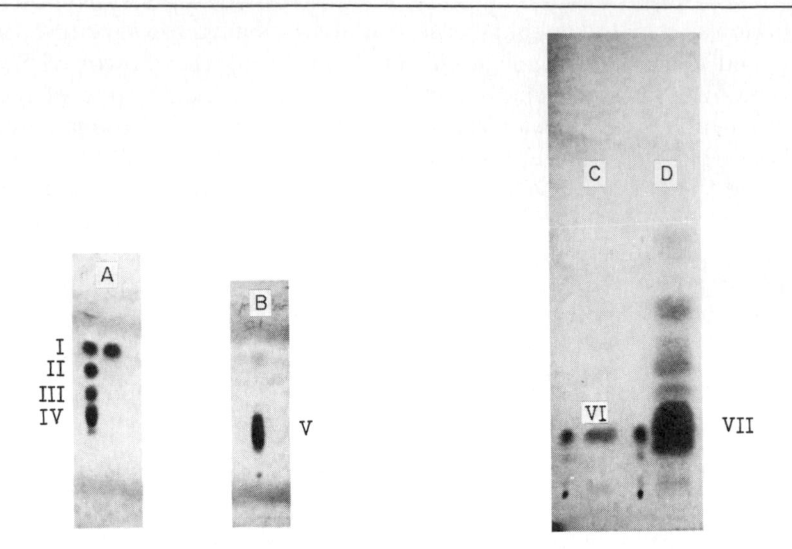

FIG. 1. (A) Resolution by reverse-phase TLC on silanized silica gel (RP-2) in acetone–water (30%) of a reaction mixture containing the following components: I, HO-dbzC-OBz; II, HO-dbzA ∓ bzC-OBz; III, $(MeO)_2$TrdbzA-ClPh; IV, $(MeO)_2$TrdbzA ∓ bzC-OBz. (B) Resolution on silica gel TLC in chloroform–methanol (5%) of the same reaction mixture as in (A). (C) PEI TLC of the deblocked tetranucleotide GGCAP-ClPh after purification of the fully protected tetranucleotide with reverse-phase TLC on RP-2 plates in acetone–water (25%). (D) PEI TLC of the deblocked tetranucleotide GGCAP-ClPh after purification of the fully protected tetranucleotide on silica gel TLC in chloroform–methanol (10%).

tography of triester oligonucleotide could yield pure product at each step of synthesis.

General Methods and Materials

Silanized silica gel 60 F-254 TLC plates (RP-2) of 0.25 mm thickness (E. M. Brinkman) and KC_{18} reverse-phase (RP-18) TLC plates with a fluorescent indicator, 200 μm thick (Whatman), were purchased commercially.

Solvent Systems

Triester oligonucleotides were analyzed on (1) silica gel F-254 TLC plates in solvent A, chloroform containing 1–15% methanol and 2% formamide, and (2) silanized silica gel F-254 (RP-2) or KC_{18} reverse-phase (RP-18) TLC plates in solvent B, acetone containing 20–30% (v/v) water.

Preparation of Fully Protected Deoxymononucleotides

The preparation of fully protected deoxymononucleotides has been described previously,[12] however, the purification was carried out on Waters Prep LC/System 500 by using solvents: chloroform containing 2% methanol for T, C, and A, components, and chloroform containing 6% methanol for G components. This instrument gave excellent separation of large amounts (25 g per batch) of material in less time (about 1 hr), and the same column could be used about 10 times.

General Method for the Synthesis of Fully Protected Di- and Trinucleotides

The fully protected mono- or dinucleotide was first dried by coevaporation of anhydrous pyridine (2× 10 ml) and then dissolved in anhydrous pyridine (10–20 ml/mmol of the nucleotide component). Excess anhydrous triethylamine[13] (at least 10 molar equivalents) was added to remove cyanoethyl groups from 3′-protected phosphates, and this reaction mixture was kept at room temperature and monitored by two TLC systems based on (1) appearance of the trityl-positive spot at the origin on the silica gel plate in chloroform–methanol (10%) and (2) movement of the trityl-positive spot on RP-2 or RP-18 TLC at the solvent front in an acetone–water (20–30%) solvent system. After 2–3 hr, the reaction was complete and the mixture was evaporated to a foam to remove the excess triethylamine and acrylonitrile liberated during the deblocking reaction. The material was then washed with anhydrous ether (2× 5 ml) and mixed with the 5′-hydroxyl mononucleotide in dry pyridine and evaporated to dryness under reduced pressure. The coupling reaction was carried out (see Fig. 2) as described below.

The syrupy residue was redissolved in dry pyridine (5 ml/mmol of the nucleotide component), followed by the addition of mesitylenesulfonyl tetrazole (3–5 molar equivalents). The coupling reaction was generally over in less than 2 hr (as judged by two TLC systems based on (1) the appearance of a major faster-moving trityl-positive component on silica gel TLC in chloroform–methanol (10%) solvent and (2) the appearance of a major slower-moving trityl-positive component on RP-2 or RP-18 TLC plates in acetone–water (20, 25, and 30%, v/v). The reaction mixture was then decomposed with cold, distilled water (5 ml), and the resultant solution was evaporated to a gum *in vacuo*. The gum was dissolved in ice-cold chloro-

[13] A. K. Sood and S. A. Narang, *Nucleic Acids Res.* **4,** 2757 (1977); R. W. Adamiak, M. Z. Bariciszewska, E. Biala, K. Grezeskowiak, R. Kierzek, A. Kraszewski, W. T. Markiewicz, and W. Wiewiorwski, *Nucleic Acids Res.* **3,** 3397 (1976).

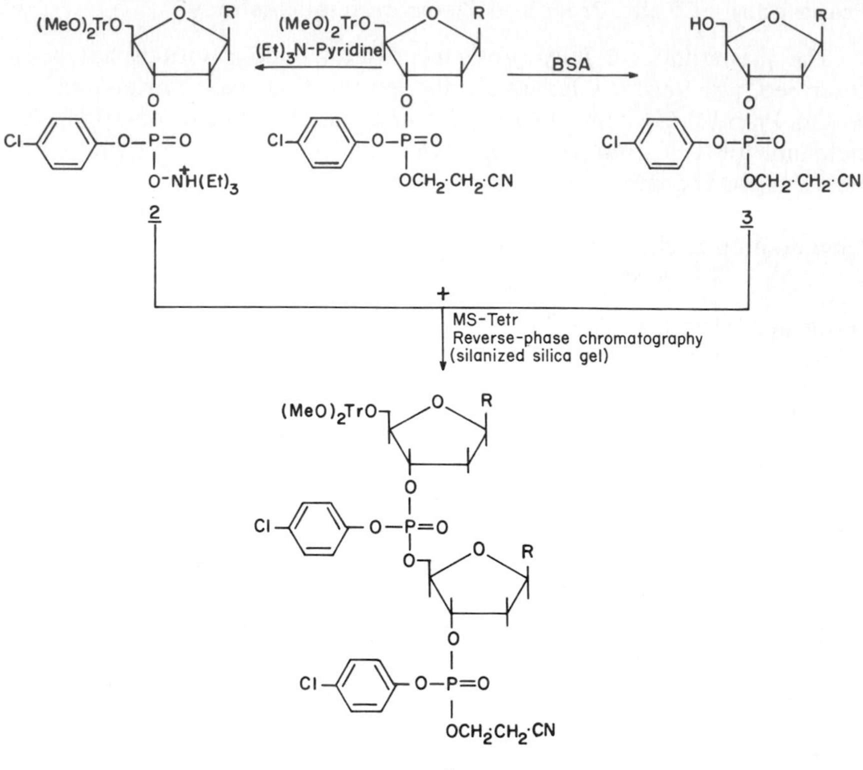

FIG. 2. Modified phosphotriester method for dinucleotide synthesis.

form (50 ml), followed by washing with 5% sodium bicarbonate (2× 25 ml) and water (1× 25 ml). The organic layer was dried over sodium sulfate. The crude reaction mixture was dissolved in 1 ml of acetonitrile–water (20% v/v) and purified by medium-pressure reverse-phase liquid chromatography (RP-8). The fraction containing the desired product was concentrated *in vacuo* until it became turbid. It was next extracted with chloroform (3× 250 ml) which was washed once with water. On drying the chloroform and then on evaporating, the product was obtained as white, foamy material in 60–80% yield.

General Method of Coupling Longer Oligonucleotides and Their Purification (Fig. 3)

The fragment containing the 5′-terminus sequence was decyanoethylated with triethylamine in anhydrous pyridine, whereas the fragment

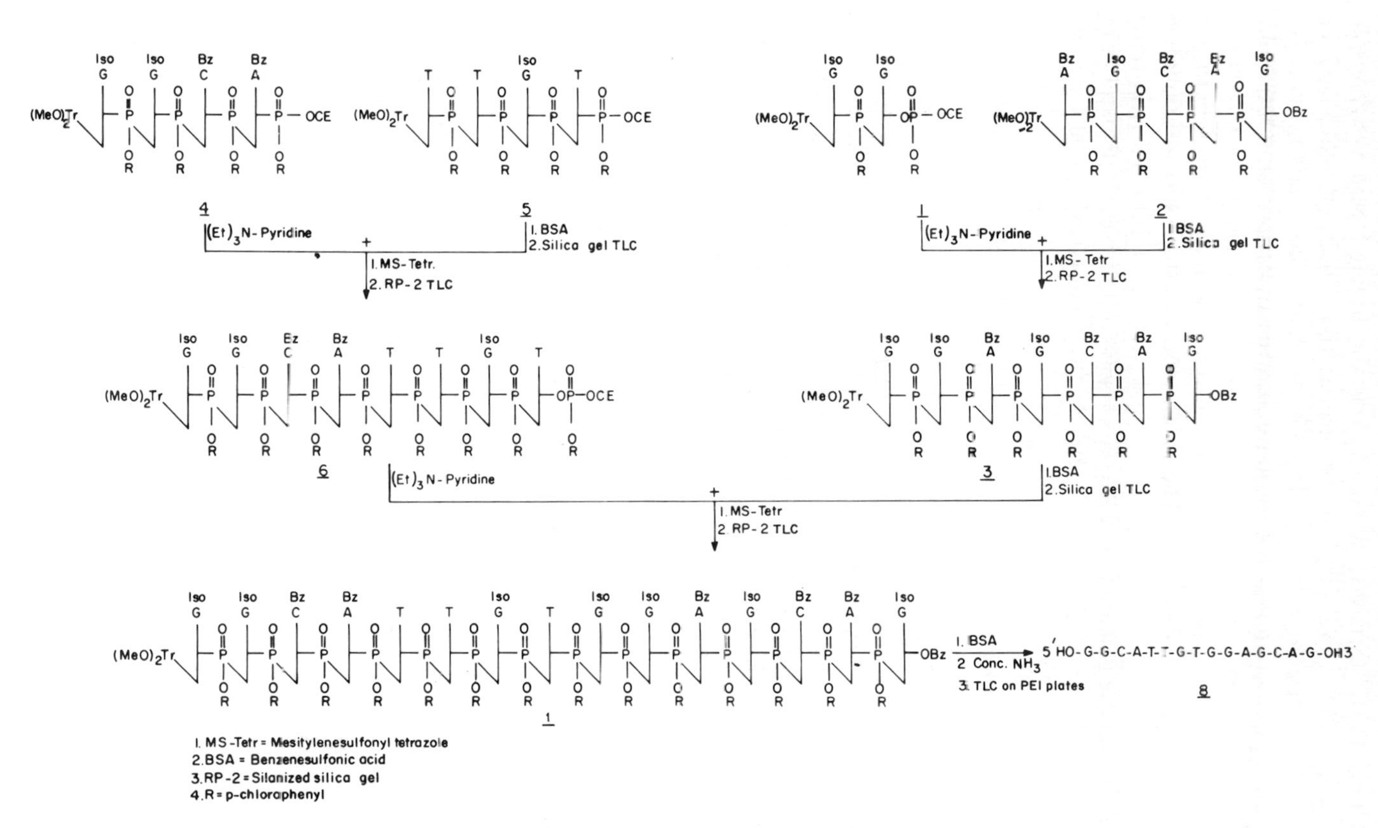

FIG. 3. Chemical synthesis of pentadecaoligonucleotide by the modified triester method.

containing the 3′-terminus sequence was detritylated with benzenesulfonic acid[7] (2%) and purified by preparative TLC on silica gel in chloroform–methanol (10%) as described earlier. The coupling reaction was carried out in anhydrous pyridine with mesitylenesufonyl tetrazole (3–5 *M* equivalents) for 2–4 hr. After the usual work-up as described above, the organic layer was passed through Whatman phase-separating filter paper[14] to remove water. The dried crude mixture was fractionated by preparative TLC (30 mg per plate) on RP-2 or RP-18 (20 × 20 cm) plates in acetone–water (20–30%, v/v) solvent. The desired band was eluted with chloroform–methanol (20%, v/v) (60 ml) in about 90% recovery.

General Method of Deblocking and Isolation of Oligomers Containing 3′-5′-Phosphodiester Linkages

The complete deblocking of each fragment was carried out by treatment with benzenesulfonic acid in chloroform–methanol to remove dimethoxytrityl groups, followed by concentrated ammonia to remove all *N*-benzoyl and *p*-chlorophenyl groups as described previously.[7] Oligomers up to 15 bases long were purified on a PEI TLC plate at 60° using an appropriate concentration of lithium chloride, pH 7.1, containing 7 *M* urea.[7] The fractionation of oligomers longer than decamers was achieved by preparative gel electrophoresis on a 20 × 40 cm slab as described below.

Preparative Gel Electrophoresis of Nonradioactive Deoxyribooligonucleotides Containing 3′-5′-Phosphodiester Linkages

A sample of oligonucleotide (~50–100 A_{260}) in 100 μl of 7 *M* urea was applied in a slot 10 cm wide and 1 cm deep on a 0.3 × 20 × 40 cm slab containing 20% (w/v) acrylamide, 0.7% (w/v) methylene bisacrylamide, 7 *M* urea, 50 m*M* Tris-borate, pH 7.5, 1 m*M* EDTA, and 3 m*M* ammonium presulfate. A mixture of dye markers, bromphenol blue and xylene cyanol, was applied in a well 1 cm deep and 1.3 cm wide on each side of the 10-cm band. Electrophoresis was carried out at 800 V/25 mA for 5 hr.

The bands on the gel were visualized under uv light by transferring them to two 20 × 20 cm silica gel TLC plates containing a fluorescent indicator covered with Saran Wrap. The desired band was sliced, crushed with a glass rod, eluted with 0.25 *M* triethylammonium bicarbonate (3 × 5 ml), and then passed through a plastic pipette containing 0.5 ml of DE-52

[14] Personal communication with Dr. J. Hackman, Collaborative Research Inc., Waltham, U.S.A.

cellulose supported on a cotton plug. It was first washed with water, and the desired compound was eluted with 2 *M* triethylammonium bicarbonate, pH 9.0, in about 70% recovery.

Sequence Determination of Synthetic Oligonucleotides

The sequence of bases in each chemically synthesized oligomer was checked by the two-dimensional mobility shift method[15] with some modification.

Oligomers Containing 3′-P-Chlorophenyl Phosphate Groups

The 5′-labeled oligonucleotide was digested at 37° in a 10-μl solution of 5 m*M* magnesium chloride, 10 m*M* Tris, pH 7.6, 10 μg calf thymus DNA,

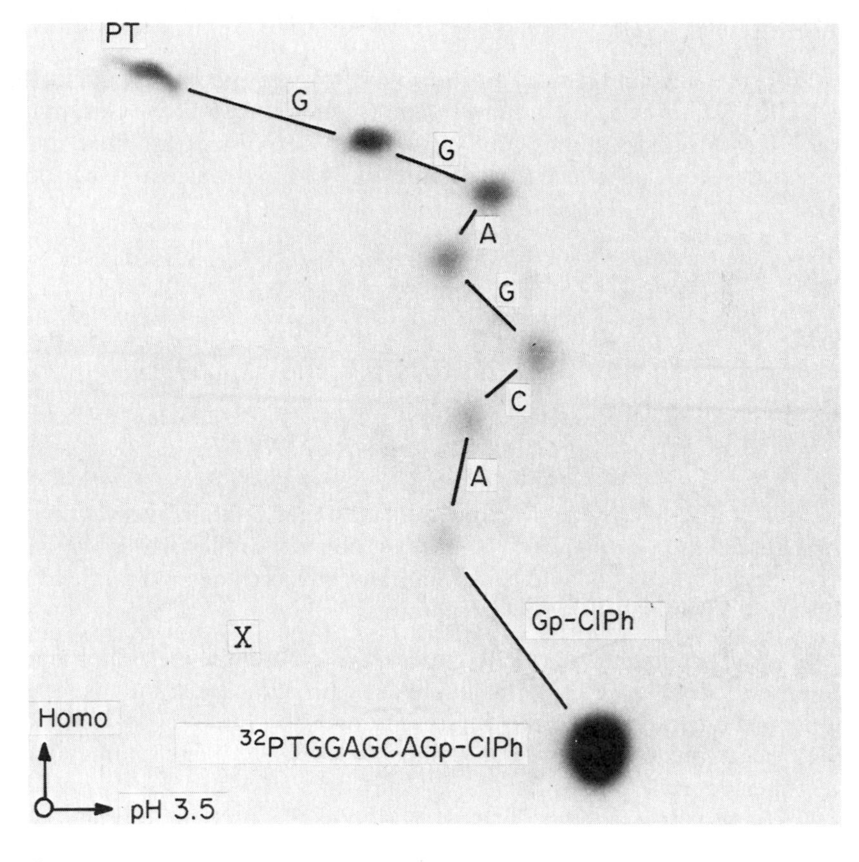

FIG. 4. Two-dimensional fingerprinting pattern of ^{32}P-labeled TGGAGCAGp-ClPh after digestion with pancreatic DNase and snake venom phosphodiesterase enzymes.

[15] C. D. Tu, E. Jay, C. P. Bahl, and R. Wu, *Anal. Biochem.* **74,** 73 (1976).

and 1 μl pancreatic DNase[16] (2 mg/ml stock, 1–2 μg) to remove 3′-chlorophenyl phosphate groups. Three-microliter samples were withdrawn after 3, 10, and 60 min into a siliconized tube and kept at 90° for 3 min (each aliquot). To the combined heated sample (11 μl) was added 2 μl of venom phosphodiesterase (1 mg/ml) and 2 μl of 300 m*M* triethylammonium bicarbonate, pH 9.2; this was incubated at 37°, and after 3, 30, and 60 min samples of 4 μl were transferred to another tube and heated each time at 90° for 3 min. Finally all the samples were combined and dried in a desiccator, dissolved in 2 μl of water, and subjected to two-dimensional chromatography (cellulose acetate electrophoresis, pH 3.5-homochromatography in solvent VI) by the standard procedure.[15] The fingerprinting patterns of 5′-P^{32} TGGAGCAGp-ClPh are shown in Fig. 4.

Concluding Remarks

With the development of an improved triester method of synthesis, arylsulfonyl tetrazoles as coupling reagents, and reverse-phase chromatography for the purification of fully protected deoxyribooligonucleotides, the synthesis of a gene has become practical within a reasonable period of time.

[16] Personal communication with Professor Ray Wu, Cornell University, Ithaca, New York.

[7] Synthetic Adaptors for Cloning DNA

By R. J. ROTHSTEIN, L. F. LAU, C. P. BAHL, S. A. NARANG, and RAY WU

Chemically synthesized oligonucleotides that contain restriction enzyme recognition sequences are useful tools for molecular biologists.[1] Three major types of adaptor oligonucleotides have been synthesized for increasing flexibility in molecular cloning:

1. A self-complementary oligomer which contains a restriction endonuclease recognition site: Forms duplexes with blunt ends; the restriction site is at the center within the duplex region.
2. Ready-made adaptors: Form short duplexes with one blunt end and one cohesive end.
3. Conversion adaptors: Form short duplexes with two different cohesive ends.

[1] R. Wu, C. P. Bahl, and S. A. Narang, *Prog. Nucleic Acid Res. Mol. Biol.* **21,** 101 (1978).

ISBN 0-12-181968-X

Self-complementary oligomers have been used for a number of interesting studies other than directly in the cloning of DNA fragments. The synthetic octamer pT-G-A-A-T-T-C-A was prepared for studying the interaction of the *Eco*RI restriction endonuclease and modification methylase with short duplex DNA.[2] Interactions of the *Hae*III and *Hpa*II restriction endonucleases[1,2] and methylases[3] were studied using the self-complementary decamer pC-C-G-G-A-T-C-C-G-G which contains an internal *Hpa*II site and generates a *Hae*III site when the duplex is ligated into a tandemly repeating polymer (i.e., 20-mer, 30-mer, etc.). A blunt-ended synthetic decamer, pC-C-G-A-A-T-T-C-G-G, was also used as a substrate for studying the reaction conditions necessary for bacteriophage T4 DNA and RNA ligases to join blunt-ended molecules.[4]

A synthetic octamer containing the *Eco*RI restriction site, pG-G-A-A-T-T-C-C, was used as a mutagen.[5] This was accomplished by randomly introducing nicks into a circular DNA molecule using pancreatic DNase. The DNA molecules were linearized by heating and made blunt-ended using the polymerization and exonuclease activities of DNA polymerase I. The octamer was then ligated onto both blunt ends. The resultant molecule was digested with *Eco*RI, creating cohesive ends which were subsequently religated. The resultant population of circular molecules contained random octamer insertion mutations which could be mapped physically by the presence of the *Eco*RI restriction site. Thus genetic functions could easily be correlated with physical map positions.

Restriction site adaptors can be built into a segment of synthetic DNA for cloning. For example, a chemically synthesized *lac* operator DNA containing *Eco*RI (5′ pA-A-T-T) cohesive ends was cloned. The synthetic fragment expressed biological function in *Escherichia coli*.[6] A gene that codes for the hormone somatostatin was synthesized with an *Eco*RI cohesive end at one terminus and the *Bam*HI sequence at the other. After introduction of the gene into *E. coli* via a plasmid vehicle, expression of the gene product was detected by radioimmunoassays.[7] The entire *E. coli* suppressor tyrosine tRNA gene together with its regulatory elements and a protruding 5′ d(pA-A-T-T) sequence at each end were synthesized and

[2] P. J. Greene, M. S. Poonian, A. L. Nussbaum, L. Tobias, D. E. Garfin, H. W. Boyer, and H. M. Goodman, *J. Mol. Biol.* **99**, 237 (1975).

[3] M. B. Mann and H. O. Smith, *Nucleic Acids Res.* **4**, 4211 (1977).

[4] A. Sugino, H. M. Goodman, H. L. Heyneker, J. Shine, H. W. Boyer, and N. R. Cozzarelli, *J. Biol. Chem.* **252**, 3987 (1977).

[5] F. Heffron, M. So, and B. J. McCarthy, *Proc. Natl. Acad. Sci. U.S.A.* **75**, 6012 (1978).

[6] K. J. Marians, R. Wu, J. Stawinski, T. Hozumi, and S. A. Narang, *Nature (London)* **263**, 744 (1976).

[7] K. Itakura, T. Hirose, R. Crea, A. D. Riggs, H. L. Heyneker, F. Bolivar, and H. W. Boyer, *Science* **198**, 1056 (1977).

cloned in *E. coli*. The biological activity of the cloned suppressor tyrosine tRNA was demonstrated.[8,9] A portion of the RNase gene containing protruding *Eco*RI cohesive ends has been synthesized for cloning.[10]

Perhaps the most flexible and general use of synthetic adaptors is for cloning a large variety of DNA fragments. For this approach, blunt-ended adaptors are first ligated onto the termini of a blunt-ended DNA fragment and subsequently cut with the appropriate restriction enzyme to generate the cohesive ends. This adapted DNA can then be cloned easily. With the use of this general method, successful cloning experiments have been conducted with DNA fragments prepared by (1) chemical synthesis, e.g., *lac* operator using an *Eco*RI adaptor,[11] a *Bam*HI adaptor,[6,12] or a *Hin*dIII adaptor,[6,12] (2) cDNA made from enriched messages, e.g., rat insulin gene,[13] rat growth hormone gene,[14] human chorionic somatomammotropin gene,[15] and chicken lysozyme gene,[16] and (3) gene banks created by random shearing and S1 nuclease treatment of DNA and by limit digestion of DNA with enzymes that give blunt ends.[17] We illustrate here the methodology of this approach with the cloning of synthetic *lac* operator DNA.

Synthetic adaptors can also be used to introduce new restriction sites into cloning vectors,[18,19] augmenting their versatility. As model systems, we describe here the methodology used to insert various restriction sites into the phage vector M13mp2 and the *E. coli* plasmid pBR322. We demonstrate the use of a combination of two different blunt-ended adaptors, and the scheme by which conversion adaptors can be utilized for increasing flexibility in molecular cloning.

Ready-made adaptors and conversion adaptors[20] have not been fully

[8] E. L. Brown, R. Belagaje, H. J. Fritz, R. H. Fritz, and K. Norris, *Fed. Proc., Fed. Am. Soc. Exp. Biol.* **36,** 732 (1977).

[9] E. L. Brown, R. Belagaje, M. J. Ryan, and H. G. Khorana, this volume [8].

[10] M. S. Poonian, W. W. McComas, and A. L. Nussbaum, *Gene* **1,** 357 (1977).

[11] H. L. Heynecker, J. Shine, H. M. Goodman, H. W. Boyer, J. Rosenberg, R. E. Dickerson, S. A. Narang, K. Itakura, S. Lin, and A. Riggs, *Nature (London)* **263,** 748 (1976).

[12] C. P. Bahl, K. J. Marians, R. Wu, J. Stawinski, and S. A. Narang, *Gene* **1,** 81 (1976); C. P. Bahl, R. Wu, and S. A. Narang, *Ibid.* **3,** 123 (1978).

[13] A. Ullrich, J. Shine, J. Chirgwin, R. Pictet, E. Tischer, W. J. Rutter, and H. M. Goodman, *Science* **196,** 1313 (1977).

[14] P. H. Seeburg, J. Shine, J. A. Martial, J. D. Baxter, and H. M. Goodman, *Nature (London)* **270,** 486 (1977).

[15] J. Shine, P. H. Seeburg, J. A. Martial, J. D. Baxter, and H. M. Goodman, *Nature (London)* **270,** 494 (1977).

[16] A. E. Sippel, H. Land, W. Lindenmaier, M. C. Nguyen-Huu, T. Wurtz, K. N. Timmis, K. Giesecke, and G. Schutz, *Nucleic Acids Res.* **5,** 3275 (1978).

[17] T. Maniatis, R. C. Hardison, E. Lacy, J. Laver, C. O'Connell, and D. Quon, *Cell* **15,** 687 (1978).

[18] R. Rothstein and R. Wu, *XIth Internat Congress Biochem.* **29** (1979).

[19] L. Lau and R. Wu, in preparation.

[20] C. P. Bahl, R. Wu, R. Brousseau, A. K. Sood, H. M. Hsiung, and S. A. Narang, *Biochem. Biophys. Res. Commun.* **81,** 695 (1978).

employed in cloning experiments to date. This is mainly because of their low efficiency of ligation to DNA. We will discuss possible solutions to this problem.

Introduction of Cohesive Ends at the Termini of DNA

A general method has been independently developed[6,11,12] for introducing double-stranded DNA molecules into cloning vehicles at different restriction enzyme sites. For this method, a chemically synthesized oligodeoxynucleotide adaptor (8 to 10 residues in length), containing a specific restriction enzyme recognition sequence, is joined by T4 DNA ligase to blunt ends of the DNA to be cloned (Fig. 1, step I). The resulting new duplex DNA is digested with the restriction enzyme specific for the site within the adaptor to generate cohesive ends (step II). The DNA to be cloned (passenger DNA) now carries the proper cohesive ends which can be joined to the same cohesive ends of the cloning vehicle (step III) to give the hybrid DNA.

As an example,[12] we joined a synthetic *Bam*HI adaptor (10 pmol) and a

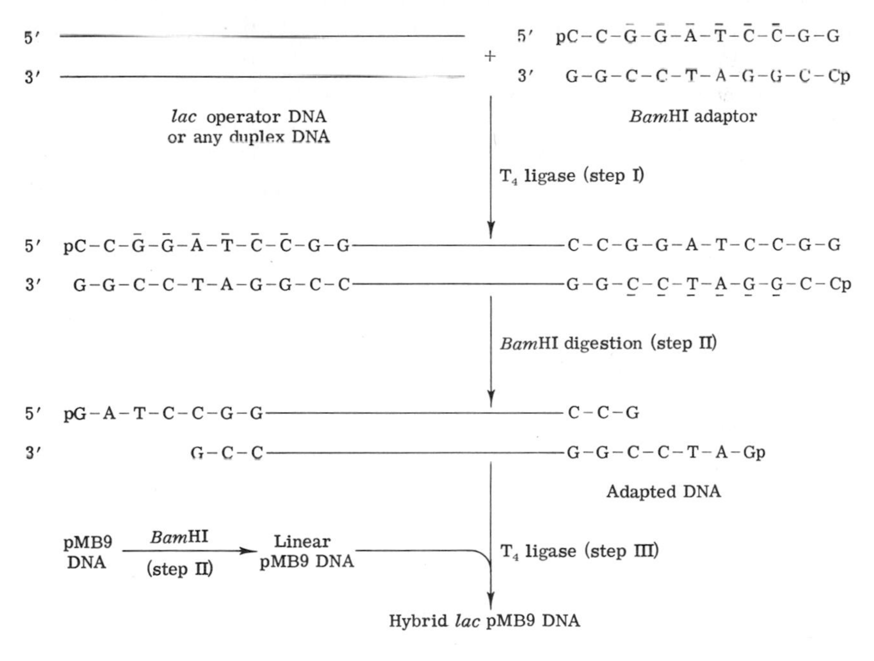

FIG. 1. A general scheme for introducing cohesive ends at the termini of any duplex DNA molecule for molecular cloning.[12] *Lac* operator DNA (21 base pairs in length) is used here as an example for a blunt-ended passenger DNA. Short dotted lines represent the recognition sequence of *Bam*HI enzyme.

synthetic 21-nucleotide-long *lac* operator (0.6 pmol, passenger DNA) end to end by incubation with 3 units of T4 DNA ligase. The incubation mixture (50 μl) also contained 20 m*M* Tris-HCl, pH 7.5, 10 m*M* dithiothreitol (DTT). 10 m*M* $MgCl_2$, and 35 μ*M* ATP. After 6 hr at 20°, the mixture was heated at 70° for 5 min to inactivate the ligase and cooled slowly to room temperature. The DNA product was precipitated with 2 volumes of ethanol and dissolved in 50 μl of a solution containing 6.6 m*M* Tris-HCl, pH 7.5, 6.6 m*M* $MgCl_2$, and 1 m*M* DTT. To this solution was added 1 μg of pMB9 DNA (vehicle) and 2 units of *Bam*Hl enzyme (Fig. 1, step II). The mixture was incubated at 37° for 12 hr to produce *Bam*Hl cohesive ends in both the *lac* operator fragment and the pMB9 DNA. The adapted DNA and the vehicle were then ligated (step III) to give hybrid *lac* pMB9 DNA which was used for transformation. We found that higher concentrations of adaptor molecules increased the efficiency of blunt-end ligation.

Introduction of a New *Hin*dIII Restriction Site into M13mp2[21]

M13mp2 is a single-stranded DNA phage which contains a cloned DNA fragment coding for the first 145 amino acids of the *E. coli* β-galactosidase.[22] This protein fragment is capable of complementing certain deletions near the N-terminus of the β-galactosidase gene to produce a functional enzyme. This complementation can be detected by cleavage of the indicator 5-bromo-4-chloroindoyl-β-D-galactoside, which results in blue plaques. Gronenborn and Messing[23] have created an *Eco*RI restriction site by mutation near the amino terminus (residue 5, Asp → Asn). This mutation does not alter the complementation properties of the strain. If an *Eco*RI restriction fragment is cloned into this site (using double-stranded RFI DNA), colorless plaques are produced after transfection, indicating that an insert has been introduced.[23] In order to make this cloning vehicle more versatile, we have introduced a *Hin*dIII restriction site into the phage, while leaving the blue color complementation system intact. The scheme for this engineering is outlined in Fig. 2. The end product includes the insertion of 18 base pairs containing a new *Hin*dIII site:

```
5′  A-A-T-T-C-C-A-C-A↑A-G-C-T-T-G-T-G-G
3′              G-G-T-G-T-T-C-G-A↓A-C-A-C-C-T-T-A-A
```

[21] After submitting the manuscript for this chapter we learned that J. Messing (Messing and Gronenborn, in preparation) had already used *Eco*RI octamers to clone a *Hin*dIII decamer into the *Eco*RI site of his M13mp2 phage which is referred to as M13mp5. However, their *Hin*dIII site was introduced with a DNA fragment longer than ours. Messing and Jones (Submitted for publication) have also used M13mp5 to clone cDNA of the major storage protein maize.

[22] J. Messing and B. Gronenborn, *Proc. Natl. Acad. Sci. U.S.A.* **74,** 3642 (1977).

[23] B. Gronenborn and J. Messing, *Nature (London)* **272,** 375 (1978).

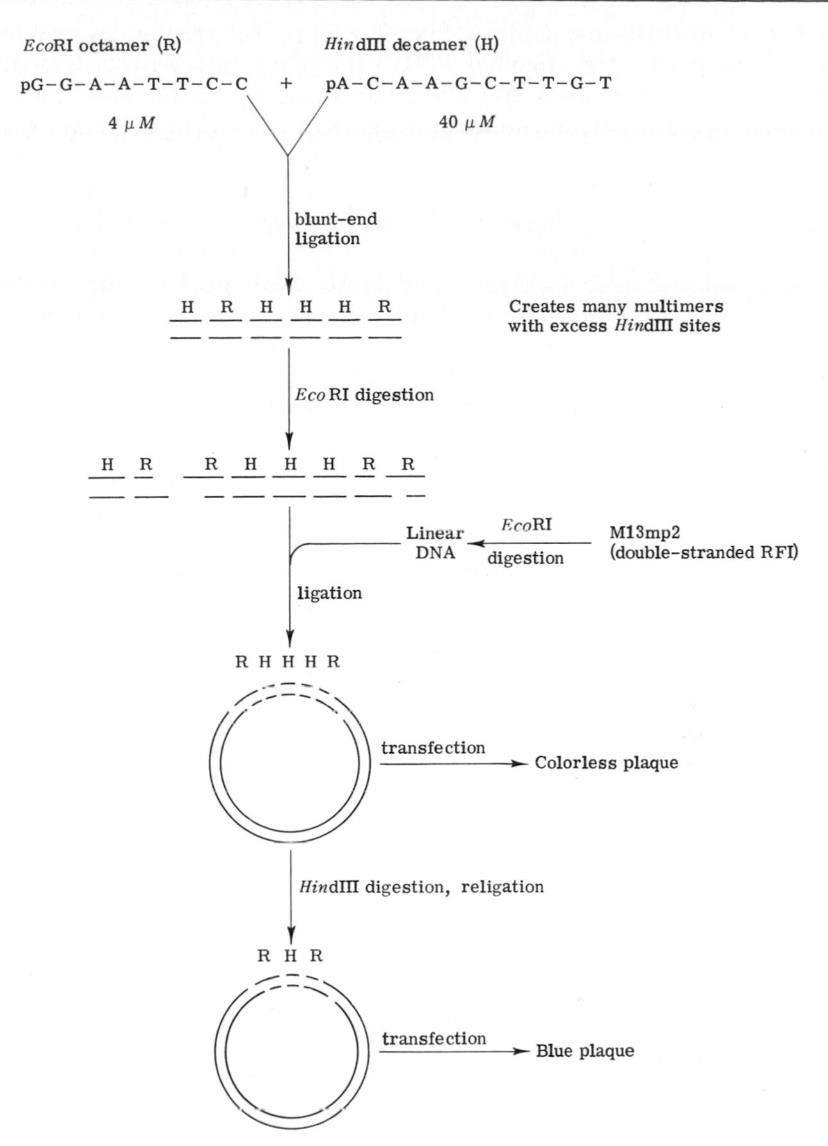

FIG. 2. Introduction of a *Hin*dIII site (H) into phage M13mp2 at the *Eco*RI site (R). The double-stranded DNA or oligonucleotide molecules shown here, and in part in Fig. 3b, are drawn with gaps to emphasize the source and location of individual adaptors in the ligated products (the gaps represent the points where ligation occurred).

This new *Hin*dIII-containing M13mp2 will be referred to as mWJR3 (where the prefix "m" signifies a M13 phage vector). First a fragment which contains several *Hin*dIII adaptors is cloned at the *Eco*RI site and is detected by the resultant colorless plaque. RFI DNA is then isolated from this phage and digested with *Hin*dIII and religated. This procedure trims out the extra *Hin*dIII fragments, and therefore the final insert into the *Eco*RI site contains a single *Hin*dIII adaptor and it is 18 nucleotides in length. This in-frame insertion of six amino acids near the amino terminus of the β-galactosidase fragment in M13mp2 does not alter the complementation as demonstrated by blue plaque formation. The resultant vector mWJR3 carries a *Hin*dIII site into which *Hin*dIII fragments can be cloned, and colorless plaques can be used to identify such clones.

The blunt-end ligation of a *Hin*dIII decamer[12] and an *Eco*RI octamer (from Collaborative Research Company) was performed at an effective ratio of 10:1. This ensured the formation of multiple-adaptor molecules with several *Hin*dIII inserts between flanking *Eco*RI adaptors. (This octanucleotide *Eco*RI adaptor was less efficient in the ligation. To achieve an effective ratio of 10:1, equimolar amounts of the two adaptors were used.) The final effective concentration of phosphorylated adaptor was 4 μM for the *Eco*RI octamer and 40 μM for the *Hin*dIII decamer. The adaptors (in 50 mM Tris-HCl, pH 7.4, 10 mM $MgCl_2$) were heated to 80° and slow-cooled to 4°. ATP and DTT were added to 0.1 mM and 10 mM respectively. Bacteriophage T4 DNA ligase (Bio-Labs, Beverly, Massachusetts 01915) was added to a final concentration of 20 units/ml. The reaction was carried out in 5 μl at 4° for 24 hr. Small samples were taken and electrophoresed on 15% polyacrylamide gel to follow the progress of ligation. We found that over 80% of the starting *Hin*dIII adaptor has been ligated to give larger products. A portion of the ligated mixture was also digested with *Eco*RI and *Hin*dIII restriction enzymes to observe the cleavage of longer DNA to shorter fragments.

To prepare the adaptors for cloning, we digested the total ligation mixture with *Eco*RI restriction endonuclease (5 units/220 pmol of starting material) in 10 μl of 100 mM Tris-HCl, pH 7.4, 50 mM NaCl, 6 mM $MgCl_2$, 6 mM β-mercaptoethanol, and 100 μg/ml bovine serum albumin for 2 hr. The reaction mixture was extracted once with an equal volume of chloroform–isoamyl alcohol (24 : 1) and ethanol-precipitated. The entire mixture (220 pmol) was then ligated with 0.5 pmol of *Eco*RI-digested M13mp2. The ligation buffer was as described above. The ligation as determined by agarose gel electrophoresis was >90% complete after 24 hr at 4°. To avoid the problem of restriction of the synthetic fragment, a restriction-deficient, modification-proficient *E. coli* strain, LE392,[24] was transfected with the ligation mix. The plates were also seeded with an *E.*

[24] P. Leder, D. Tiemeier, and L. Enquist, *Science* **196,** 175 (1979).

coli strain 71-18 containing an F′ plasmid.[22] Since M13 is a male-specific bacteriophage, the F′-containing strain is necessary for visible plaque formation. Approximately 5 × 10^4 plaques were recovered after transfection with one-half of the ligation mixture (1 μg of M13mp2).

The phages were pooled and used to infect *E. coli* 71-18 on X-Gal plates (10^{-4} *M* IPTG[24a] was added for induction). It was found that 3% of the plaques were colorless or light blue. Several colorless plaques were purified, and double-stranded RFI DNA was isolated. Digestion with *Hin*dIII restriction enzyme produced a linear molecule, confirming the cloning of the *Hin*dIII adaptor. All additional *Hin*dIII adaptor sites were also cut and eliminated after the digestion. Linear molecules were purified directly from the agarose gel (by the sodium perchlorate method described in this volume [10]). Ligation of the *Hin*dIII-cut linear DNA molecules and subsequent transfection lead to the production of blue plaques. Phage DNA was isolated and analyzed for the presence of a new *Hin*dIII site. DNA sequencing confirmed that of the inserted 18-mer.[18] In a parallel experiment, we have removed the only *Bam*HI site in M13mp2 and introduced a new *Bam*HI site adjacent to the *Eco*RI site by a similar procedure. The resulting new vector, mWJ43, is more useful in cloning *Bam*HI digested DNA since we can apply the same color change screening technique.

Introduction of *Eco*RI–*Bam*HI Conversion Adaptors into M13mp2 and pBR322

A conversion adaptor carrying an *Eco*RI site and a *Bam*HI site has been synthesized by Bahl *et al.*,[20] which consists of two decamers:

```
HO-A-A-T-T-C-C-C-G-G-G
                  G-G-G-C-C-C-C-T-A-G-OH
```

In principle, this adaptor can be used to clone DNA fragments with *Bam*HI cohesive ends into the *Eco*RI site of a cloning vector, or vice versa. This adaptor also includes a *Sma*I or *Xma*I site. The functionality of this conversion adaptor was demonstrated[20] by cloning a 21-nucleotide synthetic *lac* operator with *Eco*RI cohesive ends into the *Bam*HI site of pBR322. However, attempts to clone high-molecular-weight DNA fragments using this conversion adaptor directly have not yet been successful. It is likely that, because of the short duplex and relative instability of the adaptor, high concentrations of the substrates and DNA ligase are required for efficient ligation.[4,25] Such concentrations of substrates (10–20 μ*M*) are not always practical in most cloning experiments. High levels of DNA ligase demand that the enzyme is free of contaminating ex-

[24a] IPTG stands for isopropylthiogalactoside.

[25] H. G. Khorana *et al.*, *J. Biol. Chem.* **251**, 565 and 634 (1976).

onucleases. However, we are able to achieve effective cloning of high-molecular-weight DNA fragments employing the conversion adaptor in a stepwise manner (Fig. 3).

Each preparation of DNA ligase must be checked for the possibility of contaminating exonucleases.[28] A convenient substrate for testing single-strand- and double-strand-specific exonucleases is a [5′-^{32}P]oligonucleotide (such as the *Bam*HI blunt-end adaptor) in the denatured or annealed form. After incubation with DNA ligase, the reaction mixture is denatured and fractionated either by homochromatography or by gel electrophoresis (20% acrylamide gel) to see if shorter oligonucleotides are liberated. Polynucleotide kinase should be checked in the same way for contaminating exonucleases.

Double-strand-specific exonuclease can also be detected by incubating DNA ligase with *Eco*RI-digested λ DNA fragments which have been repaired with high specific activity [α-^{32}P]dATP in the presence of *E. coli* DNA polymerase[28] or AMV reverse transcriptase. After 1–5 hr of incubation at 20° or 37°, carrier salmon sperm DNA (100 μg) is added. The percent trichloroacetic acid-soluble [^{32}P]nucleotide is a measure of the contaminating 3′-exonuclease activity.

The first step (Fig. 3a) is to anneal and polymerize (by ligation) the conversion adaptor decamers. This step occurs efficiently, since the adaptor concentration can be made sufficiently high. When the polymer is subjected to restriction by *Bam*HI (Fig. 3b), various-sized fragments are generated that contain *Bam*HI cohesive ends and internal *Eco*RI sites. These fragments, at least 20 nucleotides in length, ligate to high-molecular-weight DNA at much higher efficiencies than the cohesive-ended decamer with a six-nucleotide-long duplex. The ligated adaptor–DNA complex (adapted large DNA) can then be digested with *Eco*RI to generate the *Eco*RI cohesive ends for cloning into a specific site of a desired vector. Any internal *Eco*RI sites of the passenger DNA can be rendered resistant to cleavage using the *Eco*RI modification methylase prior to ligation to the conversion adaptor.[17]

The following experiment illustrates the efficiency of ligation of the polymerized conversion adaptor to high-molecular-weight DNA. The 5′-phosphorylated decamers were annealed by slow cooling after heating to 75°, and polymerized at 10 μ*M* each with 0.2 units of T4 ligase in a reaction volume of 10 μl at 4° for 16 hr. Polymerized adaptors cut with *Bam*HI (see Fig. 3b) were ligated with *Bam*HI-cut pBR322 DNA and transformed to *E. coli* SK1492.[26] Transformants were screened for tetracycline sensitivity (Tet^S). Mini-plasmid DNA preparations[27] of Tet^S

[26] S. K. Kushner, *Proc. Int. Symp. Genet. Eng.*, p. 23 (1978) Elsevier/North Holland Biomed. Press.

[27] R. B. Meagher, R. C. Tait, M. Betlach, and H. W. Boyer, *Cell* **10,** 521 (1977).

[28] R. Wu, E. Jay, and R. Roychoudhury, *Methods Cancer Res.* **12,** 87 (1976).

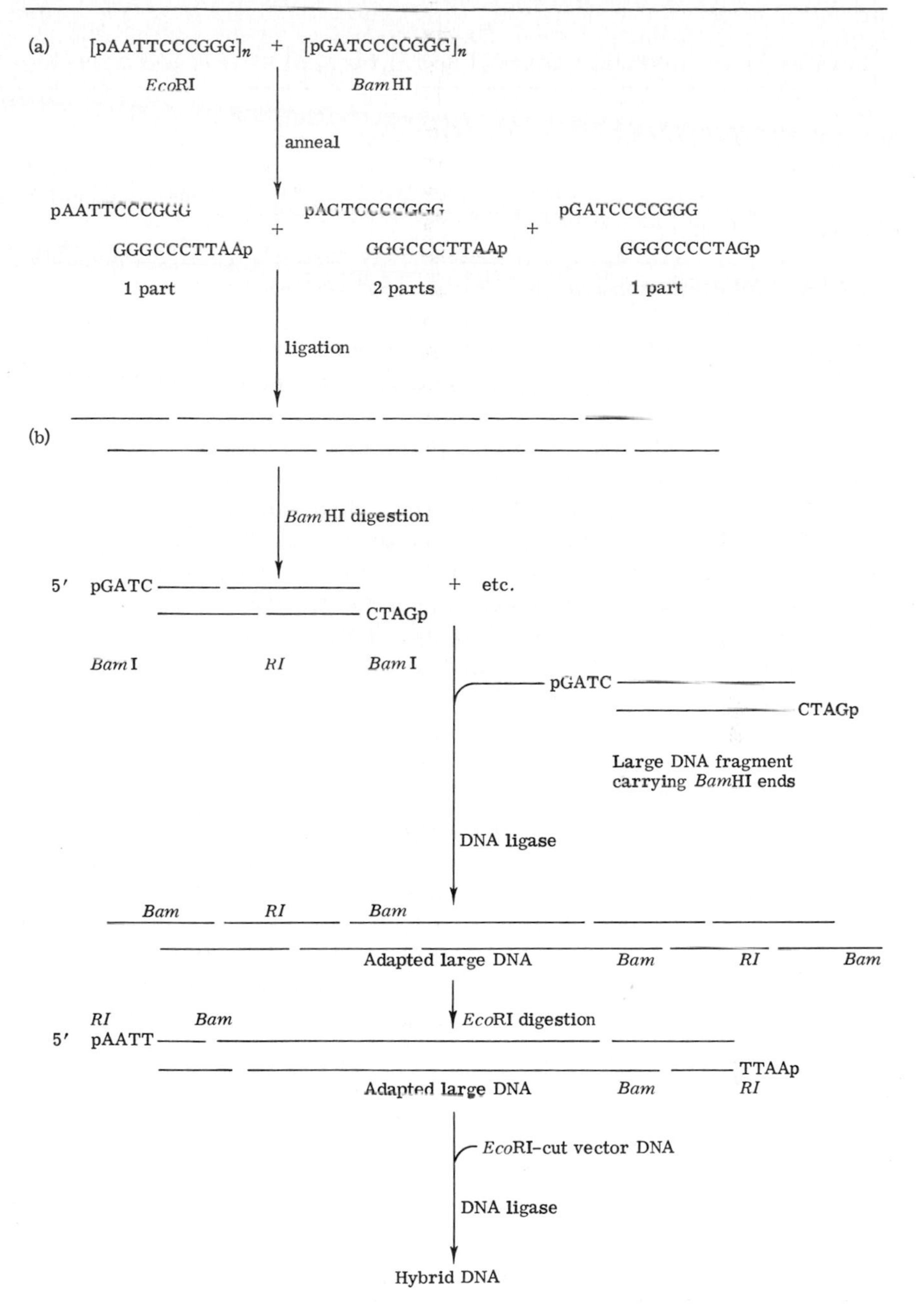

FIG. 3. Addition of *Eco*RI–*Bam*HI conversion adaptors to the termini of a large DNA fragment (passenger DNA) carrying *Bam*HI ends. The gaps between short lines represent the points where ligation occurred.

TABLE I
INSERTION OF AN *Eco*RI SITE AND A *Sma*I SITE AT THE *Bam*HI SITE OF pBR322 PLASMID

*Bam*HI-cut pBR322 DNA	*Eco*RI–*Bam*HI conversion adaptors	Transformants	Amp^R and Tet^S transformants	Insertion of *Eco*RI site	Insertion of *Sma*I site	Retention of *Bam*HI site
0.2 pmol	20 pmol	5.2×10^3	$\frac{79}{255}$ [a]	$\frac{2}{12}$	$\frac{2}{2}$	$\frac{2}{2}$

[a] Out of 255 transformants analyzed, 79 were Amp^R and Tet^S.

clones were analyzed by restriction enzymes. About 500 transformants per microgram of DNA were obtained that had the conversion adaptor cloned, inserting an *Eco*RI site and a *Sma*I site at the *Bam*HI site of pBR322. One such clone analyzed, pLL10, was shown to contain a 20-nucleotide insert at the *Bam*HI site. The results are shown in Table I.

In contrast, when the *Eco*RI–*Bam*HI conversion adaptor was ligated directly without prior polymerization to *Eco*RI–*Bam*HI-cut pBR322 DNA, less than 20 ampicillin-resistant (Amp^R) transformants per microgram of DNA were obtained. In this experiment, the background resulting from the religation of the restriction fragments of pBR322 was eliminated by exhaustive digestion with *Hin*dIII.

Polymerized *Eco*RI–*Bam*HI conversion adaptors cut with *Eco*RI were similarly ligated to *Eco*RI-cut M13mp2. The results are shown in Table II. In this experiment, a *Bam*HI site and a *Sma*I (*Xma*I) site were introduced into the *Eco*RI site of the M13mp2 phage to produce a new vector, mLL2. It is likely that the occurrence of the light-blue plaques results from the insertion of a polymerized adaptor 30, 60, or 90 nucleotides in length, yielding an in-frame insertion into the β-galactosidase gene.

From our experience with synthetic oligonucleotides, it seems that a short DNA duplex is a poor substrate for DNA ligase. While ligation of oligonucleotides with short duplex (four to six base pairs) has been achieved, very high concentrations of each substrate are required.[25] For this reason, the short adaptors[20] can be made more useful after joining to

TABLE II
INSERTION OF A *Bam*HI SITE AND A *Sma*I SITE AT THE *Eco*RI SITE OF THE M13mp2 PHAGE

*Eco*RI-cut M13mp2 DNA	*Eco*RI–*Bam*HI conversion adaptors	White plaques	Light-blue plaques	Dark-blue plaques	Insertion of *Bam*HI site	Insertion of *Sma*I site	Retention of *Eco*RI site
0.06 pmol	20 pmol	11	5	14	$\frac{2}{3}$	$\frac{2}{2}$	$\frac{2}{2}$

give longer ones (Figs. 2 and 3). Synthesis of longer adaptors, especially ready-made and conversion adaptors, is in progress.

Acknowledgments

This work was supported by Research Grants GM24904 and AM21801 from the National Institutes of Health, and from the National Research Council of Canada (NRCC No. 17371).

[8] Chemical Synthesis and Cloning of a Tyrosine tRNA Gene

By EUGENE L. BROWN, RAMAMOORTHY BELAGAJE, MICHAEL J. RYAN, and H. GOBIND KHORANA

Chemical methods for the preparation of nucleotides and derivatives were described in this series in 1963.[1] In the intervening years, methods for the chemical synthesis of short deoxyribopolynucleotides, in particular, have received emphasis. Further, by a combination of chemical synthesis and enzymatic reactions, methodology has been developed for the synthesis of macromolecular bihelical DNA of any specific sequence.[2,3] The methodology developed for the total synthesis of a given DNA containing biologically specific sequences consists of the following steps. The DNA in the double-stranded form is carefully divided into short single-stranded segments with suitable overlaps in the complementary strands. All the segments are chemically synthesized starting with protected nucleosides and mononucleotides. The 5′-hydroxyl groups of the appropriate oligonucleotides are then phosphorylated using [γ-^{32}P]ATP and polynucleotide kinase. A few to several neighboring oligonucleotides are then allowed to form bihelical complexes in aqueous solution, and the latter are joined end to end by polynucleotide ligase to form covalently linked duplexes. Subsequent head-to-tail joining of the short duplexes leads to the total DNA.

This article describes the methods used for the construction of the biologically functional suppressor tRNA gene shown in Fig. 1. The total work

[1] M. Smith and H. G. Khorana, this series, Vol. 6, p. 645.

[2] H. G. Khorana, K. L. Agarwal, H. Büchi, M. H. Caruthers, N. K. Gupta, K. Kleppe, A. Kumar, E. Ohtsuka, U. L. RajBhandary, J. H. van de Sande, V. Sgaramella, T. Terao, H. Weber, and T. Yamada, *J. Mol. Biol.* **72,** 209 (1972).

[3] H. G. Khorana, K. L. Agarwal, P. Besmer, H. Büchi, M. H. Caruthers, P. J. Cashion, M. Fridkin, E. Jay, K. Kleppe, R. Kleppe, A. Kumar, P. C. Loewen, R. C. Miller, K. Minamoto, A. Panet, U. L. RajBhandary, B. Ramamoorthy, T. Sekiya, T. Takeya, and J. H. van de Sande, *J. Biol. Chem.* **251,** 565 (1976).

ISBN 0-12-181968-X

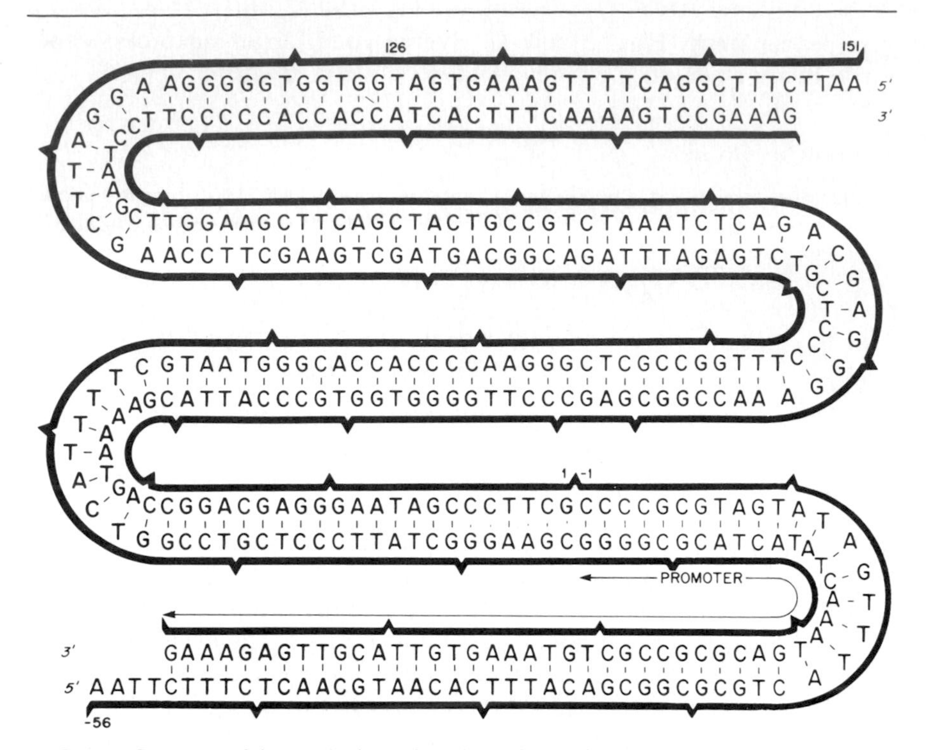

FIG. 1. Structure of the synthetic *Escherichia coli* tyrosine suppressor tRNA gene. From bottom left, following the terminal *Eco*RI endonuclease-specific sequence, the DNA contains a 51-nucleotide-long promoter region, a 126-nucleotide-long DNA sequence corresponding to the precursor RNA, and finally a 25-nucleotide-long DNA sequence. Of the latter, 16 nucleotides belong to the natural sequence adjoining the C-C-A end and the remainder is a modified sequence including the *Eco*RI endonuclease-specific sequence.

involved (1) synthesis of 126-nucleotide-long bihelical DNA corresponding to a known precursor of the tyrosine suppressor tRNA,[4] (2) sequencing of the promoter region and the distal region adjoining the C-C-A end, which contained a signal for the processing of the RNA transcript,[5] (3) total synthesis of the 208-base-pair-long DNA which included the control elements as well as the *Eco*RI restriction endonuclease-specific sequences at the two ends, and (4) full characterization by transcription *in vitro* and amber suppressor activity *in vivo* of the synthetic gene.

The preparation described in this article should be of particular interest for at least two reasons: Chemical synthesis is the only means of preparing a DNA of a defined sequence where no RNA or DNA template is

[4] S. Altman and J. D. Smith, *Nature (London), New Biol.* **233,** 35 (1971).

[5] T. Sekiya, R. Contreras, H. Küpper, A. Landy, and H. G. Khorana, *J. Biol. Chem.* **251,** 5124 (1976).

available, e.g., recent syntheses of the deoxypolynucleotides corresponding to polypeptide hormones such as somatostatin[6] and insulin.[7] Second, chemical synthesis alone permits designed and controlled changes in sequences for structure–function studies. With the recent progress and interest in DNA sequences in protein–nucleic acid interactions and a host of related studies, it is clear that synthetically available DNAs of defined sequences will play an increasingly important role.

Chemical Synthesis of Deoxyribooligonucleotides

General Principles

In the diester approach, which has been used in all this work, starting materials are protected deoxyribonucleotides, which have only the 3′-hydroxyl group free (e.g., I in Fig. 2), and protected mononucleotides (II in Fig. 2) or preformed protected di- or trinucleotides (e.g., III or IV in Fig. 4). The protecting groups used for blocking the 5′-hydroxyl of deoxynucleosides, the amino groups of cytosine, adenine, and guanine, and the 3′-hydroxyl group of mono- and oligonucleotides are shown in Fig. 2.

For condensation reactions, the 5′-phosphate groups are activated by the use of DCC[8] or, more frequently, TPS. Stepwise synthesis of an oligonucleotide chain is brought about by successive condensations of the protected mono- or oligonucleotides carrying 5′-phosphate end groups with the 3′-hydroxyl end group of the growing chain, as illustrated in Fig. 3 (V + VI → VII → VIII, etc).

[6] K. Itakura, T. Hirose, R. Crea, A. D. Riggs, H. L. Heyneker, F. Bolivar, and H. W. Boyer, *Science* **198,** 1056 (1977).

[7] R. Crea, A. Kraszewski, T. Hirose, and K. Itakura, *Proc. Natl. Acad. Sci. U.S.A.* **75,** 5765 (1978).

[8] The abbreviations used in this article are as follows. The one-letter symbols for nucleosides and the symbols for the protecting groups on the bases, sugars, or phosphates in the polynucleotides are according to the IUPAC-IUB Commission on Biochemical Nomenclature Recommendations [*J. Biol. Chem.* **245,** 5171 (1970)]. A hyphen represents internal phosphate; p represents the terminal phosphate. When placed to the left of the nucleoside initial, p indicates a phosphate group at the 5′-terminus. The symbol for the protecting group on this phosphate is placed at the left of the letter p; Ac (acetyl), *t*-$BuPh_2Si$ (*t*-butyldiphenylsilyl). These symbols are placed at the right of the nucleoside initial to indicate their location on the 3′-hydroxyl group; MeOTr (monomethoxytrityl) is placed at the left of the nucleoside initial to indicate its location on the 5′-hydroxyl group; an (*p*-anisoyl), bz (benzoyl), and ib (isobutyryl). These immediately precede the one-letter symbol for the nucleoside and represent the protecting groups on the heterocyclic rings. Other abbreviations include: CNEt, cyanoethyl; DCC, dicyclohexylcarbodiimide; DIEA, diisopropylethylamine; DIEAB, diisopropylethylammonium bicarbonate; HPLC, high-pressure liquid chromatography; TEAA, triethylammonium acetate; TEAB, triethylammonium bicarbonate; TLC, thin-layer chromatography; TPS, 2,4,6-triisopropylbenzenesulfonyl chloride; TPSE, 2-(*p*-tritylphenyl)sulfonylethyl.

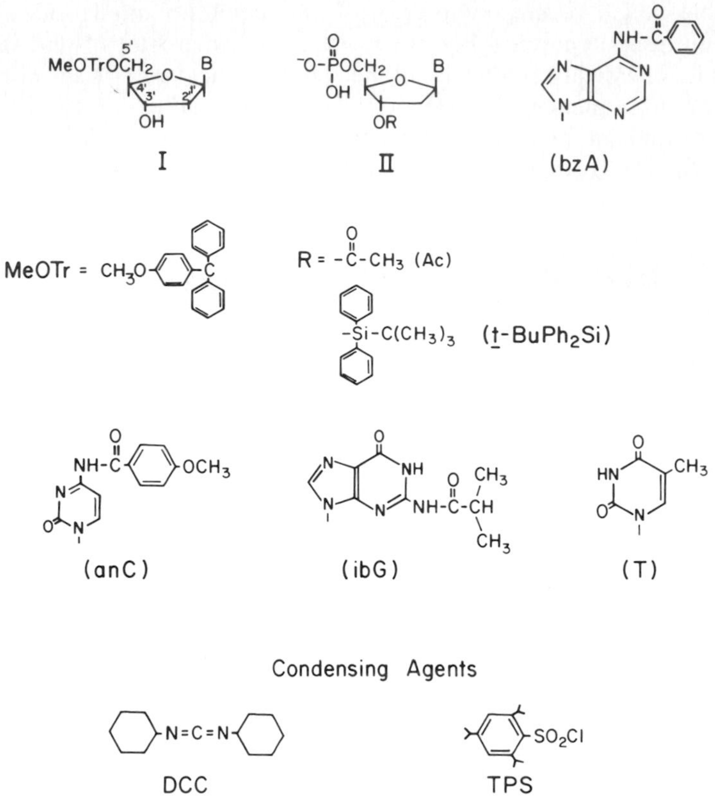

I II (bzA)

MeOTr = CH_3O-C$_6$H$_4$-C(C$_6$H$_5$)$_2$

R = -CO-CH_3 (Ac)

-Si(C$_6$H$_5$)$_2$-C$(CH_3)_3$ (*t*-BuPh$_2$Si)

(anC) (ibG) (T)

Condensing Agents

DCC TPS

FIG. 2. Protected deoxyribonucleosides and deoxyribonucleotides and condensing agents used in the synthesis of polynucleotides.

To separate the products of a condensation reaction, several methods are available. At the level of protected di-, tri-, tetra-, and sometimes pentanucleotides, the use of hydrophobic groups mentioned above, e.g., the methoxytrityl group, allows rapid separation by solvent extraction. Beyond this stage, separations are effected by anion-exchange chromatography[9] or by HPLC.[10] The latter is a powerful and extremely rapid method, and further improvements are possible by innovative use of protecting groups, e.g., the *t*-butyldiphenylsilyl group for the terminal 3'-hydroxyl group.[11]

[9] B. Ramamoorthy, R. G. Less, D. G. Kleid, and H. G. Khorana, *J. Biol. Chem.* **251,** 676 (1976), and preceding papers in this series.

[10] H.-J. Fritz, R. Belagaje, E. L. Brown, R. H. Fritz, R. A. Jones, R. G. Lees, and H. G. Khorana, *Biochemistry* **17,** 1257 (1978).

[11] R. A. Jones, H.-J. Fritz, and H. G. Khorana, *Biochemistry* **17,** 1268 (1978).

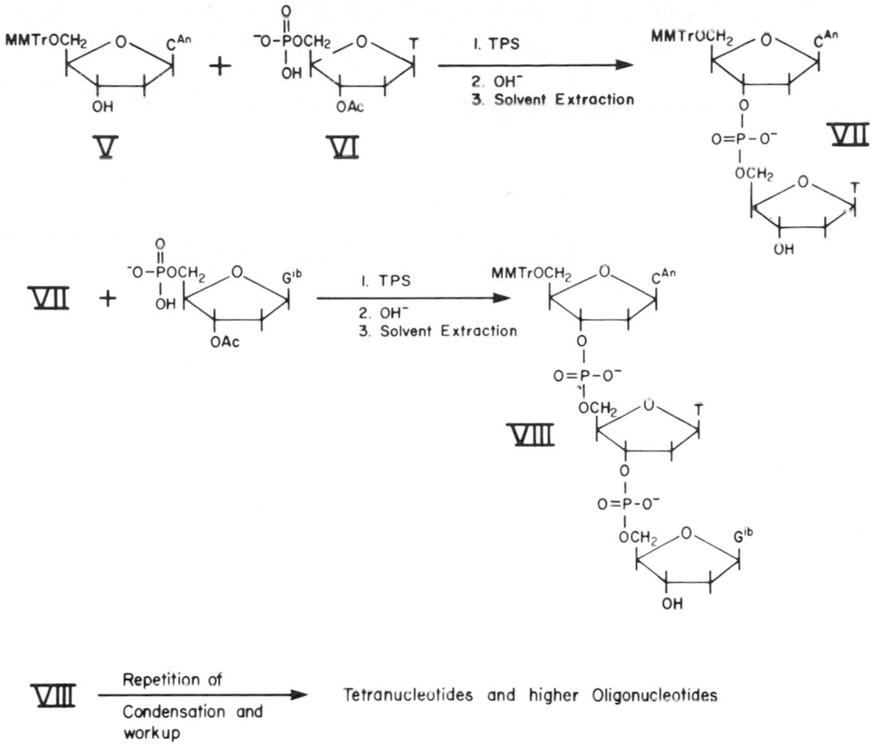

FIG. 3. Stepwise synthesis of a trinucleotide and higher oligonucleotides.

Materials and General Methods

All liquid reagents should be distilled before use. Pyridine is distilled in the presence of chlorosulfonic acid (1 g/liter), redistilled from NaOH, and stored over predried molecular sieves (Linde 4A). TPS is recrystallized from anhydrous petroleum ether before use. The deoxymononucleotides (pdA, pdG, pdT, and pdC) and deoxynucleosides (dA, dG, dT, and dC) are obtained commercially and should be carefully checked for purity by either paper chromatography as described earlier[1] or HPLC.

DEAE-cellulose column chromatography is performed at 4° using gradients of TEAB, pH 7.5. Paper chromatography is performed by the descending technique on Whatman No. 1 or 40 paper, and thin-layer TLC is carried out on precoated silica gel plates containing a fluorescent indicator (Eastman 6062 or EM 5775 silica gel F-254). Two-dimensional homochromatography is performed essentially as described by Sanger and co-workers.[12] The RNA hydrolyzates, homomixes, used for the second dimension are prepared as described in the published procedure.[13] HPLC is performed on a system consisting of the following components available from Waters Associates, Milford, Massachusetts: two M6000A solvent delivery systems, a 660 solvent programmer, a U6K injector, a 440 uv detector operating at wavelengths of 254 and 280 nm, a μBondapak C_{18} column (0.4 × 30 cm), and a Houston Instruments omniscribe TM chart recorder. The pumps and solvent programmer are operated in such a way that one pump delivers the aqueous buffer (0.1 *M* ammonium acetate or 0.1 *M* TEAA, pH 7.0) and the other acetonitrile.

Whenever HPLC is used for the separation of a condensation reaction mixture, the following general procedure is adopted. After quenching the condensation reaction mixture with water and DIEA, the mixture is evaporated to a gum and taken up in 0.2 *M* TEAB as usual. Nonnucleotidic components are partially removed by four manual extractions, two using ethyl acetate and two using ethyl acetate with 5–10% *n*-butanol. The contents of the aqueous phase are isolated by evaporation with pyridine and precipitation with dry ether. At this stage, two samples (100–300 μg each) are taken from the precipitate; one is subjected to deacetylation with 1 *M* NaOH in the usual manner. The two samples are compared by HPLC on the μBondapak C_{18} column using conditions that give good resolution of the tritylated oligonucleotides (30–35% acetonitrile in 0.1 *M* ammonium acetate). In most cases, the product peak is easily identified by its shift to shorter retention time after alkaline treatment.

The bulk of the sample, precipitated as above, is dissolved in TEAA–ethanol (1:1, v/v) and centrifuged for about 10 min to spin down all insoluble material. This solution is kept in an ice bath and applied to the preparative column as soon as possible. For preparative separations, the column and detector of the above system are replaced with a Bondapak C_{18}/Porasil B column (0.7 × 183 cm) and an Altex model 151 uv detector equipped with a preparative flow cell operating at a wavelength of 254 or 280 nm. When the fractions from preparative HPLC are concentrated, special attention is paid to ensure that a sufficient amount of pyridine is present at all times during the evaporation of solvents.

General Methods of Condensation and Workup of Reaction Mixtures

Both mono- and oligonucleotide components (the compound carrying the free 5′-phosphate group being in excess) are dissolved in dry pyridine,

[12] F. Sanger, J. E. Donelson, A. R. Coulson, H. Kössel, and D. Fischer, *J. Mol. Biol.* **90**, 315 (1974).

[13] E. Jay, R. Bambara, R. Padmanabhan, and R. Wu, *Nucleic Acids Res.* **1**, 331 (1974), and preceding papers in this series.

and the mixture is rendered anhydrous by repeated evaporation of added pyridine (four to five times), the flask being opened to the atmosphere of a dry box containing P_2O_5. During the last evaporation of the pyridine, the minimal amount of the solvent necessary for complete solubilization of the reaction components is allowed to remain, and TPS is then added inside the dry box. The reaction mixture is kept at room temperature for 4–6 hr with the exclusion of moisture. After cooling in a dry ice–ethanol bath, the reaction is terminated by adding DIEA (2 mmol/mmol of TPS used) as a 1 or 2 *M* solution in pyridine followed by water (volume equal to total volume of mixture). This solution is kept at room temperature overnight or longer.

To hydrolyze 3′-*O*-acetyl groups selectively, the solution of the oligonucleotide in pyridine–ethanol (1:1, v/v) is cooled to 0° and 2 *M* NaOH (enough to give an overall concentration of 1 *M* in the reaction mixture) is added. The alkaline mixture is kept at 0° for 5–6 min and then neutralized by adding, with shaking, an excess of pyridinum Dowex-50 ion-exchange resin. After neutralization, the resin and the solution are poured into a column and the resin washed thoroughly with 20% aqueous pyridine. The filtrate and washings are combined and either applied to a DEAE-cellulose anion-exchange column or concentrated in the presence of pyridine.

Removal of Protecting Groups

The fully protected oligonucleotides (10–20 A_{260} units) are treated with concentrated NH_4OH (1 ml) at 50° for 16–18 hr in a sealed tube. The solution is then diluted with pyridine (final concentration of 50%) and evaporated under vacuum; the pH is controlled by the addition of a few drops of 1 *M* TEAB buffer. To remove the monomethoxytrityl group, the *N*-deprotected compounds are treated with a mixture of acetic acid, pyridine, and water (14:1:3, v/v/v) at room temperature for 24–48 hr. The completely deprotected oligonucleotides are further purified by anion-exchange chromatography in the presence of 7 *M* urea or by HPLC.

Characterization of Synthetic Oligonucleotides

The standard method for characterization of fully protected deoxyribooligonucleotides is to record their uv absorption spectra and compare them with the calculated spectra as described by Büchi and Khorana.[14] To determine their purity, oligonucleotides are analyzed by paper chromatography, TLC, and reverse-phase HPLC[10]; the last-mentioned method is preferred because of its high level of sensitivity. The nucleotide se-

[14] H. Büchi and H. G. Khorana, *J. Mol. Biol.* **72**, 251 (1972).

quences in the final synthetic compounds are confirmed as follows.[15] Each of the oligonucleotide segments (10–15 pmol) is phosphorylated at its 5′-end using [γ-^{32}P]ATP and polynucleotide kinase. The phosphorylated segments are partially digested with snake venom phosphodiesterase, and the digests are analyzed by two-dimensional homochromatography.[12]

5′- and *N*-Protected Deoxyribonucleosides

Principle

Deoxyribonucleosides are acylated with an excess of an acid chloride (benzoyl chloride for deoxyadenosine and anisoyl chloride for deoxycytidine) or acid anhydride (isobutyric anhydride for deoxyguanosine). The fully acylated product is treated with alkali under controlled conditions. Mono-*N*-acyl derivatives of the deoxynucleosides are thus obtained. These derivatives are now treated with bulky reagents such as monomethoxytrityl chloride to give the *N*-acyl-5′-*O*-monomethoxytrityl deoxynucleosides.

Procedure

N-Benzoyl Deoxyadenosine.[16] A suspension of dry deoxyadenosine (1.25 g, 5 mmol) in anhydrous pyridine (15 ml) is treated with 2.5 ml of benzoyl chloride for 2 hr at room temperature. The resulting solution is then poured into ice water, and the insoluble product extracted with chloroform (3× 100 ml). The chloroform extract is washed with water and then evaporated to a gum. The gum is dissolved in a mixture of ethyl alcohol (15 ml) and pyridine (10 ml), and the solution is treated with a mixture of 2 *N* NaOH (20 ml) and ethyl alcohol (20 ml) at room temperature for 5 min. An excess of pyridinium Dowex-50 ion resin is added to remove sodium ions, and the resin is then removed by filtration and the filtrate and washings concentrated *in vacuo* to a small volume. Water (25 ml) is added, and the mixture extracted with ether (3× 50 ml). The aqueous suspension is heated, and the resulting solution cooled slowly. *N*-Benzoyl adenine separates first and is removed by filtration. The filtrate after concentration and storage yields *N*-benzoyl deoxyadenosine (1.15 g, 65%) as colorless needles, m.p. 113°–115°. The uv absorption spectrum of the product has: $\lambda_{max}^{H_2O}$ 280 nm (20,900); $\lambda_{max}^{pH\,2}$ 285 nm (22,500).

[15] R. Wu, C. D. Tu, and R. Padmanabhan, *Biochem. Biophys. Res. Commun.* **55,** 1092 (1973).

[16] H. Schaller, G. Weimann, B. Lerch, and H. G. Khorana, *J. Am. Chem. Soc.* **85,** 3821 (1963).

N-Anisoyl Deoxycytidine.[16] The *N*-anisoyl deoxycytidine is similarly prepared by using anisoyl chloride (3 ml) instead of benzoyl chloride.

N-Isobutyryl Deoxyguanosine.[14] Deoxyguanosine (2.85 g, 10 mmol) is treated with a 10% aqueous solution of tetraethylammonium hydroxide (17.5 ml). The resulting clear solution is rendered anhydrous by repeated addition and evaporation of dry pyridine. The colorless solid residue is shaken with dry pyridine (200 ml) and isobutyric anhydride (50 ml) in the dark at room temperature for 48 hr. After ethanol (70 ml) is added with cooling, the reaction mixture is kept at room temperature overnight and then concentrated to a gum *in vacuo*. Pyridine (125 ml) is added, and the solution concentrated to a volume of about 25 ml and then taken up in chloroform (500 ml). The chloroform solution is washed three times with 2 *M* $KHCO_3$ solution (200 ml) with the addition of ice, and then once with saturated NaCl solution (100 ml). The aqueous phases are backwashed with chloroform (2× 500 ml). The organic layer is rendered anhydrous by filtration through Na_2SO_4, and the solvent evaporated. The crude product, which is obtained as an oil, is dissolved in ethanol (100 ml), and then 2 *M* NaOH (100 ml) is added at 0°. The slightly turbid but homogeneous solution is kept at this temperature for 15 min. Pyridinium Dowex-50 ion-exchange resin (200 ml) is added with vigorous shaking, and the pH checked by paper (about 7). The resin is removed by filtration and washed slowly with ethanol (2 liters). The total filtrate and washings are concentrated *in vacuo* and rendered anhydrous by repeated addition and evaporation of dry pyridine. The residue is taken up in pyridine (25 ml) and precipitated into a mixture of pentane–ether (2:1, v/v). The precipitate is collected by centrifugation and washed once with a ether–pentane mixture and three times with pentane. After storage in a vacuum dessicator, *N*-isobutyryl deoxyguanosine is obtained as a fine, colorless powder. The spectral properties in ethanol are: λ_{max} 255–260 nm (16,200); λ_{max} 281 nm (11,900); λ_{min} 273 nm; $\epsilon_{260}/\epsilon_{280} = 1.37$, $\epsilon_{280}/\epsilon_{300} = 1.96$.

5'-O-Monomethoxytrityl N-Protected Deoxynucleosides. N-Protected deoxynucleoside (10 mmoles) is dissolved in dry pyridine, and the solution rendered anhydrous by repeated addition and evaporation of dry pyridine. To the residue is added dry pyridine (25–40 ml) and recrystallized monomethoxytrityl chloride (12–13 mmol). After the reaction mixture is shaken at room temperature in the dark for 16–18 hr, ethanol (5 ml) is added and the solution is kept at room temperature for 2–4 hr. This solution is diluted with chloroform (1 liter) which has been washed with 0.1 *M* NH_4HCO_3 and then extracted with the same aqueous buffer (3× 300 ml). After the aqueous phase is backwashed with chloroform (300 ml), the organic layers are combined, dried over Na_2SO_4 and evaporated *in vacuo*. The product is purified by separation on a silica gel column with chloroform containing 5–10% ethanol as the eluant. The fractions containing the desired product are pooled and evaporated with the addition of pyridine. The residue is taken up in dry pyridine (10 ml), and the product is precipitated by dropwise addition to an excess of *n*-pentane or petroleum ether (800 ml passed through an alumina column). The precipitate is collected by centrifugation, washed three times with *n*-pentane, and dried *in vacuo*. The yield is in the range of 70–80%.

Preparation of *N*-Protected Mononucleotides

Principle

Mononucleotides as their pyridine salts are treated with acylating agents in dry pyridine. The fully acylated derivatives are isolated and treated with mild alkali under controlled conditions. Only the expected mono-*N*-acyl deoxynucleoside 5′-phosphate is obtained.

Procedure

Approximately 3.9 g of the ammonium salt of deoxynucleoside 5′-monophosphate (pdA, pdG, and pdC) is dissolved in 100 ml of water. The clear solution is slowly passed (1.5–2 ml/min) through a column of pyridinium Dowex-50W-X8 resin (100 ml). The resin is washed with 5% pyridine (300 ml). The total effluent is made up to approximately 70% pyridine by adding more pyridine and evaporated *in vacuo* to a volume of 30 ml. Pyridine (50 ml) is added twice, and each time the solution is evaporated to a powder. Likewise, anhydrous pyridine (50 ml) is added three times, and the solution is again evaporated to a powder. To the dry powder are added 100 ml of dry pyridine and a 10- to 12-fold molar excess of reagent-grade acid chloride (benzoyl chloride for pdA, anisoyl chloride for pdC; however, isobutyric anhydride is used for pdG[14]). The flask is shaken for 2 hr in the dark at room temperature. As the reaction proceeds, white granules of pyridinium hydrochloride are observed as an insoluble precipitate. The reaction is then kept at 4° and checked for unreacted starting material by paper electrophoresis at pH 7.1 or HPLC. If any starting material remains, the reaction is allowed to proceed for an additional 1 hr at room temperature.

While the reaction mixture is at 4°, water (300 ml) is added to dissolve the pyridinium hydrochloride. The nucleotide is extracted from the mixture with reagent-grade chloroform (3× 200 ml). The combined chloroform extracts are filtered through anhydrous Na_2SO_4 (~100 g) to remove water. The solution is evaporated *in vacuo* to remove all the chloroform. Pyridine (50 ml) is added, and the solution evaporated *in vacuo* to 25 ml. The residue is quickly taken up in 50% pyridine (125 ml). The flask is placed in an ice bath, and 300 ml of precooled 2 *N* NaOH is added. The solution should be homogeneous. If a clear solution does not result, either water or pyridine is added depending on whether NaOH or mononucleotide has come out of solution. After verifying that the pH is 14, the clear solution is allowed to stand at room temperature for 20 min. Then pyridinium Dowex-50W-X8 resin (400 ml) is quickly added into the flask and swirled until the pH drops to 7. This is indicated by a color change from red-orange to yellow.

The resin is filtered off, and the filtrate is then passed (3–4 ml/min) through a column of pyridinium Dowex-50W-X8 resin (125 ml). The resin is washed with 50% pyridine (375 ml). The total effluent is made up to approximately 50% pyridine and evaporated to a volume of 100 ml in a rotary evaporator. Water (100 ml) is added, and the mixture is extracted four times with 100 ml of reagent-grade ethyl ether to remove the carboxylic acid (benzoic or anisic acid). The aqueous layer is made up to approximately 70% pyridine and evaporated to a minimum volume of 50 ml. Pyridine (50 ml) is added twice, and the solution again evaporated to 10 ml. Then, the last time, 50 ml of dry pyridine is added, and the solution evaporated to a thick syrup. The residue is taken up in 100 ml of dry pyridine, and the solution precipitated into 3 liters of anhydrous ethyl ether. [Pyridinium *N*-isobutyryldeoxyguanosine 5′-phosphate is precipitated from ethyl ether–chloroform (1 : 1).] The precipitate is collected by centrifugation, washed three times with ether, and finally dried *in vacuo* for 2 hr. The yield is in the range of 4–5.0 g, and the purity is checked by either paper chromatography, electrophoresis, or HPLC. The uv spectral properties at pH 7.0 are: for d(pbzA), λ_{max} 280 nm (18,300); for d(pibG), λ_{max} 258 nm (16,700); and for d(panC), λ_{max} 302 nm (22,500).

3′-*O*-Protected Derivatives of Mono- and Oligonucleotides

The 3′-hydroxyl groups of mono- and oligonucleotides carrying 5′-phosphate groups need to be protected in order to avoid self-condensation. A hydrophobic 3′-hydroxyl protecting group such as the *t*-butyldiphenylsilyl group can be particularly advantageous in rapid separation of the desired product by HPLC.

Acetyl Derivatives

Acetylation of the 3′-hydroxyl group of mononucleotide and oligonucleotide blocks is carried out by the following procedure.[17] A mixture of the dry nucleotidic compound (pyridinium or triethylammonium salt) and acetic anhydride (a 50-fold molar excess) in anhydrous pyridine (four

[17] H. Weber and H. G. Khorana, *J. Mol. Biol.* **72,** 219 (1972).

times the volume of acetic anhydride) is kept at room temperature for 4 hr. The flask is cooled to 4°, and methanol (twice the volume of acetic anhydride used) is added. After 1 hr at room temperature, water (same volume as pyridine used) is added, and the mixture is left overnight at room temperature. After the addition of an equal volume of pyridine, the solution is concentrated and rendered anhydrous by repeated evaporation of added dry pyridine. The concentrated pyridine solution is added dropwise to anhydrous ether (25–100 volumes). The precipitate is collected by centrifugation, washed three times with ether, and dried *in vacuo*. It is stored in a desiccator over $CaCl_2$ at −20° and reprecipitated from pyridine–ether just before use. The 3′-*O*-acetyl compounds are checked for purity by paper chromatography or HPLC. The yield is usually quantitative.

t-Butyldiphenylsilyl Derivatives

Silylation of the 3′-hydroxyl group of mononucleotides is carried out by the following procedure.[11] To a solution of *N*-protected deoxynucleoside 5′-phosphate (10 mmol) dried by evaporation from pyridine and 4.0 g (59 mmol) of crystallized imidazole in dry pyridine (150 ml) is added 13 ml (50 mmol) of *t*-butyldiphenylchlorosilane. After stirring for 16–18 hr at room temperature, water (150 ml) is added and the mixture kept overnight. The reaction mixture is transferred to a separatory funnel and extracted with two 250-ml portions of petroleum ether. The aqueous pyridine solution is evaporated *in vacuo* to a thick syrup. The syrup is dissolved in 500 ml of chloroform or ethyl acetate containing 25% *n*-butanol and extracted with two 100-ml portions of 0.2 *M* TEAB, pH 7.6. The organic layer is evaporated and rendered anhydrous by repeated addition and evaporation of dry pyridine. Finally, the residue is dissolved in dry chloroform, and the product is precipitated by dropwise addition to an excess of petroleum ether. The precipitate is collected by centrifugation, washed three times with petroleum ether, and finally dried in a vacuum. The yield is in the range of 85–90%. The product is obtained as a white powder which is homogeneous by TLC (35 or 50% methanol in ethyl acetate, v/v) and HPLC.

Synthesis of Di-, Tri- and Tetranucleotides Carrying 5′-Phosphate Groups

Principle

Di-, tri-, and tetranucleotides carrying 5′-phosphate end groups, which are required for the elongation of a chain, are prepared by condensing a suitable 5′-phosphate-protected mono- or oligonucleotide with a 3′-

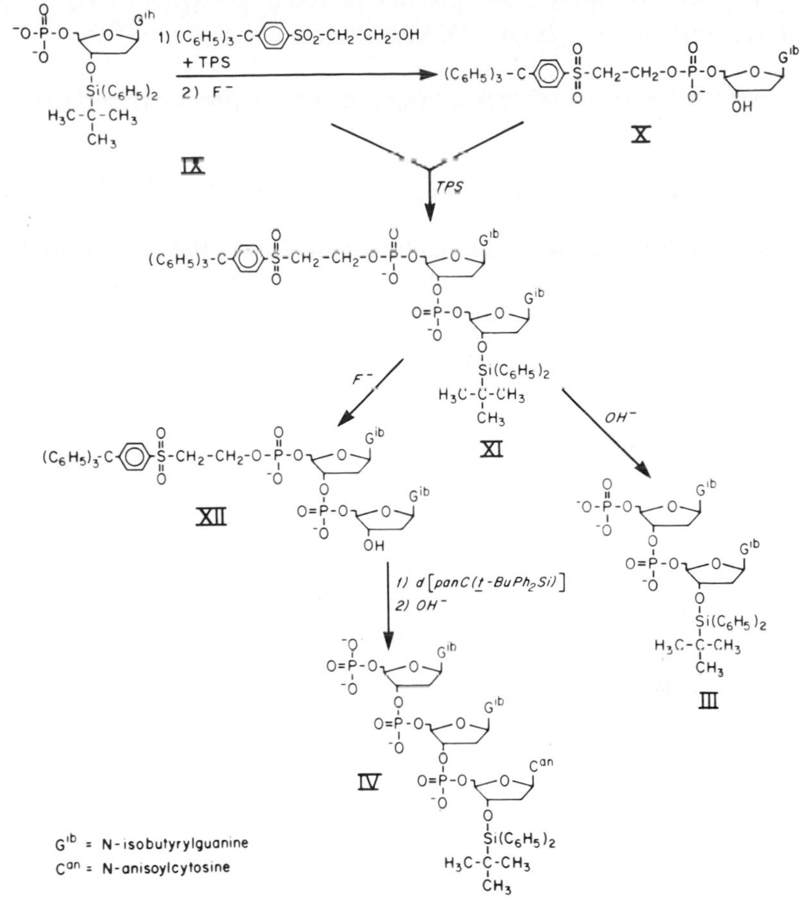

G^{ib} = N-isobutyrylguanine
C^{an} = N-anisoylcytosine

FIG. 4. The use of 3′-*O*-*t*-butyldiphenylsilyl-*N*-isobutyryldeoxyguanosine 5′-phosphate (IX) in the synthesis of suitably blocked di- and trinucleotides (III and IV).

hydroxyl-protected mono- or oligonucleotide carrying a 5′-phosphate end group. This is illustrated in Fig. 4. A number of phosphate-protecting groups have been used.[18,19] The approach shown in Fig. 4 uses lipophilic protecting groups for the 5′-phosphate and 3′-hydroxyl groups, with the advantage that both di- and trinucleotide blocks can be readily isolated and purified by solvent extraction or HPLC.[11] In the case of 5′-CNEt-

[18] K. L. Agarwal, M. Fridkin, E. Jay, and H. G. Khorana, *J. Am. Chem. Soc.* **95,** 2020 (1973), and references cited therein.

[19] K. L. Agarwal, Y. A. Berlin, H.-J. Fritz, M. J. Gait, D. G. Kleid, R. G. Lees, K. E. Norris, B. Ramamoorthy, and H. G. Khorana, *J. Am. Chem. Soc.* **98,** 1065 (1976).

protected oligonucleotides, the reaction products are separated by anion-exchange chromatography on DEAE-cellulose.

General Procedure for the Preparation of Deoxynucleoside 5'-[2-(p-Tritylphenyl)sulfonylethyl] Phosphate[19]

To an anhydrous pyridine solution (50 ml) of the *N*-protected mononucleotide (10 mmol) is added 2-(*p*-tritylphenyl)sulfonylethanol[19] (20 mmol), pyridinium Dowex-50 (1–2 g), and DCC (100 mmol), and the reaction mixture is shaken in the dark at room temperature for 16–26 hr. The reaction is followed by TLC in solvent acetonitrile–water (80/20, v/v). Water (50 ml) is added, and the reaction mixture is kept at room temperature overnight. The dicyclohexylurea is filtered off and washed with 50% aqueous pyridine. The filtrate and washings are combined and extracted with diisopropyl ether (3× 200 ml) to remove unreacted TPSE. The aqueous solution is concentrated, dissolved in 0.2 *M* TEAB, and extracted twice with ethyl acetate containing 20% *n*-butanol. The combined organic extracts are backwashed with 0.2 *M* TEAB, the organic layer is concentrated to an anhydrous pyridine solution, and the product precipitated by dropwise addition to an excess of ether–petroleum ether (1:1, v/v). Yields range from 70 to 80%. The product can also be isolated from the reaction mixture by separation on a silica gel column using acetonitrile containing 10–20% water as an eluant.

The Dinucleotide d(pT-T)[19]

An anhydrous pyridine solution (2.0 ml) of pyridinium d[pT(Ac)] (0.2 mmol) and d[(TPSE)pT] (0.1 mmol) is reacted with TPS (0.4 mmol) for 5.5 hr at room temperature. After the usual workup, the aqueous pyridine solution is concentrated, dissolved in 0.2 *M* TEAB, and extracted with ethyl acetate in a continuous solvent-solvent extractor to remove unreacted d[(TPSE)pT]. The aqueous phase is extracted with methylene chloride–*n*-butanol (7:3, v/v; 3× 5 ml). The combined organic extracts are washed with 0.2 *M* TEAB (2× 3 ml) and concentrated to an anhydrous pyridine solution; the product is precipitated by dropwise addition to an excess of dry ether. The yield of d[(TPSE)pT-T(Ac)] is approximately 75%.

The TPSE group is removed by treating a solution of the dinucleotide in pyridine–ethanol (1:1, v/v; 1.6 ml) with a 2 *M* NaOH solution (1.6 ml) at 0° for 5 min after the reaction mixture is neutralized with excess pyridinium Dowex-50 and exacted with methylene chloride–*n*-butanol (7:3, v/v) to remove the TPSE-elimination product, the aqueous phase is evaporated, and the residue as a pyridine solution is precipitated into ether. The yield of d(pT-T) is approximately 75%.

The Trinucleotide d[pibG-bG-anC (t-BuPh$_2$Si)]

Step 1. The dinucleotide d[(TPSE)pibG-ibG] is prepared as described below. An anhydrous pyridine solution (35 ml) of 3′-*O-t*-butyldiphenylsilyl-*N*-isobutyryldeoxyguanosine 5′-phosphate (IX in Fig. 4, 4.5 mmol)[11] and *N*-isobutyryldeoxyguanosine 5′-[2-(*p*-tritylphenyl)sulfonylethyl] phosphate (X in Fig. 4, 3 mmol) is reacted with TPS (12 mmol) for 17 hr at room temperature. After the usual workup, the condensed dinucleotide d[(TPSE)pibG-ibG(*t*-BuPh$_2$Si)] (XI in Fig. 4) is isolated by extraction with ethyl acetate containing 20% *n*-butanol, followed by purification on a silica gel column. The yield is 41%.

Step 2. An anhydrous solution of d[(TPSE)pibG-ibG(*t*-BuPh$_2$Si)] (XI in Fig. 4, 0.23 mmole) and tetra-*n*-butylammonium fluoride (1 mmole) in dry pyridine (0.25 *M* pyridinium fluoride, 4 ml) is allowed to react for 17 hr. The mixture is then passed over a 1-ml column of Dowex-50 pyridinium form resin, and the eluant is diluted with water (20 ml) and washed with two 25-ml portions of diethyl ether. The product is then extracted into one 50-ml and two 25-ml portions of ethyl acetate containing 50% *n*-butanol. The combined ethyl acetate layers are evaporated, and a concentrated pyridine solution of the residue precipitated by dropwise addition to 300 ml of diethyl ether. This gives 300 mg (94%) and product (XII in Fig. 4) homogeneous by TLC in 50% methanol in ethyl acetate and HPLC; A_{260}/A_{280} = 1.63, calc. 1.45; A_{280}/A_{300} = 2.96, calc. 2.50.

Step 3. An anhydrous pyridine solution (3 ml) of d[(TPSE)pibG-ibG] (XII in Fig. 4, 0.1 mmol), 3′-*O-t*-butyldiphenylsilyl-*N*-anisoyldeoxycytidine 5′-phosphate (0.5 mmol) and TPS (1 mmol) is allowed to react for 16.5 hr. After the standard workup, water (20 ml) is then added, and the mixture washed with a 30-ml portion of diethyl ether. The excess of the mononucleotide is removed by extraction with a 25-ml portion of ethyl acetate followed by a 25-ml portion of ethyl acetate containing 10% *n*-butanol. The product is then extracted into a 25-ml portion of ethyl acetate containing 25% *n*-butanol. This is evaporated, and a concentrated pyridine solution of the residue precipitated by dropwise addition to 40 ml of diethyl ether to give 150 mg of d[(TPSE) pibG-ibG-anC(*t*-BuPh$_2$Si)] (70% yeild) homogeneous by TLC in 50% methanol in ethyl acetate and HPLC; A_{260}/A_{280} = 1.17, calc. 1.15; A_{280}/A_{300} = 1.42, calc. 1.26.

Step 4. To a solution of 75 mg (0.035 mmol) of d[(TPSE)pibG-ibG-anC(*t*-BuPh$_2$Si)] in 1.7 ml of a mixture of methanol, pyridine, and water (30 : 65 : 5) at 0° is added 0.3 ml of 1 *M* NaOH in the same solvent. After 3 min, the reaction is neutralized with Dowex-50 pyridinium form resin and filtered. Water (5 ml) is then added, and the mixture is washed with a 10-ml portion of diethyl ether, a 5-ml portion of ethyl acetate, and a 5-ml portion of ethyl acetate containing 20% *n*-butanol. The aqueous layer is then evaporated, and a concentrated pyridine solution of the residue precipitated by dropwise addition to 10 ml of diethyl ether to give 50 mg (80% yield) of trinucleotide IV (Fig. 4), which is homogeneous by HPLC; $A_{260}/A_{280} = 1.17$, calc. 1.15; $A_{280}/A_{300} = 1.17$, calc. 1.26.

The Tetranucleotide d(panC-bzA-bzA-bzA)

An anhydrous pyridine solution (15 ml) of d[(CNEt)panC-bzA] (0.77 mmol) and pyridinium d[pbzA-bzA(Ac)] (0.83 mmol) is treated with TPS (4.03 mmol) for 5.5 hr at room temperature. After the usual workup and alkaline hydrolysis, the compound in 2% pyridine, 0.05 *M* TEAB, and 10% ethanol (500 ml) is applied to a DEAE-cellulose (bicarbonate form) column (2.4 × 80 cm). After pyridine removal with the buffer containing 0.05 *M* TEAB and 10% ethanol, a linear gradient is started with 3.0 liters of 0.05 *M* TEAB in 10% ethanol in the mixing vessel and 3.0 liters of 0.3 *M* TEAB in 10% ethanol in the reservoir. Fractions of 16 ml are collected. The peak that contains the desired tetranucleotide is pooled, and the product isolated as a white powder by precipitation of its anhydrous pyridine solution (12 ml) into ethyl ether (1.0 liter). The yield of the tetranucleotide is 24% as determined spectroscopically.

Chemical Synthesis of d(A-A-T-T-C-T-T-T-C)

Principle

An example of a stepwise synthesis of a nonanucleotide is given to illustrate the synthetic procedures and methods of separation using solvent extraction, anion-exchange chromatography on DEAE-cellulose, and HPLC. Finally, a typical example involving the use of the lipophilic *t*-butyldiphenylsilyl group in the rapid separation of polynucleotides[11] is also described. The sequence of steps involved in the synthesis of d(A-A-T-T-C-T-T-T-C) is:

$$\text{d(MeOTr)bzA} \xrightarrow{\text{d[pbzA(Ac)]}} \text{d[(MeOTr)bzA-bzA]} \xrightarrow{\text{d[pT(Ac)]}}$$

$$\text{d[(MeOTr)bzA-bzA-T]} \xrightarrow{\text{d[pT-anC(Ac)]}}$$

$$\text{d[(MeOTr)bzA-bzA-T-T-anC]} \xrightarrow{\text{d[pT-T(Ac)]}}$$

$$\text{d[(MeOTr)bzA-bzA-T-T-anC-T-T]} \xrightarrow{\text{d[pT-anC(Ac)]}}$$

$$\text{d[MeOTr)bzA-bzA-T-T-anC-T-T-T-anC]} \xrightarrow[\text{2. H}^+]{\text{1. NH}_3} \text{d(A-A-T-T-C-T-T-T-C)}$$

Condensation of the protected nucleoside d(MeOTr)bzA[15] with the protected nucleotide d[pbzA(Ac)] followed by condensation of the resulting dinucleotide with d[pT(Ac)] gives the trinucleotide d[(MeOTr)-bzA-bzA-T]. The latter is next condensed with the dinucleotide d[pT-anC(Ac)] to give the pentanucleotide d[(MeOTr)bzA-bzA-T-T-anC] which is purified by anion-exchange chromatography (Fig. 5). Condensation of the pentanucleotide with d[pT-T(Ac)] gives the required heptanucleotide. This product is purified by HPLC using a Bondapak C_{18}/Porasil B column (Fig. 6). Because of the small lipophilic contribution of the thymine residues relative to the standard hydrophilic contribution of the two phosphate dissociations of the added dinucleotide, d[pT-T(Ac)], there is a relatively large difference in retention time between the product and unreacted starting material. This permits isolation of the heptanucleotide completely free of the pentanucleotide. The heptanucleotide is finally condensed with d[pT-anC(Ac)], and the nonanucleotide is isolated by anion-exchange chromatography (Fig. 7). It is further purified, after complete deprotection, by chromatography in the presence of 7 *M*

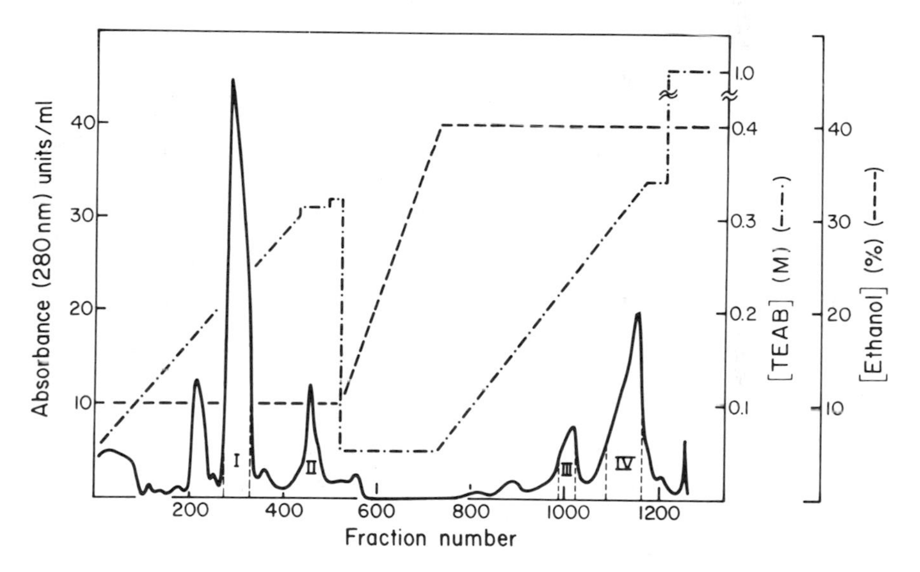

FIG. 5. Chromatography of the reaction products in the preparation of the pentanucleotide d[(MeOTr)bzA-bzA-T-T-anC]. Peak I was excess dinucleotide, and peak II was the pyrophosphate of the dinucleotide. Peak III contained the trinucleotide along with an unidentified compound, and peak IV contained the pentanucleotide.

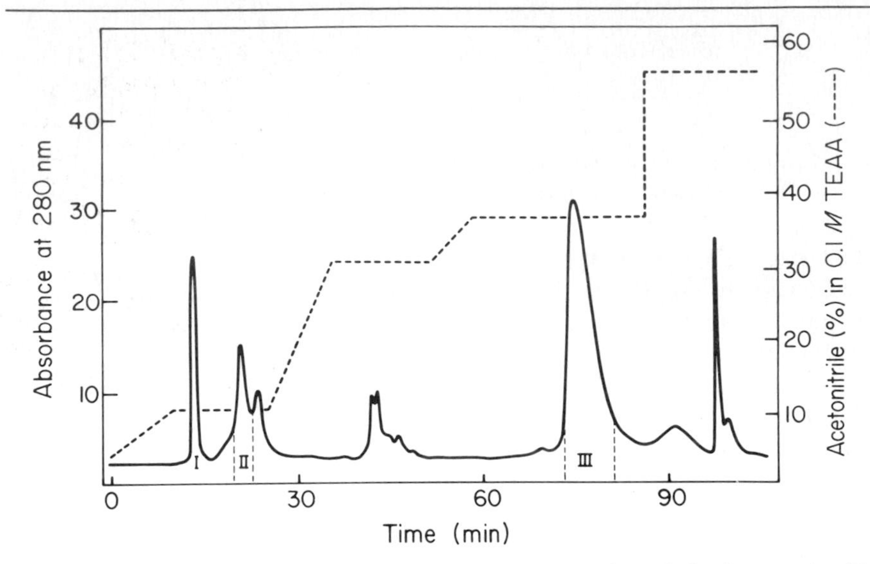

FIG. 6. HPLC of the reaction products in the preparation of the heptanucleotide d[(MeOTr)bzA-bzA-T-T-anC-T-T]. The separation was on a column (0.7 × 305 cm) of Bondapak C_{18}/Porasil B. The reaction products were fractionated with gradients of acetonitrile in 0.1 *M* TEAA (pH 7.0). Peaks I and II contained the dinucleotide and its pyrophosphate. Peak III contained the pure heptanucleotide.

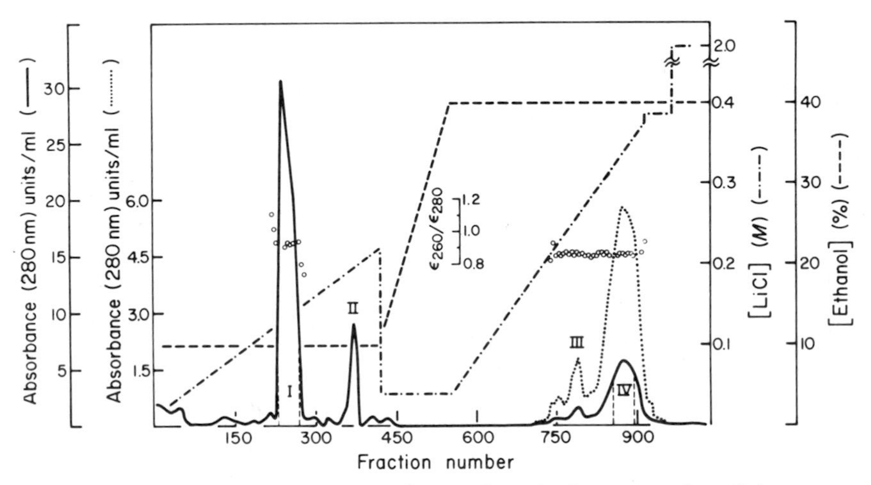

FIG. 7. Chromatography of the reaction products in the preparation of the nonanucleotide [(MeOTr)bzA-bzA-T-T-anC-T-T-T-anC]. The separation was on a column (2.0 × 90 cm) of DEAE-cellulose (chloride form) equilibrated at 4° with 0.01 *M* LiCl, 0.02 *M* Tris-HCl, pH 7.8, and 10% ethanol; the reaction products were then fractionated with gradients of LiCl and ethanol. Fractions of 10 ml were collected. Peaks I and II contained the excess dinucleotide and its pyrophosphate. Peak III contained unreacted heptanucleotide, while peak IV contained the desired nonanucleotide.

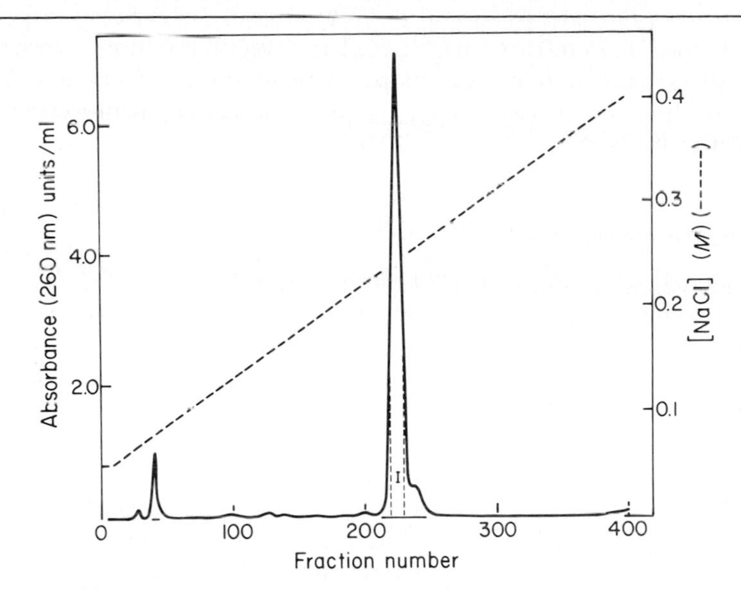

FIG. 8. Chromatography of the nonanucleotide d(A-A-T-T-C-T-T-T-C) after removal of the protecting groups. A DEAE-cellulose column (Whatman 52, chloride form) (1.1 × 115 cm) preequilibrated at room temperature with 0.05 *M* NaCl containing 7 *M* urea and 0.02 *M* Tris-HCl, pH 7.5. Elution was carried out using a linear gradient of NaCl in the presence of 7 *M* urea and 0.02 *M* Tris-HCl, pH 7.5. Fractions of 2.0 ml were collected. Peak I contained the nonanucleotide.

urea (Fig. 8). Analysis by the fingerprinting method after ^{32}P-phosphorylation of the 5′-hydroxyl group and by HPLC using an analytical μBondapak C_{18} column shows the product to be pure.

The Dinucleotide d[(MeOTr)bzA-bzA]

An anhydrous pyridine solution (35 ml) of d(MeOTr)bzA (62,500 A_{280}, 3.41 mmol), d[pbzA(Ac)] (107,000 A_{280}, 5.84 mmol) and TPS (2.80 g, 9.26 mmol) is allowed to react at room temperature for 5¼ hr. The reaction is terminated by the standard procedure. The reaction mixture as a mobile oil is taken up in 0.2 *M* TEAB (150 ml) and extracted with isopropyl ether in a solvent extractor for 16 hr at 4°. The aqueous TEAB solution is next extracted with ethyl ether–ethyl acetate (6:4; v/v; 1 × 100 ml) and finally with dichloromethane–carbon tetrachloride (5/5; v/v; 4× 150 ml). The latter four extracts contain the required dinucleotide as judged by silica gel TLC (acetonitrile–water, 85:15, v/v). Combined extracts are backwashed with 0.2 *M* TEAB (1× 35 ml) and concentrated in the presence of pyridine. The dinucleotide (3.46 g) is isolated by dropwise addition of its pyridine solution to an excess of dry ether and shown to be 95% pure by HPLC analysis. The precipitated d[(MeOTr)bzA-bzA(Ac)] is dissolved in a mixture of pyridine, ethanol, and water (4:3:3, v/v, 56 ml) and treated with 2 *M* NaOH (56 ml) for 5 min at 0°. The reaction mixture is worked up by the standard procedure, and its pyridine solution is precipitated into ether. The dinucleotide (94,800 A_{280}, 2.60 mmol) is obtained in 77% yield. Its spectral properties are: λ_{max} 281 nm; λ_{min} 259 nm; $\epsilon_{260}/\epsilon_{280} = 0.63$, calc. 0.59.

The Trinucleotide d[(*MeOTr*)*bzA-bzA-T*]

An anhydrous pyridine d[pT(Ac)] solution (25 ml) of d[(MeOTr)bzA-bzA] (88,000 A_{280}, 2.40 mmol), d[pT(Ac)] (44,500 A_{280}, 7.20 mmol), and TPS (3.96 g, 13.1 mmol) is allowed to react at room temperature for $5\frac{1}{2}$ hr. After the usual workup and alkaline hydrolysis, the reaction mixture, as a mobile oil, is dissolved in 0.2 *M* TEAB (175 ml) and extracted with ethyl acetate in a solvent extractor for 16 hr at 4°. This solution is successively extracted with ethyl acetate–*n*-butanol (95:5, v/v, 2× 75 ml), (93:7, v/v, 5× 75 ml), (90:10, v/v, 5× 75 ml) and dichloromethane–*n*-butanol (95:5, v/v, 2× 75 ml) until TLC indicates no d[(MeOTr)bzA-bzA] in the aqueous phase. The desired trinucleotide is then extracted into dichloromethane–*n*-butanol (90:10, v/v, 2× 75 ml) and (85:15, v/v, 4× 100 ml). The latter six extracts are backwashed with 0.2 *M* TEAB (1× 80 ml) and concentrated in the presence of pyridine. The precipitated trinucleotide (75,000 A_{280}, 1.75 mmol) is obtained in 73% yield. Its spectral properties are: λ_{max} 278 nm; λ_{min} 267 nm; $\epsilon_{260}/\epsilon_{280} = 0.75$ (calc. 0.71).

The Pentanucleotide d[(*MeOTr*)*bzA-bzA-T-T-anC*]

An anhydrous pyridine solution (20 ml) of d[(MeOTr)bzA-bzA-T] (26,000 A_{280}, 0.6 mmol), d[pT-anC(Ac)] (59,000 A_{280}, 2.44 mmol), and TPS (1.84 g, 6.06 mmol) is allowed to react at room temperature for $5\frac{3}{4}$ hr. After the usual workup and alkaline hydrolysis, the reaction mixture in 0.05 *M* TEAB–10% ethanol (1000 ml) is applied to a DEAE-cellulose (bicarbonate form) column (3.3 × 105 cm) preequilibrated with 0.05 *M* TEAB–10% ethanol. After removal of pyridine from the column with the same buffer, the reaction products are then fractionated with gradients of TEAB and ethanol as shown in Fig. 5. Fractions of 18 ml are collected. Fractions 1091–1163 (peak IV) contain the desired pentanucleotide. After the standard isolation procedure, the pentanucleotide (17,600 A_{280}, 0.24 mmol) is obtained in 44% yield. Its spectral properties are: λ_{max} 278 nm; λ_{min} 242 nm; $\epsilon_{260}/\epsilon_{280} = 0.82$, calc. 0.84.

The Heptanucleotide d[(*MeOTr*)*bzA-bzA-T-T-anC-T-T*]

An anhydrous pyridine solution of d[(MeOTr)bzA-bzA-T-T-anC] (10,000 A_{280}, 0.149 mmol), d[pT-T(Ac)] (91,200 A_{280}, 0.88 mmol), and TPS (797 mg, 2.63 mmol) is allowed to react for $6\frac{1}{4}$ hr at room temperature. After the usual workup and alkaline hydrolysis, the reaction mixture as a mobile oil is dissolved in 0.2 *M* TEAB and extracted with ethyl acetate (3× 100 ml) and ethyl acetate–*n*-butanol (90:10, v/v, 2× 100 ml) to remove most of the sulfonic acid. After the combined organic extracts are backwashed with 0.2 *M* TEAB (2× 75 ml), the aqueous extracts are combined and concentrated in the presence of pyridine. After drying with anhydrous pyridine, a small amount of nucleotidic material precipitates. This is removed by filtration through a glass-wool plug. The bulk of the material is isolated as a dry powder (1.39 g) by precipitation into anhydrous ether (1 liter) from an anhydrous pyridine solution (20 ml). The insoluble residue is dissolved in pyridine–DIEA (75:25, v/v, 50 ml), concentrated, and precipitated into dry ether. An additional 300 mg of the crude reaction mixture is obtained. The crude product (1.39 g) is dissolved in 0.1 *M* TEAA (pH 7.0)–ethanol (50:50, v/v, 7.0 ml) and separated in five injections using a Bondapak C_{18}/Porasil B column (0.7 × 305 cm). The tracing from one injection is shown in Fig. 6. The effluent corresponding to peak III contains the desired heptanucleotide. After concentration and precipitation, the heptanucleotide (5500 A_{280}, 69 μmol) is obtained in 46% yield. Its spectral properties are: λ_{max} 274 nm; λ_{min} 240 nm; $\epsilon_{260}/\epsilon_{280}$ = 0.92, calc. 0.86.

The Nonanucleotide d[(MeOTr)bzA-bzA-T-T-anC-T-T-T-anC]

An anhydrous pyridine solution (3 ml) of d[(MeOTr)bzA-bzA-T-T-anC-T-T] (4500 A_{280}, 56 μmol), d[pT-anC(Ac)] (12,800 A_{280}, 0.53 mmol), and TPS (467 mg, 1.54 mmol) is allowed to react at room temperature for 6¼ hr. After the usual workup and alkaline hydrolysis, the reaction mixture as a mobile oil is dissolved in 0.1 *M* Tris-HCl, pH 7.8 (50 ml), and extracted with isopropyl ether (2× 40 ml). The combined organic extracts are backwashed with 0.1 *M* Tris-HCl–10% ethanol (500 ml) and then applied to a DEAE-cellulose column. Details of chromatography and the elution profile are shown in Fig. 7. The fractions (peak IV) containing the desired nonanucleotide are pooled, diluted with 0.02 *M* DIEAB (final volume 1000 ml), and concentrated to 50 ml by membrane filtration. The concentrate is desalted by repeated addition of 0.02 *M* DIEAB (750 ml). After the retentate is diluted with pyridine, concentrated, and dried, the product is isolated as a dry powder by precipitation into anhydrous ether. The nonanucleotide (2200 A_{280}, 21 μmol) is obtained in 38% yield. Its spectral properties are: λ_{max} 274 nm; λ_{min} 240 nm; $\epsilon_{260}/\epsilon_{280}$ = 0.92, calc. 0.87).

The Unprotected Nonanucleotide d(A-A-T-T-C-T-T-T-C)

A portion (200 A_{260}, 2.25 μmol) of the fully protected nonanucleotide d[(MeOTr)bzA-bzA-T-T-anC-T-T-T-anC] is completely deprotected as

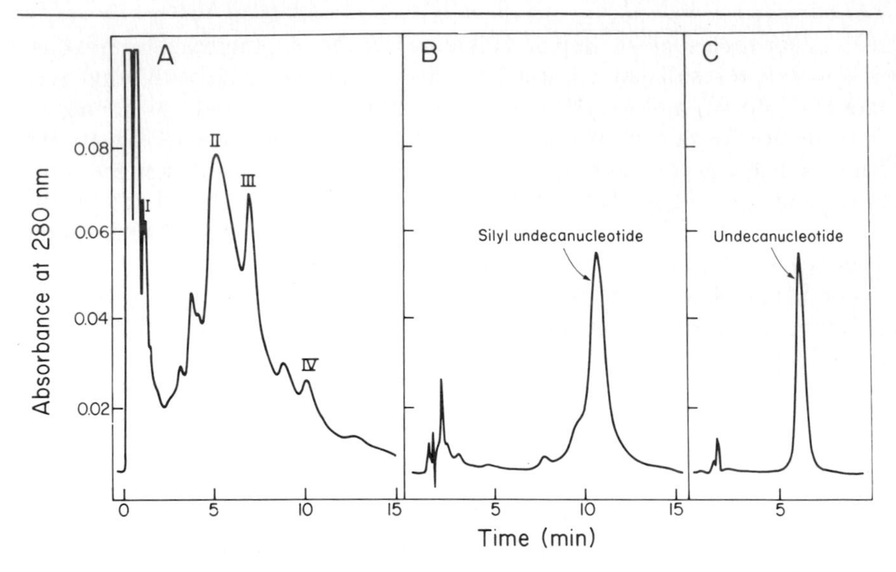

FIG. 9. (A) HPLC, at 44% acetonitrile concentration, of an aliquot from the total reaction mixture obtained after condensation of the octanucleotide d[(MeOTr)ibG-ibG-bzA-bzA-ibG-anC-ibG-ibG] with the trinucleotide d[pibG-ibG-anC(*t*-BuPh$_2$Si)]. The octanucleotide is in peak I, the trinucleotide in peak II, the pyrophosphate of the latter in peak III, while the required undecanucleotide is in peak IV. (B) HPLC, at 44% acetonitrile concentration, of the undecanucleotide after isolation by preparative HPLC. (C) HPLC (37% acetonitrile concentration) of the undecanucleotide obtained after removal of the 3′-*O*-silyl group and purification by preparative HPLC.

described in the general methods, and the product purified by anion-exchange chromatography in the presence of 7 *M* urea. The elution pattern is shown in Fig. 8. Fractions 219–229 (peak I) which contain the desired nonanucleotide are pooled and then desalted on a DEAE-cellulose column (0.8 × 10 cm). The eluant of this column is lyophilized to afford the unprotected nonanucleotide (110 A_{260}, 1.31 μmol) in 58% yield. Its spectral properties are: λ_{max} 262 nm; λ_{min} 236 nm; $\epsilon_{260}/\epsilon_{280}$ = 1.76, calc. 1.75. The nonanucleotide is phosphorylated with T4 polynucleotide kinase and [γ-^{32}P]ATP, and a portion (25 pmol) of this sample is partially digested with snake venom phosphodiesterase and analyzed by two-dimensional homochromatography.

The Undecanucleotide d[(*MeOTr*)*ibG-ibG-bzA-bzA-ibG-anC-ibG-ibG-ibG-ibG-anC*(t-*BuPh$_2$Si*)]

An anhydrous pyridine solution (0.3 ml) of the octanucleotide d[(MeOTr)ibG-ibG-bzA-bzA-ibG-anC-ibG-ibG][11] (120 A_{260}, 1 μmol) and d[pibG-ibG-anC(*t*-BuPh$_2$Si)] (36 mg, 20 μmol) is reacted with TPS

(70 μmol) for 17 hr at room temperature. After the standard workup, as described in the general methods, water (10 ml) is added and the mixture extracted with a 10-ml portion of diethyl ether, a 10-ml portion of ethyl acetate, and two 5-ml portions of ethyl acetate containing 20% *n*-butanol. Evaporation and precipitation of the aqueous layers yield 47 mg of the crude product which is analyzed by HPLC (Fig. 9A). The total precipitated reaction mixture is separated by HPLC in one run on a μBondapak C_{18} column using a 30-min gradient from 35–40% acetonitrile in 0.1 *M* TEAA. This gives 18.5 A_{260} of the desired undecanucleotide d[(MeOTr)ibG-ibG-bzA-bzA-ibG-anC-ibG-ibG-ibG-ibG-anC(*t*-BuPh$_2$Si)] (Fig. 9B).

The fully protected undecanucleotide (6 A_{260}) obtained above is dissolved in 110 μl of dry pyridine to which tetra-*n*-butylammonium fluoride (15 μmol) is added. The reaction mixture is kept at room temperature for 19 hr (a sample removed after 2 hr shows approximately two-thirds desilylation by HPLC) and then evaporated to a thick residue. The residue is dissolved in 100 μl of 0.1 *M* TEAA and purified in three runs on a analytical μBondapak C_{18} column to give 3 A_{260} units of the desired undecanucleotide d[(MeOTr)ibG-ibG-bzA-bzA-ibG-anC-ibG-ibG-ibG-ibG-anC] (Fig. 9C).

The product recovered from the HPLC column was completely depro-

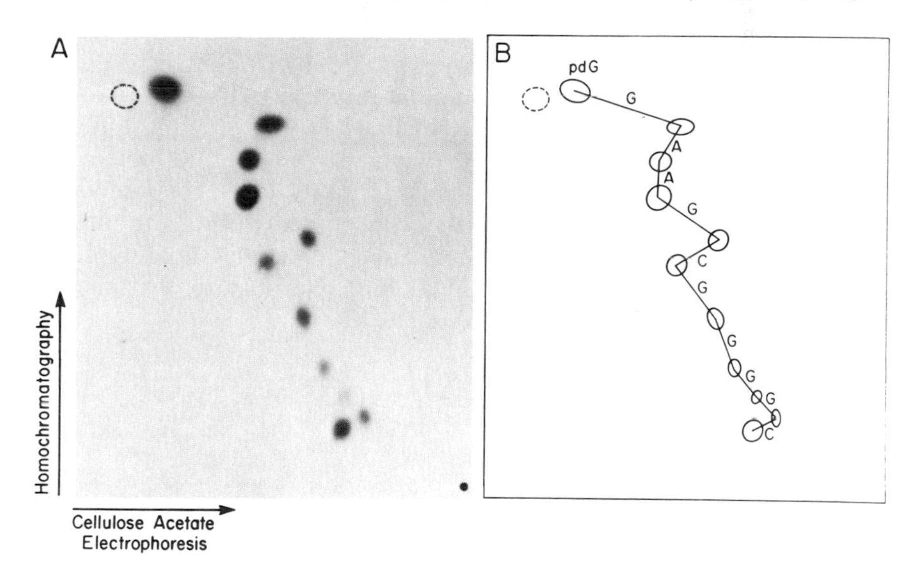

FIG. 10. (A) A two-dimensional fingerprint of a partial snake venom phosphodiesterase digest of the undecanucleotide d(G-G-A-A-G-C-G-G-G-G-C). (B) An artist's conception of the two-dimensional fingerprint shown in (A). The dashed circle in (A) indicates the position of the dye marker, xylene cyanol.

tected by the standard method. The resultant undecanucleotide, d(G-G-A-A-G-C-G-G-G-G-C), without any purification is phosphorylated by polynucleotide kinase using [γ-^{32}P]ATP and subjected to the two-dimensional fingerprinting procedure. The pattern obtained is shown in Fig. 10.

Enzymatic Joining of Oligonucleotides to Form Double-Stranded DNA

General Principles

Three to seven chemically synthesized deoxyribooligonucleotides are grouped together to afford a uniquely ordered bihelical complex. Gaps in the phosphodiester backbone are then sealed by T4 polynucleotide ligase. The resulting duplex is isolated by gel filtration or gel electrophoresis and then characterized by resistance to bacterial alkaline phosphatase and degradation to 3′- and 5′-mononucleotides. The synthetic segments are grouped together, and some are phosphorylated by means of T4 polynucleotide kinase so as to provide, whenever possible, a DNA duplex with single-stranded protruding ends bearing 5′-hydroxyl groups. To facilitate the monitoring of the joining reaction and characterization of the formed duplex, the constituent polynucleotides, except for those that will form the 5′-protruding ends, are phosphorylated with [γ-^{32}P]ATP of the same specific activity. In this way it is possible to prepare, characterize, and store relatively large quantities of the duplexes representing different parts of a larger DNA. At a later time, when the duplexes are to be covalently joined to one another, they can be readily phosphorylated with [γ-^{32}P]ATP at a much higher specific activity than that of the internal labels. Finally, the kinetics of the joining reaction is measured by determining either the percent of bacterial alkaline phosphatase resistance or percent of joining as a function of time.

Materials and Methods

Materials

[γ-^{32}P]ATP is prepared according to the procedure of Glynn and Chappell[20] using carrier-free $H_3{}^{32}PO_4$ purchased from New England Nuclear Corporation. DEAE-cellulose paper (DE-81 paper) is obtained from Whatman, Inc., and Sephadex and BioGel A gels are purchased from

[20] I. M. Glynn and J. B. Chappell, *Biochem. J.* **90,** 147 (1964).

Pharmacia Fine Chemicals and Bio-Rad Laboratories, respectively. Electrophoresis-grade acrylamide and bisacrylamide can be obtained from Bio-Rad Laboratories. Polyacrylamide slab gels are run in a buffer containing 90 m*M* Tris-borate, pH 8.3, and 4 m*M* EDTA.[21]

Enzymes

Bacterial alkaline phosphatase, micrococcal nuclease, spleen phosphodiesterase, snake venom phosphodiesterase, and pancreatic DNase are obtained from Worthington Chemical Company.

T4 DNA Ligase and T4 Polynucleotide Kinase

These enzymes are prepared generally according to published procedures.[22,23] On occasion, contaminating nucleases are observed in both enzyme preparations. The levels of these nucleolytic activities can be substantially reduced by the chromatographic procedures described below.

T4 DNA Ligase

While attempting to purify this enzyme further, it was noted that the concentration of KCl required to elute T4 DNA ligase from phosphocellulose could be reduced by adding ATP to the gradient buffers. It was then determined that this observation could be used to reduce the level of contaminating nuclease markedly. The procedure described here can be considered a form of specific elution by substrate, a topic previously covered in this series.[24] It is found that T4 DNA ligase, in its storage buffer (0.02 *M* KPO_4, pH 7.6, 0.01 *M* β-mercaptoethanol, 50% glycerol) could be adsorbed onto a small column of phosphocellulose (2.8 cm, 1.5 cm diameter) equilibrated in buffer A (0.01 *M* KPO_4, pH 7.6, 0.01 *M* β-mercaptoethanol) plus 0.1 *M* KCl. This column is washed in succession with 20 ml of buffer A plus 0.1 *M* KCl and 20 ml of buffer A plus 0.35 *M* KCl. T4 DNA ligase is then eluted with 20 ml of buffer A plus 0.35 *M* KCl plus 0.1 m*M* ATP. This fraction is immediately diluted with an equal volume of buffer A plus 0.1 *M* KCl and applied to a second small phosphocellulose column (1.7 cm, 1.5 cm diameter). After this column is washed with 10 ml of buffer A plus 0.1 *M* KCl, the adsorbed enzyme is eluted with 10 ml of buffer A plus 0.6 *M* KCl and then concentrated by dialysis against storage buffer. The T4 DNA ligase present in this final dialyzed

[21] A. C. Peacock and C. W. Dingman, *Biochemistry* **7**, 668 (1968).

[22] B. Weiss, A. Jacquemin-Sablon, T. R. Live, G. C. Fareed, and C. C. Richardson, *J. Biol. Chem.* **243**, 4543 (1968).

[23] C. C. Richardson, *Proced. Nucleic Acid Res.* **2**, 815 (1971).

[24] B. M. Pogell and M. G. Sarngadharan, this series, Vol. 22, p. 379.

fraction stored at $-20°$ represented a net 40–50% recovery of enzyme carried through this procedure.

T4 Polynucleotide Kinase

This enzyme is applied to a 1.5-cm-diameter, 45-cm column of Sephadex G-100 equilibrated and run in a solution of 50 m*M* Tris-HCl, pH 7.6, 0.1 *M* KCl, 0.01 *M* β-mercaptoethanol, 0.1 m*M* ATP, and 5% glycerol. One-milliliter fractions are collected and assayed for polynucleotide kinase activity. This enzyme, which elutes at the void volume of this column, is found to be essentially free of DNase activity. No loss of enzymatic activity is observed in this procedure.

Phosphorylation of 5'-Hydroxyl End Groups of Chemically Synthesized Deoxyribooligonucleotides

Oligonucleotides are phosphorylated with [γ-^{32}P]ATP and polynucleotide kinase in the following way. The reaction mixture containing 20–50 m*M* Tris-HCl buffer, pH 7.6, 10 m*M* $MgCl_2$, spermidine (1 μg/ml), and 1–50 μ*M* oligonucleotide is heated at 95° for 2 min to destroy any secondary structure or self-aggregation of the oligonucleotide. After the reaction mixture is cooled to 37°, 10 m*M* dithiothreitol, [γ-^{32}P]ATP (2–5 molar equivalents relative to 5'-hydroxyl end group of oligonucleotide) and T4 polynucleotide kinase (20–50 units/ml) are added. The reaction is carried out at 37°. The kinetics of the phosphorylation reaction are followed by the DEAE-cellulose paper assay[25] (see below). Under these conditions the 5'-hydroxyl end group is usually phosphorylated in 15–20 min. However, the reaction mixture is heated again to 95° for 2 min, dithiothreitol and polynucleotide kinase are added, and incubation is resumed at 37° for another 20 min. Phosphorylation is judged to be complete when no further incorporation of the label from [γ-^{32}P]ATP is observed after a heating and kinase incubation cycle. The phosphorylated product is isolated by gel filtration through a column of Sephadex G-50 with 50 m*M* TEAB as the eluant. After the appropriate fractions are pooled, the TEAB is removed by repeated lyophilization, and the residue is taken up in 10 m*M* Tris-HCl, pH 7.6. All TEAB must be removed, since it will inhibit the subsequent polynucleotide ligase-catalyzed joining reaction. This problem can be circumvented by using the phosphorylation reaction mixture directly in the joining reaction. In this case the kinase is first inactivated by heating at 95° for 2 min.

[25] A. Falaschi, J. Adler, and H. G. Khorana, *J. Biol. Chem.* **238,** 3080 (1963).

DEAE-Cellulose Paper Assay[25]

The kinetics of a phosphorylation reaction are measured in the following way. Samples (0.5–3 μl) are removed at various times and spotted about 8 cm from one end of a DE-81 strip (3 × 23 cm). The point of sample application is pretreated with 50 μl of 50 m*M* Na_2EDTA containing 50 m*M* sodium pyrophosphate. Development is by descending chromatography in 350 m*M* ammonium formate. The phosphorylated oligonucleotide stays at the origin, whereas ATP moves away with an R_f of about 0.5–0.6. The strips are scanned for radioactivity using a Packard radiochromatogram scanner. The regions containing radioactivity are cut out and counted in a scintillation spectrometer.

T4 Polynucleotide Ligase-Catalyzed Joining Reaction

A typical reaction mixture contains Tris-HCl, pH 7.6 (20–50 m*M*), $MgCl_2$ (10 m*M*), dithiothreitol (10 m*M*), oligonucleotides (2–40 μ*M* each), ATP, and T4 polynucleotide ligase (200–400 units/ml). The molar concentration of ATP is dependent on the concentration of the oligonucleotide components. If the concentration of each oligonucleotide is 5 μ*M* or greater, the joining reaction is performed in the presence of 2–3 molar equivalents of ATP relative to the total concentration of the 5′-phosphate end groups. In a small-scale reaction, where the concentration of the components is lower than 5 μ*M*, the total concentration of ATP has to be 50 μ*M* or more to achieve ligase-catalyzed joining. Before the addition of dithiothreitol, ATP, and ligase, the segments are annealed by heating the mixture at 95° for 2 min and slowly cooling to 20° (15–30 min) and then to 5°. For some joinings, the annealing process is deliberately slowed down (~18 hr) by placing the reaction mixture in a Dewar flask. The reaction is carried out at 5–10°. The kinetics of the joining reaction can be followed in two ways as described below.

The joined product is isolated either by gel permeation chromatography on a BioGel A-0.5m column[26] or by slab gel electrophoresis on a 10–15% polyacrylamide gel. The latter method is preferred, because it will resolve the desired duplex from those lacking one or more segments. The gel band containing the desired product is cut out, transferred to a polypropylene test tube, crushed to a fine powder with a glass rod, and then soaked in 2 volumes of 2 *M* TEAB. After 18 hr at 5°, the gel slurry is spun down, the supernatant is removed, and the gel is washed twice with 1 volume of water. The extracts are combined and evaporated to dryness. The residue thus obtained is dissolved in a minimal volume (<1 ml) of 50 m*M* TEAB and passed through a column (0.8 × 25 cm) of Sephadex G-50 to

[26] V. Sgaramella and H. G. Khorana, *J. Mol. Biol.* **72,** 427 (1972).

remove inorganic salts. Elution is carried out with 50 mM TEAB. After the appropriate fractions are pooled, the aqueous solution is repeatedly evaporated or lyophilized until the white residue of TEAB has disappeared. It is essential that all TEAB be removed. The residue is taken up in 10 mM Tris-HCl, pH 7.6, containing 10 mM $MgCl_2$ (10–100 μl).

The kinetics of a joining reaction are measured by either of the following methods: (1) Aliquots (0.5–1 μl) are withdrawn at various times and then analyzed by gel electrophoresis on a 15% polyacrylamide slab gel; bands are cut out, and radioactivity is determined by measuring Cerenkov radiation. (2) Aliquots (0.5–1 μl) are withdrawn and added to 50 μl of 100 mM Tris-HCl, pH 8.0, containing 0.5 mM dpT. After heating at 95° for 2 min, the samples are transferred to a 70° bath and 1 μl of bacterial alkaline phosphatase solution (1 mg/ml) is added. The reaction is carried out for 30 min and then analyzed by the DEAE-cellulose paper assay described above. The joined product stays at the origin, while the ^{32}P-labeled inorganic phosphate travels close to the solvent front.

Characterization of Joined Products

Degradation to 3′-mononucleotides is performed essentially as described in the published procedure.[26,27] To the joined product containing a few thousand counts per minute (1–10 μl) is added 25 μl of micrococcal nuclease digest mix [70 mM sodium glycinate, pH 9.2, 3 mM $CaCl_2$, calf thymus DNA (400 μg/ml), micrococcal nuclease (30–175 μg/ml)]. After incubation at 37° for 3 hr, 3 μl of spleen phosphodiesterase digest mix [167 mM NH_4OAc, 33 mM KPO_4, pH 6.5, 67 mM HOAc, spleen phosphodiesterase (1.7 mg/ml)] is added, and the incubation is continued for an additional 3 hr at 37°.

Degradation to 5′-mononucleotides has been described[26] and is carried out in the following way. To the joined product containing a few thousand counts per minute (1–10 μl) is added 25 μl of DNase digest mix [20 mM Tris-HCl, pH 7.6, 10 mM $MgCl_2$, calf thymus DNA (275 μg/ml), pancreatic DNase (255 μg/ml)]. After incubation at 37° for 30 min, 10 μl of SVD digest mix [150 mM sodium glycinate, pH 9.2, 10 mM KPO_4, pH 9.2, snake venom phosphodiesterase (450 μg/ml)] is added, and the incubation is continued for an additional 1 hr at 37°.

After the appropriate mixture of the four mononucleotide markers is added, the digestion mixture is separated by paper chromatography on Whatman no. 1 paper using the following solvent system: $(NH_4)_2SO_4$ (60 g), 0.1 M sodium phosphate, pH 6.8 (100 ml), and n-propanol (2 ml). In this solvent system the chromatographic mobility of the nucleotides, in order of fastest to slowest, is pdC, pdT, pdG, pdA (or dCp, dTp,

[27] R. D. Wells, T. M. Jacob, S. A. Narang, and H. G. Khorana, *J. Mol. Biol.* **27**, 237 (1967).

dGp, dAp). The nucleotides can also be fractionated by electrophoresis on Whatman no. 1 paper. The buffer is 0.5 M sodium citrate, pH 3.5, and electrophoresis is performed at 35 V/cm for 2–3 hr. The electrophoretic mobility of the nucleotides, in order of fastest to slowest, is pdT, pdC, pdA, pdG (or dTp, dCp, dAp, dGp).

Products having external 5′-^{32}P labels can be further characterized by treating the joined product first with bacterial alkaline phosphatase followed by degradation to 3′-nucleotides. In these cases, to the sample in Tris-HCl, pH 7.6 (5–10 μl), and 10 mM $MgCl_2$ is added 0.5 μl of bacterial alkaline phosphatase (1 mg/ml). After incubation for 30 min at 60°, 0.5 μl of 5 M NaOH is added (pH > 12), and incubation is continued for 15 min at 37°. The reaction mixture is neutralized with 0.5 μl of 5 M HOAc (the pH must be checked), and then the appropriate enzyme digest mix for degradation to 3′- or 5′-mononucleotides is added.

Joining of Short Deoxyribooligonucleotides by T4 Polynucleotide Ligase

Many DNA duplexes have been prepared for total synthesis of the synthetic suppressor gene. Three examples are presented to illustrate variations in the general procedure.

Preparation of Duplex [IVb] (Fig. 11)

The preparation of duplex [IVb] illustrates how the structure of an individual segment can influence a joining reaction. The synthesis is best carried out with a two-step approach as described.[28] Segment 22 is joined to segment 24 in the presence of segment 23. Segment 21 is then added, the mixture is heated at 95° for 2 min, and the joining reaction is resumed at 10° by the addition of polynucleotide ligase. If segment 21 is heated alone prior to its addition to the reaction mixture, no reaction takes place. This is probably due to the fact that segment 21 can form a duplex structure with itself. This is not the usual approach for duplex formation. Previous experience has shown that, in general, the use of five segments is better than the use of four segments, and the latter number is better than the use of three segments.

Procedure

For the first step, the reaction mixture (200 μl) contains 5′-^{32}P segment 24 (44 μM), segment 23 (52 μM), 5′-^{32}P segment 22 (40 μM), Tris-HCl,

[28] T. Sekiya, P. Besmer, T. Takeya, and H. G. Khorana, *J. Biol. Chem.* **251,** 634 (1976). Because a new nomenclature system has been adopted, duplex [IVb] is referred to as [I]′ in this article. Also, segments 21–24 are labeled as segments 5–2, respectively.

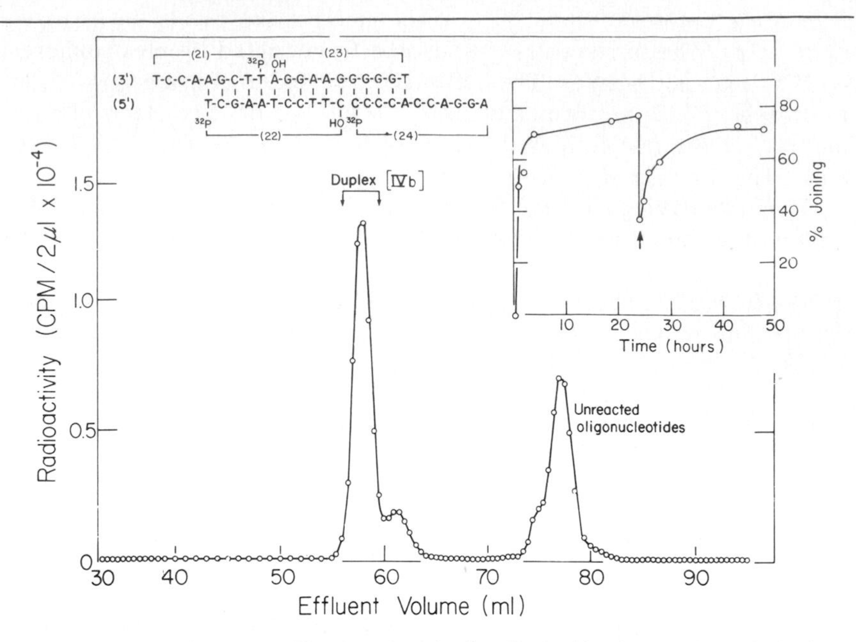

FIG. 11. Preparation and purification of duplex [IVb]. The kinetics are shown in the inset. The percent joining recorded is based on the joining of segment 24.

pH 7.6 (20 m*M*), $MgCl_2$ (10 m*M*), dithiothreitol (10 m*M*), ATP, and T4 polynucleotide ligase (100 units). After incubation at 10° for 24 hr, 5'-^{32}P segment 21 is added to a final concentration of 72 μ*M* and the total mixture heated at 95° for 2 min. After cooling to 10°, the second step of the reaction is started by the addition of dithiothreitol, ATP, and T4 ligase in the same amounts as used in the first step. The kinetics of the reaction are shown in the inset in Fig. 11. At the end of the reaction, the mixture is subjected to gel permeation chromatography. A column (1 × 150 cm) of BioGel A-0.5m (200–400 mesh) equilibrated with 50 m*M* TEAB is used. Fractions of 0.5 ml are collected every 15 min and pooled as shown in Fig. 11. The characterization of the isolated duplex is given in Table I.

Preparation of Duplex [*III*] (*Fig. 12*)

The synthesis of duplex [III] illustrates the "more is better" principle for enzymatic joining of oligonucleotides. For this and related duplex preparations, optimal yields are obtained in one-step reactions consisting of six to eight segments.[29] The extent of ligase-catalyzed joining in duplex

[29] P. C. Loewen, R. C. Miller, A. Panet, T. Sekiya, and H. G. Khorana, *J. Biol. Chem.* **251,** 642 (1976). In this article duplex [III] is labeled as [IIb]. Also, segments 14–20 are designated as segments 12–6.

TABLE I
CHARACTERIZATION OF DUPLEX [IVb]

Phosphatase Resistance

			Molar ratio	
			Found	Expected
^{32}P (cpm)				
Resistant (4107)			2.12	2
Sensitive (1930)			1.00	1

5′-Nucleotides [a]

	pdA	pdG	pdT	pdC
^{32}P (cpm)	14	39	6801	3285
Molar ratio				
Found	—	—	2.06	1.0
Expected	—	—	2	1

3′-Nucleotides [a]

	dAp	dGp	dTp	dCp + pdTp
^{32}P (cpm)	2800	17	115	5130
Molar ratio				
Found	1.00	—	—	1.84
Expected	1	—	—	2

[a] These nucleotides are separated by paper chromatography [$(NH_4)_2SO_4$ (60 g), 0.1 *M* sodium phosphate, pH 6.8 (100 ml), *n*-propanol (2 ml)].

[III], consisting of seven segments is 48%; for a related duplex consisting of eight segments (the eighth component base pairs at the 5′-end of segment 14) the joining went to 52%. This latter duplex has two recessed 5′-hydroxyl groups which cannot be quantitatively phosphorylated. Thus, duplex [III] is synthesized and used in the total synthesis of the tyrosine tRNA gene. The phosphorylation of the 5′-hydroxyl group on one protruding single-stranded end presents no difficulty. Kinetic analysis of the joining of segments 14–20 (Fig. 12B) shows that the addition of the terminal segments is slower than that of the internal segments. This can be corrected, in part, by using an excess of the terminal segment and longer reaction times.

Procedure

The reaction mixture (300 μl) containing unphosphorylated segment 14 (33 μ*M*), 5′-^{32}P segments 15–20 (23–25 μ*M* each), Tris-HCl, pH 7.6 (20 m*M*), and $MgCl_2$ (10 m*M*) are heated to 95° for 2 min and then cooled to room temperature in 15 min and then to 5°. The ligase (580 units/ml),

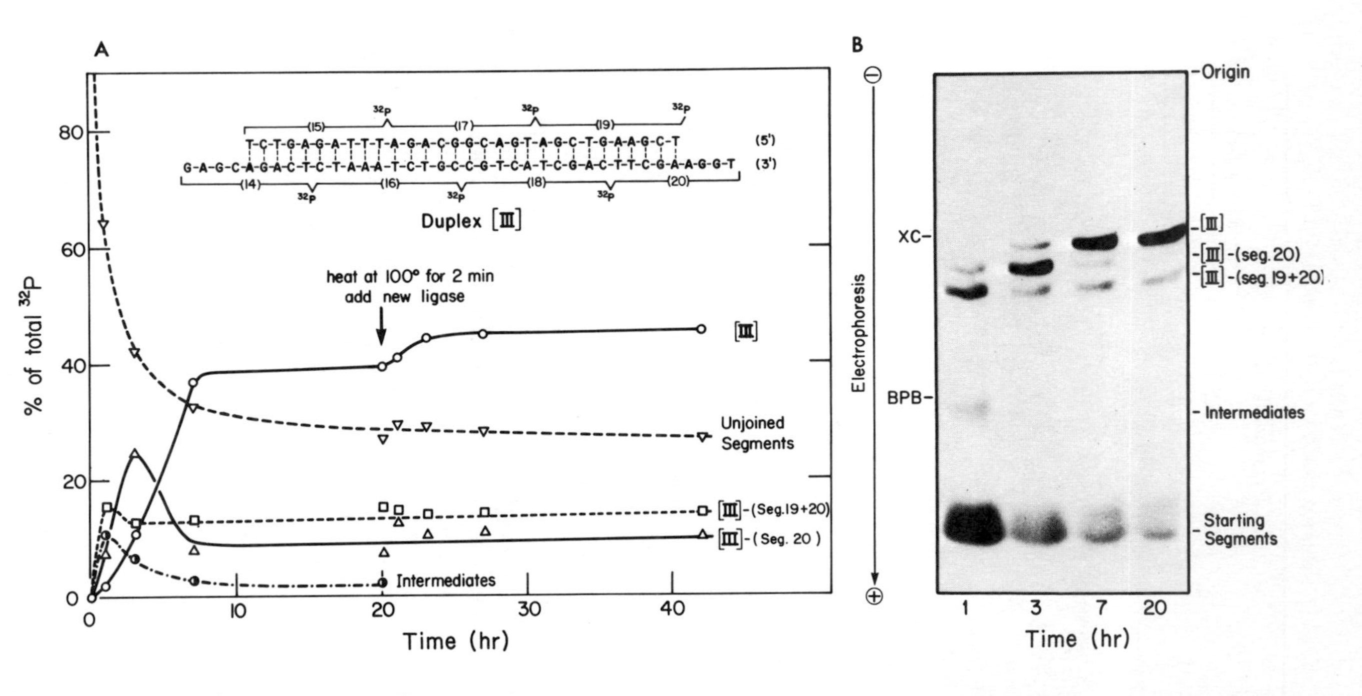

FIG. 12. Kinetics of the joining of segments 14–20 to give duplex [III] (A) as assayed by gel electrophoresis on a 15% polyacrylamide gel (B).

dithiothreitol (10 mM) and ATP (200 μM) are added, and the reaction is kept at 5°. After a 20-hr incubation, the mixture is heated again to 95° for 2 min and then cooled to 5°, and the same amounts of ligase and dithiothreitol and one-fourth of the above amount of ATP are added. The mixture is incubated at 5° for another 22 hr. Duplex [III] is isolated by preparative slab gel electrophoresis using a 15° polyacrylamide gel (20 × 20 × 0.3 cm).

Preparation of Duplex [P_{1-10}] *(Fig. 13)*

To obtain optimal yields, duplex [P_{1-10}] is prepared in four steps as described by Sekiya *et al.*[30] and outlined in Fig. 13. Because of the palindromes in this DNA[31] and the restrictions of chemical synthesis, two segments, P-2 and P-5, were prepared that had considerable sequence homology at their 5′-ends. Therefore, to obtain error-free joinings, segment P-2 is grouped with segments P-1 and P-3, and segment P-5 is grouped with segments P-4, P-6, and additional segments beginning with P-7. An additional consideration that led to the four-step plan is the following. The four-component system consisting of segments P-4 to P-7 gives essentially a quantitative yield of the expected duplex, but the seven-component system (segments P-4 to P-10) gives very little of duplex [P_{4-10}] (Fig. 13).

Procedure

Step 1. To a polypropylene test tube is added unphosphorylated segment P-1 (500 μl, 9.5 nmol), 5′-^{32}P segment P-2 (43 μl, 4.8 nmol), 5′-^{32}P segment P-3 (40 μl, 4.9 nmol), 1.0 M Tris-HCl, pH 7.6 (10 μl), and 1.0 M $MgCl_2$ (5 μl). This solution is heated at 95° for 2 min and then cooled to room temperature (15 min). After cooling to 5°, 1 M dithiothreitol (5 μl), 10 mM ATP (3 μl), and T4-polynucleotide ligase (7 units/μl, 25 μl) are added. The final volume is 630 μl. The reaction is carried out at 5° for 23 hr. Solid urea (final concentration 7 M) is added to the reaction which is heated at 95° for 2 min and then quickly chilled on ice. Tracking dyes are added, and segment P-(1 + 3) is isolated by slab gel electrophoresis on a 24% polyacrylamide gel containing 7 M urea.

Immediately before use, segment P-(1 + 3) is phosphorylated using [γ-^{32}P]ATP with a specific activity at least 50-fold higher than that of the internal [^{32}P]phosphate group. To a 1.5-ml polypropylene test tube is added segment P-(1 + 3) (45 μl 1.9 nmol), 10 μl of 10X kinase buffer [500 mM Tris-HCl, pH 7.6, 100 mM $MgCl_2$, spermidine (10μg/ml] and

[30] T. Sekiya, E. L. Brown, R. Belagaje, H.-J. Fritz, M. J. Gait, R. G. Lees, M. J. Ryan, H. G. Khorana, and K. E. Norris, *J. Biol. Chem.* **254,** 5781 (1979).

[31] T. Sekiya, H. van Ormondt, and H. G. Khorana, *J. Biol. Chem.* **250,** 1087 (1975).

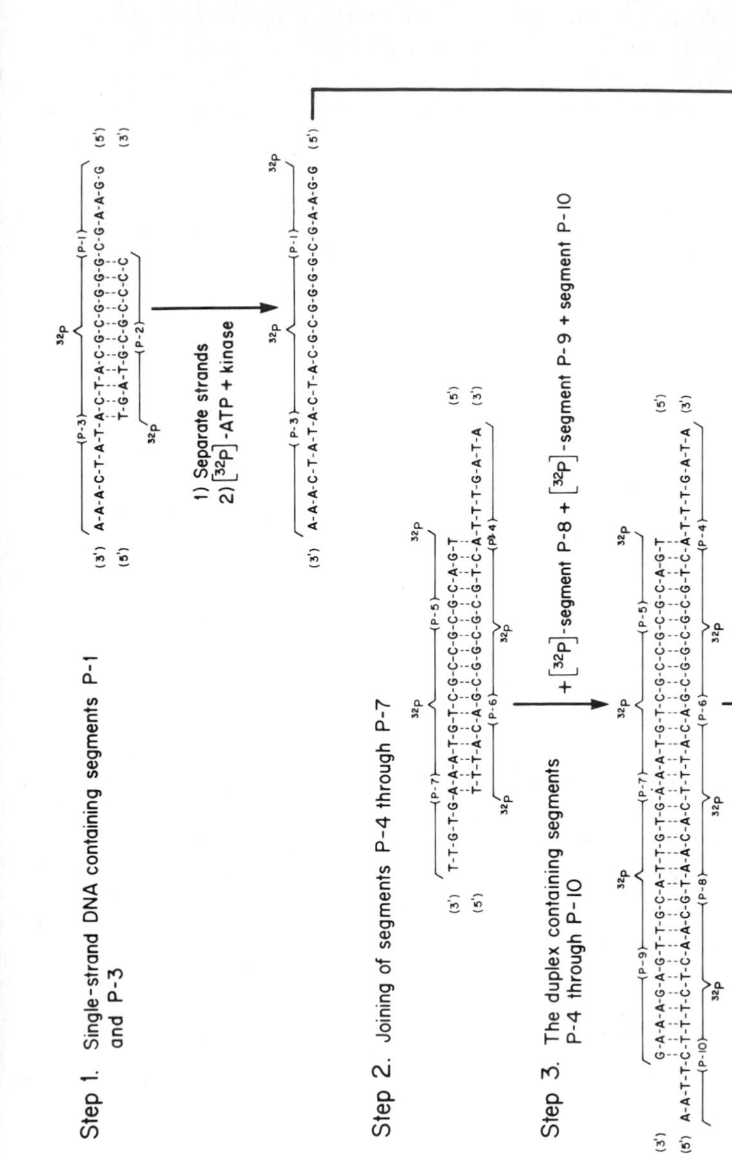

FIG. 13. Synthetic plan adopted for the stepwise joining of segments P-1 to P-10 to form duplex [P_{1-10}].

water (25 μl). This solution is heated at 95° for 2 min and then cooled to room temperature (15 min). 1 *M* dithiothreitol (1 μl), [γ-^{32}P]ATP (20 μl, 6.1 nmol; specific activity 1.7 × 10^5 cpm/pmol) and T4 polynucleotide kinase (4 units/μl, 1 μl) are added. The reaction is carried out as described in Materials and Methods. After gel filtration through a column of Sephadex G-50 and lyophilization, 5′-^{32}P segment P-(1 + 3) is taken up in 10 *M* Tris-HCl, pH 7.6 (100 μl). The concentration is 17 μ*M*.

Step 2. Segments P-4, P-5, P-6, and P-7 are phosphorylated at the 5′-end with [γ-^{32}P]ATP (specific activity is the same as that used to phosphorylate segment P-3) and, after heating at 95° for 2 min to inactivate the polynucleotide kinase, the phosphorylation reaction mixtures (100 μl each), which contain 50 m*M* Tris-HCl, pH 7.6, 10 m*M* $MgCl_2$, and 100 μ*M* ATP, are used directly in the following joining reaction. To the four phosphorylated segments (5.5–6.6 nmol of each) is added 1.0 *M* Tris-HCl, pH 7.6 (30 μl), 1 *M* $MgCl_2$ (6 μl), and water (510 μl). This solution is heated at 95° for 2 min and then cooled to room temperature (15 min). After cooling to 5°, 1 *M* dithiothreitol (10 μl), 10 m*M* ATP (6 μl, final concentration 100 μ*M*), and ligase (10 units/μl, 50 μl) are added. The final volume of the reaction mixture is 1.0 ml. The reaction is carried out at 5°. The kinetics are followed by electrophoresis on a 15% polyacrylamide gel using 0.5-μl aliquots from the mixture. The results are shown in Fig. 14A. After incubation for 9 hr, the reaction mixture is applied to a column of Sephadex G-100 (1.2 × 90 cm) at 4°. Elution is performed using 50 m*M* TEAB (pH 7.8) as the eluant. Half-milliliter fractions are collected every 10 min. After the appropriate fractions are pooled and lyophilized, and duplex [P_{4-7}] is taken up in 10 m*M* Tris-HCl, pH 7.6, containing 10 m*M* $MgCl_2$ (200 μl). The concentration of duplex [P_{4-7}] is 24 μ*M*.

Step 3. A solution of duplex [P_{4-7}] (100 μl, 2.4 nmol) and a second solution containing 5′-^{32}P segment P-8 (43 μl, 4.0 nmol), 5′-^{32}P segment P-9 (65 μl, 5.8 nmol) and unphosphorylated segment P-10 (210 μl, 11.1 nmol) are heated separately at 95° for 2 min and then slowly cooled (18 hr) to room temperature. After the two mixtures are cooled to 5° and combined, 1 *M* dithiothreitol (4 μl), 10 m*M* ATP (5 μl, final concentration 100 μ*M*), and T4 ligase (4 units/μl, 20 μl) are added. The reaction is carried out at 4°. The kinetics are followed by electrophoresis on a 15% polyacrylamide gel. The results are shown in Fig. 14B. After 24 hr, solid sucrose (final concentration ~30%) and tracking dyes are added, and duplex [P_{4-10}] is isolated by preparative slab gel electrophoresis on a 15% polyacrylamide gel. Isolated duplex [P_{4-10}] is dissolved in 10 m*M* Tris-HCl, pH 7.6, containing 10 m*M* $MgCl_2$ (50 μl) to give a final concentration of 11 m*M*.

Step 4. 5′-^{32}P segment P-(1 + 3) (38 μl, 590 pmol) in its phosphoryla-

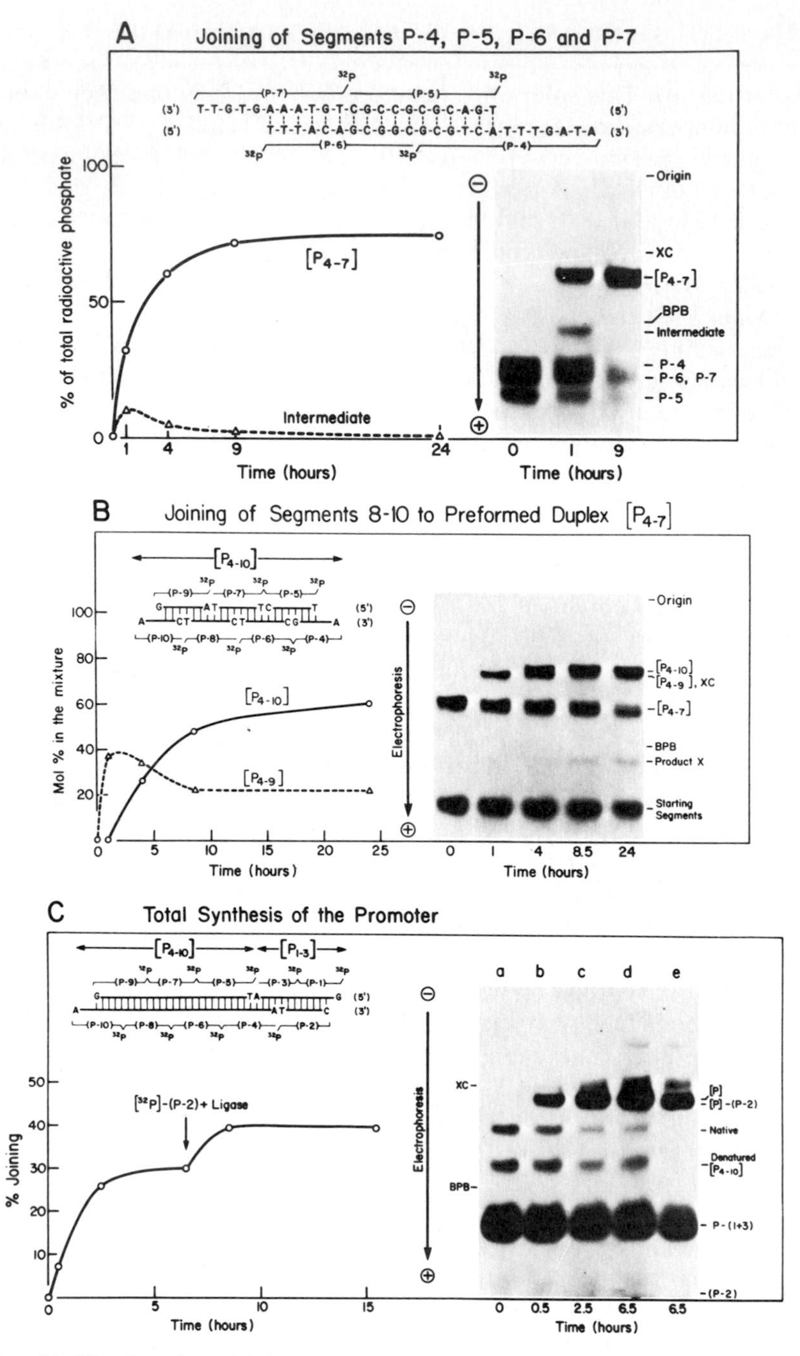

FIG. 14. Kinetics of the joining of segments P-4 to P-7 as determined by gel electrophoresis (A). Kinetic analysis of the formation of duplex $[P_{4-10}]$ (B). Kinetic analysis of the formation of duplex $[P_{1-10}]$ (C).

tion reaction mixture, which contains ATP (1390 pmol), is heated at 95° for 2 min. Then duplex $[P_{4-10}]$ (49 μl, 550 pmol), $5'$-^{32}P segment P-2 [25 μl, 1780 pmol; this phosphorylation reaction mixture contains ATP (3900 pmol)], 1 *M* Tris-HCl, pH 7.6 (2.5 μl), 1 *M* dithiothreitol (1 μl), 10 m*M* ATP (0.5 μl, final concentration 83 μM), and T4 polynucleotide ligase (8 units/μl, 5 μl) are added. The final reaction volume is 120 μl. Without heating, the reaction is incubated at 5° for 6.5 hr. Then a second batch of $5'$-^{32}P segment P-2 (25 μl, 1780 pmol) and ligase (8 units/μl, 5 μl) is added. The kinetics are followed by electrophoresis on a 15% polyacrylamide gel, and the results are shown in Fig. 14C. After 24 hr, tracking dyes in 60% glycerol (100 μl) are added, and duplex $[P_{1-10}]$ is isolated by preparative slab gel electrophoresis on a 10% polyacrylamide gel. Isolated duplex $[P_{1-10}]$ is dissolved in 10 m*M* Tris-HCl, pH 7.6, containing 10 m*M* $MgCl_2$ (50 μl). The concentration of duplex $[P_{1-10}]$ is 3.4 μ*M*.

Enzymatic Joining of Preformed Duplexes

General Principles

Short, preformed duplexes are aligned in a head-to-tail fashion by virtue of base pairing between single-stranded protruding ends. Gaps in the phosphodiester backbone are then sealed by T4 polynucleotide ligase to afford the intact total DNA. The joined product is isolated by gel filtration or gel electrophoresis. Unlike the enzymatic joining of synthetic oligonucleotides, this step is rapid and very efficient. Where appropriate, the terminal 5′-hydroxyl groups of the short duplexes are phosphorylated enzymatically with $[\gamma$-$^{32}P]$ATP immediately before use. As in the formation of the short intermediate duplexes, the ^{32}P radiolabel aids the kinetic analysis of the ligase-catalyzed joining reaction and the characterization of the total DNA.

Materials and Methods

Materials and enzymes are the same as those described in the previous section.

Phosphorylation of the Terminal 5′-Hydroxyl End Groups of the Preformed Duplexes

These polynucleotide kinase-catalyzed phosphorylations are performed as described in the previous section. The molar excess of ATP used varies with the nature of the terminal hydroxyl groups. To phosphorylate duplexes with protruding 5′-hydroxyl groups, a fivefold molar excess (relative to each 5′-hydroxyl group) is used. However, for DNAs with recessed 5′-hydroxyls, a 10-fold or greater molar excess gives the best results. Also, the specific activity of [γ-^{32}P]ATP used for phosphorylation should be at least 50-fold higher than that of the [^{32}P]phosphate groups at the internal joinings. The phosphorylated duplexes can be isolated by gel permeation chromatography on a column of BioGel A-0.5m[32] or used directly in the ligase-catalyzed reaction. In the latter case, the kinase reaction mixtures are first treated at 100° for 2 min and cooled slowly to room temperature (15–30 min).

T4 Polynucleotide Ligase-Catalyzed Joining of Preformed Duplexes

These reactions are carried out as described for the enzymatic joining of synthetic segments. Two variations need to be noted. (1) The reaction is usually complete in 2–3 hr. (2) In joinings where only phosphorylation reaction mixtures are used, a heating and slow-cooling step for annealing is not necessary.

T4 Polynucleotide Ligase-Catalyzed Joinings of Preformed Duplexes in the Total Synthesis of a Tyrosine Suppressor tRNA Gene

Seven preformed duplexes shown in Fig. 15, ([P_{1-3}], [P_{4-10}], [I]-[IV], and [Vb]), were joined to give the total DNA. Although several approaches are available for the final joinings, it was decided to prepare intermediate duplexes [P-I] and [III-IV-Vb] and then to join these to duplex [II] in a two-step reaction.[33] The advantage to this approach is that none of duplex [II], the limiting component in this stage of the total synthesis, is lost in a subsequent joining reaction as part of an intermediate.

The Joining of Duplexes [P-I], [II], and [III-IV-Vb] (Fig. 16)

Two procedures for this joining have been described by Sekiya *et al.*[33]

Procedure A. To a 1.5-ml polypropylene test tube containing water (3.5 μl) is added duplex [P-I] (1 μl, 2.3 pmol), duplex [II] (0.25 μl, 1.7 pmol), 0.5 μl 10X ligase buffer (500 m*M* Tris-HCl, pH 7.6, 100 m*M* $MgCl_2$, 100 m*M* dithiothreitol, 500 μ*M* ATP), and T4 ligase (8 units/μl,

[32] R. Kleppe, T. Sekiya, P. C. Loewen, K. Kleppe, K. L. Agarwal, H. Büchi, P. Besmer, M. H. Caruthers, P. J. Cashion, M. Fridkin, E. Jay, A. Kumar, R. C. Miller, K. Minamoto, A. Panet, U. L. RajBhandary, B. Ramamoorthy, N. Sidorova, T. Takeya, J. H. van de Sande, and H. G. Khorana, *J. Biol. Chem.* **251**, 667 (1976).

[33] T. Sekiya, T. Takeya, E. L. Brown, R. Belagaje, R. Contreras, H.-J. Fritz, M. J. Gait, R. G. Lees, M. J. Ryan, H. G. Khorana, and K. E. Norris, *J. Biol. Chem.* **254**, 5787 (1979).

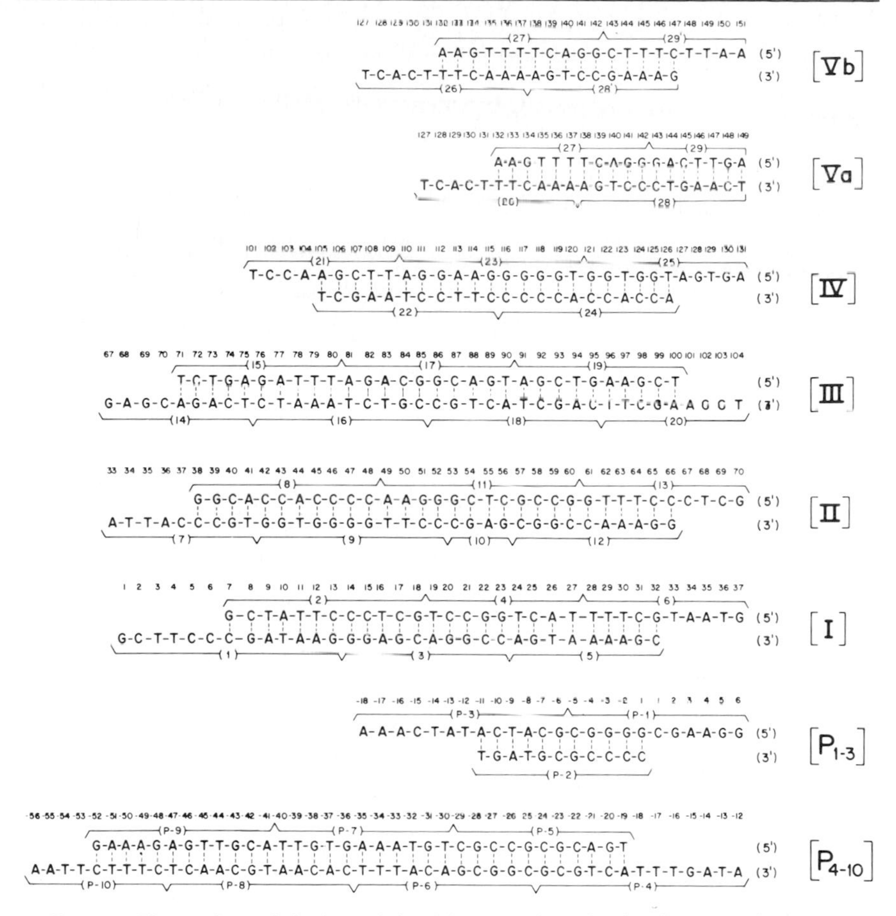

FIG. 15. The preformed duplexes joined by T4 polynucleotide ligase to give the functional synthetic suppressor tRNA gene.

0.25 μl). After incubation at 5° for 1 hr, duplex [III-IV-Vb] (0.5 μl, 2.8 pmol) is added. Incubation is continued at 4° for 19 hr. Kinetics are followed by electrophoresis on a 10% polyacrylamide gel. The results are shown in Fig. 16A. In a similar fashion, a reaction mixture (100 μl) containing duplex [P-I-II][34] (67 pmol) and duplex [III-IV-Vb] (144 pmol) is in-

[34] As described for the small-scale reaction, the first step, consisting of duplex [P-I] (115 pmol) and duplex [II] (90 pmol) in a 100-μl reaction mixture, afforded duplex [P-I-II]. However, upon addition of duplex [III-IV-Vb], no new DNA formed. After the mixture was passed through a column of Sephadex G-50 (0.8 × 30 cm), the fractions containing the duplexes were evaporated to dryness. To this residue was added the standard ligase reaction mixture (100 μl) and T4 polynucleotide ligase (8 units/μl, 5 μl). The mixture was incubated at 5° for 13 hr.

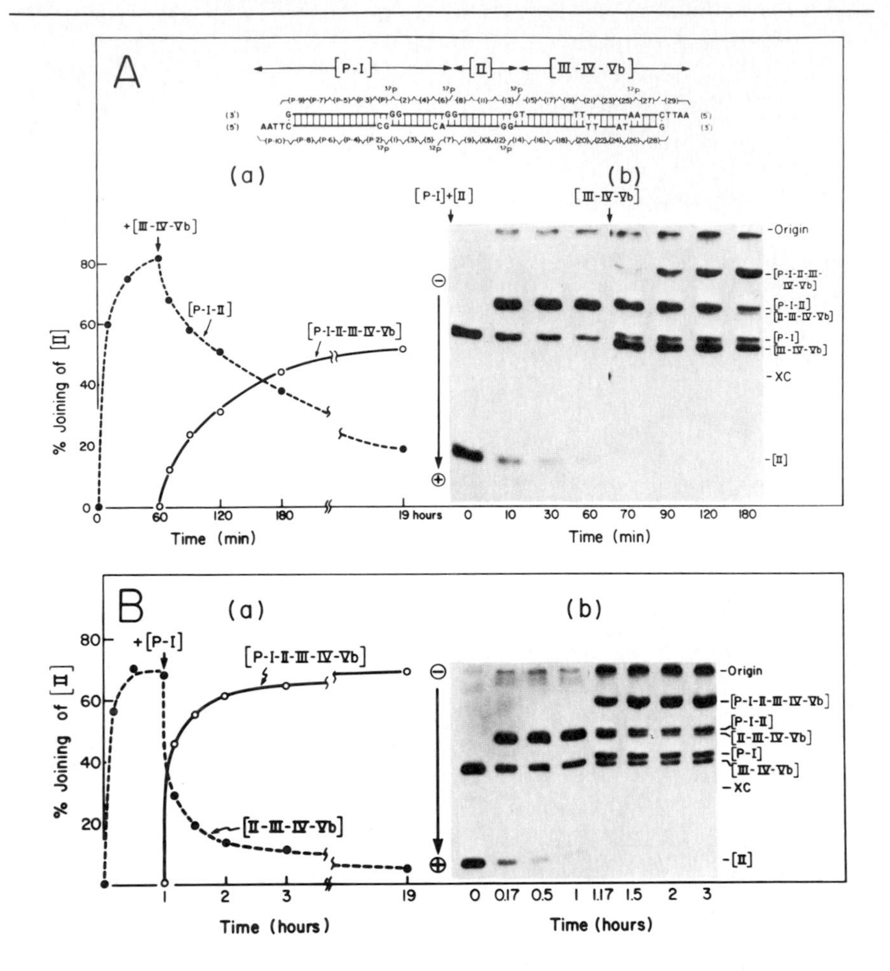

FIG. 16. Kinetics of joining duplex [P-I] to duplex [II] and then duplex [P-I-II] to duplex [III-IV-Vb] (A). Kinetics of joining duplex [III-IV-Vb] to duplex [II] and then duplex [P-I] to duplex [II-III-IV-Vb] (B).

cubated at 5° for 13 hr. After preparative slab gel electrophoresis, the synthetic gene (26 pmol) is obtained.

Procedure B. A second 5-μl reaction is carried out as described in procedure A except duplex [II] is first joined to duplex [III-IV-Vb] which is then fused to duplex [P-I]. The kinetics of both joining reactions are shown in Fig. 16B. Although the rates of both joinings are higher than those observed for procedure A, the final yields of the synthetic gene in the two procedures are essentially identical.

Biological Activity of the Synthetic Gene for a Precursor of the Tyrosine Suppressor tRNA

Principle

The correct functioning of this synthetic gene *in vivo* is established by cloning the chemically synthesized DNA and demonstrating the supression of amber mutations. These experiments were facilitated by including the 5′-AATT single-stranded termini produced by the *Eco*RI restriction endonuclease at both ends of this synthetic molecule. These ends are used to join the synthetic DNA to *Eco*RI-digested cloning vectors to yield recombinant DNA molecules which are then inserted into *Escherichia coli*. Both colE$_1$ Amp^R plasmid[35] and Charon 3A bacteriophage[36] vectors are employed. When the plasmid is used as the cloning vehicle for the synthetic suppressor gene, the host bacterium carries amber mutations in genes required for the biosynthesis of histidine and tryptophan. Therefore, colonies harboring the desired recombinant DNAs carrying the functional synthetic suppressor gene are selected as ampicillin-resistant prototrophs. The phage vector used (Charon 3A) has amber mutations in genes *A* and *B* which are required for capsid formation. Here the correct expression of the synthetic gene integrated within the phage chromosome permits suppression of the amber mutations in genes *A* and *B*, as evidenced by the formation of viable, plaque-forming phage after transfection of a nonsupressing host bacterium. The general procedures used are described below.

Methods

EcoRI Digestion of Vector DNA

Approximately 0.5–1 pmol of each vector DNA is digested for 1 hr at 37° in a 60-μl reaction containing 85 m*M* NaCl, 10 m*M* $MgCl_2$, 100 m*M* Tris-HCl, pH 7.6, and 20 units of *Eco*RI (New England Biolabs). The digestions are terminated by heating them at 65° for 10 min.

[35] M. So, R. Gill, and S. C. Falkow, *Mol. Gen. Genet.* **142,** 239 (1975).

[36] F. R. Blattner, B. G. Williams, A. E. Blechl, K. Denniston-Thompson, H. E. Faber, L.-A. Furlong, D. J. Grunwald, D. O. Kieter, D. D. Moore, J. W. Schumm, E. L. Sheldon, and O. Smithies, *Science* **196,** 161 (1977).

5′-Phosphorylation of Synthetic DNA

The 5′-hydroxyl ends of this synthetic gene (0.2 pmol) are phosphorylated by T4 polynucleotide kinase (11 units[37]). The incubation is carried out at 37° for 1 hr in 150 μl of a mixture of 50 m*M* Tris-HCl, pH 7.6, 10 m*M* dithiothreitol, 180 μ*M* ATP, and 5 m*M* $MgCl_2$. After this time, the reaction is placed directly on ice.

T4 Polynucleotide Ligase-Catalyzed Reactions

Each reaction consists of 10 μl of the above *Eco*RI-digested vector DNA solution, 50 μl of the 5′-phosphorylation reaction containing the synthetic suppressor gene (Fig. 1), and salts as required to bring the final volume to 100 μl with each component present at the same concentration as in the 5′-phosphorylation reactions. The joinings are allowed to proceed for 22 hr at 5° in the presence of 20 units[37] of T4 polynucleotide ligase.

Transfection[38] and Transformation[39]

Escherichia coli LS340 (*his* amber *trp* amber) is grown at 37° in 30 ml of LB medium (per liter: 10 g tryptone, 5 g yeast extract, 10 g NaCl) to a density of 75 Klett units (green filter). The culture is centrifuged, resuspended in 10 ml of ice-cold 10 m*M* NaCl, recentrifuged, and resuspended in 15 ml of ice-cold 30 m*M* $CaCl_2$. After a 20-min incubation on ice, this suspension is centrifuged and finally resuspended with 3 ml of ice-cold 30 m*M* $CaCl_2$. Each transformation reaction contains 0.2 ml of this cell suspension, 3 μl of 1 *M* $CaCl_2$, and the appropriate ligase reaction described above. The recombinant DNA is allowed to adsorb to the cells for 80 min at 0°. The tubes are heated at 42° for 5 min and then placed at room temperature for 10–20 min. Three milliliters of LB medium (containing 0.2 ml of 10 m*M* $MgCl_2$ and 10 m*M* $CaCl_2$ when the vector is Charon 3A) is then added, and the cultures are incubated with shaking at 37° for at least 4 hr. After this time, the cultures are either titered for plaque-forming units or plated on minimal plates containing ampicillin.

Results

As shown in Table II, both bacterial and bacteriophage amber mutations were suppressed in the presence of the cloned synthetic gene for a

[37] A. Panet, J. H. van de Sande, P. C. Loewen, H. G. Khorana, A. J. Raae, J. R. Lillehaug, and K. Kleppe, *Biochemistry* **12,** 5045 (1973).
[38] M. Mandel and A. Higa, *J. Mol. Biol.* **53,** 159 (1970).
[39] S. N. Cohen, A. C. Y. Chang, and L. Hsu, *Proc. Natl. Acad. Sci. U.S.A.* **69,** 2110 (1972).

TABLE II

In vivo SUPPRESSION OF PHAGE AND BACTERIAL AMBER MUTATIONS BY THE SYNTHETIC SUPPRESSOR tRNA GENE[a]

Vector DNA	Synthetic suppressor gene	Plaque forming units/ml
Charon 3A (0.8 μg)	−	0
Charon 3A (0.8 μg)	+	1.8×10^7
Vector DNA	**Synthetic suppressor gene**	**Amp^R prototrophs/ml**
ColEl Amp^R (0.5 μg)	−	0
ColEl Amp^R (0.5 μg)	+	47

[a] Each vector DNA is digested with the restriction endonuclease *Eco*RI and aliquots are treated with T4 polynucleotide ligase in the presence or absence of 0.01 μg of the chemically synthesized tyrosine suppressor tRNA gene. The amount of each vector DNA given in parentheses is the level present in the ligase reaction mixture used for the transformation assay (Transfection and transformation section). Phage are titered using *E. coli* CA274 (*lac*Z amber, *trp* amber) as the host strain. Ampicillin-resistant prototrophs of LS340 were detected by growth on M9 minimal plates containing 35 μg/ml of the antibiotic, 40 μg/ml L-methionine and 0.2% glucose.

precursor of the tyrosine suppressor tRNA. The recombinant DNA molecules formed in these experiments were characterized after amplification and purification. As expected, each of the molecules could be shown to have incorporated a new, low-molecular-weight *Eco*RI restriction fragment. This piece of DNA, unique to molecules shown to code for suppressor activity, was purified and shown (1) to be the same size as the chemically synthesized gene and (2) to serve as a source of suppressor activity in further transformation experiments. Furthermore, *in vitro* transcription of one of the recombinant plasmids carrying the cloned synthetic suppressor tRNA gene indicated that the synthetic promoter was recognized by the *E. coli* RNA polymerase and that the nucleotide sequence of the structural gene was accurate.

Acknowledgment

This work has been supported by Grant No. CA11981, awarded by the National Cancer Institute, DHEW; Grant No. PCM73-06757, awarded by the National Science Foundation, Washington, D.C.; Grant No. NP-140 from the American Cancer Society, Inc.; and by funds made available to the Massachusetts Institute of Technology by the Sloan Foundation.

[9] Gel Electrophoresis of Restriction Fragments

By EDWIN SOUTHERN

Analytical Procedures

There are many designs of apparatus that can be used for the electrophoresis of DNA in agarose or polyacrylamide gels.[1-3] Some of them are very easy to make from glass or Perspex, and a simple apparatus can give excellent results (Fig. 1). For both agarose and polyacrylamide gels, glass has advantages over Perspex: Perspex and other plastics have a higher coefficient of thermal expansion than glass and distort more when molten agarose is poured into them; Perspex and many other plastics inhibit the polymerization of acrylamide. Agarose gels thinner than about 3 mm are fragile and difficult to handle, but polyacrylamide gels can be used at thicknesses of less than 1 mm.

Horizontal slab gels are normally used for agarose gels below 0.5%. Some workers prefer to use horizontal gels for all applications, because they are easier to use, but in our hands vertical gels give sharper bands.

Agarose gels are simply prepared by adding the powder to cold electrophoresis buffer and bringing the solution to a boil while stirring or shaking. The solution is cooled to a temperature a few degrees above the setting point and poured into the apparatus or mold. Several buffers are employed; Tris-acetate or -phosphate at about 50 m*M* and pH 7.5–8 with 1 m*M* EDTA can be used, as can sodium phosphate, provided the buffer is circulated between the anode and cathode compartment during the run. Discontinuous buffer systems and stacking gels are not normally used. Denatured DNA can be separated in polyacrylamide gels containing formamide,[4] or in alkaline agarose gels.[3]

Before loading restriction digests of DNA the reaction is usually stopped by adding EDTA and/or heating the mixture to 65°. A dense solvent and a marker dye are usually also added. Glycerol or sucrose is often used to increase, the density, but these low-molecular-weight solutes cause streaming of the sample up the sides of the loading slot, which leads to U-shaped bands. Ficoll added to 1–2% avoids this effect. Bromphenol blue or orange G can be used as a visible marker. The latter travels faster and causes less quenching of the ethidium bromide (EtBr) fluorescence.

[1] F. W. Studier, *J. Mol. Biol.* **79,** 237 (1973).
[2] T. M. Shinnick, E. Lund, O. Smithies, and F. R. Blattner, *Nucleic Acids Res.* **2,** 1911 (1975).
[3] M. W. McDonell, M. N. Simon, and F. W. Studier, *J. Mol. Biol.* **110,** 119 (1977).
[4] T. Maniatis, A. Jeffrey, and H. van de Sande, *Biochemistry* **14,** 3787 (1975).

ISBN 0-12-181968-X

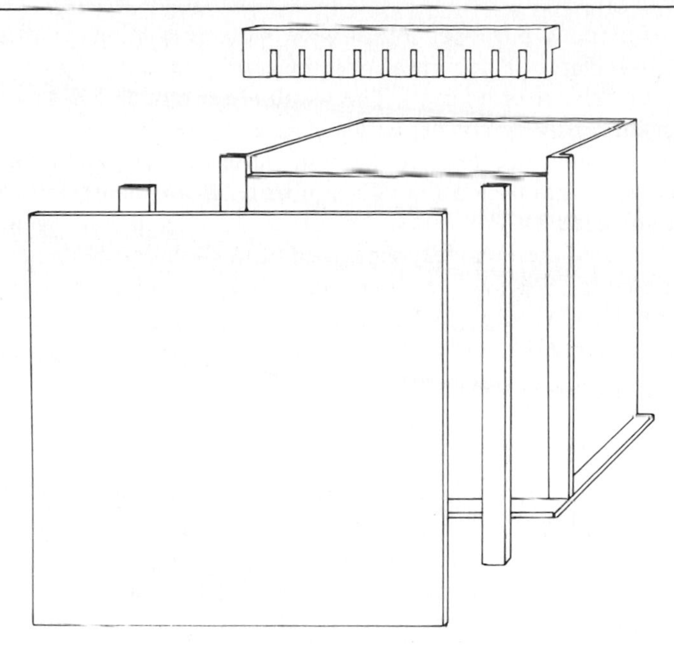

FIG. 1 Vertical gel electrophoresis apparatus. This design is typical of many based on the original apparatus of Studier.[1] The box is made of Perspex, apart from the front, which is best made of glass. The "comb" used to cast loading slots is also made of plastic, as are the spacers; many plastics inhibit polymerization of acrylamide, but polyvinyl chloride can be used with these gels. The front glass plate is held about 3 mm above the base and clamped in position. Molten agarose is poured into the gaps to a height of 1–2 cm, while the apparatus is tilted back. When this has set, the rest of the molten agarose, or acrylamide monomers, is poured in, and the comb inserted into the top to mold the slots. The box is filled with electrophoresis buffer and placed in a vessel containing the same buffer. Electrodes made of platinum wire are placed one inside the box and the other outside along the front.

The maximum load of DNA that can be applied to the gel depends on the average size of the fragments and whether the mixture is simple or complex. As a rough guide, 50 $\mu g/cm^2$ of an *Eco*RI or *Hin*dIII digest of a complex DNA is close to the maximum that can be separated with high resolution on agarose gels. Higher loading is possible with lower-molecular-weight DNA, with single-stranded DNA, and with polyacrylamide gels. Lower loading might be necessary with simple DNAs such as digests of viral DNA, or with mixtures of higher average molecular weight.

Gel Dimensions

For rapid analysis gels about 80 mm long may give adequate separation and can be run in approximately an hour at a voltage gradient of ca.

10 V/cm; for many purposes 3-mm-wide slots give bands that are wide enough. However, for accurate size measurement and high resolution 400-mm-long gels are often used. The width of the slot also has a considerable effect on resolving power; wider slots provide the eye with more information to discern fine detail in the band pattern. Thus, 20-mm slots give noticeably better resolution than 10-mm slots. Both separation and resolution are affected by the voltage gradient. Low-molecular-weight fragments diffuse and are thus best separated at fairly high-voltage gradients. Large fragments, however, diffuse very slowly, and best resolution is achieved by a low-voltage gradient.[3] It may be necessary to use a voltage gradient less than 0.5 V/cm and carry out the separation for several days to resolve large DNA fragments.[5]

Detection of DNA Fragments

EtBr may be included in the gel (added while the gel is molten) and running buffers; it has only a small effect on the mobility of DNA. However, it is not advisable to include it with high loads of DNA, because the faster-moving DNA can mop up all the EtBr, leaving the slow-moving molecules unstained. For accurate quantitative work it is advisable to stain the DNA after electrophoresis is complete. About ½ hr using a concentration of 0.5 μg/ml is standard practice.[6] However, these conditions do not saturate the DNA in a 3-mm thick gel, and longer staining may give higher sensitivity. Background fluorescence due to unbound EtBr in the gel can be reduced by destaining, but extensive washing removes the ethidium bound to the DNA.

Photography

Fluorescence of the DNA–EtBr complex is stimulated by illumination with uv light of 254, 300, or 366 nm.

The gel may be illuminated from the side, with an angle of incidence of about 5°–10°. Placing the gel on black Perspex gives a cleaner background than placing it on glass or clear Perspex which may itself fluoresce in the uv light. Alternatively, the gel may be illuminated from below (transillumination). With 254-nm lamps the gel is placed directly on the uv pass filter that covers the lamp, but with 300- and 366-nm lamps the gel can be placed on a sheet of uv transparent Perspex (Plexiglas 218, Rohm and Haas). This material is transparent to wavelengths above 300 nm but is opaque to 254 nm. Transillumination gives higher intensity than incident

[5] W. L. Fangman, *Nucleic Acids Res.* **5,** 653 (1978).

[6] P. A. Sharp, B. Sugden, and J. Sambrook, *Biochemistry* **12,** 3055 (1973).

illumination, but more light from the lamps has to be eliminated by filters on the camera.

The short wavelength gives high sensitivity and is most commonly used, but it causes damage to the DNA during the period of photographic exposure. Less damage is caused by 300-nm lamps which also give high sensitivity,[7] but visible light emitted by these lamps is more difficult to filter out than that from 254-nm lamps and requires an interference filter on the camera to remove red light. If the interference filter cuts off both above and below the emission maximum of 590 nm, it can be used without a red absorption filter. Depending on the power of the lamp, the concentration of DNA, and the sensitivity of the film, exposure times may vary from a few seconds up to half an hour. Exposures longer than a few minutes lead to reciprocity failure in film response. This has the effect of suppressing the background, giving a cleaner appearance to the picture, but faint bands are also underrepresented and may be missed under these conditions. Reciprocity failure can be overcome by prefogging the film with a brief exposure from an attenuated electronic flash, sufficient to give a fog level of about 0.05 OD_{600} on the developed film.

Lamps with an output at 366 nm give about $\frac{1}{10}$ the excitation of shorter wavelengths[7] but, as they cause little damage to the DNA, they are used to illuminate gels from which DNA is to be recovered. They can also be used to illuminate gels still contained in the electrophoresis apparatus.

Large-format film gives better results than miniature film, and most laboratories use either Polaroid, Ilford FP4, or Kodak Tri X. Because the wavelength is close to the red end of the visible spectrum, the film's sensitivity to EtBr fluorescence is not directly related to its speed rating, and we have found that fast emulsions are no more sensitive than those listed here. Sensitivity can be increased by prefogging as described above and by doubling the recommended development time. For quantitative work prefogging improves the linearity of film response to faint images, but standard development should be used.

A standard giving a variety of band sizes and intensities, such as a restriction digest of a simple DNA, should be included in gels used for quantitative analysis, and the intensity of the bands in the standard should span the full range of those to be measured. The film density in the bands is measured in a microdensitometer such as the Joyce–Loebl. A calibration curve is plotted from peak areas of the bands in the standard, against their size. Ideally this should be a straight line. If it is not, corrections must be applied to measurements of unknowns. Of course, measurements taken from gels are rarely used to give absolute amounts but are used to measure relative amounts of fragments in a mixture.

[7] C. F. Brunk and L. Simpson, *Anal. Biochem.* **82,** 455 (1977).

Film Detection of Radioactive DNA

Radioautography of ^{3}P and ^{125}I. The wet gel may be placed against the x-ray film, with a thin plastic film such as Saran Wrap in between. For long exposures the DNA should be fixed by immersing the gel in 5% acetic acid. Drying the gel gives sharper resolution. Dilute agarose gels are easily dried to a thin film supported by a paper backing. The gel is laid on a piece of Saran Wrap and covered with a piece of hard filter paper. Absorbent paper towels weighted down with a heavy plate are laid on top of this. Most of the liquid is squeezed out in about an hour, leaving the DNA trapped in a thin film of gel which adheres to the paper. This can be radioautographed without further drying. Both agarose and polyacrylamide gels can be completely desiccated using an apparatus (Fig. 2) that applies pressure, heat, and suction to the gel. Heat should not be applied to agarose gels before most of the liquid has been removed by suction. Suction is then continued while the apparatus is heated over a boiling water bath or hot plate. Polyacrylamide gels become brittle when dry, unless glycerol permeates the gel before drying.

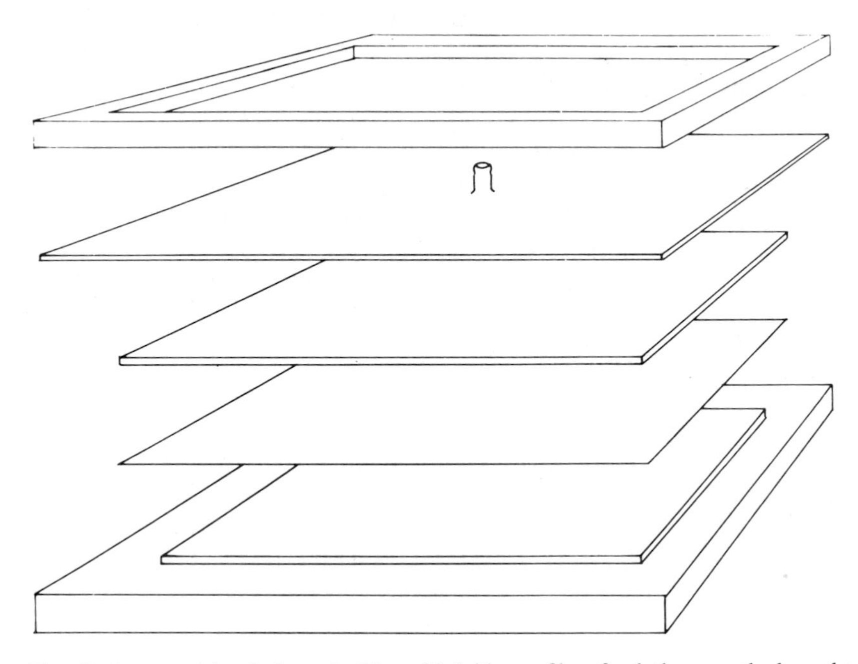

FIG. 2. Apparatus for drying gels. The gel is laid on a film of polythene on the baseplate and overlaid with a sheet of wet filter paper and a sheet of porous polythene. A sheet of rubber with a nozzle is clamped over the sandwich, and suction applied. When most of the liquid has been removed, the assembly is placed on a steam bath or hot plate until the gel has dried down to a thin film attached to the filter paper.

Indirect radioautography enhances the sensitivity to ^{32}P about 10-fold and to ^{125}I about 15 to 50-fold.[8] The x-ray film is fogged to an OD_{600} of 0.1–0.2 and laid over the dry gel with the fogged side uppermost. A calcium tungstate image intensification screen is laid over the film, and exposure is carried out at −70°. Film development is normal. After exposure, film should be unwrapped while still cold; if it is allowed to warm up, pressures that cause physical fogging may be generated. Tritium can be detected by allowing 2,5-diphenyloxazole PPO to permeate the gel after replacing the water in the gel by an organic solvent. The gel is soaked in methanol for 1 hr and then in a solution of PPO (10% w/w). The gel is then dried, without heating, and exposed at −70° to prefogged x-ray film.[9]

Detection of Specific Sequences

It is often useful to be able to identify a particular sequence in DNA fragments separated by gel electrophoresis. There are currently a number of methods available to achieve this purpose. Restriction fragments that bind to a protein may be separated as a complex by filtration through a membrane filter, and the fragment identified by electrophoresis after elution from the membrane.[10] Radioactive restriction fragments that hybridize to a particular RNA molecule can be measured by gel electrophoresis after nonhybridized parts of the fragments have been digested with the single-strand-specific nuclease S1.[11] Alternatively, nonradioactive restriction fragments, after they have been separated by agarose gel electrophoresis, can be probed with radioactive RNA or DNA to mark the position of specific sequences. Two methods have been developed for hybridizing a radioactive probe to restriction fragments after they have been separated by electrophoresis in agarose gels. Both methods start by denaturing the DNA in the gel by soaking it in alkali. (This step can be omitted if the DNA fragments were denatured before electrophoresis.) Alkali is then neutralized by soaking the gel in a neutral or slightly acidic buffer. At this point, the gel may be dried down to a paper-thin membrane, trapping most of the DNA within it, and the membrane used as a matrix for subsequent hybridization.[2] Alternatively, DNA may be transferred from the gel to a sheet of cellulose nitrate, retaining the original pattern.[12] The sheet of cellulose nitrate is laid against the gel, and solvent blotted through it by stacking absorbent paper on top. DNA is carried out of the gel by the flow of solvent and trapped in the cellulose nitrate paper, which is subse-

[8] R. A. Laskey and A. D. Mills, *Eur. J. Biochem.* **56,** 335 (1975).
[9] R. A. Laskey and A. D. Mills, *FEBS Lett.* **82,** 314 (1977).
[10] V. Pirrotta, *Nucleic Acids Res.* **3,** 1747 (1976).
[11] A. J. Berk and P. A. Sharp, *Cell* **12,** 721 (1977).
[12] E. M. Southern, *J. Mol. Biol.* **98,** 503 (1975).

quently used for hybridization using well-established methods. After hybridization, bands of radioactivity are best detected by radioautography or autofluorography.

Standard Procedure for Transfer of DNA from Agarose Gels

Step 1. Prepare the transfer apparatus (Fig. 3). This consists of a tray filled to the brim with 20× SSC, a glass plate supported on two sides of the tray, and a thick pad of filter paper soaked in 20× SSC draped over the glass plate with two ends dipping into the solution in the tray.

Step 2. Place the gel on a glass or plastic sheet and immerse it a solution of 1.5 *M* NaCl and 0.5 *N* NaOH for 15–30 min, occasionally rocking the tray.

Step 3. Carefully decant the alkaline solution, or draw it off at a water pump, and rinse the gel with water.

Step 4. Soak the gel in a neutralizing solution (e.g., 3 *M* sodium acetate, pH 5.5, or 2 *M* NaCl in 1 *M* Tris-HCl adjusted to pH 5.5) for 30 min.

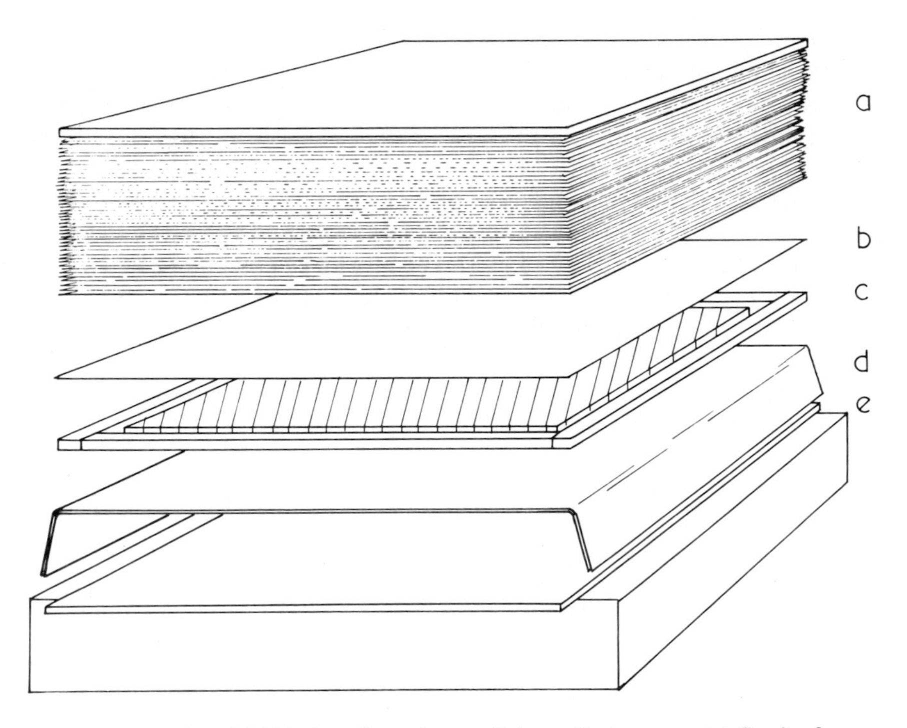

FIG. 3. Transfer of DNA from flat gels to cellulose nitrate paper. (a) Stack of paper towels weighted with a glass plate. (b) Cellulose nitrate paper. (c) Gel surrounded by plastic strips which support the edges of the cellulose paper and the towels. (d) Wad of thick filter paper that dips into the tray of 20× SSC. (e) Tray of 20× SSC with a glass plate to support the wad of wet filter paper.

Step 5. Slide the gel carefully from the plate on to the pad of filter paper, taking care to avoid trapping air beneath it.

Step 6. Cover the paper around the gel with a layer of waterproof film (e.g., thin polythene or Saran Wrap). This helps to prevent the absorbent paper, which may sag down, from becoming saturated.

Step 7. Set pieces of plastic sheet around the gel with a gap of 2 mm away from the gel. The pieces should stand about the same height as the gel.

Step 8. Squeegee excess liquid from the surface of the gel.

Step 9. Take a sheet of cellulose nitrate paper (e.g., Millipore HAWP Schleicher and Schuell BA85 or Sartorius) and wet it by floating it on water. The sheet should be at least 1 cm longer and wider than the gel. Handle the sheet with forceps or wear gloves.

Step 10. Lay the cellulose nitrate sheet on the gel, taking care not to trap air beneath it. The edges of the sheet should be supported by the pieces of plastic around the gel.

Step 11. Soak a piece of filter paper in 20× SSC and lay it on top of the cellulose nitrate, taking care to avoid trapping air beneath it.

Step 12. Stack absorbent paper (e.g., paper towels) on top of the filter paper and weigh it down lightly with a glass plate.

The rate of transfer out of the gel depends on DNA size, and gel concentration and thickness. The usual practice is to leave the transfer overnight. Large fragments (above 10 kb) may not be completely transferred from gels in this period; they can be transferred more effectively if they are broken down by irradiation with shortwave uv before denaturation. The time taken to photograph a gel in 254-nm light is enough to introduce an appreciable number of breaks.[7] Small fragments do not absorb to cellulose nitrate in low salt. However, sonicated DNA (M_r ca. 10^5) absorbs completely in 20× SSC.[12]

Hybridization

After the transfer is complete, the position and orientation of the gel are marked on the cellulose nitrate sheet with a ballpoint pen, and the sheet is thoroughly rinsed with 2× SSC. It is then baked at 80° in a vacuum oven for 2 hr. The sheet can be stored for many months after baking. Before hybridization the sheet is treated with Denhardt's solution.[13] (Some workers use 2× or even 10× usual strength, and it is an advantage to include also some denatured nonhomologous DNA.) Denhardt's solution is prepared by dissolving in 3× SSC, Ficoll (MW 400,000), polyvinylpyrollidone (MW 360,000), and bovine serum albumin, each to a concentration of 0.02% (w/v). The cellulose nitrate sheet is steeped in this solu-

[13] D. T. Denhardt, *Biochem. Biophys. Res. Commun.* **23,** 641 (1966).

tion at 65° for a few hours. The hybridization probe dissolved in the same solvent including 0.2% sodium dodecyl sulfate (SDS) is then introduced. Various methods have been devised for carrying out the hybridization of large sheets in the smallest possible volumes:

1. The sheet is wrapped around a rod which is a close fit inside a cylinder. Enough liquid is introduced to fill the space around the filter.
2. The sheet is wrapped around the inside of a cylinder or test tube which is sealed and placed on a roller or a platform that can be rotated in a horizontal plane. Enough liquid is introduced to form a continuous film over the filter as the cylinder is rotated.
3. The filter is sealed in a polythene bag, or plastic box, with enough liquid to cover it liberally. The bag or box is submerged in a water-filled shaking bath and shaken during hybridization.
4. The filter is wetted with hybridization mixture and then submerged in oil or fixed to a glass plate which is then sealed with Saran Wrap.

The aim of all these methods is to give uniform covering of the sheet with a small volume. To avoid high backgrounds it is important that the sheet be covered with a generous amount of liquid and that no areas dry out. Background is less of a problem if formamide is included in the hybridization mixture, but this solvent has the disadvantage that it slows down the rate of hybridization and also causes losses of DNA from the cellulose nitrate. Air coming out of solution can cause points of high background, which are avoided by deaerating the hybridization mixture. Sweat can cause absorption of the probes, so the sheet should not be handled with bare fingers.

The time required for hybridization depends on the concentration and sequence complexity of the probe, as well as on the temperature, solvent, and salt concentration. Usually it is not necessary to saturate all the hybridizing sequence on the filter with radioactive probe to give a readily detectable signal. Most commonly, hybridization is carried out at about 65°–70° in 2–6× SSC for a period of 1–48 hr.

Many methods of radioactive labeling can be used to make radioactive hybridization probes. The following have been used to hybridize to transfers: ^{3}H- or ^{32}P-labeled cDNA, nick-translated DNA, RNA terminally labeled with ^{32}P, ^{125}I-labeled RNA, tRNA charged with radioactive amino acids, and RNAs labeled *in vivo* with ^{3}H or ^{32}P.

After hybridization, the sheet is washed in a large volume of dilute salt solution at a high temperature for an hour or more; 0.1–2× SSC at 60°–70° are the conditions most commonly used. The high-stringency wash reduces the background of probe sticking nonspecifically to the filter and to nonhomologous sequences in the DNA. However, it should not be used in the detection of mismatched duplexes or small fragments.

RNase treatment (20 μg/ml in 2× SSC at room temperature for 20 min) can be used to reduce background when the probe is RNA; usually this treatment is not necessary.

After washing the filter it is dried and exposed to x-ray film as described for dried gels. For the detection of ^3H the sheet is first impregnated with PPO by dipping it into a 20% solution of PPO in toluene and allowing the solvent to evaporate while the sheet is held horizontal.

There are several problems associated with using the technique quantitatively. The method is rarely used to give an absolute measure of the amount of hybridization but to compare the extent of hybridization with different bands; even so there are problems. First, large fragments may not transfer completely from the gel to the cellulose nitrate. Second, tandemly repeated sequences such as satellite DNAs may reassociate in the gel to form networks that do not transfer. Third, the efficiency of hybridization of fragments falls off sharply below an M_r of 5×10^5. Fourth, the probe may "overhang" the ends of the fragment being detected and give an overrepresentation of its proportion (when the probe is RNA, such ends can be trimmed off with RNase, but enzymes specific for single-stranded DNA remove all DNA from the filter). Fifth, film detection methods other than direct radioautography do not give linear film response with very low amounts of radioactivity even with presensitized film. Many of these problems can be overcome by the use of appropriate internal standards. But in their absence, it should be apparent that the method can give only a rough measure of the amount of a fragment.

Estimation of Size from Mobility

Graphical Methods. There are two graphical methods of relating DNA size to mobility in gels. The log of the size may be plotted against mobility, or the size may be plotted against the reciprocal of the mobility. Both methods give a straight-line relationship in the small size range but a curve in the large size range. The curvature, in both cases, is more marked when the separation is carried out with high-voltage gradients. For accurate measurements in the high-molecular-weight range, separation should be carried out with low-voltage gradients.

If a plot of size L against the reciprocal of mobility $1/m$ gives a curve, this may be corrected to a straight line by plotting L against $1/(m - m_0)$, where m_0 is a factor calculated as described in the next section.

Calculation of Size from Mobility

Graphical methods of relating data introduce unnecessary errors. When the relationship can be represented by a simple equation, calculation is more accurate and often more convenient. The relationship

$$L = (k_1/m) + k_2 \tag{1}$$

is accurate over a fairly wide range provided the separation was carried out with a low-voltage gradient. When a plot of L versus $1/m$ is curved in the range of the measurements, it can be made to fit to a straight line of the form

$$L = k_1/(m - m_0) + k_2 \tag{2}$$

where m_0 is a correction factor determined by imposing the condition that the three lines joining three points all have the same slope. Three points are chosen, corresponding to size standards L_1, L_2, and L_3 with mobilities m_1, m_2, and m_3. The three values of L should span the range in which measurements are to be made. The value of m_0 that determines that these three points are joined by a straight line is given by

$$m_0 = \frac{m_3(L_2 - L_3)(m_2 - m_1) - m_1(L_1 - L_2)(m_3 - m_2)}{(L_2 - L_3)(m_2 - m_1) - (L_1 - L_2)(m_3 - m_2)} \tag{3}$$

Rearrangement gives a form that is easier to compute with a simple calculator:

$$m_0 = \frac{m_3 - m_1((L_1 - L_2)/(L_2 - L_3) \times (m_3 - m_2)/(m_2 - m_1))}{1 - ((L_1 - L_2)/(L_2 - L_3) \times (m_3 - m_2)/(m_2 - m_1))} \tag{4}$$

The terms in parentheses are the same and need only be calculated once.

To calculate the equation of the line [Eq. (2)], k_1 and k_2 are calculated from

$$k_1 = \frac{L_1 - L_2}{1/(m_1 - m_0) - 1/(m_2 - m_0)}$$

$$k_2 = L_1 - k_1/(m_1 - m_0)$$

Equation (2) can now be used to calculate the sizes of DNA fragments in the range between L_1 and L_3.

Size Standards

The sequences of some small DNAs have been completely determined, and the positions of a large number of restriction endonuclease sites are therefore precisely known. Fragments produced from these DNAs give the most accurate size standards in the range 0–5000 nucleotide pairs. Probably the most convenient of these DNAs to prepare in bulk is that of the plasmid pBR322[14] (Table II). The sizes of a number of restric-

[14] J. G. Sutcliffe, *Nucleic Acids Res.* **5,** 2721 (1978).

TABLE I
SIZES OF RESTRICTION FRAGMENTS OF PHAGE λ DNA[a]

Fragment	*Eco*RI[b]	*Hin*dIII[b]	*Eco*RI plus *Hin*dIII[b]	*Bgl*II[c]	*Ava*I[c]
A	21.8	23.7	21.8	22.8	15.9
B	7.52	9.46	5.24	13.6	8.8
C	5.93	6.75 (6.61)	5.05	9.8	6.1
D	5.54	4.26	4.21	2.3	4.6 (two fragments)
E	4.80	2.26	3.41	0.46	4.1
F	3.41	1.98	1.98		1.8
G		0.58	1.90		1.61
H			1.71 (1.57)		1.55
I			1.32		
J			0.93		
K			0.84		
L			0.58		

[a] All values are in kilobases and the full size of λ DNA is taken as 49 kb.
[b] Data from Phillipsen and Davies.[15] Values are for wild-type λ with values for λ cI857 in parentheses.
[c] Data from unpublished measurements of B. Smith on λ cI857.

tion fragments of phage λ DNA (Table I) have been measured in the electron microscope to an accuracy of about 1%. These together with the full-sized phage DNA cover the range 0.5–49 kb.[15] (Note that the strain of λ often used for bulk production, λ cI857, gives some restriction fragments differing in size from those of the wild type.) A continuous range of size standards is produced from a polymer series. Such a series can be produced by partial digestion with a restriction endonuclease of a tandemly repeated sequence such as a satellite DNA,[16] or alternatively by polymerization with ligase of a molecule that has sticky ends. Polymers of a small plasmid such as λ dV (3.1 kb) give markers in the moderate to high size range. Polymers of phage λ DNA would extend the range of accurate markers beyond 100 kb.

Measurement of Mobility

It is difficult to measure the position of bands stained with EtBr directly on the gel, and mobilities are usually taken from photographs. For the greatest accuracy, the photograph should be traced in a microdensitometer at sufficient scale expansion that the distance of the peaks from the origin can be measured accurately. Alternatively, measurements can be taken from a photographic enlargement. The accuracy of these measurements should be better than 1% if the full accuracy of the method is required.

[15] P. Phillipsen Kramer and R. Davies, *J. Mol. Biol.* **123,** 371 (1978).
[16] E. M. Southern, *J. Mol. Biol.* **94,** 51 (1975).

TABLE II
THE SIZES OF THE RESTRICTION FRAGMENTS OF pBR322[a,b]

*Hae*III	*Hpa*II	*Alu*I	*Hin*fI	*Taq*I	*Tha*I	*Hha*I	*Hae*II	*Mbo*I
587	622	910	1631	1444	581	393	1876	1374
540	527	659	517	1307	493	347	622	665
504	404	655	506	475	452	337	439	358
458	309	521	396	368	372	332	430	341
434	242	403	344	315	355	270	370	317
267	238	281	298	312	341	259	227	272
234	217	257	221	141	332	206	181	258
213	201	226	220		330	190	83	207
192	190	136	154		145	174	60	105
184	180	100	75		129	153	53	91
124	160	63			129	152	21	78
123	160	57			122	151		75
104	147	49			115	141		46
89	147	19			104	132		36
80	122	15			97	131		31
64	110	11			68	109		27
57	90				66	104		18
51	76				61	100		17
21	67				27	93		15
18	34				26	83		12
11	34				10	75		11
7	26				5	67		8
	26				2	62		
	15					60		
	9					53		
	9					40		
						36		
						33		
						30		
						28		
						21		

[a] Data from Sutcliffe.[14]

[b] These sizes (in base pairs) do not include any extension which may be left by the particular enzyme.

Mobilities can, of course, be measured directly from radioautographs of dried gels or hybridized transfers. But nonradioactive standards are often used, and their mobilities must be taken from the photograph of the ethidium stain. In this case great care must be taken, in determining the photographic reduction of the ethidium-stained gel, to convert the mobilities of the standards to the same scale as the radioautograph.

Preparative Gel Electrophoresis

Recovery of DNA from Gel Slices

Method 1. Gel slices are shaken overnight in a buffer containing SDS. The liquid is filtered through a membrane filter or extracted with phenol, and the DNA is recovered by ethanol precipitation. This method is most effective with polyacrylamide gels and small DNA fragments. Agarose gels are usually crushed before treatment.

Method 2. Agarose is dissolved in about 10 volumes of 5 *M* sodium perchlorate[16] at 60° or in saturated potassium iodide solution at room temperature. Enough hydroxyapatite (HAP), as a solid or a thick slurry, is added to absorb the DNA and, after brief shaking, the HAP is centrifuged down. The HAP is resuspended in perchlorate or iodide solution to wash away traces of agarose and then washed in 0.12 *M* potassium phosphate buffer. A small (1- to 2-ml) column of Sephadex G50, swollen in a buffer that is appropriate for storing the DNA, is prepared in a disposable syringe or pipette. The HAP is layered carefully on top of the Sephadex. DNA is eluted from the HAP with 1 *M* potassium phosphate, from which it is separated as it passes through the Sephadex.[17] Radioactive DNA is readily detected in the fractions. Nonradioactive DNA can be detected by mixing a small drop with EtBr solution and examining under uv light.

Method 3. Agarose is dissolved in saturated potassium iodide. The solution should not contain more than 0.1–0.2% agarose and should be adjusted to a density of about 1.5 g/ml (n_{20}^{D} = 1.421). The solution is centrifuged in a SW rotor at about 200,000 *g* for 20–40 hr at 20°. DNA bands at about 1.465 g/ml, and agarose above 1.55 g/ml.[18] If EtBr (20 μg/ml) is included in the gradient, the band of DNA can be seen in longwave uv light. The band is drawn off (extracted three times with *n*-butanol if EtBr is present), and potassium iodide removed by dialysis.

Method 4. Slices of gel are placed in a dialysis bag containing a dilute buffer such as 100 m*M* sodium acetate and 1.0 m*M* EDTA, which is laid in a shallow layer of the buffer between two electrodes. Direct current is passed between the electrodes for about ½ hr, and the polarity is reversed for about 10 sec to detach the DNA from the surface of the membrane. The buffer is then drawn from the dialysis bag.[3]

Alternatively, elution may be carried out in a glass or plastic tube which has one end sealed with a knotted dialysis tube. The tube is held in a standard electrophoresis apparatus. After overnight elution, and a

[17] H. F. Tabak and R. A. Flavell, *Nucleic Acids Res.* **7,** 2321 (1978).

[18] N. Blin, A. V. Gabain, and H. Bujard, *FEBS Lett.* **53,** 84 (1975).

10-sec polarity reversal, the buffer containing DNA is dripped from the end of the dialysis tube.

Method 5. DNA can be eluted from horizontal slab gels without cutting the gel into pieces. A trough is cut in front of the band of DNA and filled with a suspension of HAP. The current is switched on so that the DNA runs into the HAP, where it is absorbed.[17] The HAP is then eluted as described in method 2.

Method 6. Pieces of gel are placed between two layers of Parafilm and frozen. The frozen sandwich is squeezed between finger and thumb while the gel thaws. The drop of liquid exuded by this treatment contains a high proportion of the DNA.[19]

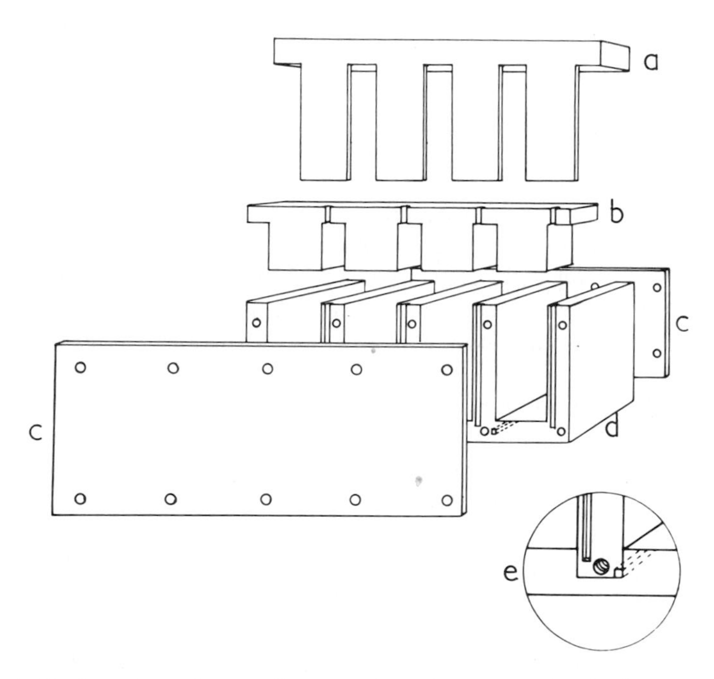

FIG. 4. Molding pieces for a small preparative gel apparatus. (a) Slot formers, which are placed behind the cover. (b) Cover, which can be placed at varying heights for gels of different sizes; this piece is left in position after the gel has set and during the run. (c) Front and back which are removed after the gel has set. (d) Main body; the grooves down the front of the uprights form seatings for the tubing to the pump used to collect fractions. (e) Detail showing the grooves down the front and the channel through the base. The channel maintains the height of the buffer in the sample collection slot at the same level as that in the rear buffer compartment. These channels are sealed off during fraction collection by a gate at the back of the apparatus (see Fig. 6).

[19] R. W. J. Thuring, J. P. M. Sanders, and P. Borst, *Anal. Biochem.* **66**, 213 (1975).

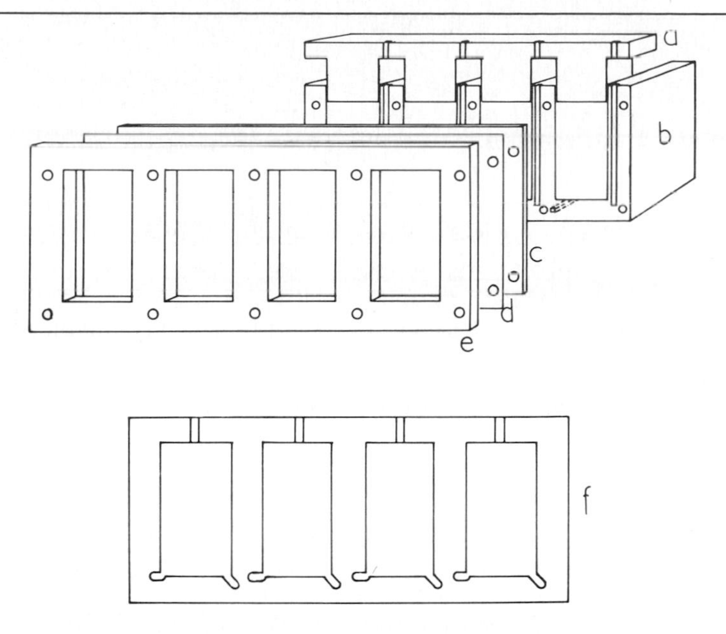

FIG. 5. Front assembly of a small preparative gel apparatus. (a) Cover, shown here in the highest position. (b) Main body. (c) Spacer, which holds the dialysis membrane 1.5–2 mm away from the gel and forms the chambers from which fractions are collected; it is grooved (see detail) with three channels in each sector—two at the bottom to connect the buffer in the chamber to the collecting tube and to the channel running through to the back of the apparatus and one at the top to let air in when the sample is pumped out. (d) Dialysis membrane. (e) Clamping frame which holds the front assembly to the apparatus; nylon screws are used to avoid corrosion. (f) Detail showing grooves in the face of the spacer; the "windows" in the spacer should be about 1 mm narrower than the width of the gel to prevent the gel from sliding forward against the membrane.

Special Preparative Gel Apparatus

Elution of DNA from pieces cut out of gels is difficult and tedious and, when a large number have to be collected or when a large amount of DNA is to be fractionated, it is better to use a special preparative apparatus. Two designs are shown here. One is for fractionation of up to 50 mg of DNA, and the other for fractionation of smaller amounts. The second apparatus can be used to fractionate several samples at the same time. Both operate on the same principle, the differences being only in the size and geometry.

Small Apparatus. The smaller apparatus (Figs. 4–6) has several channels—four in the illustration. Each channel holds a block of gel, cast *in situ,* with a loading slot molded into it. A dialysis membrane is held 1.5 mm away from the end of the gel, leaving a space from which the

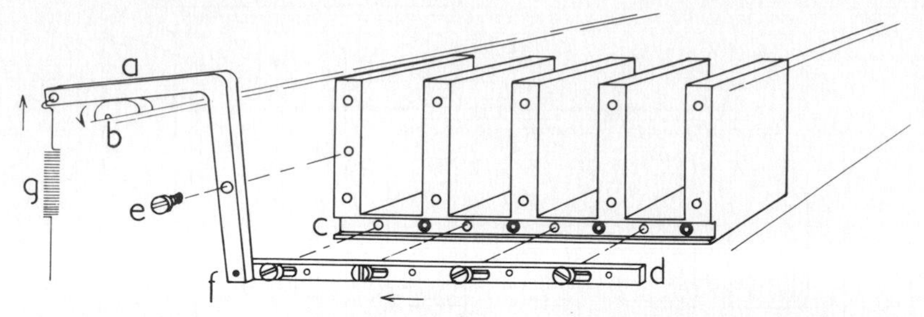

FIG. 6. Rear assembly for small preparative gel apparatus. (a) Lever which moves the gate across at each fraction collection. (c) Channel to seat the gate (d); the light circles represent threaded holes to take the nylon screws that clamp the gate to the rear of the bases. The dark circles represent rubber O rings placed in the ends of the channels that run through the base; these O rings protrude slightly to seal against the front face of the gate. (d) Gate, which is held, with light pressure, against the O rings by the nylon screws; the screws pass through slots in the gate to allow movement across the O rings; at each fraction collection, the motor-drive cam (b) raises the lever (a), moving the gate to the left; the channels are then sealed off while the fraction is pumped out of the collection chambers; when these chambers are completely emptied, the cam returns to its rest position; the spring (g) returns the gate to the right, placing the holes in register with the channels and allowing buffer from the cathode compartment to flow into the collection chambers. (e) Fulcrum of the lever. (f) Hinge holding the lever to the gate. (g) Return spring.

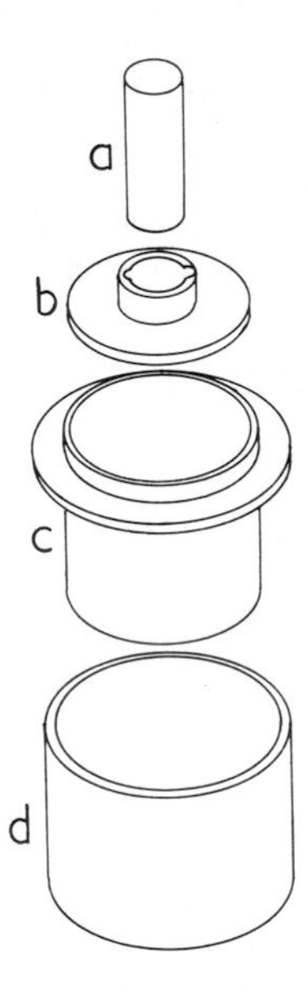

FIG. 7. Molding pieces for annular gel. The pieces are constructed of Perspex or other plastic. (a) Mold for central hole. (b) Cover which is left in position after gel has set. (c) Mold for loading slot. (d) Outer molding piece.

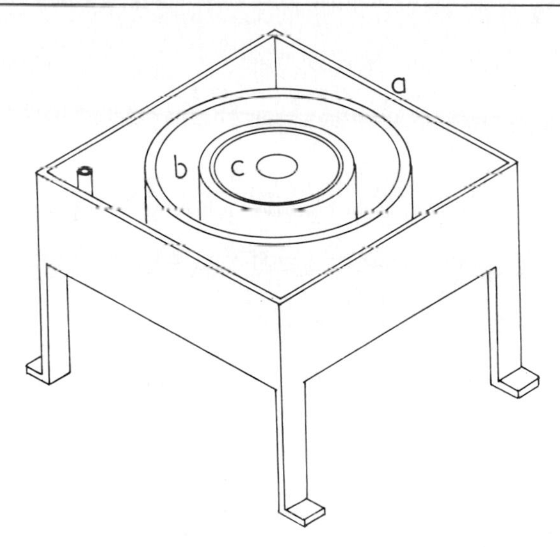

FIG. 8. Gel in position after casting. (a) Main vessel. (b) Outer electrode. (c) Gel, shown here without cover, which should be left in position.

samples are collected. This space is connected by a channel running underneath the gel to the buffer in the cathode compartment; thus the buffer in the space is maintained at a constant level. When the sample is pumped out of the space, the connection is closed off by a gate which is pushed over the ends of the grooves at the cathode side (Fig. 6). The gate is operated by a motor-driven cam and lever. The space between the gel and the membrane is also connected to a pump which removes the samples to a fraction collector.

Large Apparatus. The large-scale gel (Figs. 7–11) is cast as an annulus, with the loading slot running around the outside and a hole in the center to take the anode. The cathode is a helix of platinum wire wound on the inside of a wide tube that surrounds the gel; the anode is a helix wound around a "bobbin" that holds a tube of dialysis membrane (Fig. 9). The bobbin electrode is clamped in the base of the apparatus at its center and with a gap of about 2 mm between the membrane and the gel (Fig. 10). This space is connected to the cathode buffer by a tube that can be closed by a magnetic valve; it is also connected to the pump used to collect samples. Another pump is used to circulate the buffer continuously during the run. Buffer is pumped from a reservoir through the base of the bobbin electrode, up through the space between the membrane and the anode out through the top of the bobbin electrode, and then through a tube to circulate around the space between the outside of the gel and the cathode (Fig. 11). An overflow tube to the reservoir keeps the level of buffer constant at the same height as the gel.

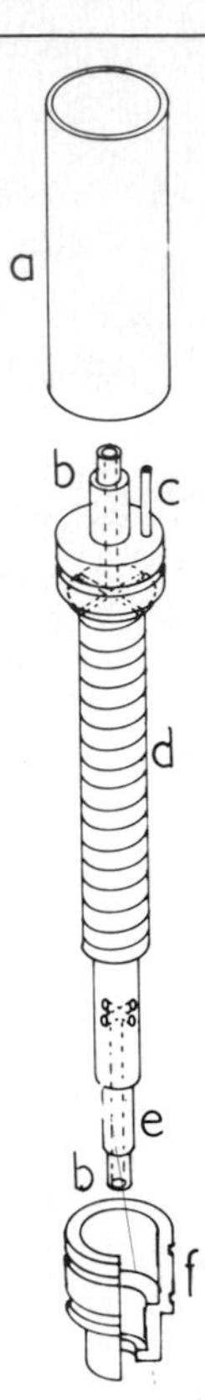

FIG. 9. Central electrode assembly. (a) Collar which crimps dialysis membrane to upper flange. (b) Nozzles for hoses carrying buffer flow. (c) Terminal. (d) Central core, made of a tough plastic such as nylon, with platinum winding. (e) Lower end, threaded to take knurled nut. (f) Lower flanges, grooved to take O rings that hold bottom end of membrane.

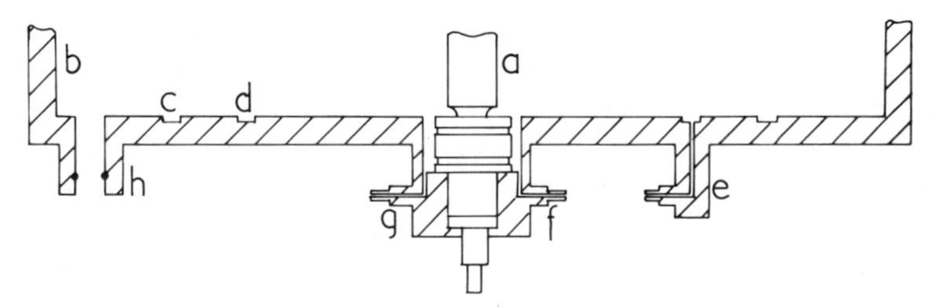

FIG. 10. Detail of base of main vessel. (a) Central electrode. (b) Side wall. (c) Groove to seat outer electrode. (d) Groove to seat outer molding piece. (e and f) Nipples to take tubing that connects outer buffer to collecting slot; the tubing is closed off by a magnetic valve during sample collection. (g) Nipple for tube connection to fraction collector. (h) Seating for buffer overflow tube.

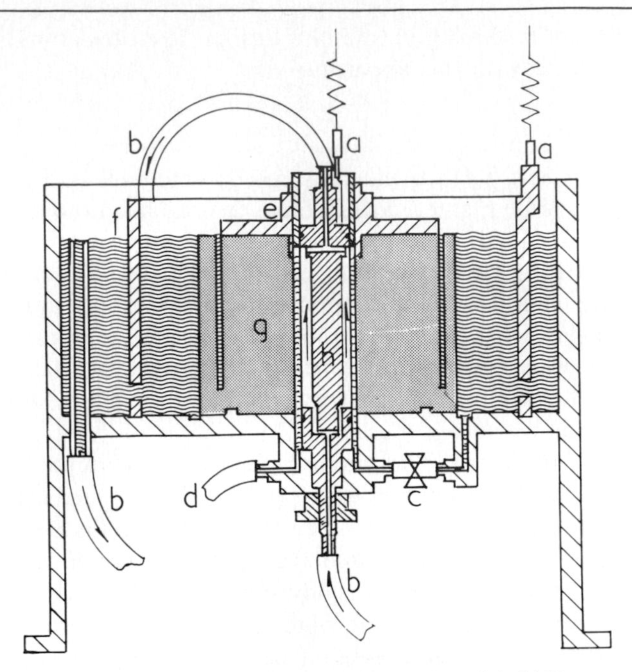

FIG. 11. Cross section through complete assembly. (a) Termini. (b) Hoses carrying circulating buffer. (c) Magnetic valve. (d) Outlet to pump to fraction collector. (e) Cover. (f) Outer electrode. (g) Gel. (h) Central electrode.

The operating cycle is controlled by the fraction collector and slave timer. The interval between fractions is set on the fraction collector. At each changeover, a pulse from the fraction collector (taken from, say, the event marker) activates the slave timer which makes the following switches over a period of 2 min:

1. The polarity of the dc supply is reversed for $\frac{1}{2}$ min to release the DNA from the membrane into the collection slot.
2. The dc supply is switched off, the magnetic pinch cock closes the tube between the outside and the inside compartments, or the motor-driven cam moves the gate across the channel, and the pump to the fraction collector is started. One minute is allowed to completely clear the sample from the collection compartment(s).
3. The pump is switched off, and the magnetic pinch cock or gate opened ($\frac{1}{2}$ min) so that the collection compartment(s) refills with buffer.
4. The cycle is restarted by switching on the dc supply with normal polarity.

The slave timer may be electronic or electromechanical. Suitable timers

are commercially available, and the design described by Brownstone[20] can also be used with this apparatus.

Casting the Gel

Small Apparatus. (1) A plate is screwed onto each end of the gel container, and the cover lowered to give the desired height to the gel (Fig. 4). (2) Agarose dissolved in electrophoresis buffer (100 m*M* Tris-acetate, pH 7.8, 10 m*M* EDTA) and cooled to just above the setting point is poured into as many channels as are needed for the separation, and any bubbles trapped beneath the cover are removed by tilting the apparatus. (3) The slot former is lowered into the gel behind the cover, and the gel left to set. (4) The slot former and the end plates are removed; any loose bits of agarose are cleared away.

Large Apparatus. (1) The molding pieces that form the outside of the gel and the central hole are placed in position (Fig. 7), and the joins sealed with a little molten agarose. (2) When this has set, gel is poured to the required height. The slot former and the cover are then lowered into position. (3) When the gel has set (it is advisable to leave it overnight), the slot former, the central rod, and the outer molding piece are removed, and loose bits of agarose cleared away. The cover is left in place.

Setting Up the Apparatus

Small Apparatus. (1) A moist dialysis membrane is sandwiched between the spacer and the clamping frames (Fig. 5). Holes are pierced in the membrane at the positions of the holes in the frame. (2) This assembly is screwed against the front of the gel container, the membrane being stretched as the screws are tightened. (3) Excess membrane is trimmed off, the gate is attached to the back, and the whole assembly placed in position in the main vessel. (4) Tubes to the pump are pushed into their grooves. (5) The vessel is filled with buffer to the height of the gel and the sample is loaded.

When the sample has run into the gel, the buffer level can be raised above the top of the gel to prevent it from drying out. Saran Wrap or a polythene sheet is floated on the surface of the buffer to stop evaporation.

Large Apparatus. (1) The collar of the central electrode and an O ring are slipped over the top. (2) A length of moist dialysis tubing is fed over the lower end of the bobbin, under the collar and O ring, and over the top end of the bobbin. (3) The O ring is forced into its seating in the lower

[20] A. Brownstone, *Anal. Biochem.* **70,** 572 (1976).

flange to clamp the lower end of the dialysis tube in place. (4) The collar is then moved up, while the dialysis tube is pulled tight, to clamp the tube against the O ring seated in the upper flange. The distance between the lower end of the collar is adjusted so that the end of the collar protrudes just below the cover on the gel. (5) Excess dialysis tubing is trimmed off, and the electrode is placed in position in the center of the gel and clamped to the base with the knurled nut. (6) Tubing to carry the circulating buffer is attached to the top and bottom of the electrode assembly. (7) With the connection between the cathode buffer and the collecting slot closed, buffer circulation is started. (8) Electric resistance between the two electrodes should be high. If it is not, it is likely that the membrane is touching the gel because it is distorted or too distended. Reducing the flow rate or refitting the membrane may break the contact. Alternatively, conduction between the electrodes may be caused by leakage through the dialysis membrane filling the gap between it and the gel. This can be checked by switching on the collection pump.

If all is well, the collection slot is filled with buffer by unclamping the connecting tube. The sample is loaded into the peripheral slot and run into the gel. The buffer level is then raised to prevent the slot from drying out.

The preparative apparatuses described here have high resolving power and, to exploit this fully, it is necessary to collect a large number of fractions. For example, if a restriction enzyme digest of a complex DNA is fractionated into about 50 fractions, there is little overlap between fractions. Resolution depends on a number of factors which include the gel concentration, path length, time between fractions, and accuracy with which the apparatus is constructed. In the annular apparatus the path length is not readily varied, and this is one disadvantage of this design. However, there is an approximately linear relationship between the molecular weight of the DNA fragment and the time required for its elution from the apparatus. This linear relationship holds over a wide range of fragment sizes and gel concentrations between 0.4 and 2.4%. But if the molecular weight is plotted against the elution time for different gel concentrations, the slope of the line decreases with increasing gel concentration. An important consequence of this is that increasing the gel concentration has the same effect as increasing the path length. Thus, to increase separation in either gel apparatus it is not necessary to increase the size of the gel; it is enough to increase the gel concentration. There is, however, an upper limit to the gel concentration that can be used, because as the gel concentration is increased the total time for a separation increases and this could become unacceptably long. A consequence of increasing resolution by increasing either gel concentration or path length is that the

DNA concentration in the fractions is reduced, and this makes recovery and analysis of the fraction more difficult.

Some guidance as to the appropriate choice of gel concentration for a given separation is provided by the data in Fig. 12.

Analysis of Fractions

The average concentration of DNA in the fraction collected from the annular apparatus is about 100 μg/ml, and from the smaller apparatus about 30 μg/ml, concentrations that are high enough to read in a spectro-

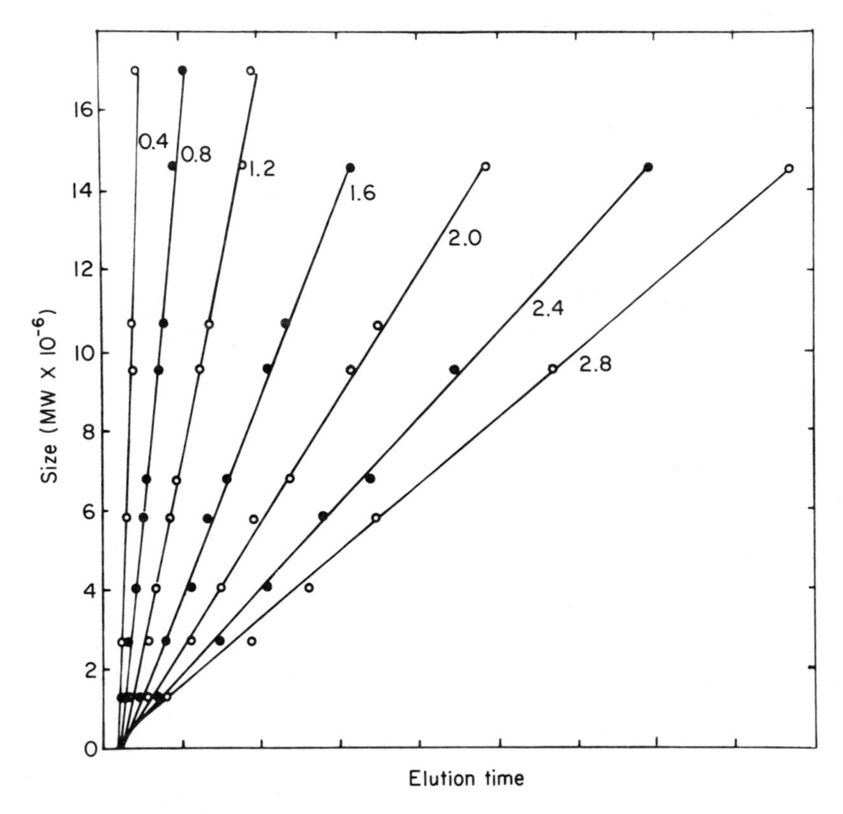

FIG. 12. Dependence of elution time on fragment size and gel concentration. The time taken for a fragment to be eluted from a preparative gel is approximately proportional to its size. Elution time for any fragment size increases with gel concentration as shown. Relative elution times only are shown here, because other factors such as gel path length and voltage gradient also affect elution time. Data for the 1.2% gel were taken from the preparative gel separation shown in Fig. 13. Data for other gel concentrations were calculated from relative mobilities in analytical gel (unpublished experiments of E. M. Southern and R. West).

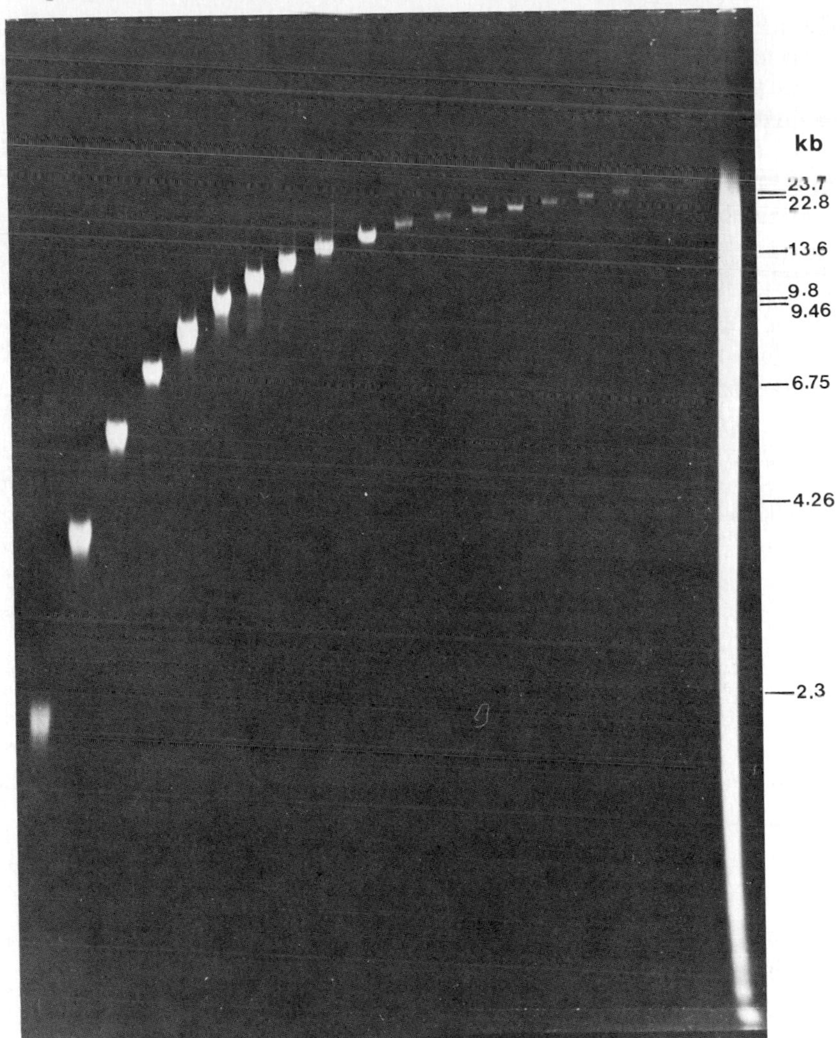

FIG. 13. Fractionation of a *Hin*dIII digest of rabbit DNA on the annular preparative gel apparatus. A *Hin*dIII digest of rabbit liver DNA (6 mg) dissolved in 100 m*M* sodium acetate, pH 7.0, containing 1% Ficoll and a little orange G was applied to a 1.2% agarose gel 4 cm high with a 12.5 cm diameter. The central hole had a diameter of 3 cm, giving a radial path length of 4.75 cm for the separation. Twenty-four volts were applied across the electrodes. Fractions (ca. 9 ml) were collected according to the following schedule: 54 fractions at 20 min, 10 at 40 min, 20 at 60 min, and at 90 min thereafter. The photograph shows an analytical agarose gel (0.8%). Fractions of 5 μl each separated by about 3 hr were applied to this gel which was stained in EtBr and photographed as normal. The track on the right shows the original sample that was applied to the preparative gel. Size standards were a *Hin*dIII and a *Bgl*II digest of λ cI857 DNA. DNA was recovered from each fraction by adding 2 volumes of ethanol and centrifuging after keeping at $-20°$ for 1 hr.

photometer. However, we have found that the uv absorption of buffers containing Tris rises throughout the run, probably because the base is oxidized. Analysis of the fractions by gel electrophoresis is more informative, and a few microliters of each fraction is enough for this analysis (Fig. 13). Where many fractions are to be analyzed it is convenient to combine them. For example, 100 fractions can be analyzed in 10 slots of a slab gel by combining portions of every tenth fraction.

Recovery and Further Purification of Fragments

Fractions may be concentrated approximately 10-fold by adding 2 volumes of *s*-butanol, removing the organic phase, and extracting the *s*-butanol from the aqueous phase with chloroform. Alternatively, the DNA may be precipitated with ethanol, provided the running buffer was sodium or Tris-acetate. Occasionally it is found that the DNA recovered from gels is resistant to the actions of enzymes such as restriction endonucleases or ligase. It is assumed that the inhibition is due to impurities leached from the agarose. A variety of methods have been used to remove these inhibitors, such as phenol extraction, HAP fractionation, and banding in cesium chloride.

[10] Elution of DNA from Agarose Gels after Electrophoresis

By ROBERT C.-A. YANG, JOHN LIS, and RAY WU

Introduction

The study of genome structure and function relies heavily on the isolation and analysis of defined DNA fragments. Gel electrophoresis is a simple, high-resolution method of separating specific DNA fragments on the basis of size.[1–5] Agarose gels at concentrations of 2.5–0.1% resolve DNA of from 150 to 880,000 base pairs, whereas acrylamide gels ranging

[1] C. W. Dingman and A. C. Peacock, *Biochemistry* **7,** 659 (1968).
[2] P. A. Sharp, B. Sugden, and J. Sambrook, *Biochemistry* **12,** 3055 (1973).
[3] M. W. McDonell, M. N. Simon, and F. W. Studier, *J. Mol. Biol.* **110,** 119 (1977).
[4] W. L. Fangman, *Nucleic Acids Res.* **5,** 653 (1978).
[5] A. M. Maxam and W. Gilbert, *Proc. Natl. Acad. Sci. U.S.A.* **74,** 560 (1977); also this series, Vol. 65, p. 499.

METHODS IN ENZYMOLOGY, VOL. 68

ISBN 0-12-181968-X

from 20 to 3% afford good resolution of fragments in the size range of 10–2000 base pairs.

Although several methods for recovering DNA from gels have been developed over the years[1–10] and have been reviewed in several articles,[11–13] no method has been universally accepted. In this article we describe and evaluate three new or revised methods for the recovery of DNA from agarose gels with which we have had considerable experience. The evaluation was based on four criteria: simplicity, speed, yield, and amenability of the purified DNA to subsequent use of enzymatic reactions. We found that all three methods met the above criteria, and they appeared to be simpler than other methods.

Materials

Agarose (Seakem standard pure powder) was obtained from Marine Colloids, Inc. (P.O. Box 308, Rockland, Massachusetts 04841) and dialysis tubing (flat diameter of 2.5–4.2 cm) from VWR Company (25225-248) or Fisher Scientific Company (8-667E). The tubings were washed by boiling in 10% Na_2CO_3 solution, rinsed extensively with distilled water containing 1 mM EDTA, and stored in 60% ethanol. Before use the bags were handled under sterile conditions by rinsing in distilled water, and they were cut to the desired size (usually 3.0 × 1.8 cm for a slot 1 cm wide); they were of single thickness.

An Eppendorf Microcentrifuge (5412) and Eppendorf centrifuge tubes (1.5-ml capacity) were obtained from Brinkmann Instruments. GF/C filters (Whatman glass fiber) and Gelman GF A-E filters were from Scientific Products. Microgranular DE52 was from Whatman Biochemicals Ltd.

$NaClO_4$ obtained from Fisher Company (purified grade) is dissolved in distilled water to give a 6M solution. To 100 ml of this solution 1 ml of 1M Tris-HCl, pH 7.4, is added. The solution is filtered through Whatman no. 1 paper and stored at −20°.

This solution also serves as the washing solution after the DNA-containing solution has been applied to the filter.

[6] N. Blin, A. V. Gabain, and H. Bujard, *FEBS Lett.* **53,** 84 (1975).
[7] R. W. J. Thuring, J. P. M. Sanders, and P. Borst, *Anal. Biochem.* **66,** 213 (1975).
[8] M. Finkelstein and R. H. Rownd, *Plasmid* **1,** 557 (1978).
[9] A. M. Ledeboer, J. Hille, and R. A. Schilperoart, *Biochim. Biophys. Acta* **520,** 498 (1978).
[10] H. F. Tabak and R. A. Flavell, *Nucleic Acids Res.* **5,** 2321 (1978).
[11] R. Wu, E. Jay, and R. Roychoudhury, *Methods Cancer Res.* **12,** 87 (1976).
[12] H. O. Smith, this series, Vol. 65, p. 371.
[13] E. Southern, this volume [9].

Results

Elution of DNA Fragments from Agarose Gels

Method 1. Electroelution of DNA into Slots. Nucleic acids can be eluted from gel slices by electrophoresis. The commonly used procedure has been to pack the slice into an open-ended tube or plastic pipette which has a dialysis membrane sack attached to one end.[2,3] The sack is immersed in the anode buffer tank, and DNA is forced out of the gel and against the walls of the sack by electrophoresis. The method has worked adequately, but it takes time to set up, and defects in the dialysis membrane or in the attachment of the membrane to the tube can result in loss of a precious sample. We find that it is faster and safer to electroelute DNA into a slot cut out of the same horizontal agarose gel on which the DNA fragments have been fractionated. The slot is made immediately in front of the DNA band to be eluted. The slot is filled with gel buffer (90 m*M* Tris-borate, 3 m*M* EDTA, pH 8.3), and the mobility of the ethidium bromide-stained band of DNA can be followed with a hand-held uv light. The DNA is removed from the slot before it reenters the gel on the other side. The elution is generally performed at 3 V/cm, and the gel buffer in a 5-mm-wide slot is collected and replaced with new buffer at 5 to 10-min intervals. The DNA can be purified from contaminating agarose and ethidium bromide by butanol or isopropanol extraction followed by ethanol precipitation. Alternatively, the DNA sample in gel buffer is directly loaded onto a small DE52 column that has been poured in 0.1*M* HCl and then equilibrated with 0.1*M* Tris-HCl, pH 7.4, 0.1*M* KCl (1 ml of DE52 can bind 25 μg of DNA). The loaded column is washed with 3 volumes of 0.1*M* Tris-HCl, pH 7.4, 0.1*M* KCl, and the DNA is eluted with 0.1*M* Tris-HCl, pH 7.4, 0.6*M* KCl. The KCl is diluted to 0.3*M* before 2 volumes of ethanol is added to precipitate the DNA.

We have measured the efficiency of recovery of DNA fragments (four different sizes) by this and the other two methods to be described. The results are listed in Table I. The yields with method 1 range from 42 to 82%. The largest fragment, 21.7 kb, was recovered at the lowest yield. Several factors make this method less than ideal for application to large DNA. Larger DNA fragments have slower mobility in the gel and therefore take longer to move completely into the slot. Moreover, large fragments have the same charge/mass ratio as small fragments and move across the slot at the same rate. Thus, slots have to be emptied at the same frequency with large and small fragments, and large fragments must be submitted to electrophoresis longer. Large fragments are also less efficiently recovered from the subsequent DE52 chromatography procedure.

Method 2. Electroelution of DNA onto Dialysis Membranes. The

TABLE I
EFFICIENCY OF ELUTION OF DNA FRAGMENTS FROM AGAROSE GELS

Method	Fragment size (kb)								
	1.9	2.2	3.5	4.8	6.6	7.5	9.4	21.7	23.7
1	—	—	82	69	—	49	—	42	—
2	95	95	—	—	95	—	90	—	90
3[a]	—	—	95	71	—	73	—	76	—

[a] We note that the efficiency (shown as a percentage) of elution by the perchlorate method is variable. Testing each batch of filters and perchlorate is helpful but does not completely eliminate potential problems.

Recently, C. W. Chen and C. A. Thomas, Jr., informed us that reduced yields can result from drying the filter too extensively after the ethanol wash and/or allowing too little time for elution of DNA from the filter. These complications can be avoided by air-drying the filters for only 3 min rather than the recommended 10 min and by eluting the DNA from the filter with a 30-min incubation in 1 m*M* Tris-HCl, pH 7.5–0.1 m*M* EDTA rather than the 1- to 10-min incubation at room temperature.

basic principle of this method is similar to that of method 1 except that a piece of dialysis membrane is placed in the slot to trap the DNA. In this way, the solution in the slot need not be withdrawn until the end of the electroelution. This method is simple, rapid, and efficient. The DNA is fractionated by electrophoresis at 2 Vs/cm in agarose gel containing ethidium bromide (0.5 μg/ml). We prefer gels 0.5–0.7 cm in thickness and use TB buffer (50 m*M* Tris-borate, pH 8.1, containing 1 m*M* EDTA). This buffer has a higher buffer capacity than Tris-acetate buffer and is therefore recommended. After electrophoretic fractionation, the DNA bands are visualized by uv illumination and marked out by cutting a slit in front of each band. Then, under room light a U-shaped slot (as shown in Fig. 1a) is cut out. For a small-scale purification (e.g., 1–5 μg DNA per slot 1 cm wide), a piece of washed dialysis membrane (about 3 × 1.8 cm) is inserted in such a way that the DNA band is surrounded with the dialysis membrane from three sides (see Fig. 1b and c). The slot is then filled with about 200 μl of the same electrophoresis buffer (TB buffer). Electroelution is carried out at 6–10 Vs/cm for 30–60 min to elute the DNA from the gel into the U-shaped slot. uv light is used briefly to check for the completeness of elution. Most of the DNA is loosely accumulated on the dialysis membrane surface. Under room light, the buffer in the slot is used to rinse the DNA from the membrane by pipetting in and out before transferring to the filtration set (see Fig. 2). A few drops (about 100–150 μl) of the buffer are added to the slot to recover the remaining DNA. The combined buffer is filtered through a cotton-plugged 1-ml tip into a 1.5-ml Eppendorf tube (see Fig. 2) by centrifugation at about 1000 *g* for 2 min to remove any

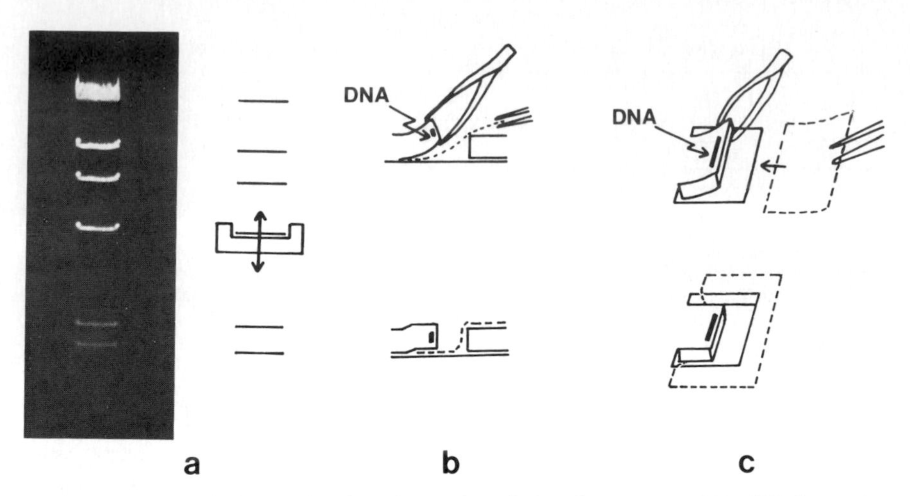

FIG. 1. (a) A typical example of an electrophoretic banding pattern of *Hin*dIII-digested λ DNA fragments (2 μg in total) ranging from 23.7 to 1.96 kb (from top to bottom) photographed under a uv lamp. The DNA bands (1 cm wide) are depicted alongside. A U-shaped slot is cut out with 0.5 cm in bank-to-bank width. The front bank is about 1.6 cm wide and 0.1 cm away from the DNA band, marked out first by cutting the agarose with a razor blade under a uv lamp. Then, under room light, cutting and removal of a U-shaped piece of gel are performed. (b) A cross-sectional view of the U-shaped elution slot along the arrow shown in (a). As indicated, one pair of blunt-ended forceps is used for lifting the gel and the other pair of forceps for inserting the dialysis membrane (dashed line). A single layer of dialysis membrane, about 3 × 1.8 cm, is cut from washed dialysis tubing. (c) A three-dimensional view of the same slot showing how the dialysis membrane is inserted. In the lower panel, the dashed line shows the part of dialysis membrane that is above the gel; the part under the gel is not shown.

contaminating gel particles. One phenol extraction is sometimes helpful in removing soluble contaminants that may inhibit subsequent enzymatic reactions. The DNA solution is mixed with 3 *M* Tris-HCl, pH 8.5, to give a final concentration of 0.5 *M*. Two volumes of isopropanol is then added. The mixture is held at −70° for 5 min and centrifuged in an Eppendorf microcentrifuge for 10 min at 0°. This precipitation step is repeated once more to remove ethidium bromide buffer. After drying for a few minutes in a vacuum desiccator, the DNA is ready for suspension in a suitable buffer for any subsequent enzymatic reaction. Preparative-scale DNA isolation (e.g., 25–50 μg DNA per slot of 5 cm width) using this method from gels have also been carried out. The yield for recovering DNA of 23.7–1.96 kb is at least 90% (see Table I).

The quality of the eluted DNA was assessed from its ability to serve as a substrate for several restriction endonucleases and for DNA ligase. The eluted DNA reacted with an efficiency indistinguishable from that of a

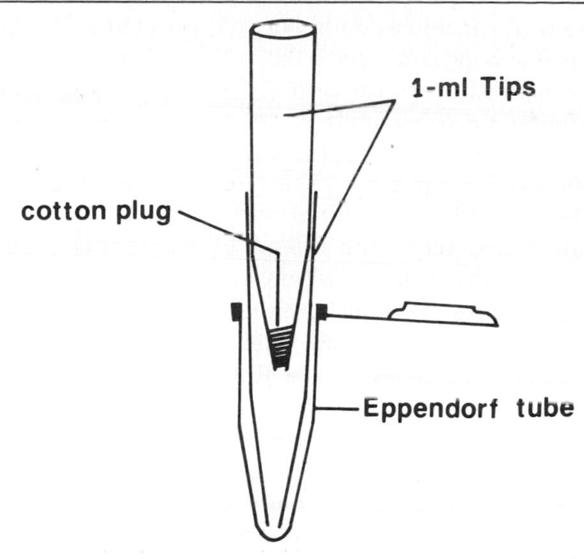

FIG 2. A filtration set composed of two 1-ml plastic pipette tips, with the cotton-plugged tip inside the other tip, as well as a 1.5-ml Eppendorf tube holding the tips. For the recovery of very small amounts of DNA (0.5 μg or less), siliconized cotton is used for the plug.

control DNA sample that was not fractionated through agarose gel electrophoresis.

Method 3. Dissolving Gel Slices in Perchlorate Solution. The method of isolating DNA from an agarose gel slice by dissolving agarose in perchlorate was described several years ago.[14,15] The DNA was separated from dissolved agarose by equilibrium density gradient centrifugation.[15] Recently, Thomas and co-workers (C. W. Chen and C. A. Thomas, Jr., personal communication) have replaced the method of recovering the DNA by time-consuming equilibrium centrifugation with a rapid filtration through glass-fiber filters. The DNA sticks to the filters in the presence of sodium perchlorate, while agarose and perchlorate pass through the filter. The DNA can be recovered efficiently from the filter in a small volume by eluting with Tris-EDTA buffer (5 m*M* Tris, 0.5 m*M* EDTA). This last method is very rapid, gives high yields for fragments at least up to λ length (70–90% yield), and the DNA is in a form that is ready for subsequent enzymatic reactions.

We have slightly modified the method of Chen and Thomas as follows. The portion of the agarose gel containing the DNA of interest is cut out and weighed. The gel is dissolved in 4 volumes of 6 *M* $NaClO_4$ at 45° for

[14] M. Fuke and C. A. Thomas, Jr., *J. Mol. Biol.* **52,** 395 (1970).
[15] N. M. Wilkie and R. Cortini, *J. Virol.* **20,** 211 (1976).

30 min.[15a] The final concentration of $NaClO_4$ is about 5 M, and agarose is 0.1–0.5% (aim for approximately 0.25%). The dissolved agarose–DNA solution is transferred dropwise with a siliconized Pasteur pipette or an Eppendorf pipette tip to a single 6- to 8-mm-diameter GF/C filter or Gelman GF A-E filter which is placed on a sintered-glass filtration set under maximum water aspiration. The filter of this size can handle up to 10 μg of DNA. The filter is washed with about 0.5 ml of $NaClO_4$-Tris washing solution and then with 0.5 ml of isopropanol, and 0.5 ml of EtOH, and then air-dried (for 3–10 min).

DNA is eluted from the filter by either of two methods. One method makes use of the fact that the 8-mm filter fits snugly into the exterior indentation of the cap of a closed 1.5-ml Eppendorf microfuge tube. Before inserting the filter, the cap is pierced several times with a 26-gauge needle; 25 μl of Tris-EDTA buffer (5 mM Tris-HCl, pH 7.4, 0.5 mM EDTA) is added to the filter, and after 1 min the cap is covered with Parafilm and centrifuged for 10 sec at 12,000 g in the microfuge. This elution step is repeated twice more. The DNA solution, about 60 μl, collects in the bottom of the microfuge tube. An alternative method is to roll up the filter and place it inside a plastic Eppendorf tip (1 ml capacity, as shown in Fig. 2, except that the cotton plug is unnecessary and only one Eppendorf tip is needed). This method can be adapted to different-sized filters more readily than the first; however, the first method appears to be more effective at keeping glass fiber out of the eluent. About 20–30 μl of Tris-EDTA (1–0.1 mM, pH 7.4) is added to the filter and after incubation for about 1–10 min at room temperature, the eluate is collected at the bottom of the tube by centrifugation at 1000 g for 2 min. Elution is repeated twice more. The DNA in this low-salt solution can be used directly for restriction enzyme digestion, ligation, or other manipulations.

After the preparation of this article, we found an article by Vogelstein and Gillespie[16] who used another chaiotropic salt, NaI, instead of $NaClO_4$ to dissolve the agarose. The DNA binds tightly to powdered glass under these conditions and can be efficiently eluted with Tris-EDTA buffer. Presumably the binding of DNA to glass-fiber filters in the presence of $NaClO_4$ is an analogous process.

[15a] With large amounts of DNA containing ethidium bromide the retention of DNA by the filter may not be as efficient. Removal of ethidium bromide by extracting the solution once with 2 volumes of isobutanol (saturated with the $NaClO_4$ washing solution) is recommended.

[16] B. Vogelstein and D. Gillespie, *Proc. Natl. Acad. Sci. U.S.A.* **76,** 615 (1979).

[11] Two-Dimensional Electrophoretic Separation of Restriction Enzyme Fragments of DNA

By STUART G. FISCHER and LEONARD S. LERMAN

Rationale

A very large change in the electrophoretic mobility of a double-stranded DNA molecule in a polyacrylamide gel is observed when part of the molecule melts.[1] Since melting takes place cooperatively as DNA is subjected to progressively stronger denaturing conditions, the transition in the electrophoretic mobility is abrupt. The onset of melting is related in a complex way to sequence, since it depends on the presence of a relatively long stretch of base pairs slightly richer in AT than the neighboring sequences. The conditions favoring the mobility transition vary strongly and characteristically among restriction enzyme fragments several thousand base pairs long.

The partial melting to which we attribute the mobility change represents an equilibrium state of the molecule in which the bases along a considerable segment remain unstacked and disordered. The configuration and flexibility of a molecule under these conditions are not the same as for a molecule that has been melted, or partially melted, and restored to an environment at a lower temperature or in a different solvent in which double helices are stable.

In the procedure described here, DNA molecules are subjected to progressively stronger denaturing conditions by electrophoretic migration into an ascending concentration gradient of a denaturing solvent at an elevated temperature. As each molecule reaches the point in the gradient where melting of some portion is initiated, its rate of migration drops so sharply that it appears to have stopped. The positions and arrangement after the last fragment has stopped are determined only weakly by the number of nucleotide pairs in each fragment and depend almost entirely on the denaturation characteristics. Since each zone containing a single fragment accumulates at the position in the gradient where its mobility drops sharply, the zones are self-sharpening and become narrow.

If restriction enzyme fragments are sorted first according to size by electrophoresis in an agarose gel, a second sorting can be effected in the perpendicular direction, resulting in a two-dimensional separation according to nearly independent properties. Reasonably clear resolution of 10^3 fragments is possible. Since the procedure described here represents

[1] S. G. Fischer and L. S. Lerman, *Cell* **16,** 191 (1979).

ISBN 0-12-181968-X

an early stage of development, the reader is urged to consult current literature or consult the authors for further improvements.

Apparatus

Gradient gel electrophoresis is carried out between $\frac{1}{4}$-in.-thick glass plates held in an acrylic frame (A in Fig. 1) that serves to clamp the plates, to provide connections to the electrodes, and to support the assembly when immersed in a 5-gal aquarium containing the anodic electrolyte. Both sides of the plates are in contact with the electrolyte, which is stirred vigorously to provide good heat exchange. We have found that fabrication from acrylic sheet is satisfactory for prolonged operation in water at 60°. Stainless steel and platinum are the only metals in contact with the electrolyte. The plate that is sealed against the upper, cathode compartment (F) is about 17.7 cm square with two tabs 2.3 × 1.5 cm at each side of the top edge (M). A tight seal against the cathode compartment is ensured by $\frac{3}{16}$-in. silicone rubber tubing (I) set into a slightly shallower, U-shaped groove that follows the contour of the tabs and the top edge of the plate. The front plate (N) is 20 × 17.7 cm, matching the corner dimensions of the rear plate. The bottom edges of the plates are set just above the bottom trough (B), ensuring that contact with the electrolyte is made by an upper rather than a lower surface to preclude obstruction by either bubbles from the electrode or those stirred into the solution. Space for the gel (C) is provided by silicone-greased acrylic spacers, $2\frac{1}{2}$ mm thick and 1 cm wide, inserted along each vertical edge. The plates and spacers are held together and the unit sealed against the upper electrode compartment by means of a series of five stainless-steel screws (D) bearing on an acrylic strip (E) along each vertical edge. The cathode chamber contains about 100 ml of electrolyte, constantly recirculated to the main tank by means of a peristaltic pump with silicone rubber tubing at 10 ml/min (J, inlet; K, overflow). The electrode (H) is a $\frac{1}{4}$-in.-diameter rod of hard, highly purified graphite, POCO grade AXF 9Q BGI (POCO Graphite, Inc., Decatur, Texas, 76234), into which a Teflon-insulated wire has been sealed by means of soft solder, an O ring, and epoxy cement. As an anode, we use (with some reservations) a platinum wire parallel to the trough. To prepare concentration gradients of denaturants in the gel, we have used a digitally controlled gradient maker in which two solution stocks are delivered at linearly varying rates from two 50-ml glass syringes. The delivery rate is unaffected by the usual differences in solution viscosity and density, and there is no contact of degassed solutions with air. In most of our work we have used Sorenson DCR 600-0.75B power supplies capable of delivering up to 600 V and/or 750 mAmp at constant voltage or constant current. As described below, each gel draws

FIG. 1. Two-dimensional DNA gel electrophoresis apparatus. Front view (a) and side view (b). A, Acrylic frame; B, trough; C, gradient gel; D, stainless-steel screws; E, screw force spreader; F, cathode buffer chamber; G, agarose gel strip; H, carbon cathode; I, rubber gasket; J, cathode buffer chamber inlet; K, cathode buffer chamber outlet; L, 90 mg/ml acrylamide plug; M, tabbed inner glass plate; N, outer glass plate; O, 10 mg/ml agarose seal.

about 200 mA, and three gels can be powered by a single supply. A comparable supply from the same manufacturer with an upper limit of 300 V can power twice as many gels.

Heating of the gel due to the current flow has been measured in a

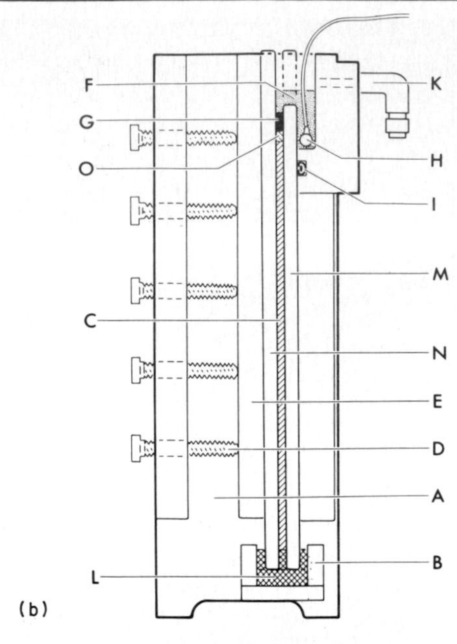

FIG. 1 (*continued*)

standard $2\frac{1}{2}$-mm slab under standard conditions by insertion of a needle-mounted thermistor into the gel from the cathode compartment immediately after switching off the high voltage. In order to minimize self-heating the thermistor is connected in series with a 100 μA constant-current source, and its resistance is determined as the voltage across the terminals by means of a high-impedance digital voltmeter. With no denaturant in the gel, its temperature was found to be 2.0° higher than the 60.0° bath at 150 V and 150 mA.

Gel Preparation

The trough is filled first with 40 ml of a degassed solution containing 90 mg/ml acrylamide in TAE buffer (0.04 *M* Tris, 0.02 *M* sodium acetate, 1 m*M* EDTA, pH 8 by addition of acetic acid) to which has been added 0.2 ml 200 mg/ml ammonium persulfate and 40 μl TEMED. All gel-forming solutions are degassed by stirring for 30 sec in a filter flask evacuated with a small, single-stage mechanical high-vacuum pump. After the plug (L) has set in the trough, hot 10 mg/ml agarose in TAE is pipetted along the outer edge of the acrylic strips and into the corners above the plug to prevent leakage of the gradient gel solution. The gel-forming solution for the gradient is delivered between the plates through a large syringe needle inserted just above the plug.

The gel-forming solutions are prepared from a stock solution containing 300 mg/ml acrylamide and 8 mg/ml bisacrylamide which has been mixed with powdered activated charcoal and filtered through a 0.5-μm membrane filter. The low-density stock solution, designated 0% denaturant, is diluted $7\frac{1}{2}$-fold to 40 mg/ml acrylamide in TAE buffer. The dense solution, designated 100% denaturant, consists of 40 mg/ml acrylamide–bisacrylamide, 7 *M* urea (BRL ultrapure), 40% (v/v) formamide [Fisher, passed through a column of Bio-Rad AG501X8 (D) mixed bed resin], and the same final salt concentration in TAE buffer. For the shallow gradients, such as those in Figs. 2 and 3, stock solutions are mixed in appropriate proportions before loading the gradient maker.

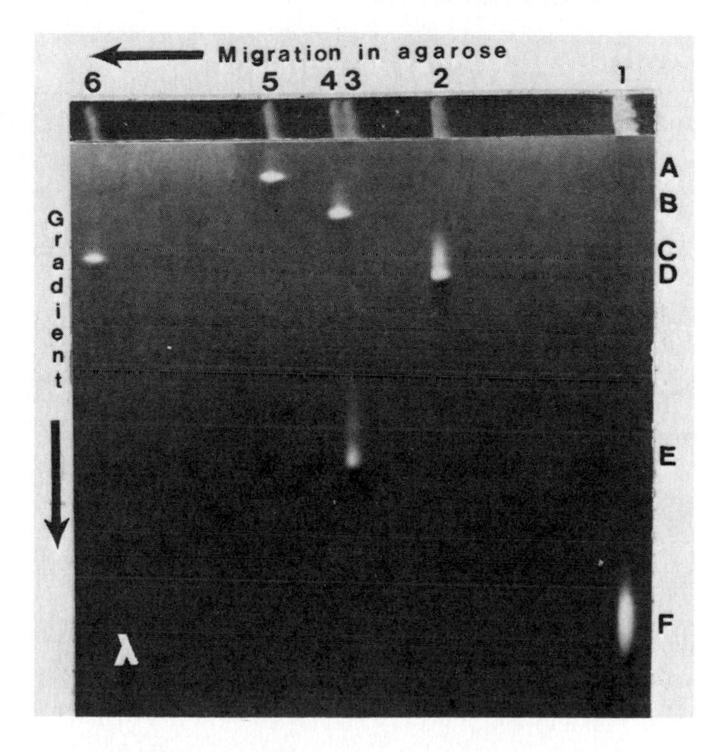

FIG. 2. Two-dimensional gel of *Eco*RI digest of λ DNA. A 10 mg/ml agarose gel of an *Eco*R1 digest of λ DNA was set across the top of a 40 mg/ml acrylamide slab gel containing a linear gradient of 4–30% formamide and 0.7–5.25 *M* urea from top to bottom; 150 V was applied for 20 hr at 60°C, and the gel was stained with ethidium. Bands 1–6 are the λ R1 fragments in order of increasing mobility in agarose under nondenaturing conditions. Spots A–F identify the final order of the fragments in the denaturing gradient gel. The photograph of the ethidium-stained agarose gel is included as a reference. Only the central region of the denaturing gradient slab gel is shown.

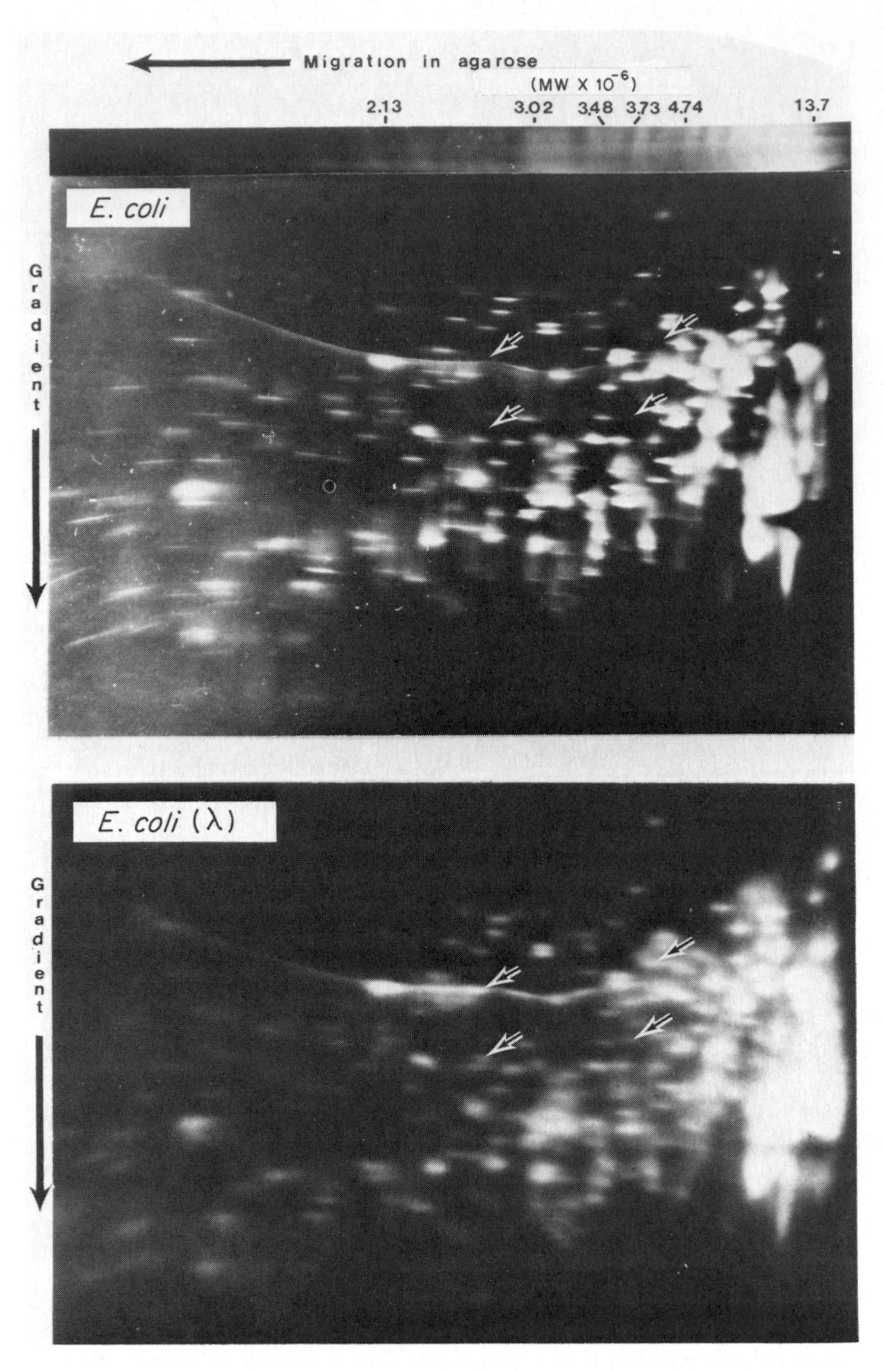

FIG. 3. Two-dimensional gel of an *Eco*R1 digest of *E. coli* and *E. coli* λ DNA. A 10 mg/ml agarose gel of an *Eco*R1 digest of a mixture of 60 μg *E. coli* DNA and 0.5 μg $^{32}PO_4$ *E. coli* λ DNA was laid across a 40 mg/ml acrylamide gel containing a vertical gradient of 4–30% form-

Immediately before casting the gel, 40 ml of each of the acrylamide solutions is degassed, and to each solution 0.2 ml 200 mg/ml ammonium persulfate and 4 μl TEMED are added. The concentrations of persulfate and TEMED are adjusted to allow about ½ hr before the onset of gelation. Delivery from the gradient maker requires about 10 min. The light solution is delivered first and displaced upward. The plates are filled to about 1 cm below the top edge of the rear plate, which is open to the cathode chamber. The top edge is sprayed with 1 mg/ml SDS to ensure a smooth, flat surface. Gels are allowed to set at least 2 hr before the SDS solution is poured off. At least 3½ hr elapse between the onset of polymerization and the beginning of electrophoresis.

The initial electrophoretic separation according to fragment length is carried out in agarose gels prepared from 10 mg/ml Seakem HGT agarose dissolved in TAE buffer. The sample slots are 6 mm wide. Since the agarose slab is the same thickness as the polyacrylamide gel (2½ mm), agarose strips fit snugly into the space between the plates. Strips about 5 mm wide and 13 cm long are excised from the agarose slab gel with a clean razor blade. Trimming the strips to a width slightly less than that of the original slot eliminates DNA retarded at the edges of each band and avoids unnecessary zone breadth in the two-dimensional gel. The strip (G) is transferred on a flexible plastic ruler and slipped into the space between the plates of the polyacrylamide gel with a gentle downward push. A space of about 3 mm is left between the bottom of the strip and the top of the polyacrylamide gel to be filled with hot 10 mg/ml agarose in TAE (O). Filling by means of a needle underneath the strip prevents the trapping of air bubbles under the agarose strip. The agarose seal is allowed to gel for about 1 hr. The assembly is immersed to the level of the agarose strip in about 14 liters of TAE buffer in a glass aquarium maintained at 60° with vigorous stirring. Electrolyte is recirculated between the cathode compartment and the tank at about 10 ml/min. An inadequate flow rate through the cathode compartment may lead to deterioration of the gel. A potential of 150 V is applied for about 20 hr, sufficient time for the slowest, most denaturation-resistant fragment to reach its mobility transition point.

Gels are stained for about 30 min in 1 μg/ml ethidium bromide in water, and fluorescence is excited with 254-nm uv light by placing the gel directly on the filter of the illuminator (Model C-61, Ultraviolet Products,

---

amide and 0.7–5.25 *M* urea and electrophoresed downward at 60° for 20 hr at 150 V. The gel was stained with ethidium (upper gel) to show *E. coli* fragments and autoradiographed (lower gel) to show *E. coli* λ fragments. The arrows indicate the positions of fragments found in gel of *E. coli* λ but not *E. coli*. The original ethidium-stained agarose gel is included at the top of the figure.

Inc.). The fluorescence is photographed on 4 × 5 in. Kodak Technical Pan film, SO 115, through a pair of filters, a Corning 3-68 glass filter and a short-wavelength-pass interference filter (650 nm cutoff) that eliminates red radiation from the transilluminator. Exposure is typically 2 min at *f*/5.6. The films are developed 7 min in Kodak D19.

Radioactive DNA was detected by autoradiography after drying the gels on a Bio-Rad Model 224 gel slab dryer and exposure to Kodak RP Royal X-omat film in a Kodak x-ray intensifier at −70°. For further data processing, the center of each spot is estimated visually, and the two-dimensional coordinates are recorded by means of a Tektronix Model 4662.

Results

Figure 2 shows a representative two-dimensional separation using *Eco*R1 fragments of bacteriophage λ DNA. The positions of the fragments in the agarose gel prior to migration into the gradient are indicated at the top. The large, heavily stained fragment, which encompasses most of the GC-rich left side of the λ genome moves the most slowly in the agarose gel and is near the right edge in this photograph. From right to left the fragments are sorted according to decreasing size (increasing electrophoretic mobility in agarose). After a long electrophoresis into the denaturing gradient gel, the final distance moved by these fragments is random with respect to their size. The largest fragment, which moved least far into the agarose gel under nondenaturing conditions, moves deepest into the denaturing gradient gel. Bands 3 and 4, which are similar in size and mobility in agarose, have much different final positions in the two-dimensional gel.

The two patterns shown in Fig. 3 derive from a single gel containing a concentration of *Escherichia coli* DNA large enough to be visualized after staining with ethidium, together with a much smaller amount ($\frac{1}{120}$) of ^{32}P-labeled DNA of an isogenic λ lysogen not large enough to be detected by staining. The two DNA samples were digested together by *Eco*R1 and run as a single sample in agarose. The stained agarose strip is shown at the top. At least 350 spots can be discerned in the original negatives. Photographs and printed reproduction of an entire gel as a single image require arbitrary printing control because of the large range of intensities; both the radioactivity and fluorescence of each spot are expected to be proportional to fragment length. The arrows indicate spots present in the pattern of the lysogen that are absent in the pattern of the parental strain.

Discussion

On migration into an increasing concentration of solutes that effect unstacking of the bases, DNA molecules undergo a cooperative change in electrophoretic mobility resulting in a final mobility 10–50 times lower than that of the intact double helix.[1] The concentration of denaturant at which the cooperative transition occurs is characteristic for each fragment and appears to depend in a complex way on the base composition and sequence. Because the mobility after the transition is so small, the position of the fragment in the gradient after the transition is almost independent of its length and original velocity and of the duration of continued electrophoresis.

Since the ratio of initial to final mobility of small fragments (those below 1200 nucleotide pairs) is smaller than that for larger fragments, the position of small fragments in the pattern is somewhat more time-dependent. Gradient gels with higher polyacrylamide concentration permit retardation of shorter fragments. The characteristic patterns suggested by clusters of nearby spots remain easily recognizable from gel to gel, despite variation in gradient steepness or length-dependent expansion. Transfer of the spots to nitrocellulose filters according to the procedure of Southern[2] is accomplished slowly and with reduced efficiency.

If 2 mm between centers in the gel is taken as the closest spacing of readily resolvable fragments and the working area of the gel is assumed to be 12×9 cm, the probability of superposition of randomly distributed spots can be estimated. Poisson statistics for 350 fragments indicate that about 308 spots correspond to single fragments, about 20 spots contain 2 fragments, and 1 spot might contain 3 fragments. However, it should be possible to determine the number of different fragments in each spot from the limit digest of a unique genome by rough quantitation of spot intensity. Fragments are produced in equimolar numbers and distributed along the x coordinate nearly monotonically by size; superposed fragments should give multiples of the prevailing local intensities.

Some of the limitations in resolution are presently discernible. Overloading the agarose gel, tilting the bands, or inclusion of the retarded edges produces excessive width in the x direction. On occasion we have observed y-axis streaking of certain fragments; the effect appears to be characteristic of some preparations and not others and is most pronounced among high-molecular-weight fragments.

Acknowledgment

This work was supported by Grant No. GM24030 from the National Institutes of Health.

[2] E. M. Southern, *J. Mol. Biol.* **98,** 503 (1975).

[12] Purification of DNA or RNA Sequences Complementary to Mercurated Recombinant DNA by Sulfhydryl Sepharose Chromatography

By SHIRLEY LONGACRE and BERNARD MACH

The rapid proliferation of DNA cloning technology using recombinant bacterial plasmids has permitted the isolation of relatively large quantities of pure DNA sequences. It is, therefore, of interest to have a procedure to exploit the potential of these recombinant plasmids as probes for the purification of any DNA or RNA containing homologous complementary sequences by preparative molecular hybridization.

There are two general approaches to this problem. The first involves hybridization to a probe covalently linked to an insoluble support, followed by melting and subsequent release of the target strands. Alternatively, hybrids formed in solution between a modified probe DNA and an unmodified target DNA are retained on an insoluble support, which binds the modified probe DNA selectively, and the unmodified target strand is eluted by *in situ* melting. Drawbacks of the first approach include long hybridization times, a lack of precise information characterizing the hybridization reaction on solid supports, and decreased efficiency when hybridization involves large DNA fragments. In addition, our experience indicates that a significant amount of covalently bound probe DNA is frequently released with most published techniques for covalent linkage.

The second approach was made feasible by the work of Dale and Ward[1] showing that mercurated DNA, which can be hybridized efficiently, is selectively retained on sulfhydryl Sepharose. In their procedure, the entire hybrid structure is then eluted with a mercaptan. This is a serious drawback in the purification of single-stranded material, since both strands of the hybrid are recovered.

The procedure described here is a modification of Dale and Ward's technique that permits selective elution of the unmodified target strand only. Both the original technique and the modification are diagramed schematically in Fig. 1. The probe polynucleotides (e.g., recombinant DNA or viral DNA) are mercurated and hybridized in solution to unpurified DNA or RNA target sequences. The hybrids are passed over a sulfhydryl Sepharose column where mercurated polynucleotides are retained together with any hybridized complementary nonmercurated sequences. As originally described, the entire hybrid, containing both mercurated and nonmercurated sequences, was eluted from the column with a mer-

[1] R. M. K. Dale and D. C. Ward, *Biochemistry* **14,** 2458 (1975).

ISBN 0-12-181968-X

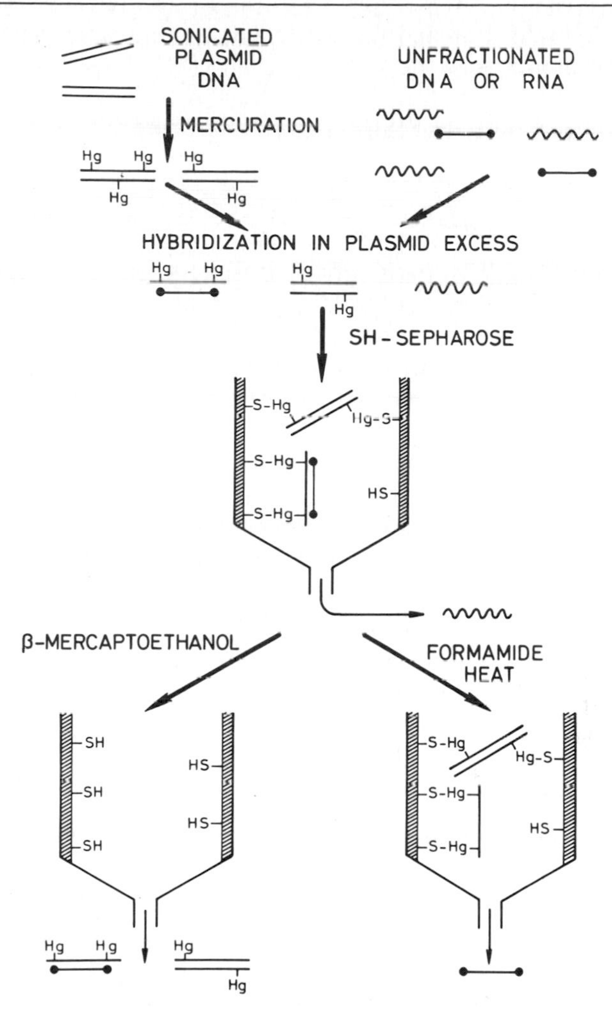

FIG. 1. A schematic representation of the purification of DNA or RNA complementary to recombinant plasmid DNA. The diagram represents, on the left, the procedure as originally proposed by Dale and Ward for the elution of mercurated hybrids with a mercaptan. Our modification for the purification of the target sequences only is illustrated on the right. In the diagram ●—● represents the DNA target sequence complementary to the mercurated probe DNA and ∿∿ represents other DNA sequences.

captan. In the modified procedure the nonmercurated target strand is melted *in situ* and selectively eluted from the sulfhydryl Sepharose, free of mercurated probe DNA.

The modification described here is important in any situation requiring the purification of a single polynucleotide strand such as single-stranded probes for hybridization and for sequences to be analyzed by hybridization or translation.

Preparation of Sulfhydryl Sepharose

Sepharose CL-4B (Pharmacia) is used as the basic solid support since it is resistant to formamide and exhibits relatively low nonspecific absorption. It is prepared following the procedure of Cuatrecasas[2] as described by Dale and Ward.[1] Wet, packed Sepharose (100 g) is washed with 4 liters of distilled water and resuspended in an equal volume of distilled water (100 ml). Under a well-ventilated hood, 20 g of cyanogen bromide (Merck) is ground and added to the stirred slurry. The pH is monitored with a pH meter and is maintained close to pH 11 by the dropwise addition of 10 *N* NaOH. The temperature is held at about 20° by the addition of crushed ice directly to the reaction mix. After the rate of pH change is substantially diminished (about 30 min), the activated Sepharose is quickly washed under suction in a sintered glass funnel with 1500 ml of water adjusted to pH 10 with NaOH. It is resuspended immediately in 100 ml of 2 *M* ethylenediamine (Merck, adjusted to pH 10 with 6 *N* HCl) and stirred for 20 hr at 4°. The Sepharose is then washed with 4 liters of distilled water and resuspended in 100 ml of 1.0 *M* $NaHCO_3$, pH 9.7. Then 16 g of *N*-acetylhomocysteine thiolactone (Fluka) is added, and the mixture is stirred for another 24 hr at 4°. The sulfhydryl Sepharose is then washed extensively with 7 liters of 0.1 *M* NaCl containing 1 m*M* 2-mercaptoethanol followed by 2 liters of 0.1 *M* NaCl. Finally, it is washed twice for 1 hr each time in 2 volumes of 97% deionized formamide in 10 m*M* Tris-HCl, pH 7.6, 100 m*M* NaCl, and 1 m*M* EDTA (TNE) at 60° and stored at 4° in TNE. The final washing step in formamide is necessary to eliminate the residual nonspecific binding observed with nonmercurated cDNA.[3] No decrease in the binding capacity of the sulfhydryl Sepharose has been noticed after a year of storage. We have had no experience with commercial preparations.

Mercuration of Probe DNA

Sonicated, double-stranded DNA has been used for mercuration, although neither sonication nor double strandedness are essential to the technique. If a sonicated probe DNA is prepared, it should be sonicated before mercuration. Mercuration is carried out in 55 m*M* sodium acetate, pH 6.0, 20 m*M* mercuric acetate (added immediately from a 200 m*M* mer-

[2] P. Cuatrecasas, *J. Biol. Chem.* **245,** 3059 (1970).

[3] S. Longacre and B. Mach, *J. Biol. Chem.* **253,** 7500 (1978).

curic acetate solution made up freshly in 50 mM sodium acetate, pH 6.0, and kept on ice to prevent precipitation) containing up to 100 μg/ml DNA. The mercuration reaction is incubated at 50° for 24–48 hr, the amount of mercury substitution being controlled as a function of time. Less extensive mercuration leads to the release of some DNA from the sulfhydryl Sepharose under melting conditions,[3] and more extensive mercuration appears to affect the stability of the hybrids formed with mercurated DNA. The mercuration reactions are terminated by the addition of 0.4 ml "quench buffer" (10 mM Tris-HCl, pH 7.6, 1.0 M NaCl, 100 mM EDTA) per milliliter of reactants. The mercurated DNA is then dialyzed three times against 1-liter changes of dialysis buffer (10 mM Tris, pH 7.6, 20 mM NaCl, 1 mM EDTA). The second dialysis solution contains, in addition, 10 μM mercaptoethanol which provides an appropriate mercury ligand.[1] The mercurated DNA can be stored as an ethanol precipitate at −20°.

Hybridization of Mercurated Probe DNA to Target Polynucleotides

The mercurated probe DNA is precipitated in ethanol and resuspended in 100% formamide (Merck, deionized for 1 hr with 1 g Amberlite IRC-50 per 30 ml) and melted at 50° for 15 min. The target DNA or RNA is added to the melted probe DNA in as small a volume as possible and incubated at 50° for another 10 min. Hybridizations are carried out at 37° in 50% formamide, 20 mM Tris-HCl, pH 7.8, 0.3 M NaCl, 1 mM EDTA. The molar ratio of probe and target sequences and the duration of hybridization must be calculated in individual cases. The extent of hybridization can be analyzed by digestion with S1 nuclease after appropriate dilution in S1 buffer (50 mM sodium acetate, pH 4.6, 0.2 M NaCl, 1 mM $ZnCl_2$, 5% glycerol). The S1 enzyme is active in S1 buffer containing up to 10% formamide. For long-term hybridizations in formamide, the use of PIPES or Tricine as buffer may be preferable to Tris-HCl.

Sulfhydryl Sepharose Chromatography

The chromatography is done in jacketed columns at least 5 × 15 mm. The column is washed extensively in TNE before loading the sample. The sulfhydryl Sepharose can accommodate at least 100 μg of mercurated DNA per milliliter of packed beads and possibly more. After hybridization the sample is diluted with TNE such that the volume is at least 1 ml and the concentration of mercurated DNA is less than 50 μg/ml. The sample can be loaded either diluted with TNE or directly in the hybridization buffer, since the presence of up to 60% formamide has no effect on binding. Samples are loaded at room temperature with a flow rate of only 1–2 ml/hr. Quantitative binding seems to depend on leaving the mercurated DNA in contact with the sulfhydryl Sepharose for a sufficient length of time.

After loading, the column is washed under gravity at room temperature with at least 6 column volumes of 60% formamide in TNE. The important factors for the elimination of nonspecific binding include the presence of 100 m*M* NaCl, the presence of formamide (at least 20%), and prewashing of the sulfhydryl Sepharose in 97% formamide in TNE at 50° as noted in the preparation of the sulfhydryl Sepharose. In order to minimize nonspecific interactions 60% formamide is used in the washing step, since this is the highest concentration compatible with well-matched hybrids (Fig. 2). However, as shown in Figs. 2 and 3, the mercurated plasmid DNA–DNA and DNA–RNA hybrids are less stable than the unmercurated controls. Therefore, in cases where it is of interest to purify homologous but nonidentical sequences (such as for the purification of human globin sequences using mercurated mouse globin plasmid) the washing step should be done with less formamide (20–40%) so as not to destabilize hybrids of interest.

Since Dale and Ward[1] have demonstrated that the mercury–carbon bond is thermolabile, conditions were determined for the melting of the hybrids with only a minimal rise in temperature. The melting profile of hybrids in 97% formamide in TNE is shown in Fig. 3A for mouse β-globin cDNA. Under these conditions even unmercurated DNA–DNA hybrids

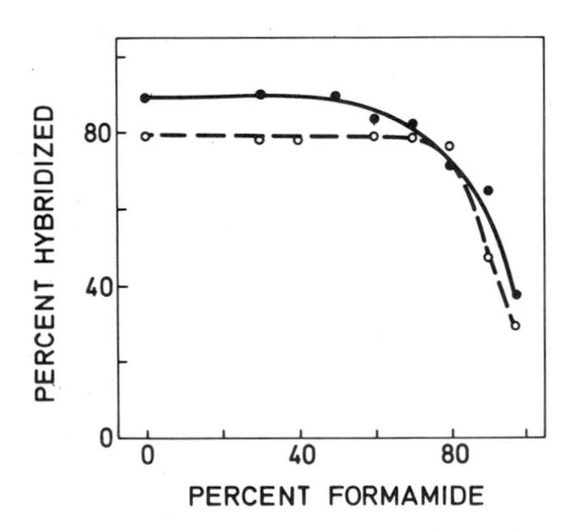

FIG. 2. Stability of DNA–DNA hybrids in formamide and TNE at room temperature. Rabbit β-globin plasmid DNA[4] was hybridized with partially purified rabbit β-globin [^{3}H]cDNA. Aliquots of the hybridization mix were precipitated separately and resuspended in TNE containing 0–90% formamide at room temperature (21°). The percentage hybridization at each formamide concentration was determined by S1 nuclease digestion.[3] ●, Unmercurated plasmid DNA; ○, mercurated plasmid DNA.

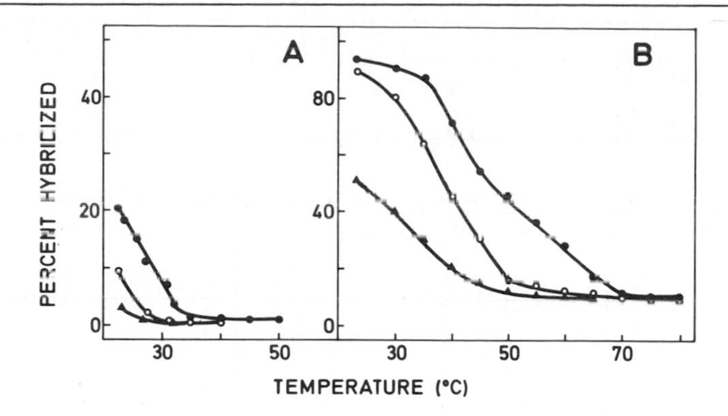

FIG. 3. Melting profiles of DNA DNA and RNA DNA hybrids in 97% formamide in TNE. (A) Melting of DNA–DNA hybrids. Mercurated or unmercurated mouse β-globin plasmid DNA[4] was hybridized to β-globin [³H]cDNA. The hybrids were precipitated in ethanol and resuspended in 97% formamide in TNE. At the indicated temperatures, aliquots were removed and the percentage hybridization was determined by S1 nuclease digestion. ●, Unmercurated plasmid DNA; ○, plasmid DNA mercurated for 2 hr; ▲, plasmid DNA mercurated for 72 hr. (B) Melting of DNA–RNA hybrids. As in (A), except that the hybrids were formed between mouse β-globin plasmid DNA and [³H]cRNA made from pCR1 plasmid DNA. Symbols are the same as in (A).

are melted at 35°. Therefore, 97% formamide in TNE and 40° are used as the melting conditions for DNA–DNA hybrids to allow a margin of safety for sequences of high GC content. The DNA–RNA hybrids are more stable in formamide and are completely denatured only at 70° for unmercurated hybrids and at 55°–60° for mercurated hybrids as shown in Fig. 3B.

For the *in situ* melting step, the column is therefore heated to 40° for DNA–DNA hybrids and to 65° for RNA–DNA hybrids, and elution is effected under gravity with 5 column volumes of 97% formamide in TNE which is preheated to the same temperature as the column. The eluted polynucleotides can be precipitated directly from 97% formamide in TNE with 2 volumes of ethanol at −20°. After precipitation it is necessary to wash the pellets twice in cold 70% ethanol.

A typical profile of sulfhydryl Sepharose chromatography is shown in Fig. 4A for the purification of mouse β-globin [³H]cDNA from total globin cDNA using a mercurated globin plasmid DNA.[4] No globin cDNA is retained when an unrelated mercurated plasmid DNA is used in the hybridizations (Fig. 4B). Recoveries of cDNA from the column are generally 80–100%.

Following melting elution of the nonmercurated target strand, about

[4] F. Rougeon and B. Mach, *Gene* **1,** 229 (1977); *J. Biol. Chem.* **252,** 2209 (1977).

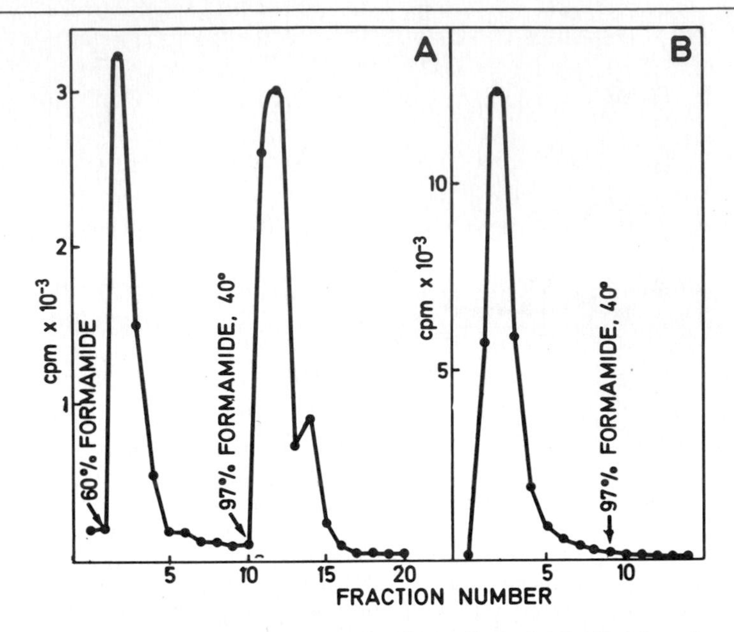

FIG. 4. Sulfhydryl Sepharose chromatographic profiles. (A) Purification of mouse β-globin [^{3}H]cDNA. Sonicated mouse β-globin plasmid DNA[4] was mercurated (36 hr) and hybridized to total mouse β-globin [^{3}H]cDNA. The hybrids were loaded on a 0.5 ml sulfhydryl Sepharose column, washed with 4 ml 60% formamide in TNE, and eluted with 5 ml 97% formamide in TNE at 40°. Fractions (0.5 ml) were collected, diluted with 1 ml water, and precipitated with an equal volume of cold 20% TCA. (B) Specificity of sulfhydryl Sepharose chromatography. Same as (A) except that mouse β-globin [^{3}H]cDNA was hybridized with an unrelated mercurated plasmid DNA (mouse myeloma light chain).

50–80% of the mercurated plasmid DNA can be recovered by the addition of 0.2 *M* mercaptoethanol. The plasmid DNA can then be reused following remercuration. Because of the incomplete release of mercurated plasmid from the sulfhydryl Sepharose, it is considered advisable to always use fresh aliquots of sulfhydryl Sepharose for each experiment.

The procedure described here is specifically designed to purify the unmercurated target sequence free of the mercurated probe sequence. To detect contamination by probe sequences a purified preparation of single-stranded target (such as cDNA or mRNA) is melted and hybridized to completion with the purified target acting as driver. The amount of target that has become double-stranded (assayed either by S1 nuclease digestion or hydroxylapatite chromatography) is used as a measure of contamination by probe DNA. Such analyses have indicated that occasionally purified target polynucleotides contain traces of probe DNA which cannot be eliminated by rechromatography on sulfhydryl Sepharose.[3] When residual probe contamination is observed, it can be eliminated either by velocity centrifugation in sucrose gradients for sequences larger than about 300 nucleotides (since the contaminants consist mainly of the smaller fragments from sonicated preparations) or alternatively by hybridization of the purified sequences followed by hydroxylapatite chromatography to recover only single-stranded material. This contamination is dependent on the conditions of mercuration of the probe DNA and, with prolonged times of mercuration (36–48 hr), usually no probe contamination is detected.

The methodology described here is potentially useful for the purification of any single-strand DNA or RNA sequence for which a pure homologous probe exists. With this method highly purified single-strand cDNA hybridization probes can be prepared for use in the sensitive quantitation of gene sequences by saturation hybridization.[3] The plus and minus strands of genomic DNA restriction fragments containing sequences homologous to a double-stranded mercurated probe can be highly enriched.[3] In addition, mRNAs or nuclear pre-mRNAs homologous to the probe can be purified for further studies on gene expression. Since mRNAs are translatable after purification by this method,[3] the translation product could be used to help characterize a previously unidentified genomic or viral probe sequence. Finally, it should be emphasized that any sequence that is sufficiently homologous with the probe to remain in hybrid form under the conditions of purification can be purified. Thus, genomic and viral sequences could be purified using probes derived from distant species providing the homology was sufficient.

[13] The Isolation of Extrachromosomal DNA by Hydroxyapatite Chromatography

By M. SHOYAB and A. SEN

Semenza[1] and Main *et al.*[2–4] were the first to report the use of hydroxyapatite (HAP) for fractionation of nucleic acids. Bernardi[5] and Miyazawa and Thomas[6] convincingly demonstrated that single- and double-stranded DNA could be separated by HAP chromatography. Since these earlier developments, HAP has been widely used in the preparation and

[1] G. Semenza, *Ark. Kemi* **11,** 89 (1957).
[2] R. K. Main and L. J. Cole, *Arch. Biochem. Biophys.* **68,** 186 (1957).
[3] R. K. Main, M. J. Wilkins, and L. Cole, *Science* **129,** 331 (1959).
[4] R. K. Maine, M. J. Wilkins, and L. Cole, *J. Am. Chem. Soc.* **81,** 6490 (1959).
[5] G. Bernardi, *Nature (London)* **206,** 779 (1965).
[6] Y. Miyazawa and C. A. Thomas, Jr., *J. Mol. Biol.* **11,** 223 (1965).

ISBN 0-12-181968-X

analysis of nucleic acids.[7,8] Britten *et al.*[9] reported a simple procedure for isolating DNA from animal tissues using HAP. Hirt[10] described a method for the separation of polyoma viral DNA from cellular DNA based on the preferential precipitation of high-molecular-weight cellular DNA in the presence of NaCl and sodium dodecyl sulfate (SDS) at 4°. During the course of our work involving the analysis of DNA–DNA and DNA–RNA hybridization mixtures on HAP columns, we consistently observed that high-molecular-weight native DNA molecules eluted very poorly in 0.4–0.5 *M* phosphate buffer (PB), pH 6.8, which is commonly used to elute the reannealed double-stranded DNA from HAP. This led us to develop a rapid and reproducible method for purifying the low-molecular-weight extrachromosomal DNA from eukaryotic cells virtually free of protein, RNA, and high-molecular-weight bulk cellular DNA.[11] We describe the procedure here using SV40 DNA in infected monkey kidney cells as an extrachromosomal DNA prototype.

Principle of the Method

HAP has a higher affinity for nucleic acid molecules with rigid and ordered structures than for those having disordered and flexible structures; this rule probably holds true for protein molecules. The interaction between phosphate groups of nucleic acids and calcium ions on the surface of HAP appears to play a major role in the adsorption of nucleic acid on HAP. When a cell lysate made in 0.24 *M* PB, pH 6.8, containing 8 *M* urea, 1% SDS, and 0.01 *M* EDTA is passed through HAP at room temperature, only double-stranded DNA is adsorbed on HAP and most other macromolecules pass through. The low-molecular-weight extrachromosomal DNA can then be eluted with 0.4–0.5 *M* PB, while high-molecular-weight bulk DNA remains trapped and adsorbed on HAP under these conditions.

Materials

1. Hydroxyapatite: DNA-grade BioGel HTP HAP can be purchased from Bio-Rad, Richmond, California, or prepared in the laboratory.[6] We recommend the use of HAP slurry made within a few hours of use in order to achieve a consistent and reasonable flow rate.

[7] G. Bernardi, this series, Vol. 21, p. 95.

[8] D. E. Kohne and R. J. Britten, *Proced. Nucleic Acid Res.* **2,** 500 (1971).

[9] R. J. Britten, M. Pavich, and J. Smith, *Carnegie Inst. Washington, Yearb.* **68,** 400 (1970).

[10] B. Hirt, *J. Mol. Biol.* **26,** 365 (1967).

[11] M. Shoyab and A. Sen, *J. Biol. Chem.* **253,** 6654 (1978).

2. PB, pH 6.8: Prepared by mixing equimolar quantities of Na_2HPO_4 and NaH_2PO_4.

3. Lysing solution: 8 *M* urea, 0.24 *M* PB, 1% SDS, 0.01 *M* EDTA.

4. Washing solution: 8 *M* urea in 0.24 *M* PB.

5. Column: The width and the length of a column depend on the amount of cell lysate to be processed.

6. Cells: Kidney fibroblasts from African green monkey (Vero line) are grown in Dulbecco's modified Eagle's medium supplemented with 10% fetal calf serum (GIBCO, Grand Island, New York). Confluent monolayers of cells are infected with SV40 at a multiplicity of 2–5 plaque-forming units per cell and maintained in media supplemented with 2% fetal calf serum. Viral DNA is preferentially labeled at 30 hr postinfection by incubating the infected cells in medium containing 2% dialyzed fetal calf serum and 100 μCi/ml of [^{3}H]thymidine for 4 hr. The total DNA of uninfected cells is labeled with [^{14}C]thymidine (10 μCi/ml) for a period of 12–14 hr.

Purification of Extrachromosomal DNA

The procedure can be outlined as follows.

Step 1. The monolayers are washed two to three times with cold phosphate-buffered saline (PBS) and harvested with trypsin–EDTA (0.1% trypsin, 0.5 m*M* EDTA solution). The cells are pelleted by centrifugation at 5000 *g* for 5 min at 4°. The supernatant is carefully drained, and the cell pellet used for DNA isolation.

Step 2. The lysing solution is added to the cell pellet (approximately 2 ml per 5 × 10^6 cells). The sample is mixed gently and allowed to stand at room temperature for 10–20 min, and the crude cell lysate is centrifuged at 80,000 *g* for 1 hr at 20°. The supernatant solution is carefully removed, and the loose pellet containing the high-molecular-weight DNA is gently resuspended in 1 ml of the lysing solution and recentrifuged as earlier. The first and second supernatant fractions are combined.[11a]

Step 3. The pooled supernatant fractions are applied to a column of HAP[11a] (1 ml bed volume) preequilibrated with 2–3 volumes of the lysing

[11a] The chromosomal DNA can be purified as follows. The loose pellet fraction containing about 99% of cellular DNA is resuspended in 1 ml of the lysing solution and sonicated at 10–12 kHz for 1 min, using an MSE ultrasonicator (the setting can be varied according to the final size of DNA desired). The total sheared DNA is purified from this mixture by HAP chromatography, as in step 3.

[11a] Solid HAP is suspended in 0.01 *M* phosphate buffer, pH 6.8, and after a few minutes fine particles are removed by decantation. The HAP suspension is then poured into a column to the desired height. The height and the diameter of the column should be approximately the same to provide a fast flow rate.

solution. The column is sequentially washed with the washing solution (8 ml) to remove proteins and RNA from HAP and then with 6 ml of 0.15 *M* PB to remove urea. The extrachromosomal DNA is finally eluted with 0.5 *M* PB (4–6 ml). The flow rate is maintained at 15–20 ml/hr.

Step 4. When only $2-4 \times 10^6$ cells are used, the crude cell lysate can be applied directly on the column without prior centrifugation to remove bulk cellular DNA; the extrachromosomal DNA can then be purified as described above. The recovery of viral DNA from infected cells without prechromatography centrifugation is virtually identical to that obtained when bulk DNA is first removed by centrifugation. The high-molecular-weight DNA remains trapped in the HAP column bed and is released by suspending the HAP in 0.5 *M* PB and ultrasonically treating the mixture. This material does not contain form I SV40 DNA. However, bulk DNA in the lysate of $>6 \times 10^6$ cells impedes the flow rate of the column, and centrifugation to remove this material is then necessary.

Step 5. Although we have detailed above the isolation of extrachromosomal DNA from only a few million cells, we have successfully used this method for the purification of extrachromosomal DNA from up to 3 ml of packed leukemic cells. In order to process large amounts of cells, approximately 20–30 ml lysis buffer is used for each milliliter of packed cells. In parallel, the HAP column bed size for such extracts is increased to $\simeq$30 ml. It is advisable to increase the diameter of the column for a desired flow rate of at least 40–60 ml/hr.

Characterization of Extrachromosomal DNA

The [^{3}H]thymidine-labeled DNA recovered in the 0.5 *M* phosphate buffer eluate of SV40-infected cell lysate was analyzed by alkaline sucrose velocity sedimentation. The majority of the DNA (about 95%) sedimented as SV40 form I DNA, while the remainder behaved as nicked viral DNA (Fig. 1A and B). Purified ^{3}H-labeled SV40 DNA was added to the crude lysate of a parallel unlabeled, infected culture. Approximately 90% of the radioactivity was recovered in the 0.5 *M* PB eluate and behaved identically to the untreated viral DNA when analyzed on an alkaline sucrose gradient (Fig. 1C). [^{14}C]Thymidine-labeled uninfected cells and [^{3}H]thymidine-labeled SV40-infected cells were also lysed directly on a parallel alkaline gradient and cosedimented (Fig. 1D). While the majority of uninfected ^{14}C-labeled cell DNA sedimented as molecules larger than 60 S, a major peak of ^{3}H radioactivity from the SV40-infected cells was seen to sediment at the position of form I SV40 DNA. When parallel batches of pulse-labeled SV40-infected Vero cells ($\sim 2 \times 10^6$) were extracted by the method of Hirt[10] or by the HAP chromatography described

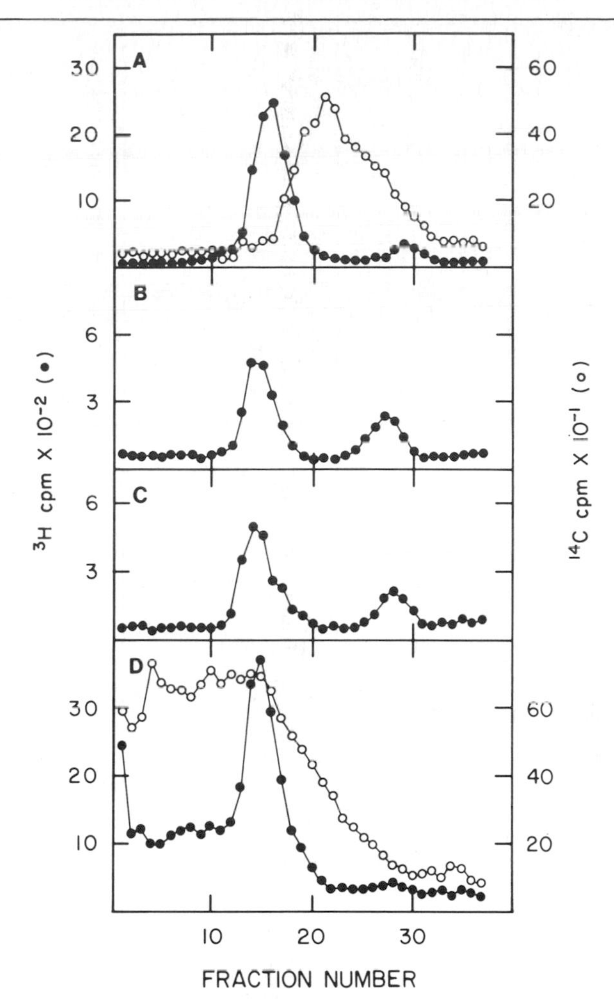

FIG. 1. Analysis of DNA in alkaline sucrose gradients. Samples (0.2 ml) were applied on 5 ml of 10–30% linear sucrose gradients containing 0.25 *M* NaOH, 0.75 *M* NaCl, and 0.01 *M* EDTA. The gradients were centrifuged at 30,000 rpm for 8½ hr at 10° in a Spinco SW 50.1 rotor. Approximately 125-μl fractions were collected from the bottom of the tubes directly on 3MM Whatman filter paper disks. The filters were dried and washed three times with 10% cold trichloroacetic acid followed by washing with acetone; the pads were then dried, and the radioactivity measured in a Beckman LS-250 liquid scintillation spectrometer. (A) [^{3}H]Thymidine-labeled extrachromosomal DNA from SV40-infected cells and [^{14}C]thymidine-labeled extrachromosomal DNA from uninfected cells. (B) ^{3}H-Labeled SV40 marker DNA forms I and II. (C) ^{3}H-Labeled SV40 DNA added exogenously to cell lysates and purified by the above method. (D) SV40-infected cells labeled with [^{3}H]thymidine, and uninfected cells labeled with [^{14}C]thymidine, lysed directly on the top of an alkaline gradient.

TABLE I

ANALYSIS OF THE ELUTION PATTERNS OF PROTEINS, RNA, AND EXTRACHROMOSOMAL DNA IN THE HAP COLUMN CHROMATOGRAPHY OF CELL LYSATE[a]

Cells and additions	Trichloroacetic acid-precipitable ^3H radioactivity in:			Trichloroacetic acid-precipitable ^{14}C radioactivity in:		
	Input (cpm)	Flow-through and washings (cpm)	0.5 *M* PB eluate (cpm)	Input (cpm)	Flow-through and washings (cpm)	0.5 *M* PB eluate (cpm)
[^{3}H]Leucine-labeled (5 μCi/ml, 8 hr) and [^{14}C]thymidine-labeled (2.5 μCi/ml, 24 hr)	2.7×10^7	2.71×10^7	ND[b]	5×10^6	ND	4.8×10^4
[^{3}H]Uridine-labeled (5 μC/ml, 4 hr) and [^{14}C]thymidine-labeled (2.5 μCi/ml, 24 hr)	1.2×10^7	1.21×10^7	3.1×10^4 (alkali-sensitive)	5×10^6	ND	4.75×10^4
Unlabeled cell extract and ^{3}H-labeled SV40 DNA	5.0×10^4	ND[b]	4.91×10^4	NA[b]	NA	NA

[a] Cells (2×10^6) labeled with either [^{3}H]leucine or [^{3}H]uridine were mixed with 2×10^6 cells labeled with [^{14}C]thymidine; the mixed cells were lysed and centrifuged, and the supernatant was chromatographed on a HAP column (~1 ml) as described.

[b] NA, Not applicable; ND, not detectable.

here, we recovered 7.1×10^5 and 7.4×10^5 cpm of [^{3}H]thymidine-labeled SV40 DNA, respectively. On the other hand, by direct lysis of cells on an alkaline gradient, only $5.6-5.8 \times 10^5$ cpm of ^{3}H radioactivity was recovered in the 53 S peak. The lower recovery with the latter method is presumably due to the trapping of viral DNA with high-molecular-weight cellular DNA in the absence of detergents and reducing agents. The data taken together suggest that (1) the majority of SV40 DNA from an infected culture is recovered in the 0.5 *M* PB-eluted extrachromosomal fraction and (2) there is no detectable degradation of DNA molecules during the purification procedure. In other experiments we have observed representative recoveries of all forms of replicating SV40 DNA by analyzing pulse-labeled (~5 min) extrachromosomal DNA from infected cells.

About 1% of the total DNA from uninfected normal fibroblasts was obtained in the extrachromosomal fraction, while the rest was recovered as bulk DNA in the prechromatography centrifugation pellet. An analysis of the size distribution of the extrachromosomal DNA from normal monkey, mouse, and human cells revealed a rather heterogeneous collection of predominantly linear molecules between 8 and 35 S (median, 26 S) in an alkaline sucrose gradient (Fig. 1A). Both the extrachromosomal DNA and the bulk DNA from normal mouse cells showed average densities of 1.7 and 1.76 g/ml in neutral and alkaline CsCl, respectively. The thermal denaturation profiles of both DNAs were also virtually identical in 0.12 *M* PB, and the thermal dissociation temperature for both was 83°.

To test the contamination of protein and RNA in the extrachromosomal DNA fractions, parallel cultures were labeled with [^{3}H]leucine or [^{3}H]uridine and mixed with [^{14}C]thymidine-labeled cells. The extrachromosomal DNA and the bulk DNA were isolated. The results presented in Table I indicate that the 0.5 *M* PB eluate did not contain any detectable [^{3}H]leucine-labeled protein and only $\geq$0.25% [^{3}H]uridine-labeled RNA. In another experiment, we found that more than 90% of labeled bacteriophage λ DNA, added exogenously to a crude lysate of *Escherichia coli* spheroplasts, eluted in the extrachromosomal fraction.

General Comments and Conclusions

The method described here for the isolation of low-molecular-weight extrachromosomal DNA is efficient and has several advantages over the procedure described by Hirt.[10] The DNA obtained by this procedure is virtually free of protein and RNA, whereas the Hirt procedure requires subsequent organic solvent extraction for the removal of proteins and equilibrium density banding for the removal of RNA. This method is rapid and simple and allows the analysis of several samples in a few hours. By avoiding the ethanol precipitation step the recovery of total DNA is considerably increased. The yield, using SV40 DNA as a prototype, is reproducibly greater than 95% as compared to the 80–85% yield commonly obtained with the Hirt procedure.[10] HAP is a stable matrix and can be used over a wide range of temperatures, as well as in the presence of salts and organic reagents. A rapid flow rate can be achieved by using positive pressure without any loss of recovery or resolution. Using the method of Hirt,[10] one obtains a widely variable fraction ranging between 2 and 15% of the total cell DNA in the low-molecular-weight DNA supernatant; therefore, it is difficult to perform meaningful analyses of the low-molecular-weight DNA species. Using the procedure described here, we have reproducibly obtained approximately 1% of the total cell DNA in the extrachromosomal fraction of normal cells from different species. Preliminary experiments have shown the presence of a higher level of labeled DNA in the extrachromosomal fraction of certain virally and chemically transformed mouse cells. It will now be possible to analyze the properties of this small fraction of cellular DNA which is present in the low-molecular-weight form in tissues or cultured cells and to study the possible correlation between the relative abundance of these molecules and the physiological states of cells and tissues.

In conclusion, this procedure represents a powerful technique for the isolation of extrachromosomal DNA from mitochondria, viruses, and plasmids (including those containing recombinant DNA) from normal and transformed eukaryotic cells, as well as for studies on the integration of viral DNA into the cellular genome.

[14] Isolation of Specific RNAs Using DNA Covalently Linked to Diazobenzyloxymethyl Cellulose or Paper

By MICHAEL L. GOLDBERG, RICHARD P. LIFTON, GEORGE R. STARK, and JEFFREY G. WILLIAMS

Substantial amounts of DNA can be linked covalently to finely divided DBM-cellulose[1,2] or to DBM paper.[3] Either form of insoluble DNA can then be used to select specific species of RNA by hybridization. If the RNA is radioactive, DNA paper can be used to analyze the fraction of RNA complementary to the insoluble DNA in a procedure very similar

[1] The abbreviations used are: ABM, aminobenzyloxymethyl; DBM, diazobenzyloxymethyl; SDS, sodium dodecyl sulfate.

[2] B. E. Noyes and G. R. Stark, *Cell* **5**, 301 (1975).

[3] G. R. Stark and J. G. Williams, *Nucleic Acids Res.* **6**, 195 (1979).

ISBN 0-12-181968-X

to the one originally developed by Gillespie and Spiegelman[4] using DNA bound noncovalently to nitrocellulose filters. Either DNA-cellulose or DNA paper can be used preparatively to select intact, specific RNAs that can be used subsequently for translation *in vitro*, as a template for reverse transcriptase, as a source of pure infectious RNA, or for any other experiment in which pure, intact RNA is required.[5] Both DNA-cellulose and DNA paper can be reused many times.

Principle

The reactions shown in Scheme 2 of Alwine *et al.*[6] are used for linking DNA covalently to cellulose or to paper. Since only single-stranded DNA reacts with DBM groups, the coupling reaction is carried out in the presence of a denaturant or after heating the DNA, under conditions where reannealing is slow. Hybridization of a specific RNA to DNA paper or DNA-cellulose is virtually complete under conditions of DNA excess and, using stringent conditions for washing, the background of RNA bound nonspecifically can be reduced to 0.01–0.02% of the input, with retention of about 80% of the specific RNA originally bound.

Materials

Preparation of DNA

For coupling, the DNA should be free of contaminating proteins, RNA, and especially amines or any other molecule that would react with DBM groups. Covalently closed, circular DNA must be nicked so that the strands can separate when denatured. Several alternative procedures may be used: cleavage with one or more restriction endonucleases, mechanical shear, limited cleavage with DNase I (which may be controlled by the

[4] D. Gillespie and S. Spiegelman, *J. Mol. Biol.* **12,** 829 (1965).

[5] It may be possible to use DNA-cellulose or DNA paper to select specific DNA fragments. Such a technique might be useful, for example, to increase the fraction of specific sequences in a population of restriction fragments prior to cloning. Noyes and Stark (see also footnote 2) were able to hybridize 96–99% of sonicated SV40 DNA (size 6 to 7 S) to a 24- to 48-fold excess of SV40 DNA-cellulose in 24 hr at 37° in buffered 50% formamide containing 0.6 *M* NaCl. However, the details of isolating DNA in this way have been explored very little. DNA–DNA reannealing in solution is much faster than hybridization of soluble DNA to complementary immobilized DNA. The reannealing problem can be overcome at least partially by using a large excess of immobilized DNA. DNA-cellulose may perform better than DNA paper in such cases because the cellulose can be dispersed much more evenly throughout the solution (see Fig. 1).

[6] J. C. Alwine, D. J. Kemp, B. Parker, J. Reiser, J. Renart, G. R. Stark, and G. M. Wahl, this volume [15].

presence of ethidium bromide), limited hydrolysis with alkali at high temperature, or cleavage mediated by uv light with ethidium bromide.[7] The DNA fragments should be relatively long (fragments in the range of 300 to 5000 base pairs have been used successfully), and it is probably a good idea to cleave DNA of low complexity (e.g., DNA from a plasmid) into pieces small enough that each single-stranded fragment will be unlikely to find a complementary neighbor after attachment to the cellulose or paper and therefore will not be able to renature during hybridization of the RNA. In the case of plasmid DNAs amplified in the presence of chloramphenicol, we have used alkaline hydrolysis (2 hr in 0.5 *M* NaOH at 37°) to cleave the RNA linkers that replace short stretches of DNA in the majority of molecules.[8]

Preparation of DNA-Cellulose

ABM-Cellulose. We have used commercially available ABM-cellulose (Miles-Yeda) most often. However, different batches of this material have varied widely in their ability to couple DNA, probably because of the instability of the ABM groups upon prolonged storage under certain conditions (see Alwine *et al.*[6] for conditions which preserve the ABM groups). It is wise to test each batch of cellulose in a preliminary experiment to determine the efficiency of coupling, using any available sample of radioactive DNA.

Alternatively, one can synthesize the ABM-cellulose readily, starting with Whatman microgranular cellulose CF-11 and following the procedure for ABM paper given in Alwine *et al.*[6] (see also Gurvich *et al.*[9]). Homemade material has given excellent results in several laboratories.

Diazotization. The procedure described is for 500 μg of DNA. First, the ABM-cellulose is converted to a finely divided form by dissolving and reprecipitating it.[10] Prepare 10 ml of ammoniacal $Cu(OH)_2$ (0.45 g $Cu(OH)_2$, 0.1 g sucrose, 6 ml concentrated NH_4OH per 10 ml of solution). Vortex the mixture in a hood and allow the excess $Cu(OH)_2$ to settle. To the supernatant solution, add 100 mg of ABM-cellulose, vortex the mixture, add another 5 ml of concentrated NH_4OH, and vortex again. Remove any undissolved material by centrifugation for 5 min at 10,000 *g* and add the supernatant solution to 90 ml of distilled water preheated to 70°. Reprecipitate the cellulose by adding 25% (v/v) H_2SO_4 dropwise to pH 6. [The blue color of the $Cu(OH)_2$ is barely visible at this point.] Col-

[7] P. A. Martens and D. A. Clayton, *Nucleic Acids Res.* **4,** 1393 (1977).

[8] D. G. Blair, D. J. Sherratt, D. B. Clewell, and D. R. Helinski, *Proc. Natl. Acad. Sci. U.S.A.* **69,** 2518 (1972).

[9] A. E. Gurvich, O. B. Kuzovleva, and A. E. Tumanova, *Biokhimiya* **26,** 934 (1961).

lect the cellulose by centrifuging the mixture for 5 min at 10,000 *g* and discard the supernatant solution. Resuspend the cellulose in 25 ml of ice-cold distilled water, transfer it to a 30-ml Corex tube, and collect it again by centrifugation. (Losses of cellulose are minimized by using a swinging-bucket rotor for all centrifugations.) Wash the finely divided ABM-cellulose twice more in ice-cold water to remove the last traces of $Cu(OH)_2$.

Resuspend the reprecipitated, washed cellulose in 33 ml of ice-cold 1.2 *M* HCl and add 0.9 ml of a freshly made solution of $NaNO_2$ (10 mg/ml). Stir this mixture for 30 min at 0°. Test for excess HNO_2 by touching a drop of the mixture to starch iodide paper. Collect the cellulose by centrifugation at 0° and wash it sequentially in the cold with 15 ml each of distilled water, buffer (see below), and a mixture of 1 part buffer plus 4 parts dimethyl sulfoxide. Work as rapidly as possible to minimize hydrolysis of the DBM groups. The cellulose usually begins to turn yellow as soon as the nitrous acid is removed (see Alwine *et al.*[6]).

Coupling of the DNA. The solution of DNA should be prepared ahead of time. Just before the diazotization is finished, suspend the DNA in 1 volume of buffer (use 1 ml for 500 μg) and heat it to 80° for 1 min to be sure that it is in solution completely. Cool the solution and then add 4 volumes of dimethyl sulfoxide.[11] Successful coupling has been achieved with the 0.2 *M* borate buffer, pH 8, used originally by Noyes and Stark,[2] but more recent work has made it clear that a lower pH is better, and we now recommend using 25 m*M* sodium phosphate buffer, pH 6.0 (*not* potassium phosphate, which forms a precipitate with the dimethyl sulfoxide) or 20 m*M* citric acid–Na_2HPO_4, pH 4.0. Resuspend the pellet of washed DBM-cellulose in 5 ml of buffer–dimethyl sulfoxide–DNA and then incubate this mixture at 4° for 24–48 hr with slow stirring to prevent the cellulose from settling. Collect the DNA-cellulose by centrifugation and wash it twice with buffer–dimethyl sulfoxide at 50°. Wash the cellulose

[10] If homemade cellulose is being used, it may not be necessary to carry out the cycle of solution and reprecipitation at all. For example, J. C. Alwine (personal communication) has obtained excellent results with microgranular CF-11 cellulose without precipitation. Precipitation at the ABM stage has sometimes caused loss of capacity, and precipitation before reaction with nitrobenzyloxymethyl pyridinium chloride has yielded intractable material. Therefore, if precipitation is needed, it is probably best to do it with the chemically stable nitro derivative of the cellulose.

[11] It is important to dissolve the DNA in buffer first and then add the dimethyl sulfoxide second, since it is extremely difficult to dissolve the DNA directly in the buffer–dimethyl sulfoxide mixture. We have used dimethyl sulfoxide without purification successfully. Use the best grade available. If coupling yields are low, it may be helpful to distill the dimethyl sulfoxide in a vacuum, to remove traces of impurities which may react with diazo groups.

four times at room temperature with 0.14 *M* NaCl–15 m*M* trisodium citrate–20 m*M* sodium phosphate, pH 7.0 and then twice with 99% formamide containing 0.1% SDS, heating the mixture in formamide to 80° before each centrifugation. Finally, wash the DNA-cellulose twice at room temperature with hybridization-storage buffer [50% formamide, 0.6 *M* NaCl, 75 m*M* trisodium citrate, 100 m*M* sodium phosphate, pH 7.0, 0.1% SDS, 200 μg/ml poly(rA), 200 μg/ml *Escherichia coli* tRNA] and then resuspend it in 5 ml of this buffer. Store the cellulose at 4°, with occasional changes of buffer.

About 40% of the DNA will be coupled in borate buffer at pH 8.0. Better yields can probably be obtained with buffers of lower pH, which have given 60–70% yields with DBM paper (see below). Electron micrographs of finely divided cellulose, with and without DNA, are shown in Fig. 1.

Preparation of DNA Paper

DBM Paper. This material is prepared exactly as described by Alwine *et al.*[6]. It is convenient to cut the paper into circles with an area of about 1 cm^2 before starting the reduction and diazotization. The circles should fit into flat-bottomed plastic or siliconized glass tubes. (The inserts commonly used with scintillation vials are very convenient.) Each circle can then be coupled to DNA individually; the tubes also serve subsequently as convenient vessels for hybridization.

Coupling of the DNA. Perform all the following manipulations in a cold room. After diazotization of the ABM paper circles for 30 min, wash the DBM circles batchwise four times with about 100 ml each of ice-cold water (total time 1–2 min) and once with 20 ml of ice-cold buffer. Remove individual filters with tweezers and blot them damp-dry with cold paper towels. Place them in individual vials and add the solution of DNA without delay. Prepare the DNA ahead of time by dissolving it in 1 volume of buffer, with heating to 80° for 1 min. Cool and then add 4 volumes of dimethyl sulfoxide. The final concentration of DNA should be high, up to 2 mg/ml. Add 10–20 μl of DNA solution to each circle, cap the vials, and allow them to stand overnight at room temperature. Wash the filters individually or batchwise several times with large amounts of water at room temperature to remove the solvent, and then wash them four times for 10 min each at 37° with 0.4 *M* NaOH, about 1 ml per circle. After a further three washes with water, blot the filters damp-dry, number them with a pencil for future identification, and store them at 4° in hybridization buffer (see below). Circles stored in this way are stable for several months at least, and probably much longer. During the NaOH washes 30–40% of the bound DNA is lost. However, the DNA remaining is held very firmly

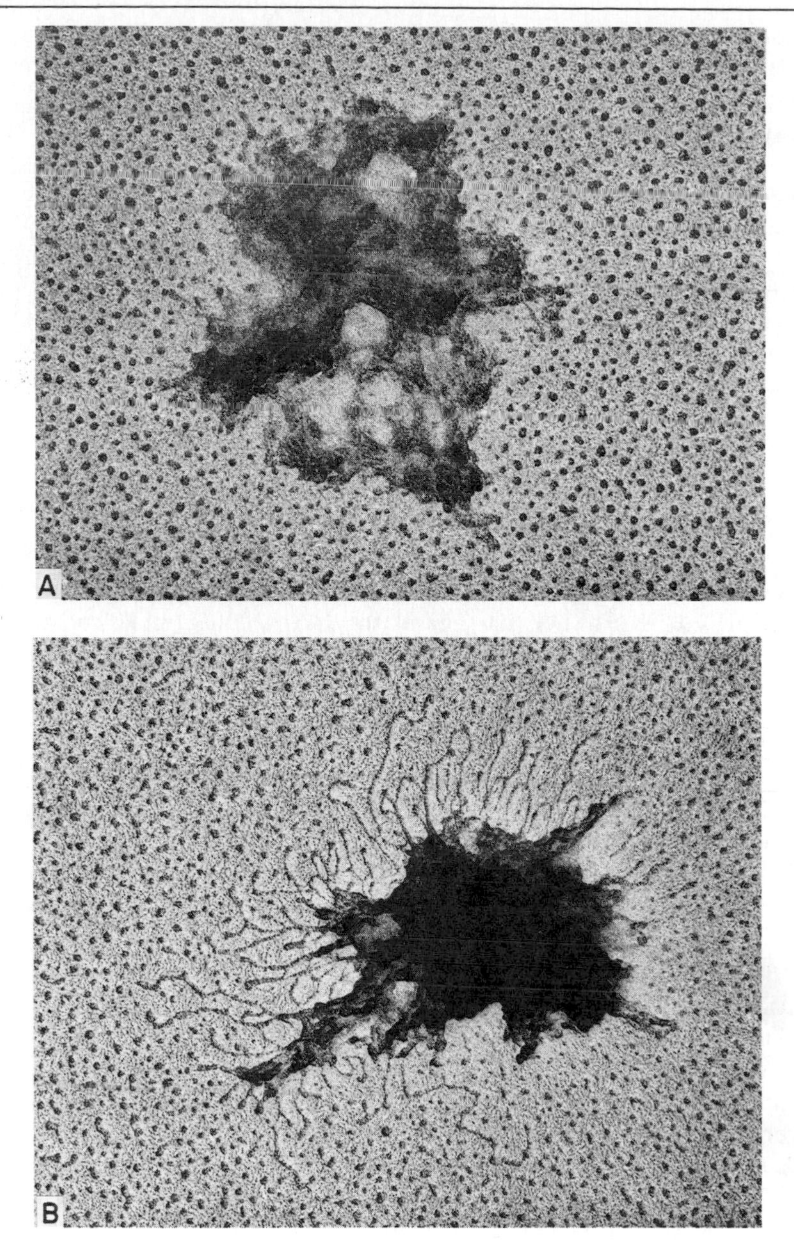

FIG. 1. Electron micrographs[2] of finely divided cellulose before and after coupling to DNA. 5.4 cm = 1 μm. (A) Control cellulose, diazotized and allowed to hydrolyze in the absence of DNA. (B) SV40 DNA-cellulose, 22 μg DNA/mg cellulose.

and survives many cycles of use. Starting with 40 μg of DNA, 25 μg can be bound stably to each 1-cm² circle. However, an even larger amount could probably be bound by starting with more DNA.

An alternative to the procedure using dimethyl sulfoxide has been used by Carl Parker (unpublished), with comparable results. Supercoiled plasmid DNA was denatured and fragmented in 0.33 *M* NaOH at 100° for 5 min. After cooling to 0–4°, an equivalent amount of 1 *M* HCl was added immediately, followed by 100 m*M* sodium phosphate buffer, pH 6.0, to a final concentration of 20 m*M*. The final volume should be no larger than 100 μl for each 1-cm² filter, and preferably should be smaller. The DNA solution was added immediately to fresh DBM paper. The papers were incubated overnight at room temperature and then washed as above.

Purification of Intact RNAs

Method Using DNA-Cellulose

Collect the DNA-cellulose from hybridization-storage buffer [50% formamide, 0.6 *M* NaCl, 75 m*M* trisodium citrate, 100 m*M* sodium phosphate, pH 7.0, 0.1% SDS, 200 μg/ml poly(rA), 200 μg/ml *E. coli* tRNA] by centrifugation and resuspend it in fresh buffer.[11a] Add enough of the suspension to the hybridization mixture to bring the DNA concentration to 5–20 μg/ml. Add the heterogeneous RNA, dissolved in a small volume of formamide or water, to bring the concentration of the specific RNA species to be purified to 1 μg/ml or less. Swirl the mixture to suspend the cellulose, heat it to 80° for 1 min, cool it to 42°, and allow annealing to proceed at this temperature for 8–20 hr, with continuous mixing. Sometimes the cellulose forms a matrix following the heating step. If this occurs, disperse the matrix by swirling the mixture vigorously.

If the final volume is 1 ml or less, a microfuge tube is a convenient reaction vessel. Fix the tube to a rack attached to a mechanical device that inverts the sample repeatedly (e.g., Labindustries Labquake). Routine care should be taken to avoid contamination with RNase. For example, the microfuge tubes should be rinsed with an aqueous solution of 0.1% diethylpyrocarbonate and dried at 65° before use. If the reaction volume is larger than 5 ml, the hybridization mixture can be placed in a 25- or 50-ml, capped Erlenmeyer flask, which is incubated in a shaking water bath to keep the DNA-cellulose in suspension. Siliconize the flask (e.g., with Dri-film SC-87) and bake it at 200° before use.

[11a] Hybridization of RNA to DNA paper can be carried out advantageously under conditions that allow RNA–DNA hybrids to form in preference to DNA–DNA hybrids (see below). Although such conditions have not been used with DNA cellulose, the results should be comparable.

When the annealing reaction is completed, collect the cellulose by centrifugation, either in the microfuge or by transferring the larger hybridization mixture to a polyallomer tube pretreated with 0.1% diethylpyrocarbonate. Spin the tube at 6000 *g* for 2–3 min in a swinging-bucket rotor. Remove the supernatant solution carefully and resuspend the DNA-cellulose pellet in 5–6 volumes of hybridization buffer. Incubate this suspension at 37°–42° for 1 min and then collect the cellulose by centrifugation. Repeat this process three to six times, followed by three to four more washes in 0.24 *M* NaCl–30 m*M* trisodium citrate–40 m*M* sodium phosphate, pH 7.0, at room temperature. If the RNA is labeled with a radioisotope, the supernatant solutions can be monitored to determine whether more washes are necessary. (Washing under more stringent conditions is also possible—see the procedure for DNA paper described below.) When washing is complete, liberate the annealed RNA by resuspending the cellulose in 99% formamide and heating the mixture to 65° for 1 min. A small volume of formamide is sufficient, just enough to wet the cellulose to a thick slurry. Collect the supernatant solution after centrifugation, taking care not to include any DNA-cellulose. Repeat this step once more. Two steps generally remove more than 95% of the bound RNA. If necessary, the pooled supernatant solutions can be centrifuged again to remove the last traces of cellulose. The RNA can be precipitated directly from the formamide by adding tRNA or another carrier to 20 μg/ml, 0.1 volume of 3 *M* NaCl or sodium acetate, and 2 volumes of ethanol, followed by cooling to −20° overnight. To recover the DNA-cellulose, heat it to 65° in a large volume of 99% formamide and then centrifuge the mixture. Resuspend the pellet in hybridization-storage buffer and store it at 4° until the next use. We have used a single batch of DNA-cellulose successfully 20 times.

Purification of the RNAs Homologous to Dm500 DNA

Dm500 is an 8-kb segment of the *Drosophila* genome generated by shear. It has been joined to the 6.7-kb bacterial plasmid ColE1 by the AT joining method.[12] This clone (cDM500) is homologous to sea urchin histone mRNA.[12] The organization of the histone genes in *Drosophila* has been determined.[13] The RNA of the K_c line of *Drosophila melanogaster* tissue culture cells[14] was labeled to a specific activity of 1×10^6 cpm/μg with [^{32}P]orthophosphate, and the total cellular RNA was purified. Hybrid-

[12] R. P. Lifton, M. L. Goldberg, R. W. Karp, and D. S. Hogness, *Cold Spring Harbor Symp. Quant. Biol.* **42,** 1047 (1978); R. W. Karp and D. S. Hogness, *Fed. Proc., Fed. Am. Soc. Exp. Biol.* **35,** 1623 (1976).

[13] R. P. Lifton, M. L. Goldberg, and D. S. Hogness, in preparation.

[13a] M. L. Goldberg, R. P. Lifton, and D. S. Hogness, in preparation.

[14] G. Escholier and A. Ohanessian, *In Vitro* **6,** 162 (1970).

ization of this RNA to an excess of cDM500 DNA bound to nitrocellulose filters[4] revealed that about 0.2% of the RNA was homologous to cDM500 DNA. Therefore, 400 μg of the RNA contains 0.8 μg of specific sequences. Since about 4 kb of one strand of cDm500 DNA is homologous to the RNA, 40 μg of total plasmid DNA contains about 5 μg of complementary sequence. A hybridization mixture (0.5 ml) with these quantities was heated to 80° and incubated at 42° for 20 hr. After annealing, the DNA–RNA hybrids were washed, and the bound RNA was eluted and precipitated as described above. Part of this RNA was passed through the hybridization-elution procedure again. The precipitated RNA was collected by centrifugation, resuspended in deionized formamide, heated to 65°, and centrifuged again, to be sure that the RNA was free of any bits of DNA-cellulose. Five discrete RNA species were purified virtually to homogeneity by two cycles of hybridization. An autofluorogram[15] of these RNAs following fractionation on an acrylamide gel in 96% formamide is shown in Fig. 2. In each hybridization, about 70% of the RNA complementary to cDM500 was recovered. The overall purification was about 400-fold. We have established that each of these RNAs encodes one of the five histones.[12] The pure RNAs retain function, since unlabeled mRNAs purified by the same procedure can be translated efficiently *in vitro* following extraction from the gel.[16]

Purification of SV40-Specific RNA for Translation[17]

Poly (A^+) mRNA from SV40-infected cells (25–50 μg) was hybridized to SV40 DNA-cellulose, containing 100–200 μg of full-length linear DNA, in hybridization buffer containing 50% formamide, 0.1 *M* Tris, pH 7.5, 1 m*M* EDTA, 0.1% SDS, 0.6 *M* NaCl, 0.5 mg/ml mouse ascites tRNA, and 0.2 mg/ml single-stranded *Escherichia coli* DNA.[18] The reaction was carried out overnight at 37° with swirling, and the cellulose was then washed six times at 37° with buffer containing 10 m*M* Tris, pH 7.5, 1 m*M* EDTA, 0.1% SDS, and 50% formamide. The hybridized RNA was eluted in 1 ml of buffer containing 10 m*M* Tris, pH 7.5, 1 m*M* EDTA, 0.1% SDS, and 98% formamide at 70° for 2 min, and the same elution was repeated. After adding carrier tRNA to the combined eluates, the RNA was precipitated with ethanol twice before it was translated in a mouse L-cell system rendered messenger-dependent by treatment with micrococcal nuclease.[19] Specific translation products could be seen on SDS gels

[15] W. M. Bonner and R. A. Laskey, *Eur. J. Biochem.* **46,** 83 (1974).
[16] M. L. Goldberg and D. S. Hogness, in preparation.
[17] E. Paucha and A. E. Smith, *Cell* **15,** 1011 (1978).
[18] In this case, single-stranded DNA was used to replace poly(rA) in the hybridization buffer of Noyes and Stark[2] because poly(rA) is inhibitory in the translational system used.
[19] E. Paucha, R. Harvey, and A. E. Smith, *J. Virol.* **28,** 154 (1978).

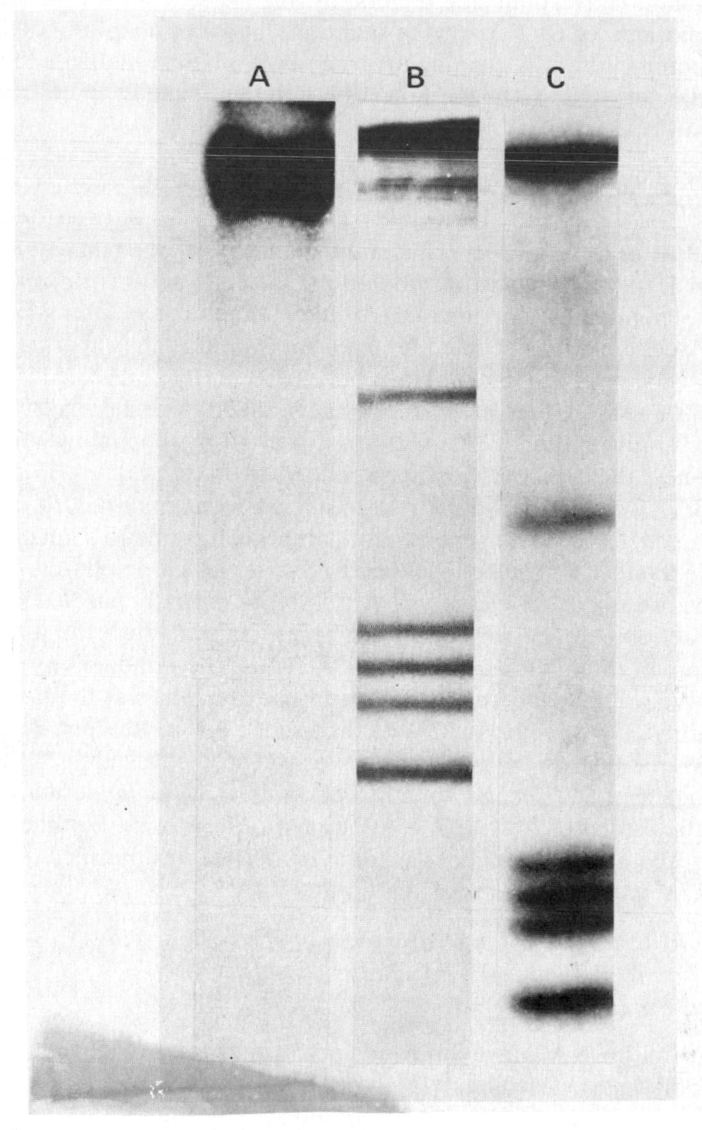

FIG. 2. Electrophoretic analysis of RNAs purified by hybridization to cDM500-cellulose or cDM500 paper. A 6% acrylamide–0.2% bisacrylamide slab gel was polymerized in 96% formamide. The RNAs were prepared as described in the text, and 10,000 cpm of each ^{32}P-labeled RNA was fractionated by electrophoresis at 15 mA, 90 V for 13 hr (A and B) or 19 hr (C). The gel was prepared for autoradiography and exposed to XR-5 x-ray film for 70 hr. (A) Total cellular RNA. (B) RNA eluted from cDM500-cellulose. (C) RNA eluted from cDM500 paper. The five lower bands in gels (B) and (C) are the individual histone mRNAs. The uppermost band is DNA, since it is sensitive to DNase but not RNase.

without immunoprecipitation. The technique was used to show that cells infected by mutants of SV40 carrying deletions between map units 0.54 and 0.59 contained little RNA coding for fragments of small-t antigen. The normal mRNA for small-t antigen could be detected readily in cells infected by wild-type virus.

Purification of Polyoma-Specific ^{32}P-Labeled RNA for Analysis of 5'-Terminal Cap Structures[20]

Total nuclear or cytoplasmic RNAs, labeled with [^{32}P]phosphate, were isolated from 15–90 m*M* plates of infected mouse cells and hybridized to 100 μg of full-length linear polyoma DNA linked to cellulose. After 4 hr at 37°, the DNA-cellulose was pelleted for 5 min at 10,000 *g*, the supernatant solution was decanted, and 3 ml of wash buffer (50 m*M* Tris-HCl, pH 7.5, 0.1% SDS, 50% recrystallized formamide) was added. The cellulose was resuspended in wash buffer at 37° for 2 min, followed by repelleting. After six such washes, the cellulose was resuspended in elution buffer (50 m*M* Tris-HCl, pH 7.5, 0.1% SDS, 99% recrystallized formamide) at 70° and again centrifuged. The pooled supernatant solutions from three sequential elutions were diluted with an equal volume of water and precipitated with ethanol in the presence of 50 μg of yeast RNA. The partially purified viral RNA was fractionated again by a second hybridization. With the use of 1.3×10^{10} cpm of total cytoplasmic RNA, 99.7% of the radioactivity was removed in the original supernatant solution and the washes, the last of which contained 3×10^6 cpm (0.02% of the input); 3.7×10^7 cpm (0.3%) was eluted from the DNA-cellulose as above, and 65% of the eluted RNA hybridized again in the second round. The final yield of twice-purified polyoma cytoplasmic RNA was 2.4×10^7 cpm (0.2% of total). For nuclear RNA, 4.3×10^7 cpm of polyoma-specific material was purified from 4×10^9 cpm, representing 1.1% of the input.

Purification of RNase-Resistant Fragments of RNA Following Hybridization to Polyoma DNA-Cellulose Made with Restriction Fragments[21]

Labeled viral mRNA was hybridized to 10 μg of restriction fragment *Hpa*II-3 of polyoma DNA (about 15% of the genome, corresponding to the late leader region), immobilized on cellulose. Proteinase K (1 μg/ml) was added to the hybridization buffer above, and the reaction was for 16 hr at 37°. The washing procedure was as just described, except that SDS was eliminated from the last three washes. The cellulose was suspended in 200 μl of buffer (500 m*M* NaCl, 10 m*M* Tris, pH 7.6, 1 m*M*

[20] A. J. Flavell, A. Cowie, S. Legon, and R. Kamen, *Cell* **16,** 357 (1979).
[21] S. Legon, A. J. Flavell, A. Cowie, and R. Kamen, *Cell* **16,** 373 (1979).

EDTA), containing 10 μg/ml of RNase T1, and incubated at 20° for 30 min. The RNA fragments released and those retained by the cellulose were obtained separately for analysis.

Attempts are being made by Paucha and Smith[17] and by Flavell *et al.*[20] to purify unique viral RNAs by hybridization to immobilized small restriction fragments containing DNA complementary to only a single RNA (e.g., the 0.54–0.59 region of SV40).

Purification of Collagen mRNA[22]

Poly (A^+) RNA derived from chick cells actively synthesizing collagen was hybridized to collagen cDNA coupled to cellulose. The cDNA fragment had been cloned in *E. coli* 1776 using the vector PBR 322. Following hybridization in a buffer containing 50% formamide (see above) and washing, the eluted RNA gave full-length collagen when translated in the nuclease-treated rabbit reticulocyte system. The hybridized cellulose was washed four times for 15 min each at room temperature with (0.15 *M* NaCl, 15 m*M* trisodium citrate, 0.2% SDS, 1 m*M* EDTA), once for 2 hr at 37° with hybridization buffer, four times for 15 min each at room temperature with wash buffer, and four times for 15 min each at 37° with hybridization buffer. The cellulose pellet was then washed once for 1 hr at 37° with 200 μl/mg of buffer (95% formamide, 0.2% SDS, 10 m*M* Tris, pH 7.5, 1 m*M* EDTA) per milligram of cellulose. Interestingly, much of the nonspecific RNA remaining was removed by this treatment, leaving a major fraction of the collagen mRNA still bound. A second wash with the same buffer under the same conditions removed 50–90% of the collagen translational activity from the cellulose. The collagen mRNA was up to 90% pure, as judged by the proportion of full-length collagen made in the messenger-dependent translational system.

Method Using DNA Paper

DNA paper is much more convenient to use than DNA-cellulose and may also provide lower backgrounds. The five histone mRNAs from *Drosophila* shown in Fig. 2C were purified using DNA paper under conditions similar to those used with DNA cellulose. DNA paper made with a cloned plasmid containing a 1.7-kb insert complementary to the 3′-region of vitellogenin mRNA (6.6 kb) was used to isolate this mRNA from *Xenopus* liver.[23] The eluted RNA gave a strong signal for full-length vitellogenin protein following translation in a cell-free system. No such product was detected using control DNA paper made with a plasmid con-

[22] S. L. Adams, J. C. Alwine, I. Pastan, and B. deCrombrugghe, *J. Biol. Chem.* **254,** 4935 (1979).

[23] D. F. Smith, P. Searle, and J. G. Williams, *Nucleic Acids Res.* **6,** 487 (1979).

taining an insert unrelated to vitellogenin mRNA. A vitellogenin cDNA paper that had been used in 25 experiments worked at least as well as a freshly made cDNA paper. DNA paper was made with a plasmid containing a cDNA insert complementary to an abundant *Dictyostelium discoideum* mRNA and used to isolate the specific mRNA at different times of development.[24] The mRNA was translated *in vitro* to yield a specific peptide and, using this assay, the time of hybridization was found to be optimal at about 4 hr, and a filter bearing 10 μg of plasmid DNA was not saturated by up to 40 ng of mRNA. Carl Parker (unpublished observations) has used DNA paper, derived from a plasmid 132E3 that contains a segment of *Drosophila* DNA complementary to a 2-kb mRNA abundant after heat shock. Starting with the total RNA made *in vitro* in isolated *Drosophila* nuclei, he isolated a labeled 2-kb transcript about the same size as the cytoplasmic mRNA found *in vivo*.

Greg Guild (unpublished) has used conditions for hybridization that favor the formation of R loops[25] to maximize the capacity of DNA paper. Using 1 cm^2 filters carrying 20 μg of stably bound plasmid DNA containing a *Drosophila melanogaster* rDNA repeat unit, sheared to 500 bases, approximately 90% of the rRNA present in a salivary gland preparation could be removed in one cycle of hybridization, using only a fourfold excess of immobilized rDNA. The quality of the purified rRNA was judged to be excellent by gel electrophoresis. The RNA sample in 1.9 μl of water was added to a mixture of 35 μl of 99% formamide, 1 μl of 10% SDS, and 0.1 μl of 0.5 *M* EDTA, pH 8. After heating to 100° for 30 sec and cooling to 4°, 2 μl of 0.5 *M* PIPES buffer, pH 6.8, and 10 μl of 5 *M* NaCl were added and 5 μl portions were applied to each of 10 filters. The filters, stacked together, were wrapped in parafilm and incubated in a humid chamber for 4–20 hr. After washing, the hybridized RNA was recovered in 99% formamide after heating to 56° for 20 min. Hybridizations under R loop conditions can probably be used to advantage with DNA cellulose and also in quantitative analyses of labeled RNAs, although these applications have not yet been tested.

Quantitative Analysis of Specific Labeled RNAs Using DNA Paper[3]

Hybridization of SV40 cRNA to DNA Paper

Circles (1 cm^2) with 12 μg each of full-length linear SV40 DNA were hybridized for 24 hr at 41° with 0.15 μg of ^3H-labeled SV40 cRNA in

[24] J. G. Williams, M. M. Lloyd, and J. M. Devine, *Cell* **17,** 903 (1979).
[25] R. L. White and D. S. Hogness, *Cell* **10,** 177 (1977).

100 μl of buffer [50% deionized formamide, 0.4 M NaCl, 20 mM PIPES, pH 6.4, 5 mM EDTA, 0.2% SDS, 1 mg/ml wheat germ RNA, and 50 μg/ml poly (rA)]. After 24 hr, 93% of the cRNA had been removed from solution. The circles were washed at 32° four times for 20 min each with 50% formamide–20 mM NaCl–8 mM trisodium citrate–0.2% SDS, and the RNA remaining was removed for counting with 0.4 M NaOH at 37° for 45 min. The overall recovery was 77% of the initial input. When ^{32}P-labeled nuclear RNA from uninfected monkey cells was included, 0.02% of the ^{32}P was found in the NaOH eluate (signal-to-noise ratio about 2×10^{-4}). When 0.6 μg of RNA per square centimeter was used, the recovery dropped to 57%. (Somewhat suprisingly, circles carrying only 1–2 μg of DNA still had the capacity to retain about 80% of an input of 0.15 μg of cRNA throughout the entire procedure. Perhaps the proportion of full-length linear SV40 DNA that reanneals is smaller when the amount of DNA coupled is smaller.) Following treatment with NaOH, the circles were washed with water and stored in 50% formamide buffered with 20 mM PIPES, pH 6.4. Some were used as many as 10 times with no change in their properties.

Hybridization of Nuclear RNA from Dictyostelium discoideum to DNA Paper[3,24]

Circles (1 cm^2) were prepared with 6 μg each of sonicated DNA derived from a plasmid containing an insert of cDNA complementary to an abundant *Dictyostelium discoideum* mRNA. As a control, comparable circles were prepared using a plasmid with an insert complementary to *Xenopus* globin mRNA. With the use of RNA made in isolated *D. discoideum* nuclei *in vitro* and labeled with tritiated UTP, hybridization was performed for 24 hr at 41° in 100 μl of buffer [50% deionized formamide, 0.75 M NaCl, 75 mM trisodium citrate, 0.2% SDS, 1 mg/ml yeast tRNA, 0.3 mg/ml Ficoll 400, 0.5 mg/ml poly (rA), and 0.02% bovine serum albumin pretreated with 0.05% diethylpyrocarbonate]. The circles were washed batchwise for 30 min at 34° using four changes of 200 ml each of 50% formamide–20 mM NaCl–8 mM trisodium citrate–0.2% SDS. The background (*Xenopus* circles) was about 0.015% of the input, whereas the signal plus background (*D. discoideum* circles) was about 0.1%.

Characteristics of the Technique

In a parallel comparison[3] of hybridization to DNA paper or DNA-nitrocellulose filters,[4] very similar results were obtained. The background was slightly lower, and the signal was slightly higher with the DNA paper.[3] An advantage of DNA paper is the ability to reuse it many times. This is particularly important when the DNA sample is precious and is a major convenience in any case. Also, using the nitrocellulose method, hybrids will be lost from the support when the RNA is complementary to the full length of the immobilized DNA; this does not occur with DNA paper. Otherwise, the two methods can give comparable efficiencies of hybridization and low backgrounds when stringent washing conditions are used[3] (see above).

[15] Detection of Specific RNAs or Specific Fragments of DNA by Fractionation in Gels and Transfer to Diazobenzyloxymethyl Paper

By JAMES C. ALWINE, DAVID J. KEMP, BARBARA A. PARKER, JAKOB REISER, JAIME RENART, GEORGE R. STARK, and GEOFFREY M. WAHL

In the technique developed by Southern,[1] fragments of DNA are separated with high resolution by electrophoresis in agarose gels. The separated fragments are then denatured, and the single strands are transferred by blotting to strips of nitrocellulose without substantial loss of resolution. Finally, specific sequences are detected by hybridization to labeled DNA or RNA probes. The backgrounds are usually quite low, so that the sensitivity of the method is limited mainly by the specific radioactivity of the probes. The use of DNA probes labeled with ^{32}P to high specific activity (about 10^8 cpm/μg) by nick translation[2] enables one to detect restriction fragments of eukaryotic DNA derived from single-copy genes.

Although Southern's technique has been of enormous value and has been very widely used, it does not work for the transfer of RNAs or small fragments of DNA (less than about 200 base pairs) since these species do not bind to nitrocellulose. These limitations can be circumvented by using chemically activated paper which reacts covalently with RNA or with single-stranded DNA. Although some procedures for linking RNA to paper covalently have been developed,[3] none of the previous methods seemed suitable for transfer of RNA from gels.

[1] E. M. Southern, *J. Mol. Biol.* **98,** 503 (1975).
[2] P. W. Rigby, M. Dieckmann, C. Rhodes, and P. Berg, *J. Mol. Biol.* **113,** 237 (1977).
[3] P. T. Gilham, *Adv. Exp. Med. Biol.* **42,** 173 (1974).

ISBN 0-12-181968-X

Principle

Noyes and Stark[4] developed procedures for linking RNA or denatured DNA covalently to finely divided cellulose through a DBM group.[5] The diazonium group reacts with single-stranded nucleic acids (double-stranded DNA does not react) to give azo derivatives, principally by coupling at the 2-position of guanosine or deoxyguanosine residues[6] and at the 5-position of uridine residues.[7] Thymidine residues react much more slowly, and no appreciable reaction is observed with polyribo- or polydeoxyribocytidine or -adenosine.[4] (More recently, reaction of polyriboinosine with DBM paper has been accomplished,[8] probably through coupling at the 2-position.)

Alwine, Kemp, and Stark[9] synthesized DBM paper and used it for the transfer of RNAs from agarose gels. The negatively charged RNA probably adheres to the positively charged diazonium groups on the paper first, through ionic interactions, followed more slowly by covalent linkage. The positive charges on the paper are lost eventually through hydrolysis, so that there is little nonspecific interaction with probe nucleic acids; i.e., specific RNAs can be detected by hybridization to specific probes, with high sensitivity and low backgrounds. In the original procedure, the RNAs were denatured with methylmercuric hydroxide and separated according to size in gels containing this compound.[10] More recently, denaturation with glyoxal[11] has been used.[12]

Small, double-stranded fragments of DNA (30–1800 base pairs) have been separated with high resolution on composite gels of agarose and polyacrylamide, cross-linked with *N*,*N*′-diallyltartardiamide instead of *N*,*N*′-methylenebisacrylamide.[13] Following electrophoresis, cleavage of the cross-links with periodic acid[14] facilitates transfer of the denatured

[4] B. E. Noyes and G. R. Stark, *Cell* **5**, 301 (1975).

[5] The abbreviations used are: ABM, aminobenzyloxymethyl; DBM, diazobenzyloxymethyl; NBPC, 1-[(*m*-nitrobenzyloxy)methyl]pyridinium chloride; NBM, nitrobenzyloxymethyl; SDS, sodium dodecyl sulfate.

[6] R. Shapiro, *Prog. Nucleic Acid Res. Mol. Biol.* **8**, 73 (1968).

[7] R. M. Acheson, "An Introduction to the Chemistry of Heterocyclic Compounds." Wiley (Interscience) New York, 1967; R. K. Robins, *Heterocycl. Compd.* **8**, 162 (1967).

[8] G. R. Stark, W. Dower, R. T. Schimke, R. E. Brown, and I. M. Kerr, *Nature* **278**, 471 (1979).

[9] J. C. Alwine, D. J. Kemp, and G. R. Stark, *Proc. Natl. Acad. Sci. U.S.A.* **74**, 5350 (1977).

[10] J. M. Bailey and N. Davidson, *Anal. Biochem.* **70**, 75 (1976).

[11] G. K. McMaster and G. G. Carmichael, *Proc. Natl. Acad. Sci. U.S.A.* **74**, 4835 (1977).

[12] L. P. Villareal, R. T. White, and P. Berg, *J. Virol.* **29**, 209 (1979).

[13] J. Reiser, J. Renart, and G. R. Stark, *Biochem. Biophys. Res. Commun.* **85**, 1104 (1978).

[14] H. S. Anker, *FEBS Lett.* **7**, 293 (1970).

fragments to DBM paper. Again, detection is accomplished with labeled probes. Large fragments of DNA can also be transferred from agarose gels to DBM paper following partial depurination with dilute acid and strand cleavage with NaOH. The efficiency of transfer is very high and independent of the size of the fragment.[15] Covalent linkage allows convenient multiple reuse of the transfers (see also Goldberg *et al.*[16]) and also allows washings to be done repeatedly under stringent conditions, helping to achieve very low backgrounds.

Materials and Reagents

Although many kinds of filter paper can probably by used, we have worked almost exclusively with Whatman 540 paper because of its excellent mechanical strength and resistance to chemicals. Schleicher and Schuell 589 WH paper also gives good results. NPBC is available commercially (British Drug Houses or Pierce Chemical Company) or can be synthesized readily at a more economical cost. ABM and NBM papers are available commercially from Schleicher and Schuell.

Synthesis of NBPC[9,17]

See Scheme 1. Bubble dry HCl gas (454 g) into a solution containing 158 g of paraformaldehyde and 200 g of *m*-nitrobenzyl alcohol in 1 liter of benzene for 2 hr at room temperature with stirring and continue the stirring overnight. After the mixture has settled, remove the upper organic phase, dry it by adding 150 g of anhydrous Na_2SO_4, and then filter the suspension. Remove the benzene under reduced pressure in a rotary evaporator. Distill the remaining yellow liquid under reduced pressure and collect the major fraction (e.g., the fraction boiling between 150° and 154° at 1.5 mm of Hg). Caution! Distillation to near dryness will probably cause an explosion. Typical yield: 216 g of yellow liquid. Add this material slowly to 750 ml of ice-cold pyridine, with stirring. Allow the pyridinium salt to crystallize, conveniently overnight. Collect the salt on a sintered-glass filter and wash it first with pyridine and then several times with petroleum ether. Dry the product under reduced pressure. Yield: 267 g (73%). Store NBPC at −20° in a desiccator.

[15] A more detailed analysis of the use of dextran sulfate to accelerate hybridization of probes to DNA paper will be found in G. M. Wahl, M. Stern, and G. R. Stark, *Proc. Natl. Acad. Sci. U.S.A.* **76,** 3683 (1979). Wetmur [*Biopolymers* **14,** 2517 (1975)] has shown previously that dextran polymers greatly accelerate the rate of DNA reannealing in solution.

[16] M. L. Goldberg, R. P. Lifton, G. R. Stark, and J. G. Williams, this volume [14].

[17] D. N. Kursanov and P. A. Solodkov, *J. Appl. Chem. USSR (Engl. Transl.)* **16,** 351 (1943).

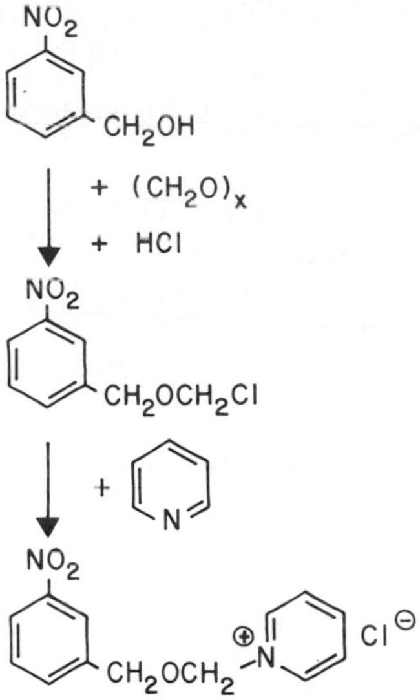

NO_2

CH_2OH

+ $(CH_2O)_x$

+ HCl

NO_2

CH_2OCH_2Cl

+ N

NO_2

CH_2OCH_2 $\overset{\oplus}{N}$ $Cl^{\ominus}$

NBPC

SCHEME 1. Synthesis of NBPC.

Preparation of NBM Paper

The complete procedure is outlined in Scheme 2. The reaction should be done in a hood to remove pyridine fumes. Cut a sheet of Whatman 540 paper to fit into the bottom of a rectangular enamel, stainless-steel, or glass pan. The size of the paper can be much larger than the size of each gel. Float the pan on a water bath at about 60°. For each square centimeter of paper, prepare a solution of 2.3 mg of NBPC (8.14 μmol, MW 280.7) and 0.7 mg of sodium acetate trihydrate in 28.5 μl of water. Using one sheet of paper at a time, pour the solution over the paper evenly and, using rubber gloves, push out any bubbles. Rub the solution evenly over the paper with a gloved hand, continuing until the paper is nominally dry. Dry one or more such papers further at 60° in an oven for about 10 min, remove them, adjust the temperature of the oven to 135°, place them back in the oven, and bake them at this temperature for 30–40 min. If the temperature is substantially below 135°, incomplete reaction can result. Several sheets may be baked at one time, with as many as three overlapping. An oven with circulating air gives the best results. Wash the papers several times with water for a total of about 20 min and three times with acetone for a total of about 20 min, and then dry them in the air. NBM paper is the most stable form and will keep for many months in the refrigerator. It is simple to activate it before each use. Alternatively, the less stable amino

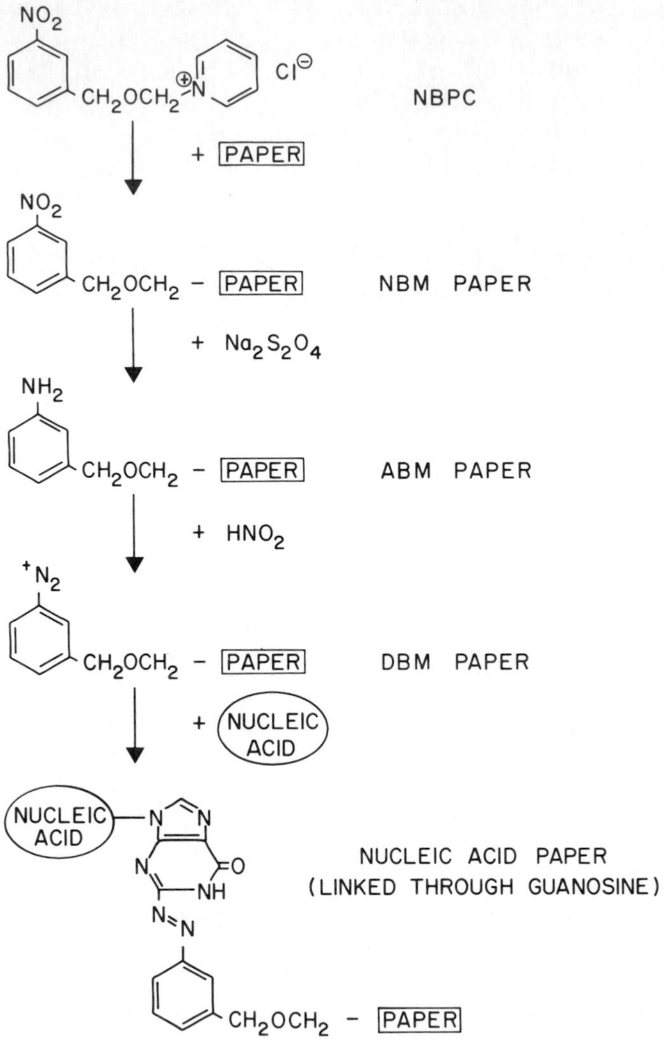

NO_2

CH_2OCH_2–$N^{\oplus}$ $Cl^{\ominus}$ NBPC

+ PAPER

NO_2

CH_2OCH_2 – PAPER NBM PAPER

+ $Na_2S_2O_4$

NH_2

CH_2OCH_2 – PAPER ABM PAPER

+ HNO_2

$^{+}N_2$

CH_2OCH_2 – PAPER DBM PAPER

+ NUCLEIC ACID

NUCLEIC ACID — N N N O NH N=N

NUCLEIC ACID PAPER (LINKED THROUGH GUANOSINE)

CH_2OCH_2 – PAPER

SCHEME 2. Synthesis of nucleic acid paper.

form (ABM paper) can be stored at 4° in a vacuum for as long as a year. DBM paper, taken directly from the $NaNO_2$–HCl mixture, frozen in liquid N_2, wrapped in Saran, and stored at −80° has worked well up to 4 months later. Of course it must be washed before use (see next section).

Preparation of DBM Paper

To reduce NBM paper, incubate it in a hood (to eliminate SO_2) for 30 min at 60° with 150 ml of a 20% (w/v) solution of sodium dithionite in water, with occasional shaking. Wash the ABM paper several times with large amounts of water for a few minutes. Be sure no odor of H_2S remains. Then wash it once with at least 100 ml of 1.2 M HCl for a 14 × 14 cm paper.[18] Transfer the wet paper directly to 0.3 ml/cm^2 of ice-cold 1.2 M HCl. For each 100 ml of HCl, add with mixing 2.7 ml of a solution of $NaNO_2$ in water (10 mg/ml) prepared immediately before use.[19] Keep the paper in this solution on ice for 30 min or a little longer, with occasional swirling. After 30 min, a drop of the solution should still give a positive (black) reaction for nitrous acid with starch iodide paper. Leave the paper in the ice-cold acid until preparation of the gel has been completed. Then pour off the acid; wash the paper rapidly twice with ice-cold water and twice with ice-cold transfer buffer (see below). The total time for these washes should be 2–3 min. Begin the transfer without delay—see below for timings relative to preparation of the gels.

Method for RNA

Preparation and Electrophoresis of the RNA

Before electrophoresis, the RNA samples should be purified, precipitated with ethanol, and dried. Large amounts of rRNA will compete with the transfer of mRNAs from overlapping regions of the gel. Hence it may be advisable to reduce this competition and to increase the concentration of a specific mRNA by selecting poly(A^+) RNAs with poly(U) Sepharose[20] or oligo(dT) cellulose[21] before electrophoresis. With two selections on oligo(dT) cellulose, very little of the RNA is ribosomal.

In order to disrupt secondary structure in the RNA completely, electrophoresis should be carried out in the presence of methylmercuric hydroxide[9,10] or after pretreatment of the RNA with glyoxal.[11,12] In either case, the distance a particular RNA migrates is directly proportional to the logarithm of its molecular weight. With rRNA and at least some (possibly all) mRNAs, it is not necessary to include dimethyl sulfoxide during treatment with glyoxal, as recommended originally.[11] Deletion of dimethyl sulfoxide makes it easier to dissolve a large amount of RNA in a small volume for loading on a single gel track.

[18] A wash with 30% acetic acid, originally recommended (see reference 9) has been deleted.

[19] A stock solution of $NaNO_2$ (10 mg/ml) can be stored for several months at 4° under an atmosphere of argon or nitrogen in a rubber-stoppered bottle, withdrawing the amount desired with a syringe.

[20] M. Adensik, M. Salditt, W. Thomas, and J. E. Darnell, *J. Mol. Biol.* **71**, 21 (1972).

[21] H. Aviv and P. Leder, *Proc. Natl. Acad. Sci. U.S.A.* **69**, 1408 (1972).

Preparation of Gels for Transfer

The denaturing agent must be removed, along with any components of the buffer that might react with the paper. Since the diazonium groups of DBM paper are relatively short-lived, it is essential to bring the RNA into contact with the paper as rapidly as possible. Low pH helps to stabilize the diazonium groups and does not affect the efficiency of the coupling reaction.

RNA from Agarose Gels Containing Methylmercuric Hydroxide. The quantities of reagents specified are appropriate for a 150-ml gel (typically 13 × 14 × 0.8 cm). Rock the gel 100 times/min for 60 min at room temperature in 250 ml of 50 m*M* NaOH containing 5 m*M* 2-mercaptoethanol. Wash the gel twice for 10 min each with 200 ml of 200 m*M* potassium phosphate buffer, pH 6.5, containing 7 m*M* iodoacetic acid at room temperature, and then twice at room temperature for 5 min each with 200 m*M* sodium acetate buffer, pH 4.0.[22,23] Reduction of the NBM paper should be started at the beginning of the NaOH wash; alternatively, diazotization of the ABM paper should be started 0.5 hr later.

RNA Pretreated with Glyoxal From Agarose Gels. Place the gel in 250 ml of 50 m*M* NaOH with or without ethidium bromide (1 μg/ml) for 1 hr at room temperature.[12,24] Neutralize the gel by washing it twice for 15 min each with 200 m*M* sodium acetate buffer, pH 4.0 (the ethidium bromide staining can now be observed).[23] Reduction of the NBM paper should be started about 0.5 hr after the NaOH wash; alternatively, diazotization of the ABM paper should be started at the beginning of the first buffer wash. Comparable signals are obtained with the methylmercuric hydroxide and glyoxal procedures (Luis Villareal, unpublished).

Transfer to DBM Paper

The procedure is essentially the one described by Southern.[1] Saturate two or three sheets of Whatman 3MM paper with the same buffer used for

[22] Methylmercuric hydroxide and its adducts with bases of nucleic acids react with 2-mercaptoethanol or other sulfhydryl compounds such as dithiothreitol. Treatment with alkali cleaves the RNAs partially and increases their rates of diffusion from the gel. Excess iodoacetic acid eliminates the mercaptans which would otherwise react with the DBM paper.

[23] We previously used 25 m*M* sodium phosphate buffer, pH 6.5, for the final wash and transfer. However, the stability of the diazonium groups is improved markedly by reducing the pH to 4.0, as now recommended. The higher concentration of buffer now used at pH 4.0 does not affect the efficiency of coupling (see also footnote 47).

[24] Treatment with base decomposes the glyoxal–RNA adducts and cleaves high-molecular-weight RNAs partially. Staining the RNA with ethidium bromide does not interfere with the subsequent coupling to DBM paper. The bands can be photographed before the transfer, using long-wavelength uv light.

the final wash of the gels, and then place them in contact with a source of additional buffer. Place the gel on top of the wet paper and place the fresh DBM paper on top of the gel, using Saran Wrap at the edges of the gel to prevent the DBM paper from touching the wet 3MM paper below. Add two or three layers of dry 3MM paper, several layers of paper towels, and a weight. Allow the buffer to blot through the gel and DBM paper overnight, either at room temperature or at 4°.[25]

Pretreatment and Hybridization

After the transfer, treat the paper with hybridization buffer *plus* 1% (w/v) glycine for 4–24 hr at 42° to remove any remaining diazonium groups and to add carrier compounds. Hybridization buffer contains 50% formamide, 0.9 *M* sodium chloride, 50 m*M* sodium phosphate, pH 7.0, 5 m*M* EDTA, 0.1% SDS, 0.02% (w/v) each of bovine serum albumin, Ficoll, and polyvinylpyrrolidone,[26] and 250–500 μg/ml of sonicated denatured salmon sperm DNA. If the transfers are not to be hybridized immediately, they can be stored at 4° in hybridization buffer plus glycine for several months or longer. For hybridization, put the paper into a plastic boiling bag (Seal-N-Save, Sears) along with 50–100 μl of hybridization buffer per square centimeter and the labeled nucleic acid probe. (The specific radioactivity of the probe needed for detection of a particular species of RNA will depend on the amount of that species in the transfer and on the time of autoradiography. For specific examples, see below.) After sealing, rock the bag gently at 42° for an appropriate time. (This parameter too depends on the amount of RNA and on the concentration and specific activity of the probe. For specific examples, see below.) At the end of the hybridization, wash the paper for 15 min each with two 250-ml portions of 0.36 *M* sodium chloride, 20 m*M* sodium phosphate, pH 7.0, 2 m*M* EDTA, 0.1% SDS at room temperature, followed by two 250-ml washes with 18 m*M* sodium chloride, 1 m*M* sodium phosphate, pH 7.0, 0.1 m*M* EDTA, 0.1% SDS for 15–30 min each at 40–50°. The optimum washing temperature should be determined experimentally, conveniently by using a hand-held monitor.[27] Blot the paper dry, cover

[25] Aromatic diazonium groups probably decompose primarily to the phenol by reaction with hydroxide ions: ⟨0⟩ $N_2^+ + OH^- \rightarrow$ ⟨0⟩ $OH + N_2$. The phenol can in turn couple with remaining diazonium groups to give azo dyes: ⟨0⟩ $N_2^+ +$ ⟨0⟩ $OH \rightarrow$ ⟨0⟩ $N{=}N$ ⟨0⟩ OH. Probably as a result of these reactions, DBM paper begins to turn yellow as soon as the HCl is washed away. The color deepens to orange with time. The final color is much more intense in alkali, probably because of ionization of the phenolic group of the dye.

[26] D. Denhardt, *Biochem. Biophys. Res. Commun.* **23**, 641 (1966).

[27] Although this procedure usually provides very low backgrounds, more stringent washes may be called for in some cases since some preparations of probe give high backgrounds.

it with Saran Wrap, and begin the autoradiography using Kodak XR-5 film exposed at −80° with a Dupont Cronex Lighting Plus intensifying screen.[28] Several exposure times should be tried with each sample.

We have found recently that the rate of hybridization can be increased dramatically without substantial effect on the background by using 10% dextran sulfate 500 (Pharmacia).[15] This procedure works well with either RNA paper or DNA paper. In an experiment with DNA paper, the intensity of signal obtained in 2 hr of hybridization with 10% dextran sulfate was about four times as great as the intensity obtained in 72 hr without dextran sulfate, with comparable backgrounds.[15]

Results

The method is illustrated with two detailed examples from our laboratory. One shows that RNAs present in very low amounts can be detected with high resolution and low background; the other shows that rough quantitation of the relative amounts of a particular mRNA in different cells can be obtained in a comparative experiment.

CV-1 monkey cells were infected at 37° with SV40 DNA by a modification[29] of the calcium phosphate technique.[30] At 9.5 hr, a very early time after infection, cytoplasmic and nuclear RNAs were collected and the poly(A^+) and poly(A^-) fractions were obtained from each. Following treatment with glyoxal, the RNAs were separated by electrophoresis (20 μg per track) in a horizontal 1.5% agarose slab gel.[11,12] Following transfer and pretreatment, the RNA papers were hybridized for 2.5 days at 42° with approximately 2×10^6 cpm per track of ^{32}P-labeled DNA probe (25 ng/ml of hybridization buffer, 1×10^5 cpm/cm^2). The probe, specific activity 5×10^7 cpm/μg, was prepared by nick-translating[2] two DNA fragments from a triple digestion of SV40 DNA by the restriction endonucleases *Hinc*II, *Bgl*I, and *Taq*I. These two fragments correspond to part of the early region of the SV40 genome (map coordinates 0.169–0.566).[31] Following hybridization, washing, and drying, the RNA papers were autoradiographed for 2 days, with the results shown in Fig. 1.

Analysis of the virus-specific transcripts early in an infection by SV40 DNA provides a stringent test of the sensitivity and resolution of the RNA transfer method. Two major early virus-specific transcripts have been de-

[28] R. A. Laskey and A. D. Mills, *FEBS Lett.* **82,** 314 (1977).

[29] E. Frost and J. Williams, *Virology* **91,** 39 (1978).

[30] F. L. Graham and A. J. Van der Eb, *Virology* **52,** 456 (1973).

[31] W. Fiers, R. Contreras, G. Haegeman, R. Rogiers, A. Van de Voorde, H. Van Heuverswyn, J. Van Herreweghe, G. Volckaert, and M. Ysebaert, *Nature* (*London*) **273,** 113 (1978).

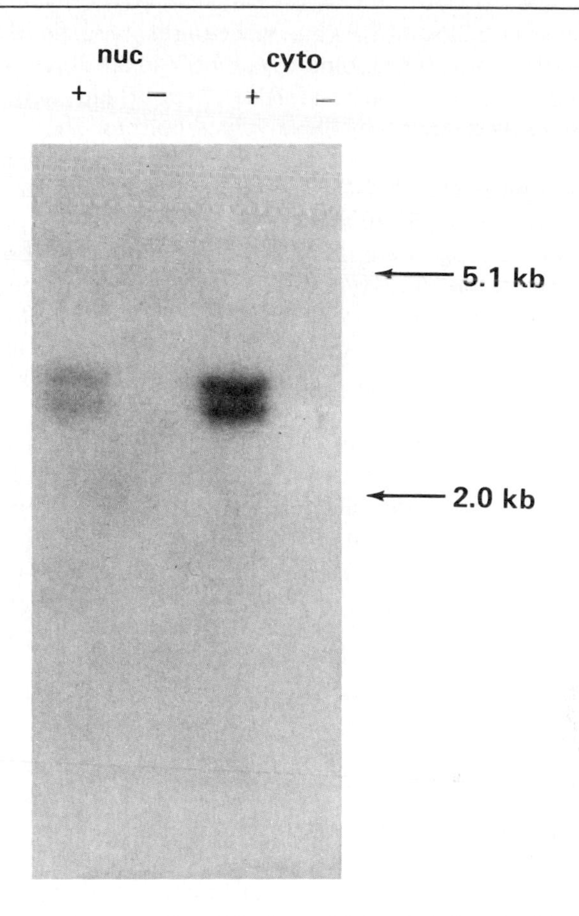

FIG. 1. Transfer of early SV40 RNAs from CV-1 cells infected with SV40 DNA for 9.5 hr.

tected before the onset of SV40 DNA replication.[32] Both have been mapped.[33] They differ in length by less than 300 bases and probably code for the big T antigen and little t antigen of SV40.[33–35] As shown in Fig. 1,

[32] Although the exact time for the beginning of SV40 DNA replication is not known, most workers suggest that it begins about 12 hr after infection. For example, see reviews by J. Tooze ("The Molecular Biology of Tumour Viruses," p. 331. Cold Spring Harbor Lab., Cold Spring Harbor, New York, 1973) and N. H. Acheson [*Cell* **8,** 1 (1976)].

[33] A. J. Berk and P. A. Sharp, *Proc. Natl. Acad. Sci. U.S.A.* **75,** 1274 (1978).

[34] L. V. Crawford, C. N. Cole, A. E. Smith, E. Paucha, P. Tegtmeyer, K. Rundell, and P. Berg, *Proc. Natl. Acad. Sci. U.S.A.* **75,** 117 (1978).

[35] E. Paucha, A. Mellor, R. Harvey, A. E. Smith, R. W. Hewick, and M. D. Waterfield, *Proc. Natl. Acad. Sci. U.S.A.* **75,** 2165 (1978).

two RNAs of length 2.6 and 2.9 kb have been resolved clearly. In an infection with SV40 virions, the amount of early SV40 RNA present at 9.5 hr is no more than 0.01% of the total RNA,[36] or about 0.5% of the total poly(A^+) RNA. Assuming that these values hold for the RNAs shown in Fig. 1, and knowing that a DNA infection is about 25 times less efficient than a virion infection under the conditions used, we can estimate that there are about 2 ng of SV40-specific poly(A^+) RNA in the cytoplasmic track and about 200 pg in the nuclear track. Therefore, about 100 pg per band of RNA is visible in 2 days, using a high concentration (25 ng/ml) of high-specific-activity DNA probe (5×10^7 cpm/μg). Since the backgrounds are so low with this technique, it should be possible to detect 10–50 pg of specific RNA in a single band after 1 week of autoradiography.

To detect specific transcripts with the low background and high sensitivity shown in Fig. 1, we had to use highly purified viral DNA for the probe, free of contamination with cellular DNA. Therefore, SV40 DNA was isolated from virions, followed by banding to equilibrium in CsCl. ^{32}P-labeled probe made from DNA isolated from infected cells according to the Hirt procedure[37] and banded once in CsCl hybridized to both cellular and viral RNAs. Cloned DNA probes would surely provide DNA of the appropriate degree of purity in other cases.

The second experiment is shown in Fig. 2. Total cytoplasmic poly(A^+) RNA pools (10 μg per track) are compared from wild-type hamster cells and from mutants that overproduce the multifunctional protein responsible for the first three steps of UMP biosynthesis[38] and the corresponding mRNA.[39] The amount of this protein or RNA in each mutant relative to the amount in the wild-type cells is shown above each track. The nick-translated DNA probe (2×10^6 cpm per track, 7×10^7 cpm/μg) is from a cloned plasmid containing a 2.3-kb insert corresponding to the 3′-end of the mRNA.[40] The wild-type cells and all the mutants contain a large (7.9-kb) mRNA which is transferred well following the partial cleavage with alkali. It is apparent from Fig. 2 that there is a rough correspondence between the known amount of 7.9-kb RNA and the signal. More careful measurement[39] reveals that the quantitative agreement is excellent. Some of the mutants contain a minor, specific cytoplasmic RNA larger than the 7.9-kb species found in the wild type. The significance of the larger RNA

[36] J. C. Alwine, S. I. Reed, and G. R. Stark, *J. Virol.* **24,** 22 (1977).

[37] B. Hirt, *J. Mol. Biol.* **26,** 365 (1967).

[38] P. F. Coleman, D. P. Suttle, and G. R. Stark, *J. Biol. Chem.* **252,** 6379 (1977).

[39] R. A. Padgett, G. M. Wahl, P. F. Coleman, and G. R. Stark, *J. Biol. Chem.* **254,** 974 (1979).

[40] G. M. Wahl, R. A. Padgett, and G. R. Stark, *J. Biol. Chem.*, **254,** 8679 (1979).

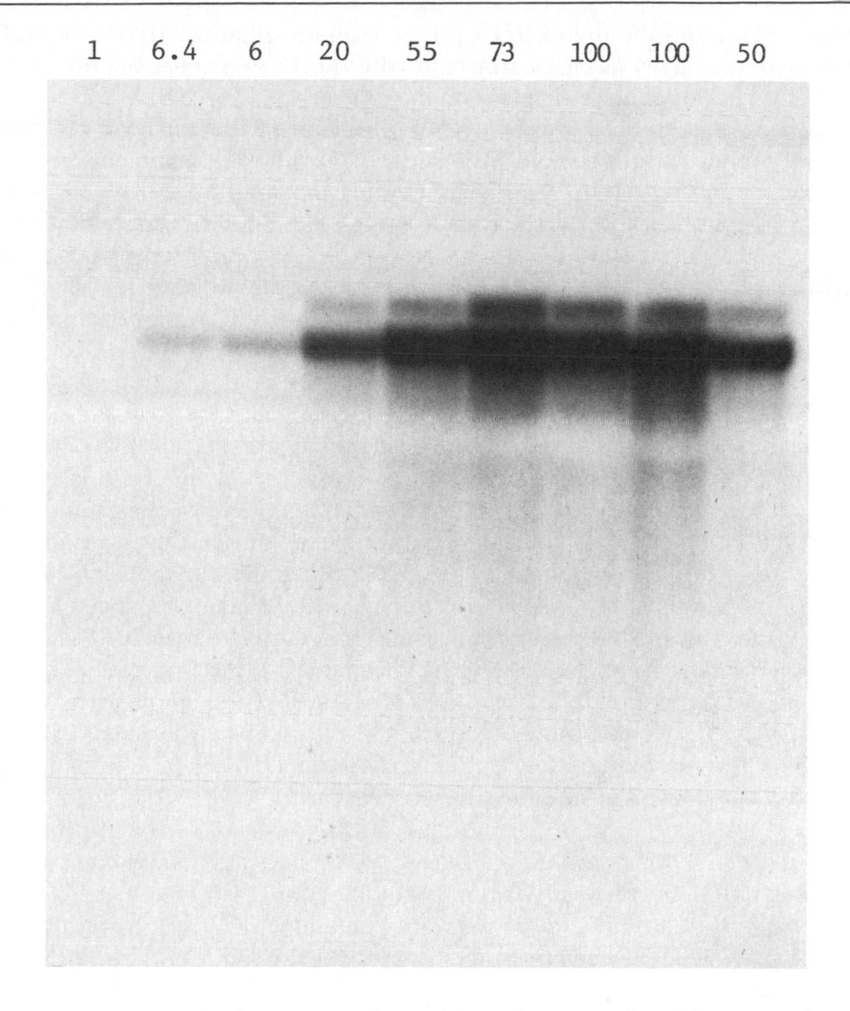

FIG. 2. Transfer of poly(A^+) RNAs from wild-type hamster cells and from several mutants that overproduce both the multifunctional protein catalyzing the first three steps of UMP biosynthesis and the corresponding mRNA. The amount of the protein in each mutant, relative to the wild-type (= 1), is shown at the top of each track.

is unclear at the moment. Note also that the autoradiographs are relatively clear in the regions corresponding to the positions of the 18 and 28 S rRNAs. [About one-half the RNA in each sample is ribosomal, since the RNAs were passed through oligo(dT) cellulose only once.] This observation is common and probably due to inhibition of binding of the probe where the amount of RNA on the paper is greatest. It may be difficult to detect mRNAs that comigrate with the rRNAs unless two rounds of poly(A) selection are employed.

We have been able to retest RNA papers with the same or with different probes, and we have also reused the hybridization buffer containing ^{32}P-labeled DNA (simply by heating the solution to 85° for 5 min and cooling it quickly). If a search is being made for two different RNAs in the same population, the ability to reprobe with a different DNA sequence allows one to do a comparative experiment with excellent internal control. By washing the RNA papers four times for 30 min each at 65° with 95% formamide, 20 m*M* HEPES buffer, pH 7.4, 1 m*M* EDTA, we were able to remove about 90% of the ^{32}P probe in an experiment where the RNA was pretreated with glyoxal. The residual 10% is not removed by further washing but can be allowed to decay before reprobing with a different ^{32}P-labeled DNA. Retention of the probe in this case is probably due to reaction with RNA from which the glyoxal has not been removed completely, as indicated by the observations that (1) none of the second probe is retained if the 99% formamide wash is repeated, and (2) none of the first probe is retained if the transfer has been made from a gel run in the presence of methylmercuric hydroxide. Some of the RNA is lost from the paper during removal of the first probe, since the second signal is less intense than the first if the same probe is used both times. Better conditions for removal of the first probe may help to alleviate this problem, but we note that some DNA is lost from DNA paper during the initial set of washes in denaturing solvents but not in subsequent washes.[16]

Villareal, White, and Berg[12] have transferred to DBM paper cytoplasmic mRNAs made in response to infection of monkey cells with wild-type SV40 or with viable mutants containing deletions or insertions in the late "leader" region. In each case, the 16 and 19 S late mRNAs, separated on agarose gels after treatment with glyoxal,[11,12] hybridized with ^{32}P-labeled DNA probes corresponding to either the leader or "body" regions, indicating that spliced messages could still be formed from mutants with altered leaders.

Mulligan, Howard, and Berg[41] transferred to DBM paper cytoplasmic mRNAs made in response to infection of monkey cells with a recombinant of SV40 in which virtually all the coding sequence for the major capsid protein VPI, but not the leader sequence, had been replaced by a cDNA that specified the entire amino acid sequence of rabbit β-globin. The transfers were probed with ^{32}P-labeled plasmid DNA containing the β-globin cDNA sequence. Two mRNA species, 1.0 and 1.8 kb, were

[41] R. C. Mulligan, B. H. Howard, and P. Berg, *Nature (London)* **277,** 108 (1979).

found, corresponding to the analogous normal 16 and 19 S late mRNAs made in an infection with wild-type SV40.

Using ^{32}P-labeled cDNA representing the total population of poly-(A$^+$) mRNAs from yeast cells grown in the presence of galactose as a probe for screening, St. John and Davis[42] have isolated recombinants of phage lambda and yeast that contain yeast DNA sequences complementary to mRNAs induced by galactose. One cloned yeast fragment hybridizes with three discrete RNA species in transfers from glyoxal gels to DBM paper of the total RNA population from cells grown in the presence of galactose. These RNAs are not detected in transfers from yeast cells grown under a variety of different conditions without galactose. It would be very laborious to obtain comparable information using any other method currently available.

Methods for DNA

General Comments

As in the procedure of Southern using nitrocellulose,[1] denaturation of DNA duplexes to single strands is essential for binding to DBM paper. Transfer of DNA is slow from the polyacrylamide gels required for resolution of small fragments. (Generally the gels contain a concentration of polyacrylamide between 5 and 12%.) Slow migration results in poor coupling to the paper. This problem is largely alleviated by using polyacrylamide–agarose *composite* gels with cleavable cross-links. Following electrophoresis, the cross-links are broken to disrupt the polyacrylamide matrix, leaving the more porous agarose matrix undisturbed to provide mechanical support for the DNA bands during transfer.[13] Following transfer, the washing and hybridization steps are exactly the same as those described above for RNA paper.

Larger DNA fragments can also be transferred from agarose gels to DBM paper, an alternative to the transfer to nitrocellulose originally described by Southern.[1] By using very low pH (4.0) with the DBM paper, transfer was achieved with excellent efficiency. We have reused a single DNA paper for different hybridizations, with low and reproducible backgrounds. Using depurination *in situ* as described below, Joseph Shlomai and Arthur Kornberg (in preparation) have found recently that DNA fragments from ϕX174 ranging in size from 1353 to 118 base pairs can be separated in 2% agarose gels (fragments of 194 and 234 base pairs are well resolved) and transferred quantitatively to DBM paper. The 194 fragment hybridizes stably with probe. Therefore the special procedure for

[42] T. P. St. John and R. W. Davis, *Cell* **16,** 443 (1979).

small DNA fragments probably needs to be used only with fragments smaller than 100–200 base pairs or where high resolution is important.

Transfer of Small DNA Fragments from Composite Gels[13]

Gel Electrophoresis. Restriction fragments are separated on polyacrylamide–agarose slab gels (23 × 14 × 0.15 cm) using Tris-acetate buffer (40 m*M* Tris-HCl, pH 7.8, 20 m*M* sodium acetate, 2 m*M* EDTA).[43] The same buffer is used in the electrode reservoirs.[44] To prepare the gels, mix 8 ml of 10× concentrated gel buffer, 59 ml of water, and 560 mg of agarose (Bio-Rad) and dissolve the agarose by boiling. To the solution cooled to 50° add an appropriate volume of 30% acrylamide stock solution [27.78 g of acrylamide plus 2.22 g of *N,N'*-diallyltartardiamide (Bio-Rad) per 100 ml] and 0.25 ml of 10% ammonium persulfate.[45] Gels containing a single concentration of acrylamide between 5 and 12% are used to resolve fragments in the size range 2000 to 10 base pairs. Polymerize the gel by adding 25 μl of *N,N,N'N'*-tetraethylenediamine (Bio-Rad). DNA samples should be precipitated with ethanol before electrophoresis. Perform the electrophoresis at room temperature, at 15–20 mA.

Preparation of the Gels and Transfer. Place the gel in 250 ml of 2% periodic acid and rock it gently for 15 min at 37° to cleave the cross-links.[14] Rinse the gel with water and put it in 250 ml of 0.5 *M* NaOH for 10 min at room temperature to denature the DNA. Rinse the gel with water and neutralize it in 250 ml of 0.5 *M* sodium phosphate buffer, pH 5.5, for 10 min at room temperature and then put it in 250 ml of ice-cold 50 m*M* sodium phosphate buffer, pH 5.5, until the DBM paper is ready (no longer than 15 min).[46] Diazotization of ABM paper should start at the same time as the treatment with periodic acid; alternatively, reduction of NBM paper should start about 0.5 hr sooner. Do the transfer as described in the procedure for RNA, except use 50 m*M* sodium phosphate buffer, pH 5.5, at 4°.[47]

[43] U. E. Loening, *Biochem. J.* **102,** 251 (1967).

[44] Tris–borate gel systems should be avoided, since borate reacts with *cis*-diols and would protect the cross-links from cleaving.

[45] We have obtained similar results with ethylenediacrylate, which is reversed by treatment with alkali [D. F. Cain and R. E. Pitney, *Anal. Biochem.* **22,** 11 (1968)].

[46] Only single-stranded DNA will bind to DBM paper, so care must be taken to prevent renaturation of fragments of low complexity. In some cases, the washes might be done with buffer of even lower ionic strength, to retard renaturation. Lower pH, successfully used in transfers of DNA from agarose gels (see below) would also help retard renaturation.

[47] The efficiency of transfer depends mainly on the stability of the diazonium groups and on the rate at which the DNA fragments diffuse from the gel to the paper. The best combination of conditions seems to be low pH (pH 5.5) and low temperature (4°). These conditions

TABLE I
EFFICIENCY OF DNA TRANSFER TO DBM PAPER[a]

Fragment	Base pairs	DNA remaining in the gel after transfer (%)	DNA transferred (% of control)
A	1790	75	11
B	1120	69	17
C	760	58	18
D	590	50	18
E	490	64	20
F	235	33	42
G	110	21	80
H	75	24	80

[a] ^{32}P-labeled SV40 DNA, 2.6 μg per slot, 14,300 cpm/μg, was digested with *Hin*fI and electrophoresed on a 5% polyacrylamide–agarose gel at 20 mA for 12 hr. The gel was pretreated and then cut into two strips. One was transferred for 32 hr at 4°, and the other was used as a control.

Results. Preparation of the gel for transfer leads to some loss of DNA. The larger SV40 *Hin*fI fragments A through E (1790 to 490 base pairs) were retained well in the gel, whereas the smaller fragments F, G, and H (235 to 75 base pairs) were retained less well.[13] The gel cross-links are removed efficiently by periodic acid, since the gel dissolves completely when the agarose is melted at 100°. Extending the periodic acid treatment to 60 min and the base treatment to 30 min caused greater losses.

The efficiency of transfer was assessed quantitatively by measuring DNA in a gel before and after transfer and by determining the amount of DNA bound to the paper (Table I). The efficiency depends on the sizes of the fragments. Only 11% of fragment A (1790 base pairs) was bound, whereas 42–80% of fragments F, G, and H (235 to 75 base pairs) were bound. The sum of the DNA remaining in the gel and the DNA transferred to the paper is not always as great as the input (60–95%). The losses may be due to partial renaturation of DNA. Since the rate of renaturation depends on the concentration of complementary strands, renaturation may be less of a problem when much lower concentrations of specific sequences are present, as in the transfer of DNA of higher complexity.

The losses during pretreatment of the gel and the efficiency of transfer have opposite size optima; i.e., larger fragments are retained better during pretreatment but are transferred less well, so that the amount

also work very well for transfer of RNA. Note also that transfer at pH 4.0 has given excellent results with larger DNA fragments from agarose gels and presumably would in this case as well.

transferred is a relatively constant fraction of the input, with slightly better yields for the smaller fragments. The bands are sharp after the transfer of labeled DNA fragments (Fig. 3A and B), indicating that there is little diffusion during the whole procedure. Bands due to the small *Hae*III fragments are broader than bands due to the larger ones because of the low percentage of acrylamide in the gel, not because of diffusion during transfer. The use of gels containing 10 or 12% acrylamide gives improved resolution of fragments smaller than 50 base pairs.

Unlabeled SV40 fragments coupled to DBM paper were detected by hybridization with DNA labeled with ^{32}P (Fig. 3D). The efficiency of hybridization to the smallest fragments I and J (28 base pairs each) was very poor, and these bands were hard to detect even after long exposure times (compare Fig. 3C with Fig. 3D).

The stability during hybridization of DNA coupled to DBM paper was also determined (Table II). Between 35 and 78% of the DNA remains bound to the paper after the complete procedure (prehybridization wash, hybridization, and washes). Since an appreciable fraction of the DNA is

TABLE II
STABILITY OF DNA FRAGMENTS BOUND TO DBM PAPER[a]

		DNA remaining bound to the paper (% of control)	
Fragments	Base pairs	A (full protocol)	B (99% formamide wash)
A	1790	71	55
B	1120	61	60
C	760	43	47
D	590	78	43
E	490	62	38
F	235	35	50
G	110	46	36
H	75	38	50

[a] ^{32}P-labeled SV40 DNA, 2.6 μg per slot, 14,300 cpm/μg, was digested with *Hin*fI and electrophoresed on a 5% polyacrylamide–agarose gel at 20 mA for 10 hr. After transfer, the paper was cut into three strips corresponding to individual gel tracks and dried overnight at room temperature. One strip (A) was treated for 22 hr at 42° with prehybridization buffer and for 58 hr at 42° with hybridization buffer. Finally it was washed for 12 hr at 37° with 50% formamide, 0.75 *M* sodium chloride, 0.075 *M* trisodium citrate. A second strip (B) was incubated for 7 hr at 80° with 99% formamide. The third strip was used as a control.

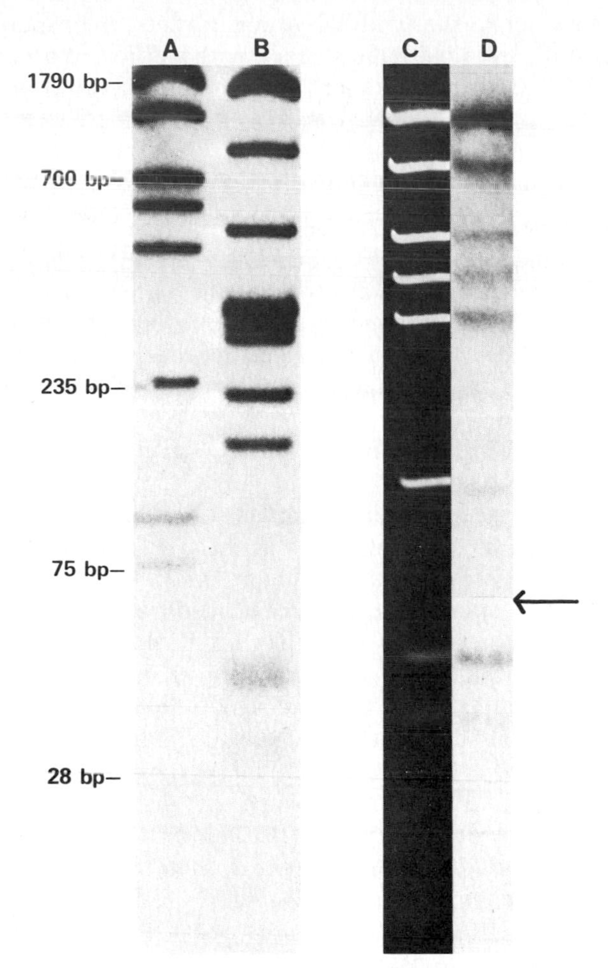

FIG. 3. Transfer of SV40 DNA fragments to DBM paper. (A and B) ^{32}P-labeled SV40 DNA, 2.3 μg per slot, 14,300 cpm/μg, was digested with *Hin*fI or *Hae*III, and the fragments were separated by electrophoresis on a 7% polyacrylamide–agarose gel. Transfer was for 29 hr at 4°. (A) *Hin*fI digest. (B) *Hae*III digest. (C and D) SV40 (I) DNA, 0.8 μg per slot, was digested with *Hin*fI, and the fragments were electrophoresed on a 5% polyacrylamide–agarose gel. One track (C) was stained with ethidium bromide (0.5 μg/ml) and photographed under uv light. The other track (D) was transferred and hybridized for 22 hr at 42° with ^{32}P-labeled SV40 DNA (3×10^7 cpm/μg, 5×10^3 cpm/cm^2). After washing, the strip was autoradiographed; the upper part (see arrow) was exposed for 6 hr and the lower part for 70 hr at −70°.

retained even after washes with DNA denaturants, the paper can be used for hybridization again, with the same or with a different probe. See Goldberg *et al.*[16] for further results and a discussion of the stability of DNA paper.

Transfer of Large DNA Fragments from Agarose Gels

Gel Electrophoresis. Restriction fragments are separated on agarose slab gels containing ethidium bromide (0.5 μg/ml) in both the gel and the buffer reservoirs. Use the electrophoresis buffer described in the previous section. Add the ethidium bromide to the molten agarose just before pouring the gel. Perform the electrophoresis at room temperature until the bromcresol purple dye marker has migrated about 12 cm (8–12 hr).

Preparation of Gels and Transfer of DNA to DBM Paper. The following protocol is designed for a 150-ml (14.5 × 13.5 × 0.8 cm) agarose gel and should be changed accordingly for gels of different thickness. It is advantageous to use bromcresol purple as the tracking dye during electrophoresis, since it provides a convenient indicator for monitoring pH changes during the later washes. All the procedures are done at room temperature. Place the gel in an enamel pan and shake it 100 times/min with two 250-ml portions of 0.25 *M* HCl for 15 min each. Decant the acid, wash the gel briefly with distilled water, and shake the gel with two 250-ml portions of 0.5 *M* NaOH–1.0 *M* NaCl for 15 min each. Decant the NaOH–NaCl solution and shake the gel with two 250-ml portions of 1 *M* sodium acetate buffer, pH 4.0, for 30 min each. Wash the diazo paper with ice-cold 1 *M* sodium acetate buffer, pH 4.0, just before transfer and perform the transfer in 1 *M* sodium acetate buffer, pH 4.0.[48] The DNA leaves the gel rapidly and transfer is complete in 2–4 hr.

Results. Southern[1] has already provided a most useful method for transferring large fragments of DNA to nitrocellulose. However, in view of the advantages provided by covalent linkage of nucleic acids to paper in the other cases described above, it seemed worthwhile to optimize transfer of large DNA fragments to DBM paper and to compare the efficiencies of transfer and hybridization for both methods.

[48] The rationale for attempting the transfer at pH 4.0 is as follows. (1) The lower the pH, the more stable the DBM groups. (2) The principal site of coupling for DNA is deoxyguanosine, which must remain unprotonated to react. The pK_a for the ring of dGMP is 2.9 (H. A. Sober, ed., "Handbook of Biochemistry," p. J93. Chem. Rubber Publ. Co., Cleveland, Ohio, 1968). (3) The corresponding pK_a for dCMP, however, is 4.4 (*ibid.* p. J81). Therefore, at pH 4.0, extensive protonation of the dCMP residues will help to keep the DNA denatured, with little protonation of the reactive dGMP residues.

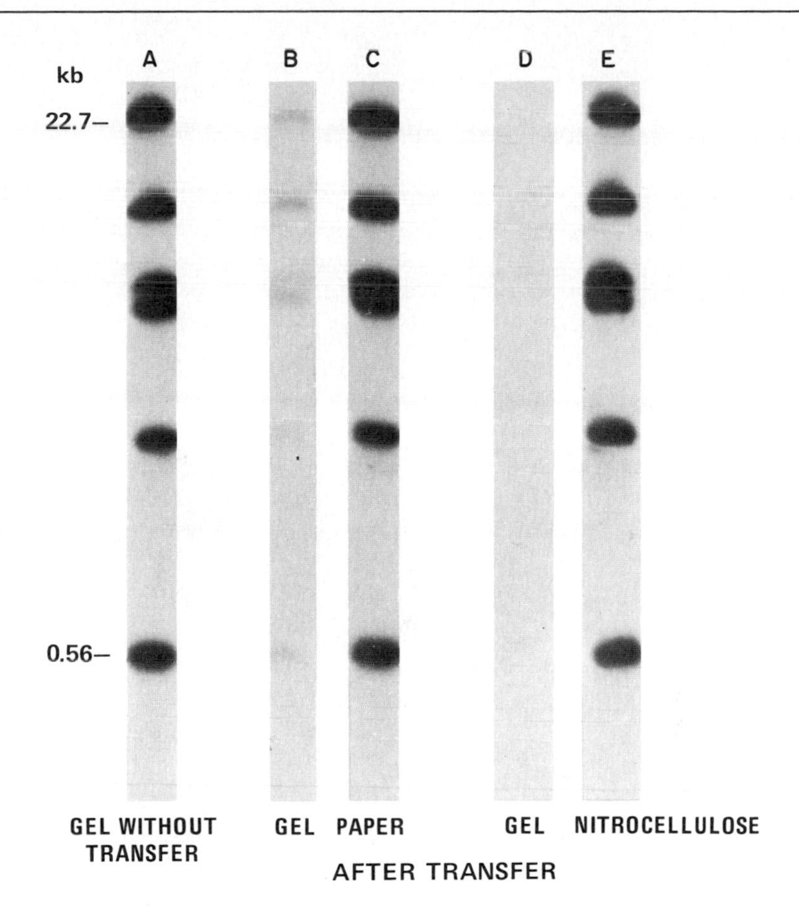

FIG. 4. Transfer of partially depurinated restriction fragments from agarose gels to diazo paper or nitrocellulose. Bacteriophage lambda (*Jam⁻ Zam⁻ Vir*)[49] DNA was cleaved with endonuclease *Hin*dIII, and the staggered ends were filled in with ^{32}P-labeled nucleotides using reverse transcriptase.[5] The labeled DNA fragments were then fractionated on a 0.7% agarose gel, depurinated partially with 0.25 *M* HCl, and processed further and transferred as described in the text. (A) Gel dried immediately after treatment. (B) Gel after transfer to diazo paper. (C) Paper obtained following transfer from gel shown in (B). (D) Gel following transfer to nitrocellulose. (E) Nitrocellulose obtained after transfer from gel shown in (D).

A quantitative comparison of the transfer of ^{32}P-labeled DNA fragments to either nitrocellulose or DBM paper is shown in Fig. 4. The majority of the DNA is transferred from the gel to DBM paper or to nitro-

[49] D. J. Donoghue and P. A. Sharp, *Gene* **1**, 209 (1977).

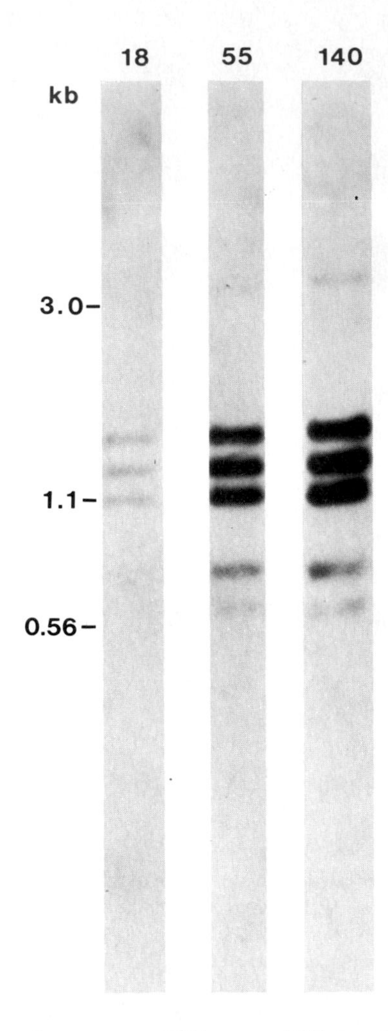

FIG. 5. Detection of restriction fragments on DNA paper. PALA-resistant Syrian hamster cell DNA was digested with either endonuclease *Pvu*II (A) or *Eco*RI (B). The restriction fragments were fractionated according to size on 1% (*Pvu*II digest) or 0.5% (*Eco*RI digest) agarose gels, depurinated, and transferred to diazo paper. DNA papers were then hybridized for 2 hr in hybridization buffer plus 1% glycine but lacking dextran sulfate, SDS, and probe, and then for an additional 12 hr in hybridization buffer lacking SDS and containing 10% dextran sulfate and 2×10^6 cpm of nick-translated probe (3×10^7 cpm/μg, 25 ng/ml). The papers were washed and autoradiographed for 18 hr using DuPont Kronex film (Fig. 5A) or for 12 hr with Kodak XR-5 film (Fig. 5B) and a Kodak Lightning Plus intensifying screen at $-70°$. The number of CAD genes in the PALA-resistant cells relative to the number in the PALA-sensitive cells is indicated above each track. The values are accurate to within a factor of about two.

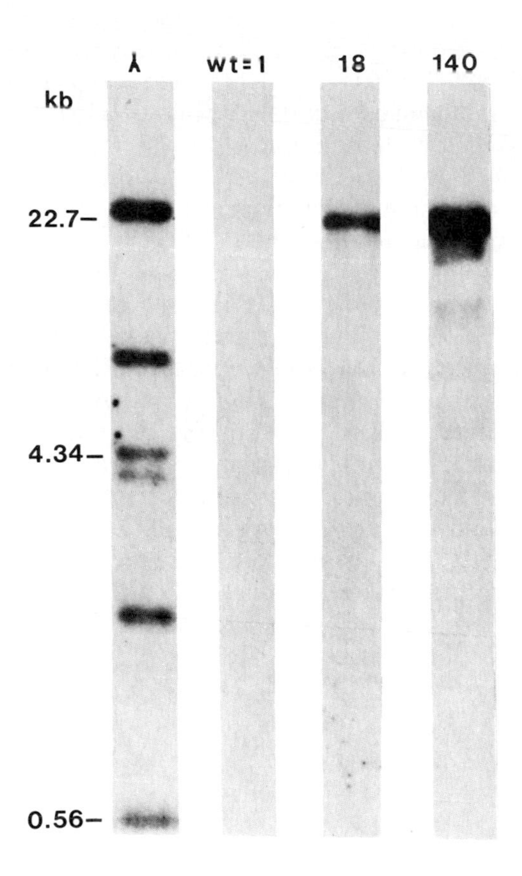

cellulose.[50] The efficiency of transfer is constant using HCl concentrations from 0.20 to 0.50 *M* (data not shown). The use of appreciably lower concentrations of HCl results in the production of fragments too large to be transferred efficiently, whereas higher concentrations produce fragments too small to bind efficiently to nitrocellulose (although DBM paper can bind very small fragments, as noted above). Since small DNA fragments will also form less stable hybrids with labeled probes, we de-

[50] For a more complete discussion of the efficiency of transfer of large DNA fragments from gels to diazo paper and to nitrocellulose, see G. M. Wahl, M. Stern, and G. R. Stark (in reference 15).

purinate with 0.25 *M* HCl and cleave with alkali to generate DNA fragments approximately 1–2 kb long.[50]

The transfer and subsequent detection of genomic restriction fragments ranging in size from approximately 0.3 to 19 kb is shown in Fig. 5A and B. The DNA used for these experiments was obtained from mutant Syrian hamster cells which overproduce the multifunctional protein CAD, which catalyzes the first three steps of UMP biosynthesis (carbamyl-P synthetase, aspartate transcarbamylase, dihydroorotase). The mutants contain approximately 18–140 times the number of CAD genes present in the parental cells.[40] CAD-specific DNA was detected using a nick-translated probe prepared from the recombinant plasmid described above (see RNA results). As shown in Fig. 5A, cleavage of DNA from mutant cells with endonuclease *Pvu*II produces five major bands of sizes approximately 0.7–1.6 kb which hybridize with the probe. The smallest CAD-specific fragment produced by digestion of the recombinant plasmid with this enzyme is approximately 0.4 kb and is detected easily (data not shown). Figure 5B shows that the major CAD-specific fragment produced by *Eco*RI digestion is 19 kb long. No DNA could be detected in either gel following transfer (data not shown). Therefore, partially depurinated genomic restriction fragments are transferred and covalently bound to diazo paper with high efficiency independently of size. The use of DNA paper allows detection of restriction fragments that vary greatly in size with equal and high sensitivity. DNA papers have been cleaned and exposed to different probes several times with little or no loss of hybridization efficiency.[51]

[51] Note: Brian Seed, California Institute of Technology, has recently developed a simpler procedure for making diazo paper. 1,4-Butanediol diglycidyl ether is used to activate Whatman 50 or 540 paper in the presence of NaOH and $NaBH_4$. After washing with NaOH, the paper-bound epoxide is reacted with 2-aminothiophenol in acetone, then washed and diazotized as usual. The results are comparable to those obtained with DMB paper.

Section IV

Vehicles and Hosts for the Cloning of Recombinant DNA

[16] Plasmids of *Escherichia coli* as Cloning Vectors

By FRANCISCO BOLIVAR and KEITH BACKMAN

The essence of molecular cloning or recombination *in vitro* is the joining together *in vitro* of two or more DNA fragments. One fragment, called the vector or vehicle, is capable of replication in some host organism; and the other(s), referred to as the cloned or passenger fragment(s), can be passively replicated when joined to the vector. Hybrid molecules are then put into a host organism by a process called transformation, and clones are isolated that contain various arrangements of the joined fragments. We summarize procedures that can be used for the generation of DNA fragments and present methods for joining DNA fragments. Then we discuss the properties of plasmid cloning vectors of *Escherichia coli* as well as techniques that can be used to select from a population of bacterial cells those carrying recombinant DNA molecules of interest.

Generation of DNA fragments

Four methods have been used to generate DNA fragments for cloning: chemical synthesis, enzymatic copying of RNA, controlled shearing, and restriction endonuclease digestion. DNA fragments to be cloned may be blunt-ended (having no unpaired bases at the termini) or sticky-ended (having a number of unpaired bases at either the 5′- or 3′-terminus). A DNA fragment may be purified prior to cloning or, if a clone carrying the desired DNA fragment can be recognized among a large number of clones that do not, the fragment may cloned without prior purification, a procedure referred to as "shotgunning."

Chemical synthesis of oligodeoxyribonucleotides was pioneered by Khorana[1] and has become significantly more facile since the introduction of the triester method of synthesis.[2] Chemical synthesis has been used to produce synthetic *lac* operators,[3,4] linkers[4-5] (about which more be said

[1] H. G. Khorana, K. L. Agarwal, H. Buchi, M. H. Caruthers, N. K. Gupta, K. Kleppe, A. Kumar, E. Ohtsuka, U. L. Raj Bhandary, J. H. van de Sande, V. Sgaramella, T. Terao, H. Weber, and T. Yamada, *J. Mol. Biol.* **72,** 209 (1972).

[2] K. Itakura, N. Katagari, C. P. Bahl, R. H. Wightman, and S. A. Narang, *J. Am. Chem. Soc.* **97,** 7327 (1975).

[3] K. Marians, R. Wu, J. Stawinski, T. Hozumi, and S. A. Narang, *Nature (London)* **263,** 744 (1976).

[4] H. L. Heyneker, J. Shine, H. M. Goodman, H. W. Boyer, J. Rosenberg, R. E. Dickerson, S. A. Narang, K. Itakura, S. Lin, and A. D. Riggs, *Nature (London)* **263,** 748 (1976).

[4a] C. P. Bahl, K. J. Marians, R. Wu, J. Stawinski, and S. A. Narang, *Gene* **1,** 81 (1976).

ISBN 0-12-181968-X

below) and, recently, a synthetic gene for the peptide hormone somatostatin.[6] Chemical synthesis presents the possibility of generating and cloning DNA sequences that may not occur in nature.

Enzymatic copying of RNA[7-12] has been used to clone many eukaryotic genes. Since many of these genes have been shown to include spacers in their genomic state, cDNA copies of mRNA may not always contain all the sequences present in the genomic gene. In practice, oligo(dT) bound to the poly(A) tails of eukaryotic mRNAs is used as a substrate for reverse transcriptase, which synthesizes a single-stranded DNA chain complementary to the RNA. The RNA is removed by alkali treatment, and a second DNA strand synthesized using DNA polymerase I. The duplex formed, after various enzymatic steps, is then joined to a vector. This method has been used to clone cDNA copies of genes for various β-globins,[7-10] ovalbumin,[11] insulin,[12] and other proteins.

Controlled shearing of DNA[13] is particularly useful for generating large (>10,000-base-pair) DNA fragments. The fragments produced by shearing have end points that are random with respect to the genome and are therefore useful in generating clone banks or "libraries," which are sets of clones that contain DNA fragments representing the entire genome of an organism.[14]

Restriction endonuclease digestion (described in detail by R. Roberts in this volume [2]) is currently the most widely used method of generating DNA fragments to be cloned and is an essential step in linearizing a plasmid vector to accept a passenger fragment. Type II restriction endonucleases cleave DNA at or near specific four- to seven-base-pair sites, producing families of fragments. These fragments can be separated by

[5] R. H. Scheller, R. E. Dickerson, H. W. Boyer, A. D. Riggs, and K. Itakura, *Science* **196,** 177 (1977).

[6] K. Itakura, T. Hirose, R. Crea, A. Riggs, H. L. Heyneker, F. Bolivar, and H. W. Boyer, *Science* **198,** 1056 (1977).

[7] F. Rougeon, P. Kourilsky, and B. Mach, *Nucleic Acids Res.* **2,** 2365 (1975).

[8] T. H. Rabbits, *Nature* (*London*) **260,** 221 (1976).

[9] T. Maniatis, S. G. Kee, A. Efstratiadis, and F. C. Kafatos, *Cell* **8,** 163 (1976).

[10] R. Higuchi, G. Paddock, R. Wall, and W. Salser, *Proc. Natl. Acad. Sci. U.S.A.* **73,** 3146 (1976).

[11] B. W. O'Malley, S. L. C. Woo, J. J. Monahan, L. McReynolds, S. E. Harris, M. J. Tsai, S. Y. Tsai, and A. R. Means, *in* "Molecular Mechanisms in the Control of Gene Expression" (D. P. Nierlich, W. J. Rutter, and C. F. Fox, eds.), p. 309. Academic Press, New York, 1976.

[12] A. Ullrich, J. Shine, J. Chirgwin, R. Pictet, E. Tischer, W. J. Rutter, and H. M. Goodman, *Science* **196,** 1313 (1977).

[13] D. S. Hogness and J. R. Simmons, *J. Mol. Biol.* **9,** 411 (1964).

[14] L. Clarke and J. Carbon, *Proc. Natl. Acad. Sci. U.S.A.* **72,** 4361 (1975).

agarose or polyacrylamide gel electrophoresis according to size (see Additional Methods), or by chromatography on RPC 5 columns.[14a]

When the size of the restriction fragment that carries a genetic region of interest is unknown, it can often be determined by a method developed by Southern[15] and is discussed in detail in this volume [9]. In brief, DNA fragments are separated according to size by agarose gel electrophoresis and transferred from the gel to a sheet of nitrocellulose paper. The DNA on the nitrocellulose paper is then hybridized to a radioactive probe RNA. Examination of the pattern of hybridization allows one to determine what DNA fragment or size class of fragment corresponds to the information contained in the RNA probe.

When a genetic region to be cloned contains internal cleavage sites for a particular restriction endonuclease, partial digestion may be performed[16,17] to produce fragments spanning that region. The production of partial digestion products may be enhanced by digestion in the presence of low concentrations of an intercalating dye or drug.[18,19] The amount of dye that gives partial digestion products is usually determined in a pilot reaction.

Joining of DNA Fragments

Jensen *et al.*,[20] Jackson *et al.*,[21] and Lobban and Kaiser[22] developed a method of adding single-stranded extensions to DNA fragments using the enzyme terminal deoxyribonucleotide transferase. When complementary single-stranded extensions are put on two different fragments, the fragments can associate by base pairing of these extensions. Fragments with complementary homopolymeric tails anneal to each other and not to themselves. This is especially important in shotgun cloning experiments because it eliminates the formation of concatemers of the passenger DNA fragments. In order to add single-stranded extensions to duplex molecules, it is often necessary to first treat the molecules with λ exonuclease, which produces short, single-stranded 3′-extensions which serve as a substrate for terminal transferase.

[14a] J. E. Larson, S. C. Hardies, R. K. Patient, and R. D. Wells, *J. Biol. Chem.* **254,** 5535 (1979).

[15] E. M. Southern, *J. Mol. Biol.* **98,** 503 (1975).

[16] C. Covey, D. Richardson, and J. Carbon, *Mol. Gen. Genet,* **145,** 155 (1976).

[17] K. Backman and M. Ptashne, *Cell* **13,** 65 (1978).

[18] J. Kania and T. G. Fanning, *Eur. J. Biochem.* **67,** 367 (1977).

[19] V. V. Nosikov, E. A. Baraga, A. V. Karlishev, A. L. Zhuze, and O. L. Polyanovskij, *Nucleic Acids Res.* **3,** 2293 (1976).

[20] R. H. Jensen, R. J. Wodzinski, and M. H. Rogoff, *Biochem. Biophys. Res. Commun.* **43,** 384 (1971).

[21] D. Jackson, R. Symons, and P. Berg, *Proc. Natl. Acad. Sci. U.S.A.* **69,** 2904 (1972).

[22] P. Lobban and A. D. Kaiser, *J. Mol. Biol.* **78,** 453 (1973).

Three methods of separating fragments joined by the terminal transferase method have been reported. The first involves cloning at *Pst*I restriction endonuclease cleavage sites[23] found in certain vectors (Table I). Not only is the 3′-hydroxyl of a cleaved *Pst*I site a direct substrate for the terminal transferase (without λ exonuclease treatment), but insertion of a passenger fragment with oligo(dC) tails into a *Pst*I-cleaved plasmid with oligo(dG) tails results in regeneration of *Pst*I cleavage sites at the junctions, thus permitting reisolation of the cloned fragment. The second method involves digestion of the (dA · dT) spacers with S1 nuclease under conditions of partial denaturation.[24] The third method involves exonuclease VII digestion of denatured linearized plasmid DNA.[25] Detailed procedures of methods used with tailing are given by T. Nelson and D. Brutlag in this volume [3].

Joining of DNA Fragments with Homologous Cohesive Ends

DNA ligase can be used to join covalently the short, cohesive ends generated by some restriction endonucleases.[26] *Escherichia coli* DNA ligase is used in 20 m*M* Tris-HCl, pH 8, 5 m*M* $MgCl_2$, 10 m*M* $(NH_4)_2SO_4$, 100 m*M* KCl, 100 μ*M* NAD, and 100 μg/ml bovine serum albumin (BSA). Ligase is added to 8 units/ml, and the reaction is incubated at 15° overnight.[27] T4 DNA ligase is used in 66 m*M* Tris-HCl, pH 7.6, 6.6 m*M* $MgCl_2$, 10 m*M* dithiothreitol (DTT), and 0.5 m*M* ATP at 4–12° for 2–12 hr. When DNA fragments with cohesive ends are joined, 1–5 units/ml of T4 DNA ligase is used.[28] The concentration of ends is adjusted in such a way that intermolecular rather than intramolecular joining is favored (typically, the DNA concentration is on the order of 100 μg/ml[29]). The joining reaction can be monitored by electrophoresis of an aliquot of the reaction on an agarose or polyacrylamide gel.

Joining of DNA Fragments with Blunt or Flush Ends

T4 DNA ligase (but not *E. coli* ligase) can covalently join blunt-ended DNA molecules.[28,30,31] This reaction requires a high concentration of

[23] A. Dugaiczyk (1976), cited in F. Bolivar, R. L. Rodriguez, P. J. Greene, M. C. Betlach, H. L. Heyneker, H. W. Boyer, J. H. Crosa, and S. Falkow, *Gene* **2,** 95 (1977).

[24] H. Hofstetter, A. Schambock, J. Van Den Berg, and C. Weissman, *Biochim. Biophys. Acta* **454,** 587 (1976).

[25] S. P. Goff and P. Berg, *Proc. Natl. Acad. Sci. U.S.A.* **75,** 1763 (1978).

[26] J. Mertz and R. W. Davis, *Proc. Natl. Acad. Sci. U.S.A.* **69,** 3370 (1972).

[27] A. Dugaiczyk, J. Hedgpeth, H. W. Boyer, and H. Goodman, *Biochemistry* **13,** 503 (1974).

[28] V. Sgaramella and H. G. Khorana, *J. Mol. Biol.* **72,** 427 (1972).

[29] A. Dugaiczyk, H. W. Boyer, and H. M. Goodman, *J. Mol. Biol.* **96,** 171 (1975).

[30] V. Sgaramella, J. H. van de Sande, and H. G. Khorana, *Proc. Natl. Acad. Sci. U.S.A.* **67,** 1468 (1970).

DNA ends in order to proceed at a usable rate. Blunt-ended DNA fragments are joined under the same T4 ligase reaction conditions given above, but the concentration of DNA ends should be at least 0.2 μM and approximately 50 units/ml of T4 DNA ligase is added to the reaction mixture.[4] Under conditions of low (5–10 μg/ml) T4 DNA ligase concentrations, 60 μg/ml RNA ligase can stimulate the reaction as much as 10-fold.[32]

Use of S1 Nuclease

DNA molecules with cohesive termini can be made blunt-ended by controlled S1 nuclease digestion or (when the protruding strand has a 5′-terminus) by repair synthesis with various DNA polymerases (a reaction called "filling in"). DNA fragments with cohesive termini (2–5 μg/ml) are digested with 3 units/μl of S1 nuclease in 0.03 M sodium acetate, pH 4.6, 0.3 M NaCl, and 4.5 mM $ZnCl_2$ at 22° for 30 min. The reaction is stopped by the addition of Tris base to 0.1 M, EDTA to 25 mM and tRNA to 25 μg/ml; alternatively, extraction with 50% phenol–50% $CHCl_3$ after the Tris base and EDTA addition may be used.

Filling in with T4 DNA Polymerase

T4 DNA polymerase reactions are carried out in 66 mM Tris-HCl, pH 8.8, 6.7 mM $MgCl_2$, 16.6 mM $(NH_4)_2SO_4$, 0.167 mg/ml BSA, 10 mM 2-mercaptoethanol, 6.7 mM EDTA, and 5×10^{-5} M of each dNTP. Reaction temperatures from 12° to 37° give satisfactory results. T4 DNA polymerase reactions can tolerate up to 30 mM NaCl. The reactions are initiated by the addition of 1 unit of T4 DNA polymerase for each microgram of DNA. The reactions are terminated by phenol extraction and ethanol precipitation.[33]

Linkers

Blunt-ended DNA fragments can also be cloned by endowing them with cohesive ends[4–5]: Short, synthetic DNA molecules (linkers) containing specific restriction endonuclease cleavage sites can be covalently joined via blunt-end ligation *in vitro* to other blunt-ended DNA molecules. Cleavage of the resulting molecules with the endonuclease that cleaves the linkers produces DNA fragments with cohesive ends. It should be

[31] V. Sgaramella, *Proc. Natl. Acad. Sci. U.S.A.* **69**, 3389 (1972).

[32] A. Sugino, N. R. Cozarelli, J. Shine, H. L. Heyneker, H. W. Boyer, and H. M. Goodman, *J. Biol. Chem.* **252**, 3987 (1977).

[33] A. Panet, J. H. van de Sande, P. C. Loewen, H. G. Khorana, A. J. Raae, J. R. Lillehaug, and K. Kleppe, *Biochemistry* **12**, 5045 (1973).

TABLE I
COMMONLY USED PLASMID VECTORS OF *Escherichia coli*[a]

Plasmid	Replicator (type or source)	Molecular weight (Md)	Genetic markers	Unique restriction sites	Reference
pSC101	Stringent	5.6	Tc^R	*Eco*RI, *Hin*dIII (Tc^R), *Bam*I (Tc^R), *Sal*I (Tc^R)	*b*, *c*
ColE1	Relaxed, multiple copy	4.2	Colicin immunity, colicin production	*Eco*RI (CP), *Sma*I (CP)	*d*
Mini-ColE1 (pVH51)	ColE1	2.1	Colicin immunity	*Eco*RI	*d*, *e*
RSF2124	ColE1	7.4	Colicin production, colicin immunity, Ap^R	*Eco*RI (CP), *Sma*I (CP), *Bam*I	*f*
pCR1	ColE1	7.4	Colicin immunity, Km^R	*Eco*RI, *Hin*dIII (Km^R)	*g*
pMB9	pMB1, ColE1-type replicator	3.5	Colicin immunity, Tc^R	*Eco*RI, *Hin*dIII (Tc^R), *Bam*I (Tc^R), *Sal*I (Tc^R), *Hpa*I, *Sma*I	*h*
pBR313	pMB1	5.6	Colicin immunity, Ap^R, Tc^R	*Eco*RI, *Hin*dIII (Tc^R), *Bam*I (Tc^R), *Sal*I (Tc^R), *Hpa*I	*i*
pBR322	pMB1	2.8	Tc^R, Ap^R	*Eco*RI, *Hin*dIII (Tc^R), *Bam*I (Tc^R), *Sal*I (Tc^R), *Pst*I (Ap^R), *Pvu*I (Ap^R), *Ava*I, *Pvu*II	*j*
pBR324	pMB1	5.5	Colicin immunity, colicin production, Ap^R, Tc^R	*Eco*RI (CP), *Sma*I (CP), *Hin*dIII (Tc^R), *Sal*I (Tc^R), *Bam*I (Tc^R)	*k*

pBR325	pMB1	3.7	Ap^R, Cm^R, Tc^R	*Eco*RI (Cm^R), *Pst*I (Ap^R), *Pvu*I (Ap^R), *Hin*dIII (Tc^R), *Bam*I (Tc^R), *Sal*I (Tc^R), *Ava*I	*k*
pACYC177	P15A, relaxed multiple-copy replicator compatible with ColE1	2.46	Ap^R, Km^R	*Pvu*I (Ap^R), *Pst*I (Ap^R), *Hinc*II (Ap^R), *Bam*I, *Hin*dIII (Km^R), *Sma*I (Km^R), *Xho*I (Km^R), *Pvu*I (Km^R)	*l*
pACYC184	P15A	2.65	Cm^R, Tc^R	*Eco*RI (Cm^R), *Hin*dIII (Tc^R), *Bam*I (Tc^R), *Sal*I (Tc^R)	*l*
pKB166	pMB1	1.6	λ immunity	*Eco*RI, *Bgl*II	*m*

[a] Vectors are listed with their replicators, molecular weights, and genetic markers. Unique restriction sites that can be used for cloning are listed and, when cloning at a site insertionally inactivates a gene, that gene is indicated in parentheses. Plasmids pSC101, pMB9, pBR313, and pBR322 are approved EK2 vectors in conjunction with *E. coli* χ1776. CP, Colicin production; Tc^R, tetracycline resistance; Ap^R, ampicillin resistance; Km^R, kanamycin resistance; Cm^R, chloramphenicol resistance. Plasmids pBR325 and pACYC184 and their derivatives which retain an intact chloramphenicol resistance gene can be amplified with spectinomycin (see text).

[b] S. N. Cohen, A. C. Y. Chang, H. W. Boyer, and R. B. Helling, *Proc. Natl. Acad. Sci. U.S.A.* **70,** 3240 (1973).

[c] S. N. Cohen and A. C. Y. Chang, *J. Bacteriol.* **133,** 807 (1977).

[d] V. Hershfield, H. W. Boyer, C. Yanofsky, M. A. Lovett, and D. R. Helinski, *Proc. Natl. Acad. Sci. U.S.A.* **71,** 3255 (1974).

[e] V. Hershfield, H. W. Boyer, L. Chow, and D. R. Helinski, *J. Bacteriol.* **126,** 447 (1976).

[f] M. So, R. Gill, and S. Falkow, *Mol. Gen. Genet.* **142,** 239 (1976).

[g] C. Covey, D. Richardson, and J. Carbon, *Mol. Gen. Genet.* **145,** 155 (1976).

[h] R. L. Rodriguez, F. Bolivar, H. M. Goodman, H. W. Boyer, and M. C. Betlach, *in* "Molecular Mechanisms in the Control of Gene Expression" (D. P. Nierlich, W. J. Rutter, and C. F. Fox, eds.), p. 471. Academic Press, New York, 1976.

[i] F. Bolivar, R. L. Rodriguez, M. C. Betlach, and H. W. Boyer, *Gene* **2,** 75 (1977).

[j] F. Bolivar, R. L. Rodriguez, P. J. Greene, M. C. Betlach, H. W. Heyneker, H. W. Boyer, J. H. Crosa, and S. Falkow, *Gene* **2,** 95 (1977).

[k] F. Bolivar, *Gene* **4,** 121 (1978).

[l] A. C. Y. Chang, and S. N. Cohen, *J. Bacteriol.* **134,** 1141 (1978).

[m] K. Backman, D. Hawley, and M. J. Ross, *Science* **196,** 183 (1977).

pointed out that this strategy assumes the absence of cleavage sites for the particular restriction endonuclease within the blunt-ended fragments being cloned. In some cases, such internal cleavage sites can be protected by treatment with an appropriate DNA methylase prior to joining with the linkers (see use of *Eco*RI methylase in Additional Methods). Detailed procedures for using linkers are presented elsewhere in this volume.

The direct joining of fragments by polynucleotide ligase offers some advantages over the terminal transferase method. When DNA fragments generated by restriction endonuclease digestion are joined, junctions involving similar ends and some junctions involving dissimilar ends are cleavable by a restriction endonuclease, allowing subsequent separation of joined fragments. Ease of separability plus the absence of homopolymeric spacers facilitates the engineering of expression of the cloned DNA.

Transformation of Cells

Isolation and amplification (cloning) of a desired arrangement of fragments from the mixture of recombinants produced in a joining reaction can be accomplished if the desired recombinant can replicate and specify a phenotype in a host organism. Cloning of plasmids is accomplished by transforming *E. coli* so that some cells receive recombinant molecules. (Certain plasmids containing the cos site of bacteriophage λ can be packaged *in vitro* into phage capsids and inserted into cells by transfection. Packaging *in vitro* is discussed by L. Enquist and N. Sternberg in this volume [18] and by B. Hohn in this volume [19].) This separates the various molecules in a manner that allows them to be recognized in individual clones of transformed cells selected for plasmid-specified phenotypes. The transformation procedures commonly used are minor modifications of that of Mandel and Higa.[34] DNA samples for transformation are prepared in 0.1 ml 30 m*M* $CaCl_2$. Bacteria are grown to 2×10^8 cells/ml in L broth, pelleted by centrifugation (Sorvall SS34 rotor, 8500 rpm, 4°, 5 min), and washed in $\frac{1}{2}$ volume cold 10 m*M* $NaCl_2$. The cell pellet is resuspended in $\frac{1}{2}$ volume cold 30 m*M* $CaCl_2$. After a 20-min incubation on ice, the cells are again pelleted and resuspended in $\frac{1}{10}$ volume cold 30 m*M* $CaCl_2$. Then, 0.2 ml of this suspension is added to each 0.1-ml DNA sample and incubated on ice for 60 min. Each transformation is then heated to 42° for 75 sec prior to the addition of 5 ml L broth. Transformed cultures are incubated at 37° for 2 hr to allow plasmid establishment and phenotypic expression. Then, 0.1-ml aliquots are plated on L plates containing appropriate selective agents (20–25 μg/ml ampicillin, tetracycline, chloramphenicol, or kanamycin). Plates are incubated 18–36 hr at 37° before scoring colonies and restreaking for single-colony isolates.

[34] M. Mandel and A. Higa, *J. Mol. Biol.* **53,** 159 (1970).

Transformation of Escherichia coli SK1592

A different transformation procedure has been developed for use with a specially designed *E. coli* host, SK1592, that exhibits a high transformation efficiency.[35] Strain SK1592 is streaked on Luria agar plate on the day preceding the experiment. A large single colony is used to inoculate 10 ml filter-sterilized L broth, pH 7.5. The culture is grown to 5×10^7 cells/ml (20 Klett units, No. 42 green filter), and 2 ml (1×10^8 cells) is centrifuged in a 15-ml Corex tube at 8000 rpm. The pellet is resuspended gently in 1 ml 10 m*M* morpholinopropane sulfonic acid (MOPS), pH 7.0, and 10 m*M* RbCl. The cells are centrifuged as above, resuspended in 1 ml 100 m*M* MOPS, pH 6.5, 50 m*M* $ClCl_2$, and 10 m*M* RbCl, and placed on ice for 30 min. The cells are centrifuged again. The tubes are drained thoroughly on absorbent material (turned upside down for 30 sec on several layers of Kimwipes), and the pellet resuspended gently in 0.2 ml of the buffer used in the preceding step. Then 3 μl dimethyl sulfoxide (DMSO) and 0.2 μg plasmid DNA are added. The DNA should be as concentrated as possible so that less than 10 μl is added. If larger volumes have to be added, the DNA should be in 10 m*M* MOPS, pH 7.0. The mixture is incubated for 30 min on ice. The transformations are then heated for 30 sec at 43.5° prior to the addition of 5 ml L broth, pH 7.5, and incubated for 60 min at 37° without shaking. For best results, transformants should be plated with 0.8 ml overlay agar (at 44°) containing the appropriate antibiotic.

Transformation of Escherichia coli χ1776

Escherichia coli strain χ1776 is an approved EK2 vector for use with certain plasmids (see Table I). Efficient transformation of χ1776 requires a special procedure, since the strain is easily lysed by a variety of conditions.[36]

An overnight culture of χ1776 grown in L broth plus 100 μg/ml diaminopimelic acid (DAP) and 40 μg/ml thymidine is diluted 10-fold in fresh medium and incubated 3.5 hr at 37° with shaking. Twenty-milliliter cultures are centrifuged at room temperature, and the pellet is gently resuspended in 10 ml 100 m*M* NaCl and recentrifuged at room temperature. Then the pellet is resuspended in 10 ml 10 m*M* Tris-HCl, pH 8, 0.8% NaCl, and 75 m*M* $CaCl_2$ and kept at room temperature for 20 min. The cells are again centrifuged at room temperature, gently resuspended in 2 ml of the buffer used in the preceding step, and chilled on ice in a glass

[35] S. R. Kushner, *in* "Genetic Engineering" (H. W. Boyer and S. Nicosia, eds.), p. 17, Elsevier, Amsterdam, 1978.

[36] M. Inoue and R. Curtiss, *in* "Molecular Cloning of Recombinant DNA" (W. A. Scott and R. Werner, eds.), p. 248. Academic Press, New York, 1977.

tube for 2 min. Then 0.2 ml cells is mixed with 0.1 ml DNA in 20 m*M* Tris-HCl, pH 8, and 0.8% NaCl and incubated on ice for 20 min. The mixture is heated to 42° for 1 min and kept on ice for 10 min. The cells are then plated directly onto selective plates.

Plasmid Vectors of *Escherichia coli*

Plasmids are circular DNA molecules capable of replicating in the cytoplasm of some microorganisms. Some plasmids have cleavage sites for certain restriction endonucleases in regions that are not essential for their replication. A plasmid linearized at such a site may be joined to another DNA fragment, recircularized, and put back into its host with no loss of ability to replicate. Plasmids with this property can be used as cloning vectors.

Two classes of plasmids have been used as cloning vectors in *E. coli*. One class, illustrated by the plasmid pSC101, undergoes stringent replication; plasmids of this class are generally present in a few copies per cell. The other class of plasmids, represented by ColE1, undergoes relaxed replication; plasmids of this class are generally present in many copies per cell. An expanded definition of relaxed and stringent replication is given in Novick *et al.*[37] Some relaxed plasmids can replicate in the absence of protein synthesis, which causes cessation of chromosomal replication. These plasmids can be amplified to copy numbers as high as 1000–3000 copies per cell by such procedures as chloramphenicol[38] or spectinomycin[39] treatment of a culture or amino acid starvation of a culture.[40] Amino acid starvation avoids the loss of viability that often accompanies chloramphenicol treatment and can be used on a small scale to increase differentially the rate of synthesis of the polypeptide products of cloned genes following termination of starvation.[40]

In deciding whether to use a phage or a plasmid as a vector, an investigator should consider the size of the DNA to be cloned, the method to be employed for identifying the inserted DNA, and the biological question to be addressed by cloning. Phage vectors are particularly well adapted to cloning large DNA fragments and fragments to be identified by hybridization with a nucleic acid probe. Phage vectors are quite large and therefore can have a poor ratio of passenger to vector DNA. This can result in difficulty in determining the restriction map of cloned DNA and in ob-

[37] R. P. Novick, R. C. Clowes, S. N. Cohen, R. Curtiss, III, N. Datta, and S. Falkow, *Bacteriol. Rev.* **40,** 168 (1976).

[38] D. B. Clewell, *J. Bacteriol.* **110,** 667 (1972).

[39] A. C. Y. Chang and S. N. Cohen, *J. Bacteriol.* **134,** 1141 (1978).

[40] K. Backman, Ph.D. Thesis, Harvard University, Cambridge, Massachusetts (1977).

taining DNA for sequence analysis free of vector-specific contaminants. The use of plasmid vectors, which are much smaller than phage vectors, can help to overcome these problems. The smaller size and simpler restriction endonuclease cleavage maps of plasmid vectors also make them better adapted to the engineering of expression of cloned DNA.

When plasmids seem better adapted to a particular cloning experiment, several other criteria must be assessed before selecting a specific vector. What amounts of vector, passenger, and recombinant DNA are required for the initial cloning and for the subsequent biological experiments and how readily are they prepared? What method of joining fragments is to be employed (tailing, linkers, blunt-end ligation, cohesive-end ligation) and what restriction endonuclease cleavage sites are present on the plasmid that can be used to clone a particular fragment? What selective pressures can be used to find plasmid carriers, recombinants, and particularly the desired recombinants? Since the answers to these questions differ from experiment to experiment, we tabulate the salient features of several plasmid vectors of *E. coli* (Table I) and leave matters of choice to the individual experimenter. The features of these plasmid vectors that should be considered include the mode of plasmid replication (relaxed replicators yield as much as 2 mg plasmid DNA from a liter of amplified culture), the presence of selectable markers (such as antibiotic resistance) on the plasmid, the presence of one or a number of usable sites at which DNA may be cloned without destroying either the ability to replicate or the selectability of the plasmid, the availability of sites at which DNA may be cloned to produce an identifiable phenotype (see Insertional Inactivation and Cycloserine Enrichment), the presence of regulatory elements which may be used to effect expression of a gene(s) on a cloned DNA fragment, and the presence of mutations that enhance the biological containability of the plasmid. Physical and functional maps of the most commonly employed plasmid vectors of *E. coli* are shown in Fig. 1.

Most plasmids listed in Table I are relaxed replicators; a new class of relaxed replicators derived from plasmid p15A is compatible with ColE1-derived plasmids.[39] Most plasmids utilize antibiotic resistance as a selectable marker, although various metabolic markers[41] and immunity to phage infection[42] can also be used as selectable markers. Many plasmids carry more than one selected marker; detection of cloned DNA in some of these vectors may be facilitated by insertional inactivation of one of the selectable markers (see below).

[41] V. Hershfield, H. W. Boyer, C. Yanofsky, M. A. Lovett, and D. R. Helinski, *Proc. Natl. Acad. Sci. U.S.A.* **71**, 3455 (1974).

[42] K. Backman, D. Hawley, and M. J. Ross, *Science* **196**, 182 (1977).

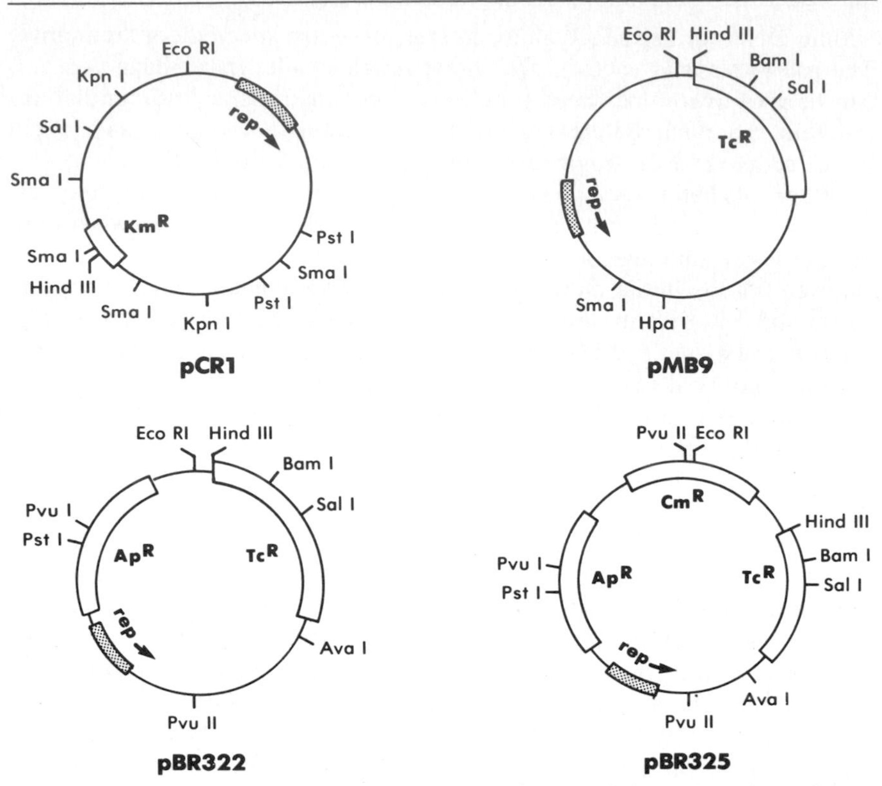

FIG. 1. Schematic representation (not to scale) of the restriction endonuclease cleavage maps and functional maps of the most commonly used plasmid cloning vectors. Indicated are the regions of genes encoding various drug resistances (Km^R, kanamycin resistance; Tc^R, tetracycline resistance; Ap^R, ampicillin resistance; Cm^R, chloramphenicol resistance) and the region essential for plasmid replication (rep). All shown vectors replicate using the ColEl mode of replication, which is unidirectional, and the direction of replication is indicated by an arrow. The map of pCR1 is based on K. A. Armstrong, V. Hershfield, and D. R. Helinski, *Science* **196,** 172 (1977); pMB9 is from F. Bolivar, R. L. Rodriguez, M. C. Betlach, and H. W. Boyer, *Gene* **2,** 75 (1977), and F. Bolivar, unpublished; pBR322 is from J. G. Sutcliffe, *Nucleic Acids Res.* **5,** 2721 (1978), and K. Backman, unpublished; and pBR325 is from F. Bolivar, *Gene* **4,** 121 (1978), and K. Backman, unpublished.

Selection of Cells Carrying Recombinant DNA Molecules

It is important to be able to select from a culture cells that have been successfully transformed, cells in which the transforming vector has acquired passenger DNA, and especially cells in which the vector has acquired the particular combination of fragments the researcher desires. Successfully transformed cells are usually selected by acquisition of a vector-determined phenotype, such as immunity to colicin E1 (see Additional Methods) or resistance to an antibiotic. Many novel schemes have been devised to allow the identification of clones in which the plasmid vector contains inserted DNA. Recombinant DNA molecules may be fractionated on the basis of size (e.g., by sucrose density gradient centrifugation or by agarose or acrylamide gel electrophoresis[40,42]) subsequent to the joining reaction and either before or after an initial round (or rounds) of cloning.

Miniscreens

There are several methods of screening individual clones for the size class of plasmid they carry.[43] We have used the following method (miniscreening). Transformants containing relaxed plasmids are grown overnight in 5 ml L broth and pelleted in 50-ml polypropylene tubes (Sorvall SS34 rotor, 8500 rpm, 4°, 5 min). The drained cell pellets are frozen, thawed, and resuspended in 1 ml cold 25% sucrose and 50 m*M* Tris-HCl, pH 8. To each sample is added 0.2 ml lysosyme (5 mg/ml in 25 m*M* Tris-HCl, pH 8) and 0.05 ml 0.25 *M* EDTA, pH 8. After incubation on ice for 15 min, 5 μl RNase (10 mg/ml in 50 m*M* sodium acetate, pH 5, boiled 10 min) is added. This is followed by the addition of 0.5 ml cold Triton lytic mix (3 ml 10% Triton X100, 75 ml 0.25 *M* EDTA, pH 8, 15 ml 1 *M* Tris-HCl, pH 8, 7 ml H_2O) and the solution is swirled gently. Samples are placed on ice for 15 min or until lysis is complete; then lysates are cleared by centrifugation (Sorvall SS34 rotor, 17,000 rpm, 4°, 35 min). Immediately, each supernatant is carefully removed, diluted with 1 ml H_2O, and phenol-extracted and ethanol-precipitated. The precipitate is collected by centrifugation (Sorvall SS34 rotor, 7500 rpm, 45 min) and resuspended in 0.05 ml 20 m*M* Tris-HCl, pH 7.5. It is sometimes necessary to repeat the phenol extraction and ethanol precipitation to remove inhibitors of some restriction endonucleases. Sufficient DNA (2–4 μg) is obtained to be analyzed before and/or after restriction endonuclease digestion by agarose or acrylamide gel electrophoresis.

Insertional Inactivation and Cyloserine Enrichment

Another method for identifying clones carrying plasmids with inserted DNA relies on the inactivation of a gene upon the insertion of foreign DNA into that gene. Many of the plasmids listed in Table I determine more than one phenotype. Insertion of a passenger DNA fragment at some sites on the plasmids inactivates a gene determining one of these phenotypes. Clones harboring recombinant plasmids can be rapidly screened for the absence of a vector-associated phenotype. If the inacti-

[43] W. M. Barnes, *Science* **195,** 393 (1977).

vated gene determines resistance to a bacteriostatic antiobiotic such as tetracycline, clones may be selected for antibiotic sensitivity via a cycloserine enrichment procedure.[44] For the selection of passenger fragments cloned in the tetracycline resistance genes of pBR322, samples of the transformation culture are diluted $\frac{1}{50}$ and, after logarithmic growth is established, ampicillin is added to 20 μg/ml. At cell saturation the culture consists entirely of ampicillin-resistant cells, i.e., plasmid carriers. The enriched culture is then diluted $\frac{1}{1000}$ and incubated for 1 hr, and tetracycline is added to a final concentration of 4 μg/ml. After a 45-min incubation in the presence of tetracycline, during which time the tetracycline-senstive cells cease growing, D-cycloserine is added to a final concentration of 100 μg/ml and the culture is incubated for 2.5 hr. Growing cells are rapidly lysed by this concentration of D-cycloserine. The culture is centrifuged and resuspended in 100 ml of L broth without antibiotics, incubated for 6–10 hr, and then plated for single colonies on L plates containing 10 μg/ml ampicillin. Between 90 and 100% of clones selected in this manner have acquired inserted DNA. In some cases, deletions within the plasmid that inactivates the tetracycline resistance genes have contributed to a percentage of isolates carrying no passenger DNA.

Use of Alkaline Phosphatase to Reduce Background of Recircularized Vector

When cloning very small amounts of passenger DNA, such as cDNA, it is possible to ensure ligation of most of the passenger DNA molecules to vector DNA by adding a molar excess of vector DNA to the ligation reaction. However, the majority of transformants obtained in this case contain recircularized vector DNA with no passenger fragment. Ullrich *et al.*[12] introduced a simple method to reduce the number of transformants containing recircularized vector plasmids. The linearized vector is treated with bacterial alkaline phosphatase in 10 m*M* Tris-HCl, pH 8.0 and 1 m*M* EDTA for 30 min at 65°. This removes the 5′-terminal phosphates from the ends of the plasmid. Alkaline phosphatase is strongly inhibited by low concentrations of inorganic phosphate, and it is usually necessary to dialyze endonuclease-cleaved DNA prior to alkaline phosphatase treatment. Ethanol precipitation is not adequate to remove inorganic phosphate. The extent of 5′-phosphate removal can be monitored as follows. An aliquot of the dephosphorylated vector is treated with ligase and analyzed by agarose gel electrophoresis. When dephosphorylation is complete, all of

[44] R. L. Rodriguez, F. Bolivar, H. M. Goodman, H. W. Boyer, and M. Betlach, *in* "Molecular Mechanisms in Control of Gene Expressions" (D. P. Nierlich, W. J. Rutter, and C. F. Fox, eds.), p. 471. Academic Press, New York, 1976.

the vector migrates as a linear molecule, and no higher forms (circles, linear dimers, etc.) are observed. As a final check, an aliquot of dephosphorylated vector can be treated with ligase and used to transform *E. coli*. Complete dephosphorylation will result in the recovery of fewer than 10^{-4} as many transformants as can be recovered from an equal amount of nonphosphatase-treated vector. This residuum is attributable to vector molecules that were not linearized by the initial endonuclease digestion. Although it is not essential to achieve 100% dephosphorylation to derive great benefit from this method, the ease of screening recombinants increases with increasing extent of dephosphorylation.

As implied in the above discussion, dephosphorylated 5′-ends are not substrates for ligase, and vector recircularization is prevented. Linear vector molecules are not proficient in transformation, perhaps because they are degraded after entering the cell. The phosphorylated 5′-ends of the passenger DNA can join to the 3′-hydroxyl ends of the vector, thus permitting the formation of plasmid–passenger nicked circles, which are proficient in transformation.

Identification of Clones Containing Specific Passenger DNA Fragments

The selection of plasmids carrying a specific fragment (or arrangement of fragments) is readily accomplished if that fragment determines a selectable phenotype. In this way, various drug resistance determinants, nutritional markers, and phage immunity genes have been cloned. In an interesting variation of the usual practice of cloning a DNA fragment on a vector having a replicator and a selectable marker, various groups have used a nonreplicating DNA fragment that determines an antibiotic resistance to clone fragments containing replicators.[45,46]

Clones carrying a specific DNA fragment in a plasmid vector can be identified by the presence of that fragment's nucleotide sequence. Grunstein and Hogness[47] have developed a method of lysing colonies on nitrocellulose filters and hybridizing radioactive probe RNA to the filters to ascertain which of the colonies contain the specific sequences desired; these clones are then isolated from a replica plate. Instead of hybridizing directly with the individual lysed colonies, it is sometimes expedient to hybridize with lysed cells from a culture grown from a small (100–1000)

[45] M. Lovett and D. R. Helinski, *J. Bacteriol.* **127,** 982 (1976).
[46] K. Timmis, F. Cabello, and S. N. Cohen, *Proc. Natl. Acad. Sci. U.S.A.* **72,** 2242 (1975).
[47] M. Grunstein and D. S. Hogness, *Proc. Natl. Acad. Sci. U.S.A.* **72,** 3961 (1975).
[48] L. H. Kedes, A. C. Y. Chang, D. Houseman, and S. N. Cohen, *Nature (London)* **255,** 533 (1975).

pool of clones.[48] When a pool containing a clone of interest is identified, that clone can be readily isolated by the Grunstein-Hogness method. Details of the colony hybridization method are presented by M. Grunstein and J. Wallis in this volume [25].

Expression of Cloned DNA

An important application of recombination *in vitro* has been the expression of cloned genetic information. Many of the genes from prokaryotic sources that have been cloned have carried with them the appropriate genetic signals to allow expression in *E. coli*. In a few cases (*his3*[49,50] and *leu2*[50] loci from yeast and dehydroquinase from *Neurospora crassa*[51]), cloned eukaryotic genes have been expressed in *E. coli,* but in the majority of examined cases, cloned eukaryotic DNA has not directed the synthesis of identifiable gene products. There may be several barriers to the expression of eukaryotic genes in *E. coli,* and some of these barriers may interfere with the expression of some prokaryotic genes in *E. coli*. Many eukaryotic genes have been discovered to contain intervening sequences or spacers in their DNA, which separate coding portions of the gene.[52-56] This poses a twofold problem: First, it can render unwieldy the size of a genomic DNA fragment that carries an entire gene; second, prokaryotic organisms are not known to possess the ability to integrate the information contained in such genes. The regulatory elements for the initiation and termination of transcription and the initiation of translation differ significantly between eukaryotes and prokaryotes; thus, a cloned eukaryotic gene may fail to be transcribed in *E. coli* for lack of a proper promoter or, being transcribed, may fail to be translated. Some eukaryotic genes may also contain sequences that cause premature termination of *E. coli* RNA polymerase. Finally, the polypeptide products of many non-*E. coli* genes may be subject to degradation in *E. coli*.

Each of these barriers to expression has been overcome in certain instances, but general solutions in all classes of barriers have not been discovered; much still depends on the demands of the particular experiment

[49] K. Struhl and R. W. Davis, *Proc. Natl. Acad. Sci. U.S.A.* **74,** 5255 (1977).

[50] B. Ratzkin and J. Carbon, *Proc. Natl. Acad. Sci. U.S.A.* **74,** 487 (1977).

[51] D. Vapnek, J. A. Hautala, J. W. Jacobson, N. H. Giles, and S. R. Kushner, *Proc. Natl. Acad. Sci. U.S.A.* **74,** 3508 (1977).

[52] J. Breathnach, L. Mandel, and P. Chambon, *Nature (London)* **270,** 314 (1977).

[53] E. C. Lai, S. L. C. Woo, A. Dugaiczyk, J. F. Catterall, and B. W. O'Malley, *Proc. Natl. Acad. Sci. U.S.A.* **75,** 2205 (1978).

[54] C. Brack and S. Tonegawa, *Proc. Natl. Acad. Sci. U.S.A.* **74,** 5652 (1977).

[55] S. M. Berget, C. Moore, and P. A. Sharp, *Proc. Natl. Acad. Sci. U.S.A.* **74,** 3171 (1977).

[56] S. M. Tilghman, D. C. Tiemier, J. G. Seidman, B. M. Peterlin, M. Sullivan, J. V. Maizel, and P. Leder, *Proc. Natl. Acad. Sci. U.S.A.* **75,** 725 (1978).

and the ingenuity of the experimenter. The barrier presented by genes in pieces has been circumvented in two ways. Chemical synthesis of a gene for the peptide somatostatin has been performed without reference to the genomic structure of the natural gene.[6] A more widely used approach is the cloning of cDNA copies of mRNAs; eukaryotic mRNAs are edited in the cell and contain an uninterrupted representation of the gene they carry. The expressibility of cDNA copies of genes in *E. coli* has been demonstrated for insulin.[57]

It has proven relatively straightforward to effect efficient transcription of cloned DNA by inserting it into a plasmid adjacent to a promoter sequence normally utilized by *E. coli.* To date, the promoter most commonly used in this manner has been the *lac* operon promoter of *E. coli.*[17,58] This promoter is subject to two kinds of regulation: positive regulation by the catabolite system[59] and repression by the *lac* repressor, which may be reversed experimentally by synthetic inducers such as isopropyl-β-thiogalactoside.[60,61] When cloned on multicopy relaxed plasmids, the *lac* promoter-operator titrates all the *lac* repressor in a cell, rendering both the plasmid *lac* promoter and the chromosomal *lac* operon constitutive.[58] The requirement for positive activation by the catabolite system may be overcome by using the UV5 mutant *lac* promoter; constitutive expression of *lac* promoters on multicopy plasmids may be repressed in host cells carrying the $lacI^{q1}$ allele, which results in 100-fold overproduction of *lac* repressor.[58] Other lower-level *lac* repressor overproducers may yield incomplete repression. The phenomenon of repressor titration and constitutive transcription is not general; for example, *E. coli* contains sufficient *trp* repressor to turn off elements of the *trp* operon cloned on multicopy plasmids.[41]

Recently, the promoter for a β-lactamase gene originally carried by Tn3 and now present in many plasmids was used to provide expression of the synthetic somatosatin gene[62] and the rat insulin gene.[57] The expression of these two genes, either from a *lac* or a β-lactamase promoter, also illustrates one method employed to effect translation of genes not normally found in *E. coli.* This method is the generation of a fusion polypeptide carrying the initial portion of a gene which is translated efficiently by *E. coli* attached in the correct reading frame to the gene whose expression is desired. The ultimate utility of this method depends on the separability

[57] L. Villa-Komaroff, A. Efstratiadis, S. Broome, P. LoMedico, R. Tizard, S. T. Naber, W. L. Chick, and W. Gilbert, *Proc. Natl. Acad. Sci. U.S.A.* **75,** 3727 (1978).

[58] K. Backman, M. Ptashne, and W. Gilbert, *Proc. Natl. Acad. Sci. U.S.A.* **73,** 4174 (1976).

[59] R. L. Perlman, B. DeCrombrugghe, and I. Pastan, *Nature (London)* **223,** 810 (1969).

[60] F. Jacob and J. Monod, *J. Mol. Biol.* **3,** 318 (1961).

[61] W. Gilbert and B. Muller-Hill, *Proc. Natl. Acad. Sci. U.S.A.* **56,** 1891 (1966).

[62] K. Backman, unpublished observations (1978).

of the desired portion of the fusion polypeptide from the portion being parasitized for its translatability. In the case of the synthetic somatostatin gene, a methionine codon was placed immediately adjacent to the portion of the gene encoding somatostatin, allowing separation of somatostatin from the carrier polypeptide by cyanogen bromide cleavage.[6]

Another method of effecting efficient translation is the "hybrid ribosome binding site" strategy.[17] This technique proceeds from the premise that an *E. coli* ribosome binding site consists of two functional units: a sequence homologous to the 3′-end of the 16 S rRNA [the Shine-Dalgarno (SD) sequence] and an initiator codon. All genes, prokaryotic or eukaryotic, are thought to have a methionine initiator codon; the strategy, then, is to juxtapose this codon (and the remainder of the gene) with an SD sequence from *E. coli* (the SD sequence from the *lacZ* gene is fortuitously followed by an *AluI* cleavage site) such that efficient translation results. Although there is a leeway of a few nucleotides in the acceptable spacing of these two elements,[17] precise juxtapositioning requires specific knowledge of the cleavage map and nucleotide sequence of the gene to be expressed.

The most formidable barrier to efficient expression of cloned genes is degradation, perhaps because it is the most poorly understood of the pertinent problems. Degradation is not a barrier to expression per se, but to recovery or utilization of the gene product after expression. Only one type of solution to this problem has been successful at this writing; that is, the use of fusion polypeptides. Attachment to the β-galactosidase protein serves to sequester the somatostatin peptide from degradation until it can be purified away from cellular proteases.[6] Attachment of rat preproinsulin to the β-lactamase protein accomplishes a similar result, with the additional wrinkle that the fusion polypeptide is secreted into the periplasmic space, as is normal β-lactamase. This serves to isolate the fusion polypeptide from cytoplasmic proteases.[57]

Additional Methods

Strains

The following derivatives of *E. coli* K12 are commonly used as recipient cells in transformation experiments and are good host strains for plasmid purification.

HB101: *pro⁻ leu⁻ thi⁻ lacY⁻ hsdR⁻ endA⁻ recA⁻ rpsL20 ara-14 galK2 xyl-5 mtl-1 supE44*

RR1: A *recA⁺* derivative of HB101

GM48: *dam-3 dcm-6 thr⁻ leu⁻ thi⁻ lacY⁻ galK2 galT22 ara-14 tonA31 tsx-78 supE44*. DNA from this strain is nonmethylated DNA and is cleaved by the restriction endonucleases *Eco*RII and *Mbo*I.

294: *endA⁻ hsdR⁻ thi⁻ pro⁻*

SK1592: *gal⁻ thi⁻ T1*^{R} *endA⁻ sbcB15 hsdR4*

χ1776: *tonA53 dapD8 minA1 minB2 supE42 Δ40(gal-uvrB) nalA25 thyA57 metC65 rfb-2 Δ29(bioH-asd) cycA1 cycB2 hsdR2*. This strain is an approved EK2 host when used with certain plasmids (see Table I).

Media

L Broth contains (per liter) 10 g tryptone (Difco), 5 g yeast extract (Difco), and 10 g NaCl. The medium is titrated to pH 7.5 prior to sterilization.

M-9 minimal medium is made as follows: 20 ml 20% casamino acids, 10 ml 0.1 *M* $MgSO_4$, 10 ml 0.01 *M* $CaCl_2$, and 840 ml H_2O are mixed and autoclaved. After cooling, 20 ml 20% glucose and 100 ml 10 × salts are added. When necessary, 2 μg/ml thiamine is added. A 10 × salt solution consists of 70 g Na_2HPO_4, 30 g KH_2PO_4, 5 g NaCl, 10 g NH_4Cl, and H_2O to 1 liter.

For plates, agar (1.5% final) is added. Antibiotics, such as tetracycline, ampicillin, kanamycin, and chloramphenicol, are added to a 20–25 μg/ml final concentration after the agar has cooled to below 50°.

Enzymes

Commercially available enzymes are adequate for most procedures described. A simple method for the purification of many restriction endonucleases is given in Greene *et al.*[63] T4 DNA polynucleotide ligase, kinase, and polymerase may be prepared as described by Panet *et al.*[33]

Phenol Extraction and Ethanol Precipitation

Removal of protein prevents the formation of insoluble precipitates following ethanol precipitation. Aqueous samples are extracted with an equal volume of cold phenol (equilibrated with 50 m*M* Tris-HCl, pH 7.5, 200 m*M* NaCl, 1 m*M* EDTA) by vortexing. The phases are separated by centrifugation. If the aqueous phase contains sufficient sucrose to be of a density similar to that of phenol, poor phase separation results. This can be corrected by the addition of a small amount of water. Loss of small oligonucleotides into the phenol phase can be reduced by extracting with

[63] P. J. Greene, H. L. Heyneker, F. Bolivar, R. L. Rodriguez, M. C. Betlach, A. A. Covarrubias, K. Backman, D. J. Russel, R. Tait, and H. W. Boyer, *Nucleic Acids Res.* **5**, 2373 (1978).

50% phenol–50% $CHCl_3$. After phenol extraction, residual phenol is removed from the aqueous phase by extraction with an equal volume of $CHCl_3$. The aqueous phase is made 0.25 *M* NaCl or 1% NaOAc, and the nucleic acid is precipitated with 2 volumes cold (−20°) ethanol. Nucleic acid is pelleted by centrifugation. If a copious precipitate occurs upon the addition of ethanol, it may be collected immediately. Otherwise, it is best to allow the precipitation to go for 2 hr in a dry ice–ethanol bath or overnight at −20° before centrifugation. Nucleic acid pellets are washed once with 70% ethanol, dried in a vacuum, and resuspended in a small volume of 10 m*M* Tris-HCl, pH 7.8, and 1 m*M* EDTA.

Selection of Colicin El-Immune Cells

Colicin E1 is prepared from *E. coli* K12 strain JC411 (ColE1). The cells are grown in L broth to 4×10^8 cells/ml and treated with mitomycin C (2 μg/ml). Twenty minutes later, the culture is pelleted, resuspended in 10 m*M* phosphate buffer, pH 7.4, and sonicated. The high-speed supernatant (Spinco 35 rotor, 30,000 rpm, 1 hr, 4°) is precipitated with 50% ammonium sulfate. The pellet is resuspended in 10 m*M* potassium phosphate, pH 7.2, 1 m*M* NaN_3, and 1 m*M* EDTA and diluted with an equal volume of glycerol for storage at −20°. To select cells immune to colicin El, 0.1 ml of a transformation culture and 1 drop of a $\frac{1}{100}$ dilution of colicin El are spread on an L plate. The optimal amount of colicin El that gives low background without killing immune cells must be determined experimentally for each colicin El preparation. Clones selected as colicin El-immune should be retested.

Use of EcoRI Methylase

*Eco*RI methylase is used in 100 m*M* Tris-HCl, pH 8, 10 m*M* EDTA, 1.1 μM *S*-adenosylmethionine (stored in 0.01 *N* H_2SO_4), 0.4 mg/ml BSA and 100 μg/ml DNA. Reactions are incubated at 37° for 20 min and stopped by phenol extraction and ethanol precipitation.[64]

Growth, Labeling Conditions, and Preparations of Amplifiable Plasmid DNA

Cultures are usually grown at 37° in M-9 minimal medium, containing [^{3}H]thymine (8 Ci/mmol) and deoxyadenosine (200 μg/ml) if plasmid is to be labeled. When the culture reaches a cell density of 2×10^8 cells/ml, chloramphenicol (170 μg/ml) or spectinomycin (300 μg/ml) is added, and the culture is incubated for 12–15 hr.

[64] P. J. Greene, M. C. Betlach, H. W. Boyer, and H. M. Goodman, *Methods Mol. Biol.* **7**, 87 (1974).

The culture is centrifuged. The pellet may be washed once with 10 m*M* NaCl. The pellet is frozen briefly and thawed (this results in improved lysis). Cells from 1 liter of culture are resuspended in 20 ml 25% sucrose and 50 m*M* Tris, pH 8.0. It is expedient to perform the subsequent steps in the centrifuge tubes to be used for clearing the lysate, since the viscous lysate produced is not easy to pour or measure, and unnecessary handling breaks the chromosomal DNA and reduces the efficiency with which it pellets. The suspension is placed on ice, and 1 ml 5 mg/ml lysozyme, 3 ml 0.25 *M* EDTA, pH 8, and 0.1 ml RNase (10 mg/ml in 50 m*M* NaOAc, pH 5, boiled 10 min) are added. After 15 min, 3 ml of cold Triton lytic mix (3 ml 10% Triton X-100, 75 ml 0.25 *M* EDTA, pH 8, 15 ml 1 *M* Tris-HCl, pH 8, 7 ml H_2O) is added, and the mixture is allowed to stand with occasional gentle swirling for 15 min. Lysates are then cleared by centrifugation (Sorvall SS 34 rotor, 17,000 rpm, 4°, 35 min, or Spinco 30 rotor, 30,000 rpm, 4°, 25 min). The supernatant is carefully removed and diluted with 1 volume of H_2O. In a good cleared lysate, more than 95% of the chromosomal DNA is removed.

The cleared lysate is phenol-extracted and ethanol-precipitated. The DNA is resuspended in a minimum volume (2 ml or less) of A50 buffer (500 m*M* NaCl, 20 m*M* Tris, pH 8.0, 1 m*M* EDTA) and applied to a 2.5 × 50 cm BioGel A50 agarose column. Absorbence at 260 nm is monitored, and the fractions under the first peak are pooled and ethanol-precipitated. The plasmid DNA is further purified by centrifugation to equilibrium (20 hr at 36,000 rpm, 20°, in a Spinco SW 50.1 rotor) in a CsCl–propidium diiodide gradient (2.2 g CsCl, 2 ml DNA solution, 0.25 ml 2 mg/ml propidium diiodide, mineral oil to fill the tube).

The DNA is visualized by uv illumination. The denser (bottom) band, which contains supercoiled plasmid DNA, is collected, and the propidium diiodide extracted by passing the DNA over a 0.5 × 6 cm Ag 50W-X8 Dowex column equilibrated with 50 m*M* Tris-HCl, pH 8.0, 1 m*M* EDTA, and 1 *M* NaCl. The eluent is dialyzed against 10 m*M* Tris-HCl, pH 7.8, and 1 m*M* EDTA, ethanol-precipitated, and resuspended in 10 m*M* Tris-HCl, pH 7.8, and 1 m*M* EDTA. DNA concentrations are determined spectrophotometrically in the above buffer; an A_{260} of 1.0 corresponds to 50 μg/ml DNA.

Agarose and Acrylamide Gel Electrophoresis

Slab gels (12 × 14 × 0.5 cm) containing 100 ml of 0.8–1.0% agarose (Seakem) are prepared by autoclaving the agarose for 10 min in Tris-EDTA-borate buffer[65] (90 m*M* Tris base, 2.5 m*M* Na_2EDTA, 90 m*M*

[65] A. C. Peacock and C. W. Dingman, *Biochemistry* **8**, 608 (1969).

H_3BO_3, pH 8.3). It is convenient to prepare a 5- or 10-fold concentrated stock solution of this buffer. After some time, a precipitate forms in 10-fold concentrated stock solutions. Not immediately, but eventually 10-fold-concentrated buffer solutions fail to give satisfactory results and must be remade. After the gels solidify, samples containing from 0.2 to 1 μg DNA are loaded onto the gel in 10–40 μl containing $\frac{1}{10}$ volume dye mix (6 *M* urea, 0.5% xylene cyanol green, 0.5% bromphenol blue). Electrophoresis is performed at 130–150 V at room temperature for 2 hr or until the bromphenol blue reaches the bottom of the gel. Slab gels (12 × 14 × 0.1 cm) of 7.5% acrylamide are prepared by mixing 2.4 ml of 10× Tris-EDTA-borate buffer with 15.6 ml H_2O, 2 ml acrylamide stock (29% acrylamide, 0.6% bisacrylamide), and 0.12 ml freshly prepared 10% ammonium persulfate. The amounts of H_2O and acrylamide stock can be varied to produce gels of from 3.75 to 20%. The acrylamide mixture is degassed, 12 μl TEMED is added, and the gel poured immediately. Acrylamide and bisacrylamide obtained from sources other than Bio-Rad may require purification to give satisfactory results. Acrylamide electrophoresis is usually performed at 120–200 V. The running buffer is Tris-EDTA-borate, described above. After electrophoresis gels are stained for 5 min in 4 μg/ml ethidium bromide solution and visualized with a short-wave uv transilluminator (Ultraviolet Products, San Gabriel, California). Gels may be photographed through a yellow No. 9. Kodak Wratten gelatin filter.

For agarose electrophoresis, the following molecular weight standards can be used: the 6 *Eco*RI-generated fragments of the λ bacteriophage genome (13.7, 4.68, 3.7, 3.56, 3.03, and 2.09 mD), and the 6 *Hin*dIII-generated fragments of the SV40 genome (1.13, 0.75, 0.68, 0.35, 0.28, and 0.13 mD). For acrylamide gel electrophoresis, the plasmid pBR322 is a good source of smaller DNA standards. For example, the 22 *Hae*III-generated fragments of pBR322 have sizes between 587 and 7 nucleotide base pairs.[66]

Elution of DNA Fragments from Polyacrylamide Gels

Maxam and Gilbert[67] have described a very simple and elegant method for eluting DNA fragments from acrylamide gels, which provides virtually complete recovery of DNA fragments less than 400 nucleotide pairs long.

The liquid volume and sizes of the plasmid cone, glass rod, and centrifuge tube specified are designed for gel slices approximately 10 × 5 × 2 mm. These dimensions may be scaled up for larger pieces.

Elution of DNA from the gel takes place inside a plastic cone plugged

[66] J. G. Sutcliffe, *Nucleic Acids Res.* **5,** 2721 (1978).

[67] A. M. Maxam and W. Gilbert, *Proc. Natl. Acad. Sci. U.S.A.* **74,** 560 (1977).

tightly with siliconized glass wool (the plug retains fragments of polyacrylamide in a later step). For small gel slices, a 1000-μl (blue) Eppendorf pipette tip is appropriate. Larger gel slices fit in larger pipette tips or conical plastic centrifuge tubes. The opening at the apex of the cone is sealed by melting the polypropylene (at the tip) with a small flame. Crush the gel in the cone with a glass rod that just fits inside it. Hand-grind the crushed gel to a not-too-fine paste. Rinse the paste off the rod in the cone with 0.6 ml gel elution solution (0.5 *M* NH_4OAc, 0.01 *M* $Mg(OAc)_2$, 0.1% sodium dodecyl sulfate (SDS), 0.1 m*M* EDTA). Introduce 20 μg tRNA in a small volume (1–10 μl) if labeling of the DNA is not to follow elution. Mix the tRNA in by stirring with the micropipette used to transfer it to the gel elution solution. Seal the top of the cone with Parafilm.

Seat the sealed cone in a siliconized, thick-walled 10 × 75 mm glass test tube. Incubate at 37° for at least 10 hr. Remove the Parafilm from the top of the cone, cut off the sealed pointed end with scissors, and insert the cone back into its tube.

The DNA has diffused out of the gel fragments and into the gel solution, which can be collected by centrifuging it through the glass wool into the tube below. A short spin (2 min) at low speed (a clinical centrifuge) is sufficient. Introduce another 0.2 ml gel elution solution into the cone and spin again. If the DNA is labeled with ^{32}P, its movement from the elution vessel into the glass tube can be followed with a Geiger counter.

The cone is discarded, and 2.0 ml 95% ethanol is added to the tube. The tube is sealed with Parafilm, and the phases are well mixed by inverting the tube at least five times. Chill at −70° in a dry ice–ethanol bath for 10 min. The DNA is sedimented at 10,000 *g* for 15 min at 4° (Sorvall SS34 angle rotor), and the supernatant is removed with a pipette, being careful not to disturb the pellet or film on the side wall of the tube.

The sedimented DNA is redissolved in 0.5 ml 0.3 *M* NaOAc, and 1.5 ml ethanol is added. The phases are mixed, chilled, and centrifuged, and the supernatant is removed. One milliliter of ethanol is introduced into the tube without mixing. After centrifugation, the supernatant is removed. The DNA is dried under vacuum for a few minutes.

The nearly salt-free DNA residue appears as a pellet or film inside the tube and is dissolved in a small volume of 10 m*M* Tris, pH 7.8, and 1 m*M* EDTA.

The eluted DNA may be absorbed to a small column of hydroxyapatite (BioGel DNA grade) and eluted with 0.4 potassium phosphate, pH 6.9, prior to using it in enzymatic reactions.

Acknowledgments

K. B. was supported by a postdoctoral fellowship from the Helen Hay Whitney Foundation. We wish to thank B. Rapagnani and P. Clausen for help in preparing the manuscript.

[17] Plasmid Cloning Vehicles Derived from Plasmids ColE1, F, R6K, and RK2

By MICHAEL KAHN, ROBERTO KOLTER, CHRISTOPHER THOMAS, DAVID FIGURSKI, RICHARD MEYER, ERIC REMAUT and DONALD R. HELINSKI

Bacterial plasmids can be used as vehicles for the stable maintenance of foreign DNA in *Escherichia coli* and other bacteria. By using *in vitro* recombination techniques, naturally occurring plasmids can be modified to facilitate the manipulation of cloned DNA and to provide a high level of biological containment of recombinant plasmids. These modifications have several main objectives: (1) to reduce the molecular weight of the plasmid cloning vehicle, (2) to decrease the probability that a recombinant plasmid will be transferred to other bacteria, (3) to construct plasmid vehicles that contain only one site for a given restriction enzyme located in a region not essential for plasmid replication, and (4) to introduce one or more phenotypic markers into the plasmid to permit the selection of cells containing the plasmid and to aid in the identification of plasmids that have incorporated foreign DNA in a way leading to an altered phenotype. In addition, it is often advantageous to use plasmids that can be obtained in large yields.

This article concentrates on cloning vehicles related to plasmids ColEl, F, R6K, and RK2. Because these replicons have different properties, cloning vehicles derived from one of them may be more suitable for a particular application than vehicles derived from another. Also, since derivatives of a single plasmid do not stably coexist, it is often desirable to have available cloning vehicles from a variety of incompatibility groups in order to study interactions between cloned fragments. The ColE1 vehicles pMK20,[1] pMK16,[2] and pMK2004[3] are very useful for general-purpose cloning, since they are maintained in high copy number and provide good selective markers for transformation. Where a low gene dosage of the insert is desirable, pDF41,[4] a derivative of the low-copy-number plasmid F′lac, can be used. Like the ColE1 vehicles, pRK353,[5] a derivative of the antibiotic resistance plasmid R6K, is present in a large number of copies per cell and can be used as a high-copy-number cloning vehicle

[1] M. Kahn and D. R. Helinski, *Proc. Natl. Acad. Sci. U.S.A.* **75,** 2200 (1978).

[2] D. Helinski, *in* "Genetic Engineering for Nitrogen Fixation" (A. Hollaender *et al.*, eds.), p. 19. Plenum, New York, 1977.

[3] M. Kahn and D. R. Helinski, in preparation.

[4] D. Figurski, unpublished observations (1977).

[5] R. Kolter and D. R. Helinski, *Plasmid* **1,** 571 (1978).

ISBN 0-12-181968-X

that will coexist with ColE1. In addition, the R6K replicon has been separated into two components that make up a functional replicon: an R6K origin of replication and a gene that codes for a trans-acting protein required for R6K replication.[6] A plasmid containing only the origin of replication can be used as a cloning vehicle restricted to hosts into which the specific R6K replication gene(s) has been inserted. Plasmid RK2 is unusual in that it is maintained in a wide variety of gram-negative bacteria,[7] and therefore RK2 derivatives may be of use as cloning vehicles in gram-negative bacteria other than *E. coli.*

Methods

Plasmid Isolation and Clone Analysis

Plasmid DNA can be separated from chromosomal DNA on the basis of its smaller size or by taking advantage of the unique properties of covalently closed circular DNA molecules. Generally, analysis of the DNA of a number of clones should be carried out only after the number of clones has been reduced as much as possible by genetic tests. It is best to choose the method of plasmid isolation according to the amount of information needed in the particular experiment. If the desired plasmid can be identified on the basis of size only, it may be convenient to use a rapid sodium dodecyl sulfate (SDS) lysis method involving small samples. If further purification of the plasmid is necessary, as is often the case when the DNA is to be cleaved with restriction enzymes, we have found it most efficient to purify plasmid DNA from as large a quantity of cells as can be handled conveniently. If large quantities of DNA of a few standard vehicles or DNA of low-copy-number plasmids is required, it may be desirable to use CsCl–ethidium bromide gradient centrifugation to purify and concentrate the covalently closed plasmid circles even though this method requires substantially more time and equipment.

Purification of Plasmids Using Equilibrium Centrifugation in Dye–CsCl Gradients. Total cell lysates can be prepared using a lysozyme–sarkosyl[8] or a lysozyme–SDS lysis.[9] In a modification of the former procedure, cells are grown at 37° in M-9 medium (6 g Na_2HPO_4, 3 g KH_2PO_4, 0.5 g NaCl, 1 g NH_4Cl, 0.1 m*M* $MgCl_2$, 0.1 m*M* $CaCl_2$ per liter) containing 1% glucose and 0.5% casamino acids, or in LB medium (5 g yeast extract, 10 g tryptone, 10 g NaCl per liter). Plasmids derived from F′ lac,

[6] R. Kolter, M. Inuzuka, and D. R. Helinski, *Cell* **15,** 1199 (1978).

[7] J. E. Beringer, *J. Gen. Microbiol.* **84,** 188 (1974).

[8] D. Clewell and D. R. Helinski, *J. Bacteriol.* **110,** 1135 (1972).

[9] P. Guerry, D. J. LeBlanc, and S. Falkow, *J. Bacteriol.* **116,** 1064 (1973).

R6K, and RK2 are isolated from fresh stationary-phase cultures. ColE1 plasmids continue to replicate in the absence of protein synthesis, and the fraction of plasmid DNA can be increased considerably by adding chloramphenicol (250 mg/liter) when the culture reaches 5 × 10^8 cells/ml (LB) or 3–4 × 10^8 cells/ml (M-9) and by continued shaking for an additional 12–16 hr.

Cells from 1 liter of culture are harvested by centrifugation and resuspended at 0° in 60 ml of 50 m*M* Tris and 20% sucrose, pH 8. Twenty milliliters of a 10 mg/ml solution of lysozyme (Sigma) dissolved in 25m*M* Tris, pH 8, is added, followed 5 min later by the addition of 28 ml of 100 m*M* NaEDTA, pH 8. Some lysozyme preparations contain small amounts of nuclease, and this can lead to low yields of supercoiled DNA. The cells are held on ice for 10–30 min, and then 30 ml of 2% sarkosyl in 50 m*M* Tris, pH 8, is added slowly and mixed thoroughly. After adjusting the volume of the lysate to 180 ml with 50 m*M* Tris, 185 g of CsCl and 20 ml of a 5 mg/ml solution of ethidium bromide (Cal Biochem) are added. The lysates are centrifuged in polyallomer tubes at 40,000 rpm for 40 hr in a Spinco Ti60 rotor (160,000 *g*) or in a Spinco Ti50 rotor (125,000 *g*). The DNA bands are visualized using long-wavelength uv light, and the upper (chromosomal DNA) band is removed from the top of the tube with a Pasteur pipette. Plasmid DNA can then be collected through the side of the tube using an 18- or 21-gauge hypodermic needle. Generally a second centrifugation is necessary to concentrate and further purify the plasmid DNA. Ethidium bromide is removed by extracting three times with isoamyl alcohol or with isopropanol that has been equilibrated with water and solid CsCl.

As an alternative especially useful for ColE1 plasmids, a cleared lysate procedure similar to that of Katz *et al.*[10] can be used. Cells from 1 liter of culture are harvested by centrifugation and resuspended at 0° in 25 ml of 50 m*M* Tris and 20% sucrose, pH 8. Three milliliters of a 10 mg/ml solution of lysozyme (Sigma) in 25 m*M* Tris, pH 8, is added, followed 5 min later by 3 ml of 250 m*M* NaEDTA, pH 8. Then 30 ml of 0.2% Triton X-100, 50 m*M* Tris, and 25 m*M* EDTA, pH 8, is added and mixed immediately and thoroughly. The cells are then placed at room temperature until the suspension has cleared noticeably, but for no longer than 15 min. The lysate is centrifuged at 20,000 *g* for 20 min, and the supernatant from the clearing spin is taken. One gram of CsCl and 0.1 ml of 5 mg/ml ethidium bromide are added for each milliliter of lysate, and the lysates are centrifuged at 40,000 rpm for 40 hr. Since much of the chromosomal DNA is removed during the clearing spin, only a single centrifugation step may be needed. Using this procedure, the yield of DNA per liter of

[10] L. Katz, D. Kingsbury, and D. R. Helinski, *J. Bacteriol.* **114**, 577 (1973).

cells is usually lower than that obtained with a sarkosyl lysate, but a lysate derived from more cells can be loaded into an ultracentrifuge rotor.

Another method that can be used to reduce the amount of chromosomal DNA in the lysate prior to ultracentrifugation relies on the higher pH required for the denaturation of the covalently closed form of DNA.[11] Cells are lysed using lysozyme or a protease and vortexed to break chromosomal DNA. The pH is slowly adjusted to 12.3 with NaOH, and then Tris-HCl is added to lower the pH to 8.5. NaCl is added to 3% w/v, and the lysate is extracted with phenol. Denatured chromosomal DNA is found at the interface and in the lower (phenol) phase, and the covalently closed plasmid DNA is in the upper (aqueous) phase. After extraction of the aqueous phase with chloroform, CsCl and ethidium bromide are added and the lysate is centrifuged as above.

Other methods using polyethylene glycol[12] or hydroxyapatite[13,14] have been used to isolate covalently closed circular DNA from cell lysates. The separation of linear from covalently closed DNA can also be accomplished using acidic phenol (pH 4.0) and low salt,[15] but this method requires relatively pure DNA.

Rapid Purification of Plasmid DNA. Relatively small cultures of cells containing plasmids derived from R6K, RK2, and ColE1 yield enough DNA for clone analysis. Forty-milliliter cultures are grown in LB medium in 25 × 200 mm culture tubes held at an angle (30°–40°) and shaken vigorously. When isolating DNA of a ColE1-derived plasmid, 10 mg of chloramphenicol is added at a cell density of 5×10^8 cells/ml. Cultures are shaken overnight, and the cells are pelleted in 50-ml polypropylene or polycarbonate centrifuge tubes. The cells are resuspended in 1 ml of 50 m*M* Tris and 10% sucrose, pH 8, by vortexing and then placed on ice. Then 0.1 ml of a 10 mg/ml solution of lysozyme (Sigma) dissolved in 25 m*M* Tris, pH 8, is added, and the mixture is incubated on ice for 5 min. Next 0.1 ml of 0.25 m*M* EDTA, pH 8, is added, followed by 1 ml of 0.2% Triton X-100, 50 m*M* Tris, and 25 m*M* EDTA, pH 8. Each tube is shaken immediately after addition of the Triton solution and then placed at room temperature. When the suspension has cleared significantly, but not longer than 15 min later, the lysates are centrifuged at 20,000 *g* for 20 min. With good lysis, the pellet is compact but jellylike, and the supernatant is not very viscous. To achieve the proper degree of lysis, it may be neces-

[11] T. C. Currier and E. W. Nester, *Anal. Biochem.* **76,** 431 (1976).

[12] G. O. Humphreys, G. A. Willshaw, and E. S. Anderson, *Biochim. Biophys. Acta* **383,** 457 (1975).

[13] W. Pakroppa, W. Goebel, and W. Muller, *Anal. Biochem.* **67,** 372 (1975).

[14] A. Colman, M. J. Byers, S. B. Primrose, and A. Lyons, in press.

[15] M. Zasloff, G. D. Ginder, and G. Felsenfeld, *Nucleic Acids Res.* **5,** 1139 (1978).

sary to vary the duration of the lysozyme digestion; digestion times of from 3 to 15 min have been used for different strains. A higher Triton X-100 concentration may also be required to obtain good lysis.

The supernatant of the clearing spin is mixed thoroughly with 3 ml of buffer-saturated phenol and centrifuged at 6000 *g* for 10 min in order to separate the phases. The use of reagent-grade phenol crystals is satisfactory. These are dissolved immediately before use in 100 m*M* Tris, pH 8.5. The final pH should be above 6.5, and the phenol should be colorless. Residual phenol is extracted from the aqueous (upper) phase by shaking it vigorously with 2–3 volumes of diethyl ether until the bottom (aqueous) phase is clear. The phases are separated by gravity, and the aqueous phase is taken. A small amount of bromphenol blue can be added to make it easier to distinguish the aqueous and ether phases. An equal weight of isopropanol is added to the DNA and mixed thoroughly. This concentration of isopropanol precipitates DNA and RNA but not the EDTA. After 30 min at 0° the heavy precipitate is pelleted in a swinging-bucket rotor centrifuged at 10,000 *g* for 10 min. The pellets are dried under vacuum and resuspended in 100 μl of 6 m*M* Tris and 6 m*M* NaCl, pH 7.4. The yield is 20–50 μg of ColE1 DNA, or 2–10 μg of R6K or RK2 DNA. The DNA is 80–95% plasmid DNA and can be used immediately for restriction or transformation. A procedure similar to this has been developed using SDS to lyse the cells.[16]

Analysis of Plasmid DNA Using Small Quantities of Cells. A method has been developed in which individual colonies are lysed and analyzed on gels.[17] An alternative to this procedure uses 2-ml overnight cultures grown in LB broth. These are pelleted and resuspended in 0.1 ml of 50 m*M* Tris, 5 m*M* EDTA, and 5% sucrose, pH 8. Ten microliters of 10% SDS is added and mixed immediately. The samples are incubated at 65° for 15 min and then centrifuged at 10,000 *g* for 15 min. The jellylike pellet is removed, and 25 μl of the supernatant is analyzed by gel electrophoresis. This quantity of DNA can also be purified further for restriction enzyme digestion.[16]

Plasmid Construction

In manipulating and rearranging DNA, a variety of selective markers carried on single restriction fragments have been used in our laboratory. Table I[1,4–6,18–23] lists these fragments. Ligation of DNA with cohesive

[16] J. A. Meyers, D. Sanchez, L. P. Elwell, and S. Falkow, *J. Bacteriol.* **127,** 1529 (1976).
[17] W. M. Barnes, *Science* **195,** 393 (1977).
[18] M. A. Lovett and D. R. Helinski, *J. Bacteriol.* **127,** 982 (1976).
[19] K. Timmis, F. Cabello, and S. N. Cohen, *Proc. Natl. Acad. Sci. U.S.A.* **72,** 2242 (1975).
[20] M. Kahn, unpublished observations (1977).

TABLE I
RESTRICTION FRAGMENTS CONTAINING SELECTIVE MARKERS

Enzyme	Selection	Size (kb)	Source	Reference
*Eco*RI	Kan^R	7.1	pML21	18
	Amp^R	6.4	pSC122	19
	Amp^R	6.0	pMK31	20
	Trp ED	7.1	pVH153	21
*Hae*II	Kan^R	1.5	pMK20	1
	Amp^R	2.0	pRK646	22
	Trp E	2.7	pMK2005	20
*Hin*dIII	Amp^R	13.8	R6K	5
	Amp^R	7.3	pRK492	6
	Sm^R	3.9	R6K	23
	TrpE Amp^R	11.9	pRK491	6
	Trp E	5.3	pDF38	4
*Bam*HI	Amp^R	6.3	pMK31	20
	Kan^R	14.4	pIF11	4
	Trp E	6.5	pDF31	4
*Kpn*I	Kan^R	5.3	pML21	4
	Trp E	10.5	pIF1	4
*Bgl*II	Kan^R	8.9	pIF11	4
	Amp^R	3.1	pRK646	22
*Sal*I	Kan^R	5.6	pDF11	4
*Pst*I	Kan^R	10.1	pDF11	4
	Trp ED	6.5	pDF71	4

ends is carried out in 66 m*M* Tris, 33 m*M* NaCl, 10 m*M* $MgCl_2$, 10 m*M* dithiothreitol, and 70 μM ATP, pH 7.4, at a DNA concentration of 40–200 μg/ml. Then 0.05 units of T4 DNA ligase (Miles) are added, and the ligation mixture is incubated at 15° for 3–8 hr.

Escherichia coli is prepared for transformation[24] by growing cells in LB medium to a density of 3–5 × 10^8 cells/ml. Cells are washed once with 0.5 volume of cold 10 m*M* $MgCl_2$ or 10 m*M* NaCl and resuspended in 0.5 volume of 100 m*M* $CaCl_2$. After 20 min at 0° the cells are pelleted and resuspended in 0.1 volume of 100 m*M* $CaCl_2$. DNA is added to 0.1 ml of the cell suspension, and the transformation mixture is held on ice for 60 min and then incubated at 42° for 2 min. If the phenotypic marker carried by the plasmid requires time to be expressed prior to selection, 1 ml of LB medium is added and the cells are shaken vigorously for the required time [2 hr when selecting for resistance to ampicillin (Amp^R), tetra-

[21] V. Hershfield, H. W. Boyer, C. Yanofsky, M. A. Lovett, and D. R. Helinski, *Proc. Natl. Acad. Sci. U.S.A.* **71,** 3455 (1974).

[22] R. Kolter, unpublished observations (1978).

[23] J. Crosa, L. K. Luttropp, and S. Falkow, *J. Mol. Biol.* **124,** 443 (1978a).

[24] E. M. Lederberg and S. N. Cohen, *J. Bacteriol.* **119,** 1072 (1974).

cycline (Tet^R), or colicin E1 immunity ($E1^{imm}$), and 1 hr when selecting for resistance to kanamycin (Kan^R) or streptomycin (Str^R)]. Selective plates contain 100 μg/ml of ampicillin or penicillin G, 50 μg/ml of kanamycin, and 25 μg/ml of tetracycline or streptomycin.

Slab gel electrophoresis is carried out in agarose or acrylamide gels using a Tris-borate buffer (10.8 g Tris base, 0.93 g EDTA, 5.5 g boric acid per liter).[25] A small volume of a bromphenol blue marker dye containing 50 μg/ml of RNase A and 20% glycerol is added to each sample. Gels are run at 10–15 V/cm, stained with ethidium bromide (1 μg/ml), illuminated with long-wavelength uv light, and photographed using Polaroid 52 film and a Wratten No. 16 color filter.

Properties of Plasmid Vehicles

Vehicles Derived from ColE1-Type Plasmids. Cloning vehicles that contain the replication region of plasmids ColE1 and pMB1 have been widely used to maintain foreign DNA in *E. coli.* Since many derivatives of these plasmids have been constructed,[26–28] a large choice of selective markers and restriction endonuclease sites is available. These plasmids are of low molecular weight and are present in high copy number. In addition, since ColE1 can replicate after host DNA synthesis has been stopped by the addition of chloramphenicol,[8] relatively large amounts of these plasmids can be obtained readily. The fact that the copy number of these plasmids can be amplified also facilitates the preparation of plasmid DNA.

Conjugal transfer of ColE1 can be promoted by the presence of a conjugative plasmid, such as F, in the same cell. This mobilization requires both a specific site on the ColE1 plasmid and ColE1-encoded proteins.[29,30] In order to confine these plasmids to biologically safe hosts, the site needed for mobilization has been deleted in all the ColE1 plasmids listed in Table II except pCR1.[31]

Each of the ColE1-type plasmids in Table II has more than one selec-

[25] P. J. Greene, M. C. Betlach, H. M. Goodman, and H. W. Boyer, *Methods Mol. Biol.* **7,** 87 (1974).

[26] F. Bolivar, R. Rodriguez, P. J. Greene, M. C. Betlach, H. L. Heyneker, H. W. Boyer, J. H. Crosa, and S. Falkow, *Gene* **2,** 95 (1977).

[27] H. Boyer, M. Betlach, F. Bolivar, R. Rodriguez, H. Heyneker, J. Shine, and H. Goodman, *in* "Recombinant Molecules: Impact on Science and Society" (R. F. Beers and E. G. Bassett, eds.), p. 9. Raven, New York, 1977.

[28] D. Helinski, V. Hershfield, D. Figurski, and R. Meyer, *in* "Recombinant Molecules: Impact on Science and Society" (R. F. Beers and E. G. Bassett, eds.), p. 151. Raven, New York, 1977.

[29] J. Inselberg, *J. Bacteriol.* **132,** 332 (1977).

[30] G. J. Warren, A. J. Twigg, and D. J. Sherratt, *Nature (London)* **274,** 259 (1978).

[31] C. Covey, D. Richardson, and J. Carbon, *Mol. Gen. Genet.* **145,** 155 (1976).

TABLE II
PLASMID CLONING VEHICLES

Plasmid	Replicon	Size (kb)	Copy number[a]	Amplify[b]	Selective markers[c]	Single sites[d]
pCR1	ColE1	13.1	20	+	Kan^R, $E1^{imm}$	*Sal*I, *Xho*I, *Hin*dIII, *Eco*RI, (*Bst*EII)
pMK20	ColE1	4.1	70	+	Kan^R, $E1^{imm}$	*Hin*dIII, *Sma*I, *Xho*I, *Eco*RI, *Pst*I, (*Bst*EII)
pMK16	ColE1	4.5	50	+	Kan^R, Tet_a^R, El^{imm}	*Sma*I, *Xho*I, *Eco*RI, *Bam*HI, *Sal*I, *Hin*cII, (*Bst*EII)
pMK2004	pBR322	5.2	40	+	Kan^R, Tet_a^R, Amp^R	*Sma*I, *Xho*I, *Eco*RI, *Bam*HI, *Sal*I, *Pst*I
pMK2005	ColE1	6.9	30	–	$E1^{imm}$, Trp E^+	*Sma*I, *Bgl*II, *Eco*RI, *Hin*dIII, (*Bst*EII), *Hpa*I
pDF41	F'lac	12.8	1–2	–	Trp E^+	*Bam*HI, *Eco*RI, *Hin*dIII, *Sal*I
pDF42	F'lac/ColE1	17.3	15	+	Trp E^+, El^{imm}, Tet_a^R, Kan^R	
pRK353	R6K	11.0	15	–	Trp E^+	*Bam*HI, *Eco*RI
pRK646	R6K	3.4	15	–	Amp^R	*Pst*I, *Bam*HI, *Bgl*II, (*Hin*dIII)
pRK248	RK2	9.6	8	–	Tet_b^R	*Sal*I, *Bgl*II, *Eco*RI
pRK2501	RK2	11.1	8	–	Tet_b^R, Kan^R	*Sal*I, *Bgl*II, *Eco*RI, *Hin*dIII, *Xho*I

[a] Plasmid copy number per *E. coli* chromosome.

[b] + indicates that the plasmid continues to replicate in the presence of chloramphenicol.

[c] Insertion in the restriction sites in parentheses inactivates these phenotypic markers: Kan^R (*Sma*I, *Hin*dIII, *Xho*I), Tet_a^R (*Hin*dIII, *Sal*I, *Bam*I, *Hin*cII), Tet_b^R, (*Sal*I), Amp^R (*Pst*I, *Hin*cII), Trp E (*Bgl*II, *Hpa*I).

[d] Insertion of DNA into sites in parentheses inactivates plasmid replication.

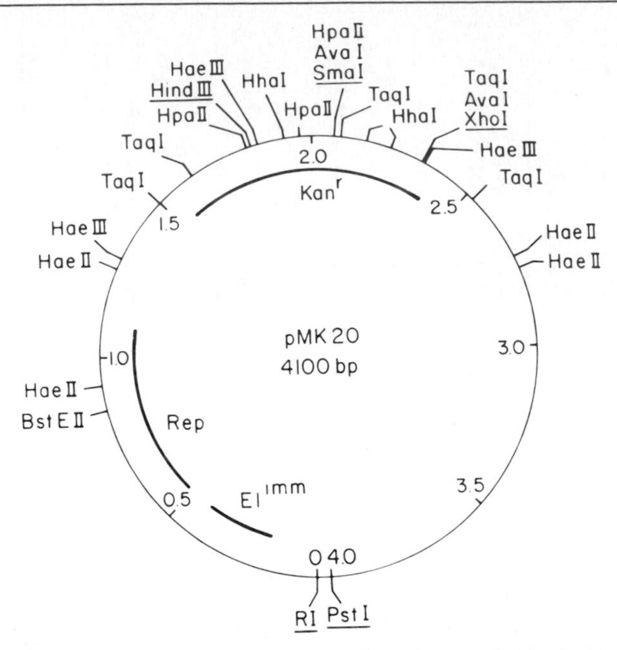

FIG. 1. Restriction map of pMK20. The restriction enzyme sites in the *Hae*II kanamycin resistance region bordered by the *Hae*II sites are from Kolter and Helinski.[5] Additional restriction endonuclease sites in the ColE1 region have been mapped.[32] Rep refers to the plasmid replication region. A restriction site is underlined if it is present only once and if insertion into it does not inactivate plasmid replication.

tive marker. Insertion of DNA into some of the single restriction endonuclease sites indicated in the table inactivates a selective marker. For instance, insertion of DNA into the *Sma* site of pMK16 leads to Tet^R Kan^S transformants, and therefore isolation of Tet^R Kan^S clones selects for recombinant plasmids. Plasmids with three selective markers such as pMK16 and pMK2004 permit one to insert DNA into one marker and then delete segments of the insert plus the other marker while retaining the third marker for selection. As an example, DNA cloned into the *Pst*I cleavage site of pMK2004 inactivates *amp*^{r}. If the insert contains several *Bam*HI sites, a set of plasmids deleted for the region between these sites and the *Bam*HI site in the *tet*^{R} region can be generated by partial digestion of the recombinant plasmid with *Bam*HI, transforming with the linear DNA, and screening for Tet^S colonies. This can be of use in determining the location of genetic information within the original insert and in mapping the insert.

Plasmid pMK20 (Fig. 1)[5,32] was constructed by ligating the Kan^R *Hae*II restriction fragment obtained from pCR1 (Table I) with pMK5, a plasmid

[32] A. Oka, *J. Bacteriol.* **133**, 916 (1978).

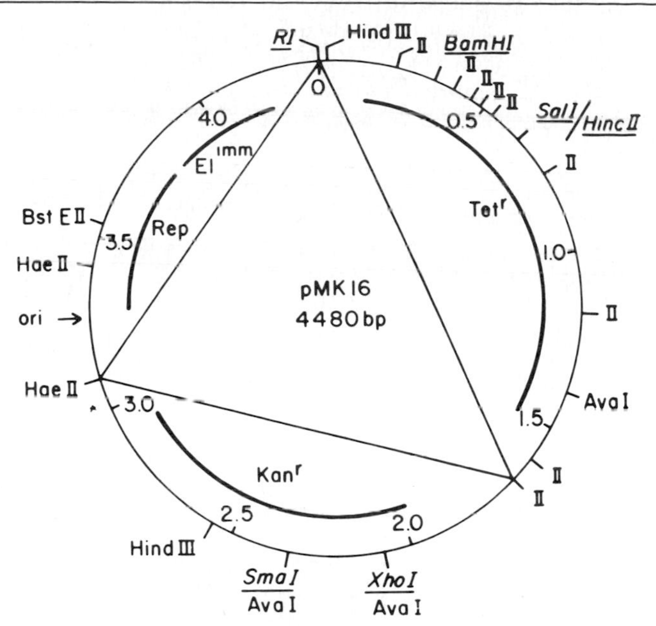

FIG. 2. Restriction map of pMK16. Additional restriction sites have been mapped in the ColE1 replication region,[32] the Kan^R fragment[5] (Fig. 1), and the tetracycline resistance-determining fragment.[26] The region of pBR322[26] from 0 to 1729 bp[33] is homologous to the corresponding region of pMK16. Rep refers to the plasmid replication region. A restriction site is underlined if it is present only once and if insertion into it does not inactivate plasmid replication. Some *Hae*II sites have been shown as II. (pMK16 is the same plasmid referred to earlier[2] as pTK16.)

that contains only the *Hae*II A and E restriction fragments of ColE1.[3] pMK20 was joined to pSC101 at the *Eco*RI site, and the hybrid plasmid was reduced using *Hae*II to give pMK16 (Fig. 2).[2,5,26,32,33] A hybrid plasmid between pMK16 and pBR322 containing both replication origins and two tetracycline resistance regions was obtained by *in vivo* recombination. This hybrid was reduced using *Hae*II to yield pMK2004 (Fig. 3).

Cloning Vehicles Derived from F'lac. F'lac is a large conjugative plasmid (93 kb) that is maintained at a low copy number (one to two copies per chromosome) in *E. coli.* A low-molecular-weight derivative of F'lac, designated mini-F, has been shown to retain this low copy number.[18,19] pDF41[4] (Fig. 4) is a low-molecular-weight derivative of mini-F that accepts inserts at several restriction sites and that carries a *trpE* gene which can be used to select transformants. It was constructed by combining mini-F and a *Trp ED Eco*RI selective fragment (see Table I) and then deleting a *Hin*dIII fragment from the recombinant.

[33] J. G. Sutcliffe, *Cold Spring Harbor Symp. Quant. Biol.* **43,** 77 (1979).

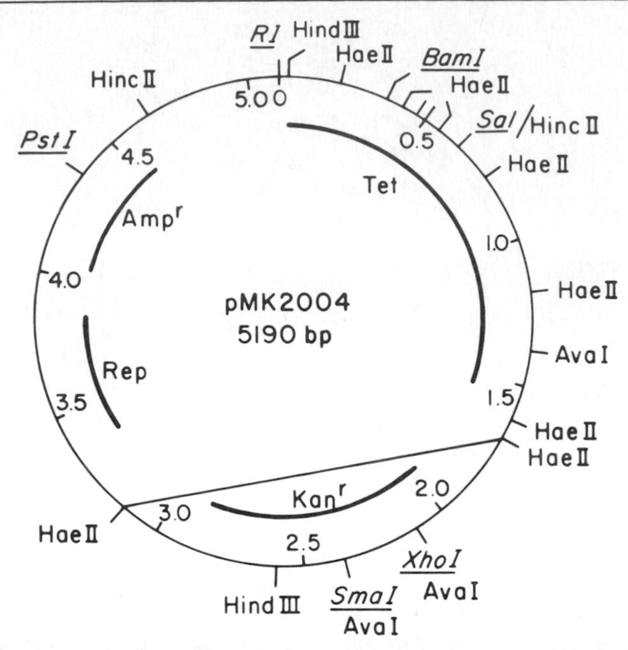

FIG. 3. Restriction map of pMK2004. A region of pBR322 containing the plasmid replication region (Rep) and the ampicillin and tetracycline resistance genes (nucleotides 2351–4729[33]) has been joined to the kanamycin resistance gene. A restriction site is underlined if it is present only once and if insertion into it does not inactivate plasmid replication.

The chief advantage of mini-F plasmids is their low copy number. This permits cloning of DNA that would be lethal to *E. coli* if it were present in many copies per cell. Mini-F derivatives can also be used to lower the gene dosage of an insert in studies of gene regulation. Mini-F plasmids are quite stable, are nonconjugative, and do not appear to be mobilizable.

It is difficult to isolate large quantities of pDF41 DNA because of its low copy number. To overcome this problem, pDF42 (Fig. 4) was constructed by joining pDF41 to pMK16 at their *Eco*RI sites. pDF42 has a high copy number and replicates in the presence of chloramphenicol. pDF41 can be removed from pDF42 by cleavage of the hybrid DNA with *Eco*RI.

Cloning Vehicles Derived from R6K. R6K is a large conjugative plasmid maintained at a relatively high copy number (10–15 copies per cell). The replicon region of this plasmid can be reduced to 2.1 kb, and this region can be divided further into a small (400-bp) replication origin and a segment that encodes for a protein required in R6K replication.[6]

pRK353[5] (Fig. 5) is an R6K plasmid that can be used as a general-purpose cloning vehicle. It retains the R6K copy number and is quite stable. In addition, it is compatible with ColE1-derived plasmids and can

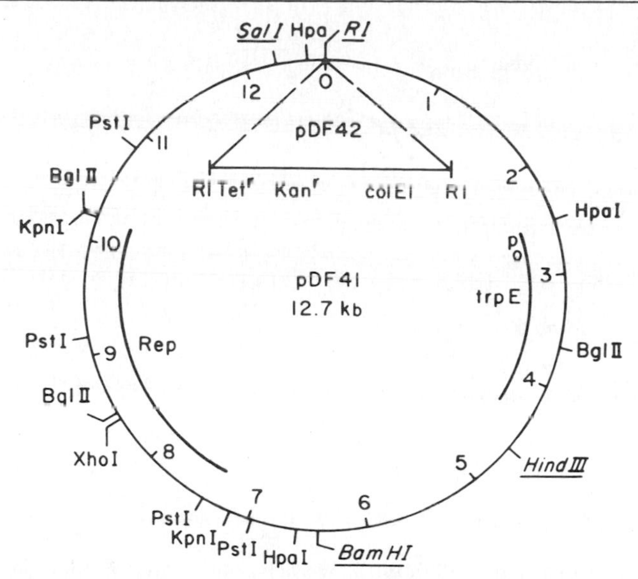

FIG. 4. Restriction map of pDF41. pDF41 was constructed by combining mini-F and the Trp E–*Eco*RI selective fragment and then deleting a *Hin*dIII fragment from the recombinant.[4] The insertion of pMK16 into pDF41 to give pDF42 is as shown. A restriction site is underlined if it is present only once and if insertion into it does not inactivate plasmid replication.

be used to study the interaction of two different inserts in the same cell.

The most novel R6K vehicle is pRK646,[22] which contains only the R6K origin fragment. This plasmid is completely dependent on the presence of a helper function provided by another fragment of R6K DNA. This helper fragment can be carried on a ColE1 plasmid or as a λ lysogen.

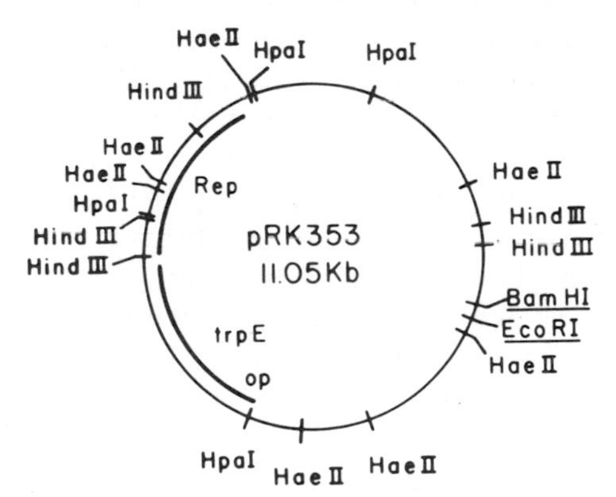

FIG. 5. Restriction map of pRK353. The construction and mapping of pRK353 have been described.[5] Rep refers to the plasmid replication region. A restriction site is underlined if it is present only once and if insertion into it does not inactivate plasmid replication.

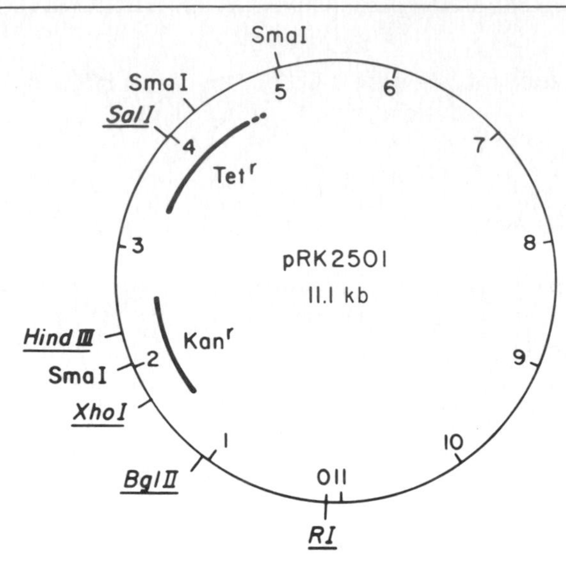

FIG. 6. Restriction map of pRK2501. A restriction site is underlined if it is present only once and if insertion into it does not inactivate plasmid replication.

This property of pRK646 makes it possible to contain pRK646 in specially constructed *E. coli* EK2 hosts that carry the R6K helper function.

Cloning Vehicles Derived from RK2. Plasmid RK2 has a broad host range and is stably maintained in *E. coli, Pseudomonas, Rhizobium, Acinetobacter,* and other bacterial genera.[7] It is thus potentially useful for cloning of DNA into species other than *E. coli.* A low-molecular-weight derivative of RK2, pRK248, has been constructed[34] that is nonconjugative and nonmobilizable. Another RK2 plasmid, pRK2501[35] (Fig. 6) contains a *Hae*II kanamycin resistance marker inserted into pRK248. It has not been possible to test whether or not the broad host range of the original plasmid has been retained in these derivatives because of the restrictions on these experiments by the National Institutes of Health guidelines on recombinant DNA research.[36]

Recently, it has been shown that a non-self-replicating fragment of RK2 containing the origin of RK2 DNA replication replicates in the presence of a fragment derived from RK2 and inserted into another replicon.[37] This result raises the possibility that the origin fragment can be used to carry bacterial DNA in genera other than *E. coli* and can be confined to strains especially constructed for this purpose.

[34] R. Meyer and D. R. Helinski, in preparation.
[35] C. Thomas, unpublished observations (1978).
[36] Anonymous, *Fed. Regist.* **41,** 769 (1976).
[37] D. Figurski and D. R. Helinski, *Proc. Natl. Acad. Sci. U.S.A.* **76,** 1648 (1979).

[18] *In Vitro* Packaging of λ *Dam* Vectors and Their Use in Cloning DNA Fragments

By L. Enquist and N. Sternberg

With recombinant DNA technology, foreign DNA fragments can be inserted efficiently into suitable bacteriophage λ vectors. However, the process of getting this DNA into *Escherichia coli* for replication and encapsidation into phage particles has veen very inefficient. For example, while the theoretical yield should be about 2×10^{10} phage plaques per microgram of λ DNA, the typical values for the commonly used $CaCl_2$ transfection method range from 10^4 to 10^6 plaques per microgram of λ DNA. In addition, when λ vector DNA is cleaved with the proper enzyme and then religated, transfection efficiency usually drops an additional 10- to 100-fold. Substantially increased λ DNA maturation efficiency can be achieved using partially assembled λ virions to package recombinant DNA *in vitro* into plaque-forming phage particles. Values of 10^7–10^8 plaques per microgram of uncleaved λ DNA are commonly obtained by *in vitro* packaging procedures.

We first describe the procedures used for *in vitro* packaging in some detail. Second, we describe the use of an amber mutation in λ gene *D* that, when carried by λ vectors, allows recognition of different size classes of hybrid phage and facilitates the isolation of deletions of the inserted foreign DNA fragment. We demonstrate the utility of this method using a specific λ *Dam* vector.

General Principles of the Packaging System

The procedures described here were initially developed by Becker and Gold (1975).[1] We modified the system so that only exogenous DNA could be efficiently incorporated into infectious λ particles.[2] The functions required for particle assembly in the test tube are provided by induction of specially constructed *Escherichia coli* λ lysogens. Two such lysogens (NS428 and NS433) are used to provide complementing extracts. The method requires two types of cell extracts: one extract is made by lysozyme freeze-thaw lysis, and the other extract is sonicated to break open the concentrated cells. The lysogens used to make these extracts carry the following prophage and host mutations.

1. Nonsense (amber) mutations in either capsid gene *A* (NS428) or *E* (NS433).

[1] A. Becker and M. Gold, *Proc. Natl. Acad. Sci. U.S.A.* **72,** 581 (1975).
[2] N. Sternberg, D. Tiemeier, and L. Enquist, *Gene* **1,** 255 (1977).

ISBN 0-12-181968-X

2. The *Sam7* mutation: In the absence of the phage gene *S* product, induced lysogens fail to lyse spontaneously, phage DNA continues to replicate, and intracellular phage products continue to accumulate for 2–3 hr beyond the normal time of lysis. These properties enable one to harvest fully induced yet intact cells and to concentrate these unlysed cells for the preparation of active packaging extracts.

3. The *b2* deletion: This phage deletion damages the λ attachment site and prevents prophage excision after induction. The induced λ *b2* prophage is effectively trapped in the *E. coli* chromosome, even though many rounds of prophage replication occur. Since trapped prophage DNA cannot be packaged into plaque-forming virions, the packaging extracts contain no endogeneous source of packageable DNA. Under these conditions, only the exogenous DNA is available for *in vitro* encapsidation.

4. The *recA* mutation in the host and the *red3* mutation in the phage: These two mutations effectively inactivate the major general recombination systems present in the packaging extracts. It was necessary to prevent recombination, because the addition of exogenous DNA to rec^+ red^+ extracts resulted in the production of a low level of phage carrying markers present in both endogeneous and exogenous DNA. Under optimal conditions, no such recombination can be detected in the rec^- red^- packaging extracts.

Methods for *in Vitro* Packaging

The method described here, which we developed with D. Tiemeier, contains several modifications of the published procedure. Many of these were worked out and communicated to us by T. Maniatis and co-workers.

Reagents

1. Buffer A: 20 m*M* Tris-HCl, pH 8.0, 1 m*M* EDTA, 3 m*M* $MgCl_2$, 5 m*M* β-mercaptoethanol
2. Buffer B: 6 m*M* Tris–HCl, pH 7.4, 15 m*M* ATP, 16 m*M* $MgCl_2$, 60 m*M* spermidine-HCl, 30 m*M* β-mercaptoethanol
3. Tris-sucrose: 50 m*M* Tris-HCl, pH 7.4, 10% sucrose
4. Nase buffer: 100 m*M* NaCl, 10 m*M* $MgSO_4$, 50 m*M* Tris-HCl, pH 7.5, 0.01% gelatin, 10 μg/ml DNase I
5. Lysozyme: 1 mg/ml egg-white lysozyme (Sigma), 0.25 *M* Tris-HCl, pH 7.4

Media

1. LB broth: 10 g Difco tryptone, 5 g Difco yeast extract, 5 g NaCl, 1 liter water. Adjust pH to 7.4 and sterilize by autoclaving. When cool, mix 10 ml 1 *M* $MgSO_4$ per liter of media.
2. Tryptone broth: Prepared as in LB broth but without the yeast extract.

3. TB plates: Prepare tryptone broth but include 11 g Difco agar per liter and pour the autoclaved media into standard petri plates. Fill the plates about half full and allow to solidify.
4. TB top agar: Prepare broth but include 7 g Difco agar per liter. Autoclave and distribute in 100-ml bottles. Melt agar prior to use and keep molten at 50°.

Bacterial Cultures

1. NS428: N205[3] (λ *Aam11 b2 red3 cIts857 Sam7*)
2. NS433: N205 (λ *Eam4 b2 red3 cIts857 Sam7*)

Growth and Induction of Lysogens

Prepare 10 ml overnight cultures of NS428 and NS433 by adding 1 drop of a −20° glycerol stock to 10 ml tryptone broth; incubate with shaking at 32° overnight. Add 1.5 ml NS428 overnight culture to 150 ml LB broth in a 500-ml flask and 7.5 ml NS433 overnight culture to 750 ml LB broth in a 2-liter flask. Shake the cells vigorously at 32° until they reach $1-2 \times 10^8$ cells/ml (0.2 absorbance at 630 nm). This usually takes about 3–4 hr. At this point, induce the lysogens by swirling them in a 90° water bath until the temperature of the culture is 42°. Monitor the temperature with an alcohol-sterilized thermometer placed directly in the culture flasks. Put the flasks in a shaker at 42° and aerate vigorously for 20 min.

Shift the cultures to 38° for an additional 70 min and again shake them as hard as possible. After 70 min, quick-chill the cultures in an ice-water bath. To test for effectiveness of induction, transfer a few milliliters of the culture to a glass tube and shake with a few drops of chloroform. The culture should clear completely in several minutes. Chloroform will not lyse a noninduced culture. The induced cultures are now ready for processing and should be used as rapidly as possible.

Preparation of Extract A (Sonic Extract, SE)

Mix 150 ml of *each* of the two chilled, induced cultures and centrifuge the mixture at 8000 rpm for 10 min at 4° in a G50 Sorvall rotor. Resuspend the pellet in 0.6 ml buffer A in a polypropylene tube (e.g., Falcon 2063). Clamp the tube securely in ice and sonicate for 10–12 2-sec bursts, allowing 15–30 sec between bursts. Use the microtip of a Bronwill Biosonik sonicator at the lowest setting. Sterilize the sonicator tip with alcohol prior to use. The culture will clear and become viscous as sonication proceeds. Sonicate until the viscosity just disappears. Another way to judge the proper amount of sonication is to centrifuge a 50-μl aliquot for 5 min at top speed in a microfuge (Eppendorf or Beckman). There

[3] Strain N205 is *recA* and *gal*$^+$.

should be little or no pellet when compared to a similar sample of nonsonicated cells. Sonication beyond this point yields extracts with lowered activity.

Add 0.15 ml cold glycerol and mix with a Pasteur pipette. Keep the sonicate on ice during mixing. Put 50-μl aliquots of this mixture into cold polypropylene tubes (e.g., Nunc or Provials) and quickly submerge them in liquid nitrogen. These extracts can be stored indefinitely under liquid nitrogen for use in future experiments. Two standard packaging reactions can be performed with one tube of extract A.

Preparation of Extract B (Freeze-Thaw Lysate, FTL)

Concentrate the chilled, induced cells from the remainder of the NS433 culture (600 ml) by centrifugation at 8000 rpm for 10 min at 4° in a Sorvall G50 rotor. Carefully pour off the supernatant and remove the last traces of fluid with a sterile cotton swab. Do this by slanting the bottle on several paper towels with the pellet up, so that the residual fluid runs to the opposite side. Resuspend the pellet in 1.2 ml Tris-sucrose buffer. Distribute 80-μl aliquots of this mixture into each of about 25 polypropylene tubes and immediately submerge in liquid nitrogen. All these tubes will be used in the following steps, but process only half the tubes at a time, keeping the remainder frozen in liquid nitrogen.

Thaw 10–12 tubes in a beaker of water at room temperature. Follow this by quick-freezing in liquid nitrogen and then rapidly thaw them in room-temperature water. To each tube add 4 μl lysozyme solution and stir into the extremely viscous lysed cell suspension using a 10-μl glass micropipette. Incubate the lysates for 30 min on ice. Mix together in a separate tube, on ice, 0.625 ml glycerol and 0.2 ml buffer B. Add 33 μl of this solution to each of the lysates, stirring with a 10-μl glass micropipette to mix. Freeze the lysates in liquid nitrogen. Many tubes of extract B can be prepared in 1 day and can be stored indefinitely in liquid nitrogen for use in future experiments. One standard packaging reaction can be done with one tube of extract B.

The Packaging Reaction

Mix in this order at room temperature in a polypropylene tube: 30 μl buffer A, 2 μl buffer B, 20 μl extract A, freshly thawed, and 5 μl DNA (5 μg/ml).

Incubate at room temperature for 15 min and then add the entire mixture to a freshly thawed tube of extract B; stir in thoroughly with a 10 μl glass micropipette. Incubate at 37° for 60 min and then add 150 μl DNase buffer. Incubate 37° for 15 min and stir occasionally with a 10-μl glass micropipette. Add 2–3 drops chloroform.

The mixture can now be treated as a conventional phage lysate. For long-term storage, the debris should be removed by a 5-min spin in the microfuge. Typically the mixtures contain about 1–5 × 10^6 phage/ml or about 2–10 × 10^7 phage per microgram of DNA if wild-type λ DNA is used. One should expect to find 10- to 100-fold fewer phage if the DNA used is a restricted and religated phage vector–foreign DNA mixture.

In general, we observe a linear relationship between the number of plaques produced and the amount of DNA added, up to about 1 μg DNA per packaging reaction. We find that plating more than 30 μl of the final packaging mixture results in a reduction in the number of plaques. For best results use *Escherichia coli* cells grown in 0.2% maltose for the plating bacteria. Maltose induces more phage receptors on the cell surface and improves the plating efficiency of the packaging mixtures.

Another variation of the Becker and Gold[1] procedure has been devised by H. Farber, D. Kiefer, and F. Blattner. The principles are the same, but rather than using NS433 to provide p*A* protein, p*A* is partially purified from an induced lysogen bearing a defective λ *gal* prophage.

Growth and Maintenance of Bacterial Strains

Extreme care must be taken to ensure the purity and viability of the bacterial strains used for *in vitro* packaging. Upon receipt of the strains, single colonies should be prepared by streaking the cultures out on TB plates and incubating them at 30°–32° for about 24 hr or until colonies are visible. The incubation temperature is critical, because the strains are λ lysogens and carry a thermoinducible prophage. Growth at intermediate temperatures (e.g., 34°–36°) inevitably leads to loss of the prophage, cell death, and selection of unwanted mutants. Pick one colony and grow a 5-ml culture in tryptone broth at 30°–32° overnight. Pellet the cells by low-speed centrifugation (5000 rpm for 10 min in a Sorvall SS34 rotor) and pour off the supernatant. Resuspend the cell pellet in 2 ml sterile 10 m*M* $MgSO_4$. Cells prepared in this way are stable for 1–2 weeks when stored at 4°. For long-term storage, add to a 5-ml screw-cap vial 2 ml sterile 80% glycerol solution (80 ml sterile glycerol plus sterile water up to 100 ml) followed by 1 ml of the cell suspension in 10 m*M* $MgSO_4$. Mix the suspension well and store at −20°. The solution should not freeze, and the cells will remain viable for at least a year. Inoculations of overnight cultures are made directly from the glycerol stocks by adding a drop or loopful to sterile medium. Even though cells remain viable for a year or more, it is prudent to make fresh glycerol stocks every 2 or 3 months.

A quick test to verify that these cultures contain a thermoinducible prophage is as follows. Streak a loopful of the cells on two TB plates and incubate one at 32° and the other at 42°. Thick growth and many single colonies will appear on the 32° plate after overnight incubation but, on the 42° plate, only a few colonies will appear at the origin of the streak. A more quantitative assay can be made by plating dilutions of the cells at 32° and 42°. The ratio of colonies at 42°: 32° should be $10^{-4}:10^{-5}$. If cell death at 42° is not observed, discard the culture.

The packaging lysogens carry a *recA* mutation. Two consequences of this mutation are (1) the lysogens become very sensitive to 260-nm uv irradiation and (2) the cells grow slower, hence form smaller colonies. Because of this latter property, it is possible by picking large colonies (a popular tendency) to enrich for *recA*$^+$ revertants. By examining the uv sensitivity of the cultures, one can verify the presence of the *recA* mutation. Set up a uv lamp (e.g., a GE germicidal bulb) on a ring stand so that a dose of about 350 ergs/mm^2 is delivered in 10 sec. Streak a loopful of culture across a TB plate in a swath ½ cm wide. Adjacent to this, streak a similar swath of wild-type *E. coli* (e.g., N99 or 594 is preferred). Remove the cover from the petri dish and cover half the plate with cardboard. Expose the plate to a dose of about 350 ergs/mm^2, remove the cardboard, recover the plate, and incubate for 18–24 hr at 32°. Although wild-type *E. coli* is resistant to this dose of uv, *recA* strains are efficiently killed. There should be no growth over the portion of the *recA* swath exposed to uv, while heavy growth will cover the remaining swath.

General Features of Lysogen Induction

A common cause for failure of *in vitro* packaging is poor induction of the lysogens. Best results are achieved when vigorous aeration is used during all phases of cell growth. We have found the baffled flasks with stainless-steel closures distributed by Bellco to be superior to standard Erlenmeyer flasks.

Prior to induction, it is important to have cells growing logarithmically at 32°. These cultures should have doubled in mass at least two or three times during growth and should be induced at cell densities not exceeding 2×10^8 cells/ml. When cultures reach 2×10^8 cells/ml, they must be rapidly and uniformly heated to 42°–45° and maintained thereafter above 37°–39° to inactivate the temperature-sensitive phage repressor. Because phage development and cell growth are poor above 40°, the temperature must be lowered to 38°–39° after the initial 10–20 min of heating at 42°–45°. Temperatures lower than 37° must be avoided, because the phage repressor can renature below this point and active repressor can curtail further phage development.

Packaging Efficiency and the Size of the Input λ DNA

When λ DNA molecules of different sizes were packaged, we discovered that the *in vitro* packaging efficiency decreased markedly as the size of the DNA decreased.[2] For example, a λ vector that was 78% the size of wild-type λ was packaged at least 200 times less efficiently than wild-type λ. This same size-related packaging efficiency is observed with the procedure developed by H. Faber *et al.*,[4] but in this method the size selection can be eliminated by using a buffer containing both spermidine and putrescine, rather than spermidine alone. We have not tested the spermidine–putrescine buffer to see if the size selection is eliminated in our method. The size selection is particularly useful when a small vector phage is used, as the packaging system then provides a strong selection for phage with inserted DNA fragments. This is convenient for vectors with only one site of insertion. Previously a selection of this sort was available only with special vectors like λ gt · λ C which only form plaques when a DNA fragment is inserted. In addition, we have found that size selection aids in the isolation of vectors carrying multiple fragments and combinations that can be missed by more conventional methods.

Preparation of the Substrate DNA

While long concatemeric molecules of λ are substrates for packaging in the cell, linear, monomeric λ DNA can be efficiently packaged *in vitro*.[5] On the other hand, circular λ DNA molecules with only one cohesive end site (e.g., a monomeric circle) cannot be packaged *in vitro* into plaque-forming particles.[5] If two or more cohesive end sites are present on a circular λ molecule (e.g., a circular dimer), they are again good substrates. The ligation conditions we prefer are chosen so that the λ vector fragments and insert fragments join by their restriction enzyme-produced cohesive ends and not by their λ cohesive ends. This ensures production of packagable linear molecules and reduces monomeric circle formation. We generally incubate the ligation reaction at 9°, a temperature unfavorable for λ cohesive-end joining but favorable for joining of *Eco*RI restriction fragments. After ligation, this mixture is heated at 70° for 10 min, quick-cooled on ice, and then packaged. It is also important to choose the optimum ratio of vector DNA to insert DNA. For most λ vectors, a two-fold molar excess of vector DNA to insert DNA generates the maximum number of plaque-forming hybrids. In practice, it is best to determine the

[4] H. Faber, D. Kiefer, and F. Blattner, personal communication (1978).

[5] B. Hohn, *J. Mol. Biol.* **98**, 93 (1976).

ratio of λ vector DNA to insert DNA that gives the highest level of recombinants. This is done by varying the insert DNA fragment with a constant amount of vector DNA, packaging the ligated mixtures, and determining the number of plaques produced.

Rules for *in Vitro* Packaging in EK2 Systems Using λ Vectors

The National Institutes of Health (NIH), Office of Recombinant DNA Activities, NIGMS, has adopted the following rules for *in vitro* packaging using λ vectors, based on the recommendations of the Recombinant DNA Advisory Committee. These rules supersede the previous packaging criteria published in the *Recombinant DNA Technical Bulletin,* Volume 1, Number 1.

1. The packaging extract must be free from viable bacteria.
2. Any packaging protocol may be used, provided that control experiments on the packaging of EK2 vector DNA meet one of the following two criteria: (a) The number of $amber^+$ phages produced must be less than 10^{-6} times the number of $amber^-$ phages. If shotgun populations are to be propagated in bulk culture or by confluent lysis methods, this measurement must be made on packaged EK2 vector DNA propagated to the same extent. (b) If the total number of $amber^-$ phages produced in a packaging experiment is less than 10^6, and if the shotgun population is not to be propagated in bulk, the number of observed $amber^+$ plaques must be zero.
3. The above tests must be done on each batch of packaging extract used.
4. Any individual clone isolated from the shotgun must be tested for retention of the safety characteristics of the vector before it can be used for bulk propagation.
5. A description of the packaging protocol should be filed with NIH for information, but NIH approval is not required provided that the above numerical criteria are met.

The Properties of the λ *Dam srI*λ*3* Vector

The DNA of this λ vector is shown in Fig. 1. While much of the following discussion deals with the cloning of *Eco*RI fragments, the vector can be used to clone *Bam*HI fragments at the *sBam*HI λ4 site,[6] *Sal*I fragments at either of the two phage *Sal*I sites, and *Xho*I fragments at the *Xho*I site. The following features of the vector have proven to be most useful both in

[6] One cannot clone into the *sBam*HI λ1 site because this apparently inactivates an essential λ function.

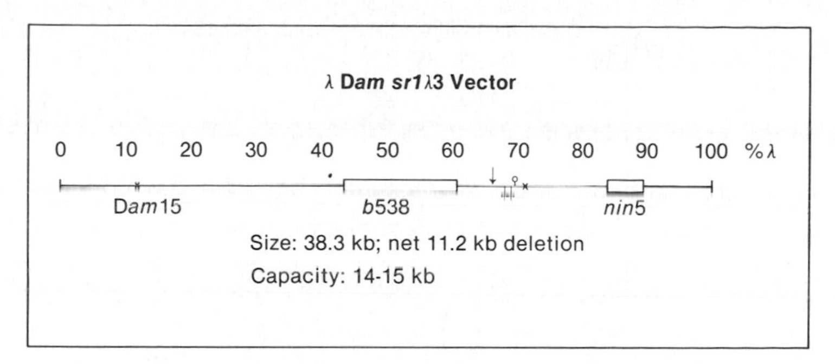

FIG. 1. A map of λ *Dam srIλ3* vector. The approximate positions of the *Dam15* mutation, the *b538* and *nin5* deletions, the *Eco*RI site, *srIλ3* (↓), the *Bam*HI sites (x), the *Xho*I site (♀), and the two *Sal*I sites (↑) on λ DNA are shown. The size of λ *wt* DNA is taken as 49.5 kb with the DNA removed by the two deletions totaling 11.2 kb. As the maximum amount of λ DNA that can be packaged is about 53 kb, the vector can accommodate 14–15 kb of foreign DNA.

the cloning of *Eco*RI restriction fragments and in their subsequent analysis.

The Presence of a Single EcoRI Cloning Site, srIλ3

The vector contains a single *Eco*RI-sensitive cleavage site, as the other four sites normally present in the DNA of phage λ have been removed either by the *b538* deletion (*srIλ1* and *srIλ2*) or by mutation (*srIλ4* and *srIλ5*). The single remaining *Eco*RI site (*srIλ3*) is located in the phage *red*α gene, a gene encoding the phage exonuclease. As the integrity of this gene is not normally necessary for phage growth, *Eco*RI fragments can be cloned into this site and the resulting hybrid phages amplified as one would the original vector. However, since red^- phage do not form plaques in a bacterial strain with either a *lig ts* or *polA* mutation, this property can be used to distinguish hybrid phages from the original λ *Dam srIλ3* vector.[7]

The Size of the Vector DNA

The λ *Dam srIλ3* vector contains two deletions of nonessential λ DNA (*b538*, a 16.7% deletion, and *nin5*, a 5.75% deletion) which remove a net 22.6% of the λ genome. As the upper limit of packageable λ DNA is about 108% that of λ wild type, the vector can accommodate a foreign DNA content equivalent to 30% that of λ wild type (~15 kb).[8]

[7] J. Zissler, E. Signer, and F. Schaefer, *in* "The Bacteriophage λ" (A. D. Hershey, ed.), p. 454. Cold Spring Harbor Press, Cold Spring Harbor, New York, 1971.

[8] J. Weil, R. Cunningham, R. Martin, E. Mitchell, and B. Bolling, *Virology* **50**, 373 (1973).

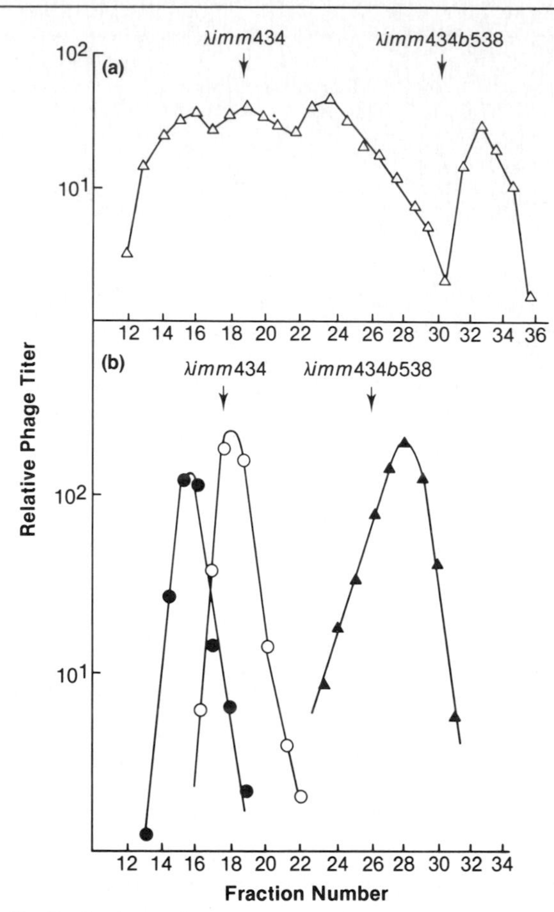

FIG. 2. The distribution in CsCl density gradients of the λ *Dam srI*λ*3* vector and various hybrid phages. (a) The density distribution of the phage in a pool of 6000 plaques is shown. The plaques were obtained by mixing λ *Dam srI*λ*3* and *E. coli* DNA, digesting that DNA with *Eco*RI, ligating the restricted fragments, and packaging the ligated DNA *in vitro*. The marker phages used in the gradient are λ *imm434* (0.97 λ *wt* DNA content) and λ *b538 imm434* (0.805 λ *wt* DNA content). The vector phage peak is located between fraction 31 and 36, and the hybrid phages are distributed between fractions 12 and 30. (b) A plaque obtained either from fraction 13 (●) or fraction 19 (○) of the gradient shown in Figure 2a was purified, lysates prepared, and an aliquot from these lysates banded in a CsCl density gradient with the two λ *imm434* marker phages. The density distribution of the λ *Dam srI*λ*3* vector (▲) is included for comparison. The figure is a composite of three gradients aligned by the marker phages used.

The packaging of λ DNA into a phage capsid is independent of the size of that DNA within the range 75–105% of the λ wild-type DNA content.[9] Thus, when foreign DNA is inserted into the vector *srI*λ*3* site and that DNA packaged *in vitro*, the increased size of the phage DNA is a direct

[9] M. Feiss, R. A. Fischer, M. A. Crayton, and C. Enger, *Virology* **77**, 281 (1977).

measure of the size of the inserted fragment(s). We can measure this increased DNA content as an increase in the bouyant density of the phage particle in a CsCl equilibrium density gradient. In fact, if the size of the fragment of interest is known, then the position of the resulting hybrid phage in a density gradient can be predicted, and the approximate size class of cloned fragments can be selected. We illustrate this technique with the following example. The density distribution of a mixture of λ *Dam srI*λ*3* hybrid phages containing fragments of *Eco*RI-digested *E. coli* DNA is shown in Fig. 2a. The phage contained in plaques made either from fractions 13 or 19 of this gradient were amplified and then banded in density gradients with appropriate λ density markers (λ *imm434* and λ *imm434 b538*). The composite gradient shown in Fig. 2b contains the two hybrid phages and the λ *Dam srI*λ*3* vector. It is clear from these results that hybrids of defined density containing a given size class of fragment can be selected by this procedure.

The Dam15 Mutation

The use of this mutation for cloning relies on the results regarding the dispensability of the normally essential λ capsid protein specified by phage gene *D*.[10] Phage λ will grow normally in the absence of gp*D* if its DNA content is no greater than 82% (40.6 kb) that of λ wild type. As the λ *Dam srI*λ*3* vector contains less than this amount of DNA, it grows equally well in bacterial strains with or without *am* suppressors. Hybrid phages containing foreign *Eco*RI fragments larger than 4% of the λ wild-type genome (1.9 kb) require gp*D* for growth and consequently only form plaques on bacterial strains with *am* suppressors. In contrast, hybrids containing fragments less than 1.9 kb form plaques on strains with or without *am* suppressors.

In addition to their value in selecting hybrid clones, the properties of *Dam* phage can be used to dissect the cloned fragments. DNA deletions can readily be isolated in λ *Dam* hybrid phages with a DNA content in excess of 82% that of λ wild type, since such deletions, if they reduce the phage DNA content below the 82% limit, allow the phage to form plaques on a bacterial strain without an *am* suppressor. Since many of the deletions should dissect the inserted DNA fragment, this procedure can be used to localize markers that might be contained on that DNA.

We can illustrate the properties of *Dam* vectors using λ *Dam srI*λ*3* hybrid phages containing cloned *Eco*RI fragments derived from the DNA of phage P1 (λ P1 hybrids).[2] Figure 3 is an agarose gel containing three *Eco*RI-cleaved, DNA-extracted λ P1 hybrid phages. The DNA shown in Fig. 3d and e contains P1 fragments 7 (5.9 kb) and 8 (5.5 kb), respectively,

[10] N. Sternberg and R. Weisberg, *J. Mol. Biol.* **117**, 733 (1977).

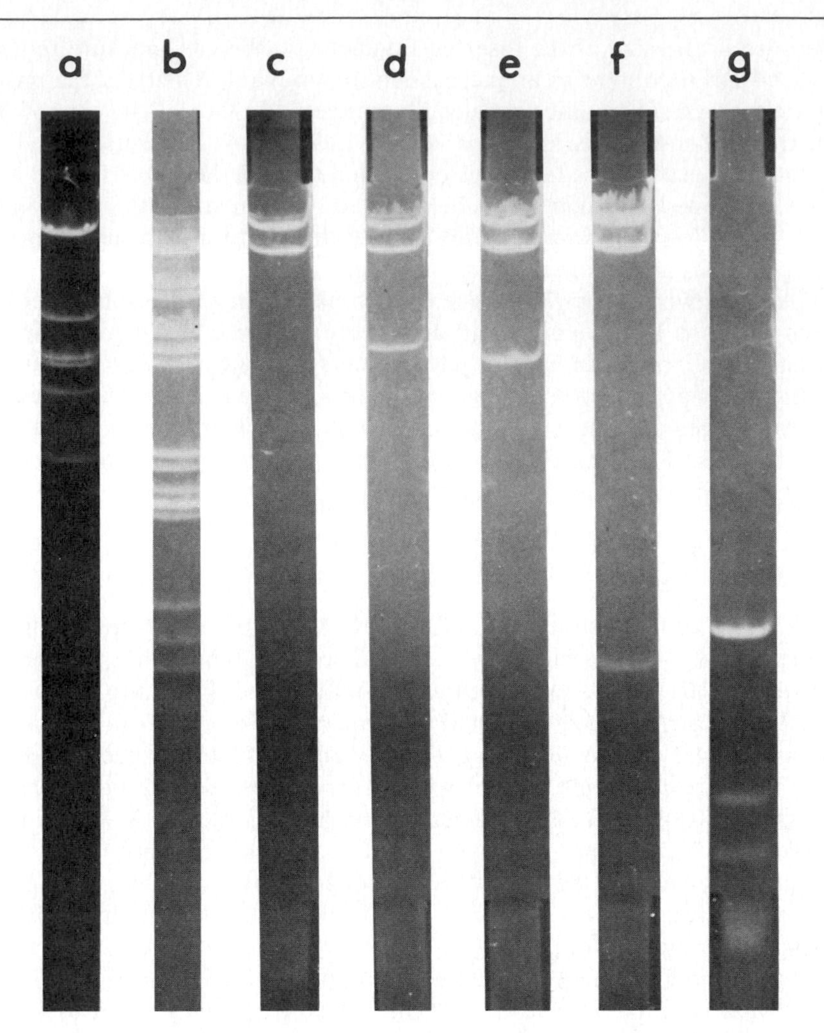

FIG. 3. Restriction enzyme analysis of λ *Dam srI*λ*3* hybrids containing fragments of P1 DNA. (a) λ*cIts857;* (b) P1 *cI. 100* r^- m^-; (c) λ *Dam srI*λ*3*; (d) λ *Dam srI*λ*3* P1:7; (e) λ *Dam srI*λ*3* P1:8; (f) λ *Dam srI*λ*3* P1:16; (g) SV40. Channels a through f contain 0.5 μg of DNA digested with *Eco*RI, and channel g contains SV40 DNA digested with *Hae*III. The digested DNA samples were subjected to electrophoresis in 1% agarose slab gels.

and the DNA in Fig. 3f contains P1 fragment 16 (1.4 kb). As expected, the λ P1:7 and λ P1:8 hybrid phages form plaques only on a strain with an *am* suppressor, while the plaque-forming ability of the λ P1:16 hybrid is independent of the amber-suppressing state of the host.

We use the λ P1 hybrid phage containing P1 fragment 3 (9.3 kb) to illustrate the use of the *Dam* vector to select deletions. The size of the λ P1:3 hybrid DNA is 98% that of λ wild type (*wt*). A deletion permitting this phage to form a plaque on a strain without an *am* suppressor must remove at least 16% of the λ P1 genome but cannot remove more than 25% of that genome. (The lower limit of packaged λ DNA compatible with plaque formation is about 73% that of λ wild type). The deletions could remain entirely within the fragment or could extend into the λ sequences on either side of the fragment. Practically, most of the deletions we isolate that extend into the λ DNA remove sequences to the left of the insertion rather than those to its right. The reason for this asymmetry is the presence of the λ *gam* gene just to the right of the site of insertion. A *red⁻gam⁻* phage, especially one with a reduced DNA content, does not grow as well as a *red⁻gam⁺* phage and so would be underrepresented among the population of deleted phages. In Figure 4 the markers present on P1 fragment 3 and those removed by 28 independently isolated deletions are shown. The exact location of the deletion end points for several of the deletions has been determined by heteroduplex analysis of the DNA of a deletion phage and the DNA of the undeleted hybrid. A deficiency of the deletion analysis described here is the limitation on the deletion sizes that can be isolated. This deficiency can be overcome by genetically altering the hybrid. For example, the *imm21* marker and the *b221* deletion can be crossed into the λ P1 : 3 hybrid, replacing, respectively, the *imm*λ marker and the *b538* deletion. The resulting λ *b221 imm21* P1 : 3 hybrid has a DNA content of 87% that of λ wild type. Deletions isolated from this hybrid range in size from 5 to 14% of the λ genome and thus

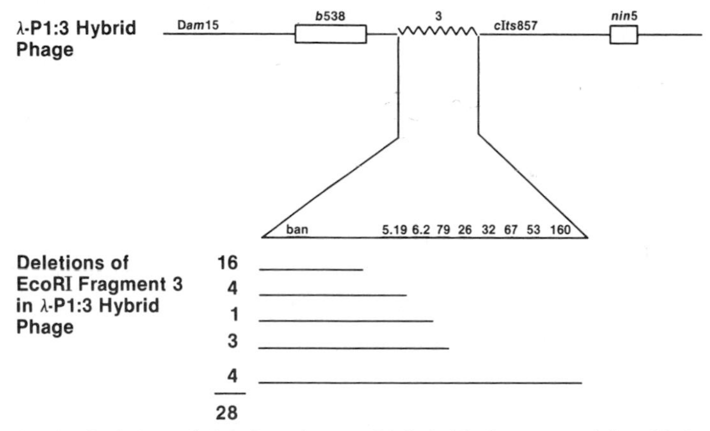

FIG. 4. The isolation of deletions in a λ P1 hybrid phage containing P1 fragment 3. Twenty-eight independently isolated deletions of λ P1 : 3 were screened for a variety of P1 markers normally contained on fragment 3. The P1 *ban* gene makes an analog of the host *dnaB* protein, and so *dnaB ts* (λ P1 : 3) lysogens express the *ban* gene and survive at 42°. By this test none of the deletions carry a functional *ban* gene. The presence of other markers normally contained on fragment 3 (*am* mutations represented by numbers) is assessed by standard marker rescue tests with the results shown in the figure.

should allow for a finer dissection of P1 fragment 3 than was possible with the original λ P1:3 hybrid.

The Temperature-Sensitive Repressor Mutation, cIt857, and the b538 Deletion

As a consequence of the *cIts857* mutation the phage can form stable lysogens at 32°, which can then be induced to initiate phage production by raising the temperature to 42°. The *b538* deletion present in the vector removes the λ attachment site (*att*) and *int* gene, thereby inactivating the λ site-specific recombination system that would normally integrate the infecting λ DNA into the host chromosome. Integration of vector or hybrid phage DNA is achieved by recombination of homologous λ DNA sequences found in the infecting phage DNA and in a cryptic λ prophage present in certain selected bacterial strains (e.g., strain RW842).

Media and Buffers

LB broth, TB broth, and TB top agar are prepared as previously described. Tris-EDTA buffer is 10 m*M* Tris-HCl and 10 m*M* EDTA adjusted to pH 7.4. TMG buffer is 10 m*M* Tris-HCl, pH 7.4, 10 m*M* $MgSO_4$, and 0.1% gelatin.

Bacterial and Phage Strains

Bacterial Strains

YMC (*supF*)[11]: Used to assay phage with *am* mutations.

594 (*supO*)[12]: Used to assay either λ *wt* or λ *Dam* phage with less than 82% of the λ *wt* DNA content.

N3098[2] = YMC (*lig ts7*): Used to detect λ red^- phage, e.g., the λ *Dam srI*λ*3* vector with a foreign DNA fragment inserted at the *srI*λ*3* site in the *red* α gene.

RW842[13] = *HfrH*(*supO*)[λ(*int-F*II)$^{\Delta}$ *galT*]: This strain contains a cryptic λ prophage inserted in the bacterial *galT* gene, which serves as a source of DNA homology for the integration of a λ *att*-deleted phage into the bacterial chromosome.

YMC (λ *imm*λ) and YMC (λ *imm434*): Used as indicator bacteria for titering gradients.

C600 (*supE*)[14]: Used to distinguish λ *Sam7* phage from λ S^+ phage.

[11] G. Dennert and V. Henning, *J. Mol. Biol.* **33,** 322 (1968).
[12] A. Campbell, *Virology* **37,** 340 (1965).
[13] L. Enquist and R. Weisberg, *Virology* **12,** 147 (1976).
[14] R. K. Appleyard, *Genetics* **39,** 440 (1954).

Phage Strains

λ *imm434:* Used as a density marker; DNA size is 0.97 that of λ *wt*.

λ *b538 imm434c Sam7:* Used as a density marker: DNA size is 0.805 that of λ *wt*.

λ *Dam srI*λ*3* and various hybrids made by inserting DNA fragments into the vector.

Techniques

Phage Amplification

The Plate Lysate Method. A phage plaque containing 10^5-10^6 phage (when plaques are small, several plaques are used) is added to a tube containing 1 drop of a fresh saturated culture of strain YMC (*supF*) grown overnight in TB broth. After 5 min for phage adsorption at 38°, 2 ml of TB broth followed by 2 ml of TB top agar are added, and the mix is poured onto TB plates. The overlay is allowed to solidify at room temperature for 5 min, and the plate is then incubated at 38° until the bacterial lawn clears, usually 5–7 hr. The top agar overlay is removed with a pipette and processed as follows: A drop of chloroform is mixed with the lysate to kill any surviving bacteria, the lysate is placed on ice for 10 min, and the remaining agar is removed by centrifugation in an SS34 Sorvall rotor at 6000 rpm for 10 min at 4°. The supernatant usually contains $10^{10}-10^{11}$ phage/ml.

The Liquid Lysate Method. The procedure is the same as for the plate lysate method through the phage adsorption step. After adsorption the infected cells are diluted 200-fold into LB broth and vigorously aerated at 38° until lysis occurs, usually 4–6 hr. Lysis is completed by shaking the culture an additional 5–10 min with several drops of chloroform. Debris is removed by centrifugation, e.g., 10,000 rpm for 10 min in the Sorvall SS34 rotor.

Selecting Individual Hybrid Clones on the Basis of Their Genetic Properties

As a consequence of the *Eco*RI restriction, ligation, and *in vitro* packaging of a mixture of 0.1 μg λ *Dam srI*λ*3* DNA and 0.2 μg foreign DNA one should expect to obtain 10^5-10^6 plaque-forming units (PFUs). To obtain individual hybrid phages an aliquot of the packaging reaction containing about 500 PFUs is added to 1 drop of a fresh overnight culture of strain YMC grown in TB broth with 0.4% maltose. The mixture is incubated for 10 min at 32°, and 3 ml melted TB top agar is added; the mix is then poured onto a TB plate. The top agar is allowed to solidify for 5 min at room temperature, and then the plates are incubated overnight at 32°.

Each plate should contain about 500 plaques. Individual plaques are purified by removing an agar plug containing the plaque with a capillary pipette (75-mm blue-tip nonheparinized hematocrit tubes, Fisher Scientific) and transferring the plug to 1 ml TMG buffer. A drop of $CHCl_3$ is then added, and the solution is vortexed to kill any surviving bacteria. The phage can be stored in this state without any loss in titer for up to 3 months at 4°. One should expect to obtain about 10^5 PFUs from a normal-sized λ plaque. In order to determine whether particular plaques contain hybrid phages, an aliquot of the TMG plaque suspension is spotted with a capillary pipette onto a series of TB plates containing top agar overlays with the following bacterial strains: YMC, N3098, and 594. The spots (~50 μl) are allowed to dry with the cover of the plate off, and the plates are then incubated overnight at 32°. Each of the spots on the YMC (*supF*) indicator plate should give 20–100 plaques. A hybrid phage containing a foreign fragment >1.9 kb in size does not form plaques on the 594 (su^-) indicator plate and forms no better than pinpoint plaques on the N3098 indicator plate. A hybrid phage with an inserted fragment whose size is less than 1.9 kb differs in its properties only in the fact that it forms plaques on the 594 indicator plate with the same efficiency as on the YMC indicator plate. The vector plaques normally on all these indicator plates. It is generally desirable to include a spot from a resuspended plaque containing the vector phage on each plate as a control. Amplification of individually selected hybrid plaques has been described above.

Selecting Hybrid Phages on the Basis of Their Density in CsCl Equilibrium Density Gradient

To a TB agar plate containing about 6000 plaques from a packaging reaction, 2 ml TB broth is added and the top agar overlay is scraped off with either a glass rod or pipette. The surviving bacteria are killed by vortexing the agar mix with 1–2 drops chloroform, and the bacterial debris and agar are removed by centrifugation in a SS34 Sorvall rotor at 6000 rpm for 10 min. Ten microliters of the supernatant, containing about 10^8 PFUs/ml, is added to a CsCl solution prepared by mixing 2.7 g CsCl with 3.5 ml 10 m*M* Tris-HCl, pH 7.4, and 10 m*M* $MgSO_4$. The resulting CsCl–phage mixture is added to a Spinco SW56 cellulose nitrate tube, the tube is filled with paraffin oil, and the solution centrifuged to equilibrium in an SW56 rotor at 30,000 rpm for 24 hr at 4°. The distribution of phage in the gradient is determined by puncturing the bottom of the tube and collecting 7-drop fractions (~60 fractions) into 0.1 ml TMG buffer. It is usually helpful to add two marker phages, λ *imm434* (DNA content 97% that of λ wild type) and λ *b538 imm434c Sam7* (DNA content 80.5% that of λ wild type), to each gradient. A typical gradient is shown in Fig. 2a.

Fractions 31 through 36 contain vector phage and fractions 12 through 30 hybrid phage with inserted fragments of different sizes. In practice, the approximate titer of the phage present in each gradient fraction is determined by spotting with a capillary, aliquots of the fractions on TB plates containing either YMC, YMC (λ *imm434*), or YMC (λ *imm*λ) as indicator. (One should never use the same capillary to transfer aliquots to both the YMC (λ *imm*λ) and YMC (λ *imm434*) plates, as phage present on either one of these indicator lawns will form plaques on the other.) The titer of appropriate fractions is assayed more quantitatively on indicator strains YMC (λ *imm*λ) or YMC (λ *imm434*), and then phage from a desired region of the density gradient is isolated by removing aliquots from appropriate fractions and plating them with strain C600 on TB plates. The plates are incubated overnight at 38°, and clear plaques are picked into 1 ml of TMG buffer with 1 drop chloroform. It should be noted that the selection of clear plaques on the C600 indicator plates incubated at 38° allows one to avoid selecting the marker phages: λ *imm434* forms a turbid plaque, and λ *b538 imm434c Sam7* does not form a plaque on strain C600 (*supF*). The genetic properties of isolated plaques are tested and the phage amplified as previously described. If desired the density of the amplified phage can be determined in a CsCl equilibrium density gradient as described in Fig. 2b.

The Isolation of Deletions in the DNA of Hybrid Phages Containing a Dam Mutation

A 1 to 5-μl aliquot of a λ *Dam* lysate with a titer on strain YMC of about 10^{10} phage/ml is plated on TB plates with indicator strain 594 (*supO*). After incubation at 38° for 8 hr about 500 plaques can be seen on the plate. About 5–10% of these plaques are the same size as λ *wt* plaques and contain phage whose *Dam* mutation has reverted to wild type. The remainder of the plaques are smaller, and the great majority of these plaques represent phage with deletions. The smaller plaques, each containing about 10^4 phage, are picked into tubes containing 1 ml TMG and mixed with a drop of chloroform. We note here that, because of the increased sensitivity to inactivation of *Dam*-deleted phage grown in a *supO* host, the following two conditions must be met in order for this isolation procedure to be successful. First, the media (agar and broth) used to grow these phage must always contain 10 m*M* $MgSO_4$. Second, plaques are best picked into TMG after 8 hr of growth on TB plates, rather than after overnight incubation, as the plaque phage titer drops drastically after 8 hr.

To confirm the presence of a deletion in the selected phage, advantage is taken of the unique properties of λ phage containing less than 82% of the λ *wt* DNA content but lacking gp*D*. First, the phage are purified once on strain 594, and then their EDTA sensitivity is tested as follows. A drop of TMG solution containing a purified plaque is spotted onto TB plates with agar overlays of either strain YMC or 594 and, after the spots have dried, the plates are incubated for 8 hr at 38°. An agar plug is taken from the center of each of the cleared spots made by the phage in the bacterial lawn and the plug resuspended in 1 ml TMG containing a drop of chloroform. Twenty-microliter aliquots of each of the two phage suspensions are diluted into tubes containing either 200 μl Tris-EDTA or 200 μl TMG, and the tubes are incubated for 20 min at 4°. The EDTA inactivation reaction is then stopped by adding 100 μl of a 200 m*M* $MgSO_4$ solution to each tube. To assay plaque-forming phage a drop of a fresh, saturated culture of YMC is added to each tube and, after incubating the tubes for 5 min at 38°, 3 ml TB top agar is added and the contents of the tube poured onto a TB plate. The plates are incubated overnight at 38°, and plaques scored. *Dam* phage carrying deletions are sensitive to EDTA treatment (even at 4°) when grown in a *supO* strain (594) but are resistant when grown in an *am*-suppressing strain (YMC).[10] Thus the titer of phage derived from plaques obtained from the 594 plate and treated with Tris-EDTA should be 100–1000× less than the titer of the same phage treated with TMG. The titer of phage picked from the YMC plate should be the same regardless of the buffer in which the phage are subsequently incubated. Once the presence of a deletion has been confirmed, phage lysates are prepared using strain YMC.

The extent of the DNA deletion and the location of deletion end points in a particular hybrid phage can be assessed by a variety of techniques including CsCl equilibrium density gradient analysis, DNA heteroduplex analysis, agarose gel electrophoresis analysis of restricted DNA, and marker rescue experiments with genetic markers known to be located on the inserted fragment. All these procedures have been used to characterize the deletions isolated in the λ P1 hybrid phage shown in Fig. 4.

Acknowledgments

We thank Tom Maniatis, Fred Blattner, Dave Tiemeier, and John Seidman for helpful advice and criticism, and M. Glasser and H. Hoeksema for help in preparing the manuscript. Supported in part by the National Cancer Institute under contract No. NO1-CO-75380 with Litton Bionetics, Inc.

[19] *In Vitro* Packaging of λ and Cosmid DNA

By BARBARA HOHN

The efficiency of cloning a desired DNA fragment in bacteriophage λ depends on a number of factors such as (1) the availability of a useful λ vector, (2) the probability of obtaining the DNA fragment in a possibly enriched form, (3) the efficiency of conversion of the ligated hybrid DNA molecules to plaque-forming particles, and (4) the availability of a suitable probe for detection of the hybrid phage.

The conversion of hybrid DNA to viable phage particles is commonly carried out by transfection of Ca^{2+} treated, permeable cells with DNA, the subsequent gene expression and multiplication of which yield plaque-forming phage.[1] Study of the morphogenetic pathway of λ phage (reviewed by Hohn and Katsura[2]) made possible an alternative of producing infectious particles from naked DNA: *in vitro* encapsidation of the DNA in question by supplying packaging proteins and preheads.

Morphogenesis of Bacteriophage λ

Structure

The phage (V in Fig. 1) consists of an icosahedral head with a radius of 30 nm and a flexible tail 150 nm long. The capsid is composed of 420 copies each of the 38,000-dalton protein p*E* (product of gene *E*) and the 12,000-dalton protein p*D*. λ DNA in its packaged form is linear and has cohesive ends, the right one of which (as defined by the genetic map) protrudes into the upper third of the tail.[3] The DNA inside the head is arranged regularly, possibly in the form of a concentric spool.[4]

The 32 disks of the tail tube consist of *V* protein, the basal part is a complex structure consisting of many different proteins, and the fiber is composed of a few copies of protein p*J*.

Assembly of the Head

The earliest precursor of the phage λ head is the "scaffolded prehead" (I in Fig. 1). It contains a central core of scaffold protein p*Nu*3, and *B* and *C* proteins as minor components. With the help of a host protein p*groE*

[1] E. M. Lederberg and S. N. Cohen, *J. Bacteriol.* **119,** 1072 (1974).
[2] T. Hohn and I. Katsura, *Curr. Top. Microbiol. Immunol.* **78,** 69 (1977).
[3] J. O. Thomas, *J. Mol. Biol.* **87,** 1 (1974).
[4] W. Earnshaw and S. C. Harrison, *Nature (London)* **268,** 598 (1977).

ISBN 0-12-181968-X

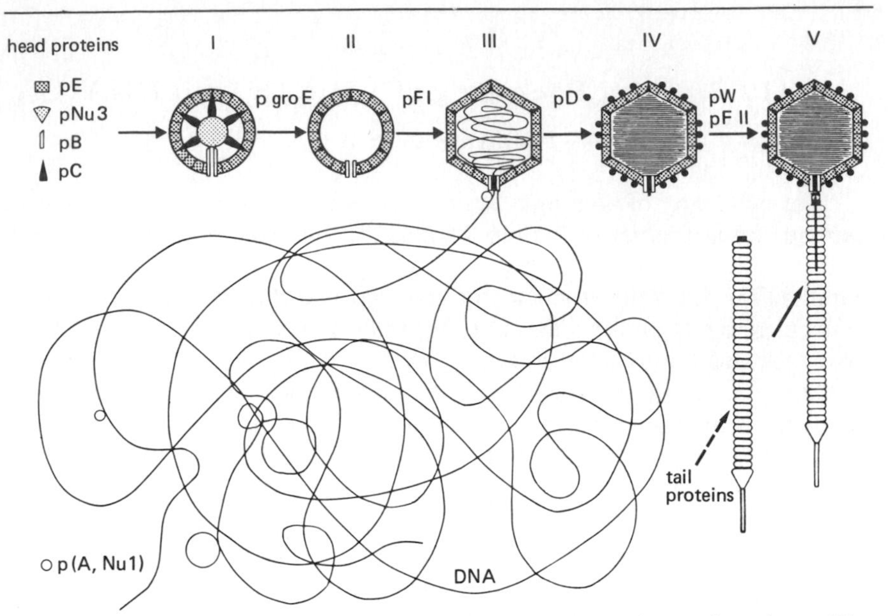

FIG. 1. Schematic representation of the bacteriophage λ morphogenetic pathway. The position of minor proteins is hypothetical. See description in the text; for a more comprehensive discussion and documentation see Hohn and Katsura.[2]

the minor components are processed and the scaffold removed. The resulting particle (II in Fig. 1) is called a prehead. It belongs, like its scaffolded precursor, to the family of *petit* λ particles as defined by their head size which is 20% smaller than the phage head.

The next step is the DNA packaging. This is a complex reaction coupling or coupled by several processes: specific selection of λ DNA by the p(*A*-*Nu*1) complex; enlargement of the prehead, possibly with the help of the protein p*FI*; cleavage of the concatenated precursor DNA at the *cos* (cohesive end) sites to produce the cohesive ends characteristic of mature λ DNA; and aggregation of *D* protein onto the capsid.

The resulting head (IV in Fig. 1) is by the successive action of proteins p*W* and p*FII* made competent to attach the independently assembled tail.

In Vitro Packaging: The DNA Precursor

The assembly pathway as outlined above was constructed after the examination of precursors accumulating in mutants and completing them *in vitro*. The availability of an *in vitro* packaging system thus enabled us to study the specificities required on the part of the DNA in order to serve as a packaging substrate.

The DNA substrate for packaging *in vivo* is concatemeric λ DNA, which is a tandem polymer of several DNA units. This molecule is packaged, *in vivo* as well as *in vitro* and cohesive ends are produced. Linear monomeric DNA is not packaged *in vivo,* because it does not exist in this form outside the phage head. *In vitro* this form of λ DNA is packaged with high efficiency.[5]

Monomeric circular DNA containing one *cos* site cannot be packaged, *in vivo* [6] or *in vitro,*[5] whereas circular DNA containing a duplicated *cos* site can.[6]

There are specific recognition regions on the DNA that are required for packaging a DNA molecule into an infectious particle: a site close to, but not on the left *cos* site or cohesive end, respectively which might be recognized by the p(*A-Nu*1) protein(s),[5,7] and of course the *cos* site as substrate for the p*A* containing[8] terminase, because only a phage particle with cohesive-ended DNA is infectious.

These specific sites are not the only requirement for DNA to be packaged; the amount of DNA *between* successive sites is also important. Only *cos* sites 38–52 kb (λ wild-type DNA has 49 kb) apart are subjected to packaging-dependent cleavage, thus yielding infectious particles. The efficiency of obtaining infectious particles with shorter or longer DNA decreases sharply.[9] This sets an upper limit on the amount of DNA that can be cloned in λ, which is 24.6 kb.[10]

It was knowledge of the physical form, required recognition regions, and length limits of the DNA that permitted development of a new type of cloning vehicles, the cosmids.[11] These plasmids, preferably small, which contain the λ *cos* site and a selectable genetic marker, e.g., antibiotic resistance. These plasmids cannot be packaged because they are circular. Opened forms of cosmids, ligated to a fragment of DNA to be cloned, resemble concatemeric DNA and are thus packaged in a λ coat. Infection with these particles yields bacteria carrying hybrid plasmids. Because of the length dependence of the DNA for packaging, as discussed above, a strong selection in favor of hybrid DNA in the range 38–52 kb occurs, the weight range of the cloned fragment being (only) smaller by the size of the cosmid used.

[5] B. Hohn, *J. Mol. Biol.* **98,** 93 (1975).

[6] M. Feiss and T. Margulies, *Mol. Gen. Genet.* **127,** 285 (1973).

[7] M. Feiss, R. A. Fisher, D. A. Siegele, B. P. Nichols, and J. E. Donelson, *Virology* **92,** 56 (1979).

[8] D. Kaiser and T. Masuda, *Proc. Natl. Acad. Sci. U.S.A.* **70,** 260 (1973).

[9] M. Feiss, R. A. Fisher, M. A. Crayton, and C. Egner, *Virology* **77,** 281 (1977).

[10] F. R. Blattner, B. G. Williams, A. E. Blechl, K. Denniston-Thompson, H. E. Faber, L.-A. Furlong, D. J. Grunwald, D. O. Kiefer, D. D. Moore, J. W. Schumm, E. L. Sheldon, and O. Smithies, *Science* **196,** 161 (1977).

[11] J. Collins and B. Hohn, *Proc. Natl. Acad. Sci. U.S.A.* **75,** 4242 (1978).

In Vitro Packaging as a Method to Produce Infectious Particles from DNA

λ and cosmid DNA, as well as *in vitro* recombinants thereof, are conveniently packaged using a combination of two lysates each of which is defective in another step of λ morphogenesis.[12,13] Empty precursor particles (II in Fig. 1) accumulate after induction in bacteria containing a prophage mutant in gene *D*. This mutation can be complemented by addition of the missing *D* protein, most effectively by supply of an induced λ E^- lysogen. The mutation in this latter strain effects the main capsid protein; in its absence all other head proteins are available in soluble form. In the presence of ATP, spermidine, and putrescine DNA is packaged, the head matures, and tails, components of both lysates, are attached. The result is a DNase-resistant infectious particle which can be stored like any *in vivo* produced phage.

The complete genotypes of the strains are N205 $recA^-$ (λ *imm434 cIts b2 red3 Eam4 Sam7*)/λ and the *Dam15* analog of it. The temperature-sensitive *imm434 cIts* repressor allows efficient prophage induction by temperature shift, and the mutation in gene *S* prevents lysis of the induced bacteria. Lysis of the concentrated, induced cells is effected by freezing and thawing. The combination of the bacterial *recA* and phage *red3* mutations and the b2 deletion drastically decreases the amount of endogenous packageable DNA. To eliminate completely background phage and possible recombination between endogenous DNA and DNA to be packaged, the induced bacteria are heavily uv-irradiated.[12]

Bacterial Strains and Media

Prophages of the strains BHB2688 [N205 $recA^-$ (λ *imm434 cIts b2 red3 Eam4 Sam7*)/λ] and BHB2690 [N205 $recA^-$ (λ *imm434 cIts b2 red3 Dam15 Sam7*)/λ] were produced by recombination of λ *cIts 857 red3 b2 Eam4 Sam7*,[14] a gift from Nat Sternberg, with λ *imm434 cIts* and *Eam4* and λ *Dam15*, respectively. Strain N205 $recA^-$, also from Nat Sternberg, was lysogenized and lysogens selected for on plates seeded with λ *imm434 cI b2 red3*.

Indicator bacteria, usually BHB2600 [803 $supE^+$ $supF^+$ r_k^- m_k^+ met^-, from Ken Murray], are grown in tryptone broth supplemented with 0.4%

[12] B. Hohn and K. Murray, *Proc. Natl. Acad. Sci. U.S.A.* **74,** 3259 (1977).
[13] B. Hohn and T. Hohn, *Proc. Natl. Acad. Sci. U.S.A.* **71,** 2372 (1974).
[14] N. Sternberg, D. Tiemeier, and L. Enquist, *Gene* **1,** 255 (1977).

maltose, centrifuged, and resuspended in 0.01 *M* $MgSO_4$. Platings are done after preadsorption of phage particles to bacteria.

L broth: 10 g Difco Bacto-tryptone, 5 g Difco yeast extract, and 10 g NaCl made up to 1 liter with H_2O, pH 7.2

LAM plates: As L broth, 10 g Difco agar, 2.5 g $MgSO_4 \times 7\ H_2O$ per liter

LAM top agar: As LAM plate agar, but with 0.6% instead of 1% Difco agar

M-9: 7 g Na_2HPO_4, 3 g KH_2PO_4, 0.5 g NaCl, 1 g NH_4Cl, 10^{-3} *M* $MgSO_4$, 10^{-4} *M* $CaCl_2$, 4 g glucose per liter

NZY^{10}: 10 g N-Z amine, 5 g yeast extract, 5 g NaCl, 2 g $MgCl_2 \times 6\ H_2O$ per liter

Packaging buffer: 0.04 *M* Tris-HCl, pH 8, 0.01 *M* spermidine × 3 HCl, 0.01 *M* putrescine × 2 HCl, 0.1% β-mercaptoethanol, 7% dimethyl sulfoxide

Nunc plates: 20 × 20 cm for uv irradiation (Nunc, Kamstrup, 4000 Roskilde, Denmark)

In Vitro Packaging of Recombinant λ and Cosmid DNA

Growth and Induction of the Packaging Strains

The packaging strains are simultaneously pregrown and tested for lysogeny (incorrectly treated strains tend to induce their prophage spontaneously): One fresh single colony of each of the strains is picked and streaked onto two tryptone or LAM plates each. One plate of each is incubated at 32° overnight for the production of inocula for the liquid cultures. On the other two plates, which are incubated at 42°, no or only a few colonies should grow because of the killing of their hosts by the temperature-induced prophages.

L broth prewarmed to 32° is inoculated separately with bacteria grown on the 32° plates to a density at 600 nm of 0.1 or below. (Growth in NZY^{10} instead of in LB may be preferable.) Incubation is continued until a density of 0.3 is reached, the doubling time being about 60 min under these conditions. Prophages are induced by incubation of the cultures at 45° for 15 min. Thereafter they are transferred to 37° and incubated for three additional hours with *vigorous* aeration. A small sample of each culture should be tested for successful prophage induction: The bacteria will lyse upon addition of a drop of $CHCl_3$, whereas uninduced bacteria or cells that have lost their prophage do not lyse under these conditions. The two cultures are mixed in a 1 : 1 ratio and can be stored at this stage for up to 2 weeks in the refrigerator.

uv Irradiation

The mixed cultures are centrifuged at 5000 rpm for 10 min, and the cells resuspended in the original volume of cold M-9. Ten- or 100-ml quantities of the suspension are transferred to empty sterile petri dishes of normal or 10× normal size (20 × 20 cm), respectively, and uv-irradiated. Conditions for irradiation must be determined for every uv lamp. Figure 2 shows the uv inactivation curve of free λ *cIts857 Sam7* phage, in M-9 medium, plated on BHB2600; the inactivation of endogenous λ DNA as measured by its ability to produce plaques upon packaging in the absence of exogenous DNA; and the capacity of the uv-irradiated cells to package exogenous DNA. In this particular experiment a Philips TUV 30-W lamp was used at a distance of 45 cm. The lamp should not be too distant, so as to ensure short irradiation times which avoid unnecessary exposure of the cells to room temperature.

An irradiation causing 20 lethal λ hits (3–4 min in the experiment shown in Fig. 2) is sufficient to eliminate completely any background of plaque-forming particles containing endogenous DNA. To satisfy requirements for EK2 biological containment conditions an irradiation time corresponding to 40 phage lethal hits has to be used (see below).

The irradiated mixture is concentrated about 500-fold by centrifugation and resuspension in packaging buffer and is made 1 m*M* in ATP.

Storage

Samples of the cell concentrate (20 μl) are pipetted into 1.5-ml Eppendorf centrifuge tubes. The mixture should be distributed on the tube walls to ensure rapid freezing when the closed tubes are placed in liquid nitrogen. (Do not use dry ice or dry ice mixtures for the freezing; packaging tubes that have been frozen in liquid nitrogen can be transported in dry ice.) The samples now can be stored at −57° or below for up to and probably more than a year without losing packaging ability.[12] The DNase and Mg^{2+}-containing packaging mix (see below) can be frozen and stored in larger aliquots, since the cells are still pipettable after thawing, because of the action of DNase.

Cells that have been stored frozen can but do not have to be placed in liquid nitrogen again before thawing in ice.

Packaging

Samples are removed from the liquid nitrogen or the deep freezer and thawed by placing them in ice for a few minutes. At this time the samples have not lysed yet, and the DNA to be packaged (up to 7 μl of any concentration) is added and carefully mixed with the concentrate and the

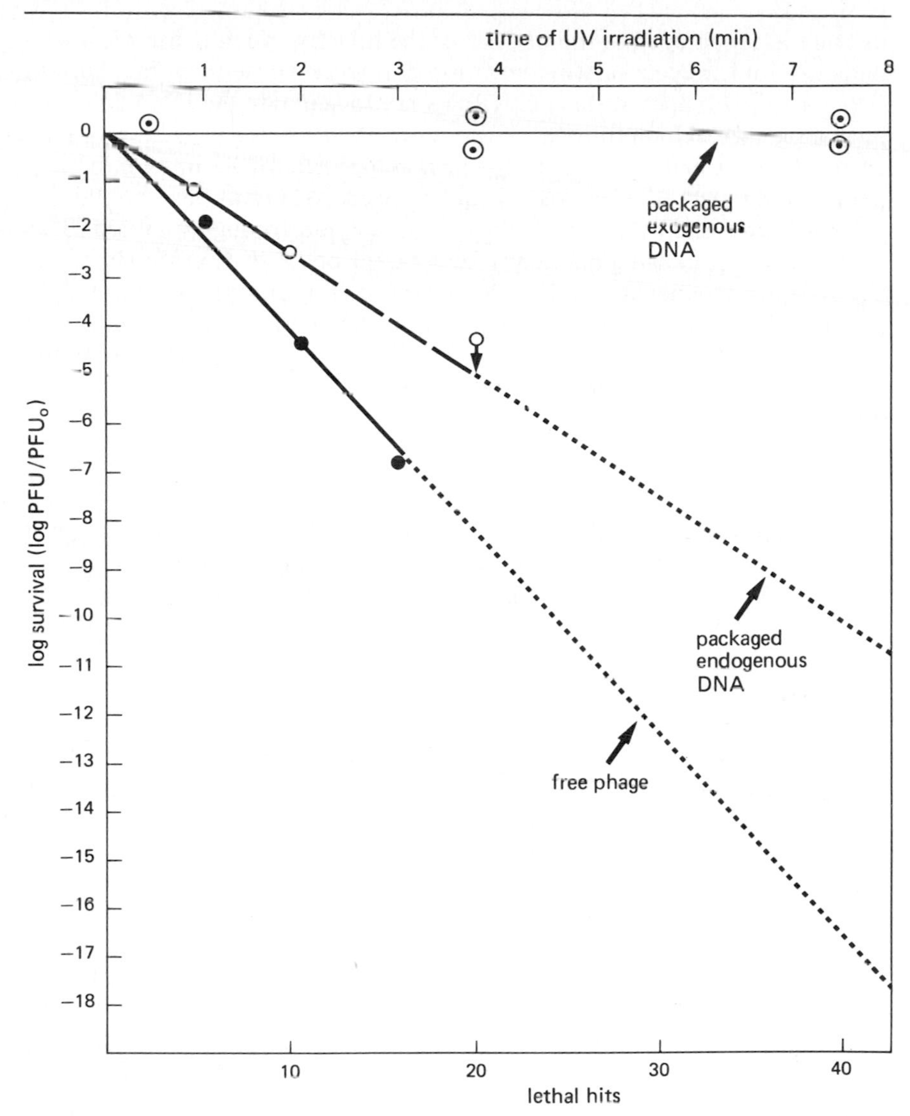

FIG. 2. Inactivation of bacteriophage λ and an *in vitro* packaging mixture by uv light. Free phage is λ *cIts857 Sam7* in M-9. ●, Measured, ---, extrapolated. Inactivation of biological activity of endogenous DNA as measured by *in vitro* packaging, in the absence of exogenous DNA, of cells that had been uv-irradiated. ○, Measured, ---, extrapolated. The effect of uv irradiation on the ability of a packaging mix to encapsidate exogenous DNA is indicated by ⊙. A Philips T uv 30-W lamp was used at a distance of 45 cm. Irradiation was performed in 20 × 20 cm petri dishes with a volume of 100 ml. All platings were performed on BHB2600. Calculation of lethal hits was done using the formula $\ln PFU/PFU_0 = -m$, at a given time t of uv irradiation (m is the mean number of hits, PFU is the number of plaque-forming units at time t, and PFU_0 is the number of plaque-forming units at time 0).

mixture is centrifuged to the bottom of the tube by spinning for a few seconds to avoid unnecessary evaporation during the subsequent incubation. (Do not spin too hard at this point so as not to separate the DNA from the packaging mix.) Upon transfer of the samples to 37°, the cells lyse and liberate all components necessary for packaging. After a 30- to 60-min incubation period, 20 μl of a frozen and thawed packaging mix containing 0.01 *M* $MgCl_2$ and 10 μg/ml DNase is added and throughly mixed. This increases the packaging efficiency by a factor of 1.5 to 5, probably by resupplying components required in postpackaging morphogenetic stages (see Fig. 1). Incubation is continued for 30–60 min and terminated by the addition of 0.5 ml of phage buffer and a drop of $CHCl_3$. The packaged phage or cosmid particles are now ready for plating or transduction, respectively.

Plating, Transduction, and Packaging Efficiency

An aliquot of packaged λ DNA is preadsorbed onto indicator bacteria, treated as described under Bacterial Strains and Media, by incubation for 20 min at 37°. Thereafter, the sample is mixed with 2–3 ml LAM top agar and plated on LAM plates. An aliquot of packaged cosmid is preadsorbed onto bacteria that have been pregrown like λ indicator bacteria, as described in Bacterial Strains and Media, in order to induce the λ receptors. After the adsorption time L broth is added and the sample incubated at 37° for an appropriate time to allow expression of the resistance marker to be used for the selection. Subsequent plating is performed, as with transformed bacteria (see this volume [20]).

If no uv irradiation is carried out, the background of phage with packaged endogenous DNA (detectable only on a *supF*$^+$ indicator such as BHB2600) is about 10^3 plaque-forming particles per packaging sample. The plaques are tiny and carry the 434 immunity. When uv irradiation is carried out for 20 or more phage lethal hits (Fig. 2), background from endogenous phage is totally eliminated (Table I, line 1). The packaging efficiency of exogenous λ DNA is 10^7–10^8 plaque formers per microgram of DNA, and that of restricted and ligated DNA accordingly less (Table I, lines 2 and 3). As an example of the packaging efficiency the cloning of mouse immunoglobulin κ genes in the EK2 vector Charon 4A[10] is shown. Details about conditions for restriction, enrichment of κ gene-containing fragments, and ligation are given in Lenhard-Schuller *et al.*[15] Lines 4 and 5 in Table I show the efficiency of cosmid and hybrid cosmid packaging. In this particular experiment, described in more detail by Collins and

[15] R. Lenhard-Schuller, B. Hohn, C. Brack, M. Hirama, and S. Tonegawa, *Proc. Natl. Acad. Sci. U.S.A.* **75,** 4709 (1978).

TABLE I
PACKAGING EFFICIENCIES[a]

	Exogenous DNA		Plaque-forming units or cosmid-containing particles		
	Vector	Insert	Per packaging sample	Per μg λ vector DNA	Per μg insert DNA
1	None	None	<1		
2	λ *b2*	None	2×10^7	1.5×10^8	
3	Charon 4A "arms"	Mouse	$0.7–5 \times 10^4$		$0.5–5 \times 10^5$
4	Cosmid pJC75-58	None	4.8×10^2	1.1×10^3	
5	Cosmid pJc75-58	*E. coli*	$0.9–6 \times 10^4$		$0.6–4 \times 10^5$

[a] Packaging efficiencies are expressed as plaque-forming units plated on BHB2600 (lines 1–3) and as ampicillin-resistant colonies after transduction of HB101 with packaged cosmids. In the experiments summarized in line 3, which were performed in a P3 laboratory, 5 μl of ligation mixture containing 60 μg/ml Charon 4A "arms" and 20–30 μg/ml mouse DNA from various sources and enriched for various sizes of *Eco*RI fragments[15] were packaged. In the experiments summarized in lines 4 and 5, 2 μl ligation mixture containing 225 μg/ml *Bgl*II cleaved and religated cosmid pJC75-58 DNA (line 4) and 225 μg/ml *Bgl*II cleaved cosmid DNA ligated to 75 μg/ml *Bgl*II-cleaved *E. coli* DNA of various sizes (line 5) were packaged.[16]

Brüning[16] and Collins (this volume [20]), packaging efficiencies expressed as cosmid-containing transducing particles of restricted and religated vector cosmid pJC75-58 (line 4) and of restricted and ligated vector cosmid plus *E. coli* DNA (line 5) are given.

Comments

As noted before,[12] recovery of *in vitro* recombinant DNA molecules by packaging is independent of the genetic constitution of the DNA, since the DNA serves only as a passive packaging substrate. The composition of the population of phage or cosmid particles produced should reflect the relative abundance of the various DNA molecules present in the packageable size range. In addition, cohesive-end ligation leading to mixed hybrid concatenated molecules is resolved into single genomes by *in vitro* packaging, since the function creating the termini is included in the packaging reaction.[13]

The efficiency of *in vitro* packaging is independent of the concentration of the exogenous DNA; i.e., the number of plaques is linearly dependent on the amount of DNA added.[5] For a cloning experiment the ligation mixture can be used directly.

In the packaging system described here, the packaging efficiency is

[16] J. Collins and H. J. Brüning, *Gene* **4,** 85 (1978).

independent of the length of λ DNA, as described for the earlier recombinant-proficient system.[12] However, if a cell-free extract is produced from the induced, uv-irradiated, concentrated cells, the packaging efficiency is strongly dependent on the size of the DNA, the longer molecules being packaged preferentially (B. Hohn, unpublished), similar to the length-dependent packaging system described by Sternberg *et al.*[14] Other methods for enrichment of longer genomes, applicable also to cosmids, are based on the positive selection for large molecules on infection of *pel*$^-$ hosts[17] or on the difference in buoyant density of particles with different DNA contents.

The efficiency of obtaining hybrid clones, per microgram of insert DNA, is on the order of $0.5-5 \times 10^5$, depending on the insert size, for both λ DNA and cosmid DNA (Table I; for more details on ligation conditions and insert sized, see Lenhard-Schuller *et al.*,[15] Collins and Brüning[16] and this volume [20]).

For λ cloning experiments carried out under EK2 biological containment conditions the following precautions, apart from the chloroform treatment, must be made (Recombinant DNA Technical Bulletin, vol. 1, no. 1, fall 1977): "1) The packaging preparation must be irradiated with ultraviolet light to a dose of 40 phage lethal hits; 2) the packaging preparation should be assayed without addition of exogenous λDNA and the ratio of plaque-forming units without addition (endogenous virus) to plaque-forming units with exogenous DNA (exogenous virus) must be less than 10^{-6}; 3) when the EK-2 vector is packaged, the ratio of *am*$^+$ phage (recombinants) to total phage must be less than 10^{-6}. These criteria must be satisfied for each packaging extract used." A dose of 40 phage lethal hits corresponds to a theoretical inactivation of *cIts857,* plated on a *recA*$^+$ host, to 10^{-17} (Fig. 2). This dose, as is also shown in Fig. 2, does not change the competence of the packaging mix to encapsidate exogenous DNA. A comparison of lines 1 and 2 in Table I shows how easily requirement 2 can be satisfied.

Requirement 3, although easily satisfied (for instance, the relevant control for the experiment shown in Table I, line 3, was $<7 \times 10^{-7}$ *am*$^+$ phage per *am* EK2 vector phage after packaging), is not very stringent. Endogenous DNA carries amber mutations as well, and any *am*$^+$ vector recombinant could only have arisen by a rare crossover in a small fraction of the λ genome, a rare reversion, or an equally rare combination of both events. Since this is an important issue in biosafety and authenticity of recovered clones, a test experiment was performed using as exogenous DNA a combination of λ 607 *Aam32 Bam1* DNA (a recombinant of vector λ 607[18] and λ *Aam32 Bam1*), restricted with *Eco*RI and ligated to

[17] S. W. Emmons, V. Maccosham, and R. L. Baldwin, *J. Mol. Biol.* **91,** 133 (1975).
[18] N. E. Murray, W. J. Brammar, and K. Murray, *Mol. Gen. Genet.* **150,** 53 (1976).

*Eco*RI-restricted *E. coli* DNA, with DNA of λ *imm434 cIts Qam21 Ram60*. Restriction and ligation of the former vector was included in this test experiment to mimic the real situation and assess the influence on recombination of the restriction and ligation conditions. Only recombinants plate on a sup^0 indicator, the reversion frequency of λ 607 $A^- B^-$ and λ *imm434 cIts* $Q^- R^-$ phage being 4.6×10^{-8} and 1.0×10^{-9}, respectively. No plaques appeared in 1.5×10^6 packaged molecules. Recombination during packaging in the $recA^-$ red^- packaging cells described in this chapter therefore, is safely excluded.

Experiments to test the safe use as EK2 vectors of various cosmids are under way.

Acknowledgments

I thank Thomas Hohn, Ken Murray, John Collins, and Susumu Tonegawa for cooperation at various stages of the development of the described system and Nat Sternberg for generously supplying strains.

Financial support for the part of the work done while the author was at the Biocenter in Basel, Switzerland, was provided by Grant No. 3,472,75 from the Swiss National Fonds.

[20] *Escherichia coli* Plasmids Packageable *in Vitro* in λ Bacteriophage Particles

By John Collins

It has recently been shown that *in vitro* λ bacteriophage packaging systems can be used for the efficient production of hybrid bacteriophage from DNA prepared by fusion of a bacteriophage vector and foreign DNA fragments *in vitro*.[1,2] The *in vitro* packaging system appears to be insensitive to the DNA built on adjacent to the small λ region required for packaging. This is illustrated by the fact that most of the λ DNA, including all regions required for λ replication and lysogeny, can be replaced by plasmid DNA and efficient packaging will still occur. The packaging is then dependent on the entire molecule being of a certain minimum size and on the presence of concatemeric forms induced by *in vitro* restriction endonuclease cleavage and ligation at high DNA concentrations[3,4] (Fig. 1).

The conditions required for efficient packaging of plasmids carrying

[1] H. Hohn and K. Murray, *Proc. Natl. Acad. Sci. U.S.A.* **74,** 3259 (1977).

[2] N. Sternberg, D. Tiemeier, and L. Enquist, *Gene* **1,** 255 (1977).

[3] J. Collins and B. Hohn, *Proc. Natl. Acad. Sci. U.S.A.* **75,** 4242 (1978).

[4] J. Collins and H. J. Brüning, *Gene* **4,** 85 (1978).

ISBN 0-12-181968-X

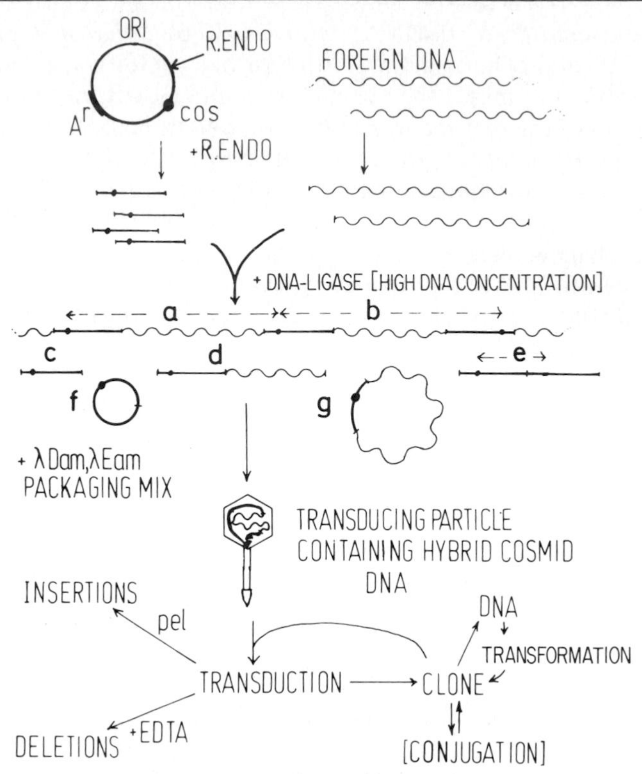

FIG. 1. A scheme illustrating a method of gene cloning with plasmids containing a selectable marker (e.g., antibiotic resistance, A^r), the cohered λ cohesive ends (cos), a replication origin (ORI), and a site (↓) for a restriction endonuclease (R. ENDO). A plasmid having such properties is called a cosmid. The cosmid is cleaved to linear form with a restriction endonuclease. The foreign DNA to be cloned is cleaved partially with an enzyme generating the same cohesive ends (fragments in the 16–30 Md size range). The DNAs are mixed and ligated. The use of high DNA concentrations during this ligation increases the formation of molecules of types a, b, and e. The DNA forms most likely to be packaged and subsequently lead to formation of a transductant are a and e, where the distance between the *cos* sites is about 30 Md. Packaging of the hybrid DNA takes place by cleavage of two consecutive, similarly oriented *cos* sites and the subsequent encapsulation of the intervening DNA in the λ particle. For small vectors the likelihood of producing transductants of the vector alone (from molecules e and f or higher polymers) is low. The packaging takes place in an *in vitro* packaging mix,[5] and the transducing particles formed are used to inject λ-sensitive su^- *E. coli*. The hybrid plasmid is injected, circularizes, and replicates as a normal plasmid without expression of any λ function. The isolated clones can now be further manipulated genetically (1) by repackaging the hybrid cosmids by superinfection with helper phages (efficiency low); (2) by conjugation, when the plasmid is mobilizable, by introduction of a sex factor; or (3) by transformation of purified supercoiled DNA. The use of repackaging could be used, presumably, in conjunction with standard methods used in genetics to select for insertions or deletions on the basis of the effect of DNA size (1) on particle stability, enhanced by EDTA to select for deletions, or (2) on penetration during transduction enhanced on infection of *E. coli pel*$^-$ hosts to select for insertions. Reproduced by permission of Elsevier North Holland Biomedical Press (Collins and Brüning[4]).

the λ *cos* site are described below, with particular emphasis on conditions leading to the selection of hybrids carrying large pieces of foreign DNA subsequent to *in vitro* recombination. The *in vitro* packaging system is based on that developed by Hohn[1,3-5] with some modifications, but it is likely that the plasmid vectors described for use with this system (cosmids)[3,4] can also function well with other λ *in vitro* packaging systems, as long as the criteria for preparation of a good DNA substrate are observed. A selection of 12 plasmid vectors that can be used for cloning with over nine restriction endonucleases that generate cohesive ends is described. Vectors of different molecular weight are also included, since they may be used for selective cloning of foreign DNA fragments in a particular size range.

Plasmids usable with this λ packaging technique are referred to below as cosmids, since the λ region necessary for recognition by the λ packaging system is called the λ *cos* site. As demonstrated below, the particular advantage of the system over the λ vectors is the small size of the vector, which consequently allows packaging of large foreign DNA fragments. The main advantages of the cosmid system over plasmid cloning vectors are (1) the high efficiency of hybrid clone formation; (2) the fact that essentially only hybrid plasmids are formed, thereby obviating the need for screening or selecting for hybrid clones or pretreating vector DNA to prevent vector ring closure; and (3) the fact that the system selects particularly for large fragments in contrast to plasmid transformation systems which give a bias toward the cloning of smaller fragments.

The development of new vectors can be rapid, since the regions required for the principal properties of the cosmid (replication origin, selective marker, single restriction site, and *cos* site) are small, and vectors may be easily developed to meet specific cloning requirements.

Principle of the Method

For DNA to be efficiently transported into the *Escherichia coli* recipient cell through *in vitro* packaging, two main criteria can be distinguished:

1. Packaging of the DNA into the λ particle, which is dependent on ATP and on the presence of suitable recognition sites distributed along the DNA to be packaged (see below).

2. Injection of the packaged DNA into the recipient cell, which is dependent on the size of the packaged DNA, being efficient in the approximate range 23–32 Md.[3,4,6]

DNA that can be used for packaging may have one of the following

[5] B. Hohn, this volume [19].

[6] M. Feiss, R. A. Fisher, M. A. Crayton, and C. Egner, *Virology* **77**, 281 (1977).

forms[7]: (1) linear molecules having λ cohesive ends (m and m′), or (2) concatemeric molecules with the fused (cohered) cohesive end site (*cos* site consisting of m plus m′) dispersed along the concatemer.

Molecules apparently not used either *in vivo*[8–13] or *in vitro*[14–16] as a efficient substrate for packaging are circular molecules with a single *cos* site. Suitable concatemeric substrates may be produced from them by recombination *in vitro*[17] or *in vivo*[11,18] but at a low frequency. The exact extent of the region in the vicinity of the *cos* site required for packaging is not well defined, but about 178 base pairs spanning this site have been sequenced[19] and show unusual palindromic regions occurring up to 60 base pairs away from the axis of the cohesive ends.

A suitable substrate for packaging is derived by the ligation of restriction endonuclease-cleaved foreign and vector DNA at high DNA concentrations (Fig. 1). The number of possible assemblies of vector and/or foreign DNA is large (molecules a–g in Fig. 1). From this random ligation molecules of forms a and e will be cleaved out at the *cos* sites and packaged *in vitro* into λ heads. This having occurred, the assembly of the bacteriophage tail and foot takes place to give a complete particle able to inject the encapsulated cosmid or cosmid hybrid DNA into an appropriate recipient *Escherichia coli*. This injection of the packaged DNA only occurs with reasonable efficiency if the DNA is in the size range 23–31 Md, i.e., a large hybrid or a multimer of the vector. This leads to automatic production of clones containing mostly large hybrid plasmids after subsequent selection for the antibiotic resistance carried on the plasmid vector. A further reduction in the percentage of nonhybrid vector (multimeric form) can be obtained by using large vectors (16–20 Md), small vectors (4–7 Md), or temperature-sensitive replication mutants with a low copy number (Table I).

On injection, the cosmid hybrid DNA, which is in a linear form, having λ cohesive ends (m and m′), circularizes by reforming the *cos* site (m plus m′). DNA replication now continues under control of the plasmid

[7] T. Hohn and I. Katsura, *Curr. Top. Microbiol. Immunol.* **78,** 69 (1977).

[8] J. Szpirer and P. Brachet, *Mol. Gen. Genet.* **108,** 78 (1970).

[9] F. W. Stahl, K. D. McMillan, M. M. Stahl, R. E. Malone, Y. Nozu, and V. E. A. Russo, *J. Mol. Biol.* **68,** 57 (1972).

[10] L. W. Enquist and A. Skalka, *J. Mol. Biol.* **75,** 185 (1973).

[11] M. Feiss and T. Margulies, *Mol. Gen. Genet.* **127,** 285 (1973).

[12] D. Freifelder, L. Chud, and E. E. Levine, *J. Mol. Biol.* **83,** 503 (1974).

[13] P. Dawson, A. Skalka, and L. D. Simon, *J. Mol. Biol.* **93,** 167 (1975).

[14] B. Hohn, B. Klein, M. Wurtz, A. Lustig, and T. Hohn, *J. Supramol. Struct.* **2,** 302 (1974).

[15] B. Hohn, *J. Mol. Biol.* **98,** 1975).

[16] M. Syvanen, *J. Mol. Biol.* **91,** 165 (1975).

[17] M. Syvanen, *Proc. Natl. Acad. Sci. U.S.A.* **71,** 2496 (1974).

[18] K. Umene, K. Shimada, and Y. Takagi, *Mol. Gen. Genet.* **159,** 39 (1978).

[19] B. P. Nichols and J. E. Donelson, *J. Virol.* **26,** 429 (1978).

TABLE I
PROPERTIES OF COSMID CLONING VECTORS

Vector cosmid	720	703ΔBglII	703ΔHindIII	703ΔPstI	720ΔBglII-3	720ΔBglII-2	720ΔBglII-2·ΔPstI	74	75-65	75-58	76	77		78	79	
MW (Megadalton)	16	7.2	11.2	7.6	13.3	10	7.2	10.5	8.2	7.6	4.0	6.3		6.8	4	
Selection	rif	E1	E1	E1	rif	rif	E1	Ap	Ap	Ap	Ap	Ap	Tc	pro⁺	Ap	Tc
EcoRI G↓AATTC				+				+	+	+		+	+	+	+	+
BamHI G↓GATCC								+	+	+					+	−
BglII A↓GATCT		+		+		+	+	+	+	+				+		
SalI G↓TCGAC		+		+			+	+				+	−		+	−
PstI CTGCA↓G				+			I							+	−	+
ClaI AT↓CGAT											+					
XmaI C↓CCGGG	+	+	+		+	+										
HindIII A↓AGCTT	+				+							+	−		+	−
KpnI GGTAC↓C						+										
Copy No.	15	15	15	15	15	15	15	15	5	5	5	30		5	30	
Temp. sensit. (ts)									ts	ts	ts			ts		
mob	+	+	+	+	+	+	+	+	+	−	−	−		−	−	

[a] Cosmid vectors of the designated molecular weights can be selected for with rifampicin (Rif), ampicillin (Ap), tetracycline (Tc), or colicin E1 (E1). Where a single cutting site for a restriction endonuclease exists this is indicated by a plus sign under the selection still available and a minus sign under the resistance inactivated by insertion at that site. The *Kpn*I site in pJC720 Δ *Bgl*II-2 destroys rifampicin resistance, leaving only E1 immunity for selection. The plasmid copy number in LB broth at 30° is indicated below the list of restriction enzymes and their cut sites on the DNA. Temperature-sensitive plasmid replication leads to loss of the plasmid at 42° and a lower copy number at 30°; both properties help reduce the background of nonhybrid clones in cloning experiments. Mobilizability (Mob^+ or mob^-) is indicated in the bottom row and refers to loss of plasmid functions required for plasmid mobilization in sex factor-promoted conjugation (but not to loss of the plasmid conjugational transfer origin). Restriction endonuclease maps of these plasmids are given in Figs. 3 and 4. pJC77, a hybrid of *Sau*I fragments from pJC75-58 and pBR322[29] is still being tested for its properties as a cloning vector. pJC78 is a vector designed to have exceptionally good containment qualities. In addition to the containment qualities detailed for pJC75-58 it has no homology with the *E. coli* chromosome or known R factors. The *proB* and *C* alleles it carries are derived from *Methylomonas*[30] pHC79 is made from a pBR322 *Sau*I-induced rearrangement, which has a single *Bg*/II site (pJC80, 2.95 Md) located between the end of the tetracycline resistance gene and the origin of replication. A 1.1-Md *Bg*/II fragment carrying the *cos* site from λ Charon 3A was inserted at this *Bg*/II site.[31] The small *cos* fragment is easily reisolatable from pHC79, and this should facilitate the construction of further cosmid derivatives. Reproduced by permission of Elsevier North Holland Biomedical Press (Collins and Bruning[4]).

replicon. Since no bacteriophage functions are expressed, no bacteriophage particles can be formed without further infection with a λ helper phage.

The *in vitro* packaging system described here[14] consists of a mixture of induced cultures of λ *Dam Sam* and λ *Eam Sam* lysogens, with suppressor minus backgrounds, which are only able to complement each other in λ prehead formation *in vitro* on lysis of the cells. Lysis is induced by warming up the cells. During the packaging of exogenous (cosmid hybrid) DNA the DNA of the bacteriophage used to produce the packaging mix is also packaged to some extent. For this reason the adsorption and injection of the packaged DNA must be carried out in a suppressor minus *E. coli* host. If the λ DNA present in the packaging mix is inactivated by uv irradiation,[5] a suppressor plus recipient strain can be used. In addition, to prevent degradation of foreign DNA cloned into the hybrids, a recipient should be used that has a defective host restriction system.

Materials and Reagents

Spermidine hydrochloride (Serva); putrescine (Merck); adenosine 5′-triphosphate (Boehringer); *Bam*HI (Boehringer); *Hin*dIII (Boehringer); DNA ligase (Boehringer); *Bgl*II (gift from W. Rüger); *Sau*I (gift from Streeck); *Sal*I, *Eco*RI, *Pst*I, and *Cla*I (gifts from H. Mayer and H. Schütte).

Preparation of DNA

Plasmid vector DNA was prepared from 1 liter L Broth (1% Bactotryptone, 0.5% yeast extract, 0.5% NaCl) cultures grown at 30° either overnight in stationary phase or after a 12-hr treatment with chloramphenicol (150 μg/ml added at OD_{550} = 1.0). This latter method of plasmid amplification in the absence of protein synthesis is possible since all vectors described here are derived from ColE1. Cleared lysates were prepared after the method of Katz *et al.*[20] using a solution of lysozyme, EDTA, and Triton X-100. The plasmid DNA was then further purified by CsCl–ethidium bromide centrifugation (Ti60 Beckman rotor) [20-ml gradients at 36,000 rpm for 36 hr at 15° followed by recentrifugation of the pooled lower (plasmid supercoil) band in a Ti50 rotor at 36,000 rpm for 36 hr at 15°]. After extraction of the ethidium bromide with three washes of isopropanol at 0° and dialysis against three changes of 10 m*M* Tris, pH 8.0, and 0.1 m*M* EDTA buffer the plasmid DNA was concentrated by precipitation with 2 volumes ethanol in the presence of 100 m*M* NaCl at −20° for 10 hr. All DNA preparations were stored at concentrations of 200–1000

[20] L. Katz, D. T. Kingsbury, and D. R. Helinski, *J. Bacteriol.* **114,** 577 (1973).

μg/ml in sterile 10 m*M* Tris, pH 8.0, and 0.1 m*M* EDTA at −20°. Repeated freezing and thawing were avoided by keeping small aliquots.

High-molecular-weight chromosomal DNAs are prepared from cells or tissues by published procedures shown to give molecular weights in excess of 30 Md. During purification RNase and CsCl ultracentrifugation are included. DNA preparations are kept unfrozen in 10 m*M* Tris, pH 8.0, and 0.1 m*M* EDTA.

Restriction Endonucleases and DNA Cleavage

The endonucleases listed in Table II,[21] which generate cohesive ends, may be used for cloning with the vectors described.[22]

Methods

The details of the λ packaging method used are described by Hohn,[5] and the application of the packaging system to a specific cloning problem using λ *vectors* is described elsewhere.[23]

The description of the packaging system as applied to cosmids is therefore confined to the vectors so far available and to variables that have an effect on the efficiency of hybrid formation.

Bacteria and Plasmids

Preparation of the plasmids pJC720 and pJC703 has been described,[24,25] as has the derivation of their deletions.[4] Preparation of the ampicillin-resistant pJC74 by cloning *Pst*I fragments from R1*drd*19 into pJC703 *Pst*I (Figs. 2 and 3) has been described.[4] Temperature-sensitive derivatives of pJC74 are shown in Fig. 3, where their formation is also outlined.[4] The formation of pJC76 is described in the legend for Fig. 4.[26] pJC77, mentioned in Table I, has not been fully mapped but is one of a number of derivatives from a pBR322[27,28] in which *Sau*I fragments from pJC75-58 have been inserted into a *Sau*I site of pBR322.

[21] Recognition sites: R. Roberts, personal communication; *Crit. Rev. Biochem.* **4,** 123 (1976).

[22] H. Mayer and G. Hobom, personal communication.

[23] R. Lenhard-Schuller, B. Hohn, C. Brack, M. Hirama, and S. P. Tonegawa, *Proc. Natl. Acad. Sci. U.S.A.* **75,** 4709 (1978).

[24] J. Collins, N. P. Fiil, P. Jørgensen, and J. D. Friesen, *Control Ribosome Synth., Proc. Alfred Benzon Symp., 9th Munksgaard, Copenhagen 1976* p. 356 (1977).

[25] J. Collins, *Curr. Top. Microbiol. Immunol.* **78,** 121 (1978).

[26] J. Collins, in preparation.

[27] F. Bolivar, and K. Backman, this volume [16].

[28] F. Bolivar, R. L. Rodriguez, M. C. Betlach, and H. W. Boyer, *Proc. Natl. Acad. Sci. U.S.A.* **74,** 5265 (1977).

TABLE II
ENDONUCLEASES USED FOR CLONING[a]

Vector cleaved with:	Foreign DNA cleaved with:
*Hin*dIII (A ↓ AGCTT)	*Hin*dIII
*Eco*RI (G ↓ AATTC)	*Eco*RI (*Eco*RI*)
*Sal*I (G ↓ TCGAC)	*Sal*I, *Xho*I (C ↓ TCGAG), *Ava*I (C ↓ PyCGPuG)
*Bam*HI (G ↓ GATCC) }	{ *Bam*HI, *Sau*I (↓ GATC)
*Bgl*II (A ↓ GATCT) }	{ *Bcl*I (T ↓ GATCA), *Bgl*II (A ↓ GATCT)
*Pst*I (CTGCA ↓ G)	*Pst*I
*Xma*I (C ↓ CCGGG)[b]	*Xma*I
*Cla*I[22] (AT ↓ CGAT)[c]	*Cla*I, *Taq*I (T ↓ CGA), *Hpa*II (C ↓ CGG)
*Kpn*I (GGTAC ↓ C)	*Kpn*I

[a] Restriction endonuclease digestions were carried out in the following buffer: 10 m*M* $MgCl_2$, 10 m*M* NaCl, 10 m*M* Tris-HCl, pH 7.8, with the exception of *Kpn*I for which 6 m*M* $MgCl_2$, 6 m*M* NaCl, 6 m*M* Tris-HCl, pH 7.5, 10 m*M* dithiothreitol, 100 μg/ml BSA was used.

[b] *Xma*I is included only on a theoretical basis, since it is currently considered highly unstable, is not commercially available, and is appreciably contaminated with nonspecific endonucleases. Any plasmid carrying a single site for any restriction endonuclease not cutting within the λ *cos* site or within the genes used for selection or for replication can be used as a packageable cloning vector (cosmid, Figs. 1 and 2, Table I). It is not known if this cloning technique works with ligation of flush-molecules as are derived from *Hin*cII or *Sma*I cleavage for example. In view of the high degree of ligation required at high DNA concentrations it is not to be expected that this would be particularly efficient.

[c] The enzyme *Cla*I recently purified from a *Caryophon* species (H. Mayer and H. Schütte, personal communication) has been shown to have the recognition sequence ATCGAT and to produce the same two base pair sticky ends as *Taq*I (T ↓ CGA) and *Hpa*II (C ↓ CGG) (H. Mayer, personal communication). The particular advantage of using "multicut" enzymes, with only 4-bp recognition sites, for the production of gene banks is discussed later.

CC309 ColE1$^+$ (*thr leu thy*) was used for colicin E1 production and carries a chromosomal mutation leading to temperature inducibility of colicin. JC411 RI*drd*16, ColE2$^+$ (*his pro arg met*) was obtained from D. Helinski. HB101 (*pro leu* r^- m^-) from B. Hohn was used as a recipient for cosmid transduction after packaging.

Properties of Cosmid Cloning Vectors

A summary of the main properties of a number of cosmids is shown in Table I.[4,29–31]

[29] F. Bolivar, R. L. Rodriguez, P. J. Greene, M. C. Betlach, H. L. Heyneker, H. W. Boyer, J. H. Crosa, and S. Falkow, *Gene* **2,** 95 (1977).

[30] J. Collins and H. Berger, in preparation.

[31] B. Hohn and J. Collins, submitted to *Proc. Natl. Acad. Sci.*

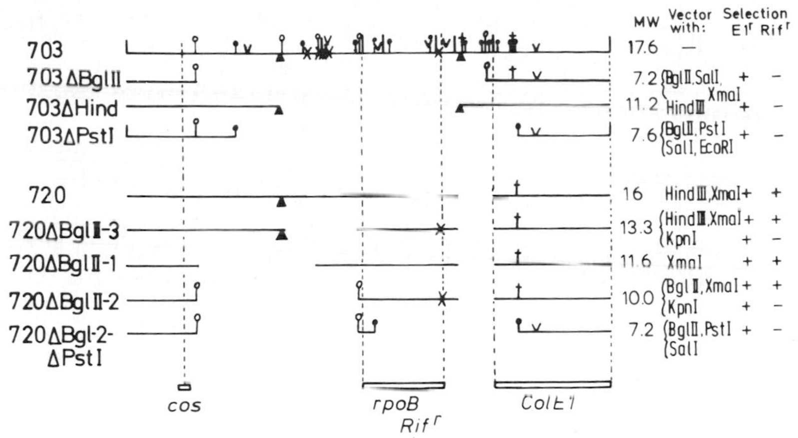

FIG. 2. Restriction maps of deletions of the ColE1 *rpoB* (rif^R) plasmids pJC703 and pJC720 usable as cosmids.[4] The vectors of the indicated molecular weight were derived by direct deletion from pJC703 or pJC720 with the indicated enzymes. The selection available (E1 immunity or rifampicin resistance) in conjunction with particular restriction enzymes is shown at the right. Reproduced by permission of Elsevier North Holland Biomedical Press (Collins and Brüning[4]).

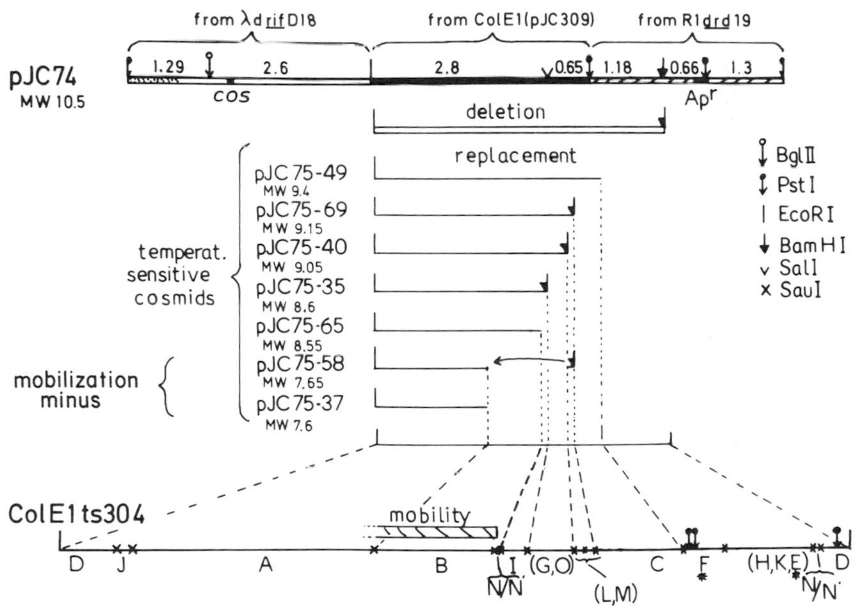

FIG. 3. The derivation of temperature-sensitive derivatives of pJC74, in which the *Eco*RI-*Bam*HI fragment of pJC74 was replaced by *Eco*RI-*Sau*I fragments from the temperature-sensitive ColE1 *ts*304.[4] The location of conjugational mobilization functions and the *Sau*I fragments of ColE1 *ts304* are shown below. *Sau*I fragment P (20 bp) has not been placed. Fragments E and F are also indistinguishable. The dotted region to the left of the *cos* site is a region of *E. coli* chromosomal DNA from λ dRifD18. Reproduced by permission of Elsevier North Holland Biomedical Press (Collins and Brüning[4]).

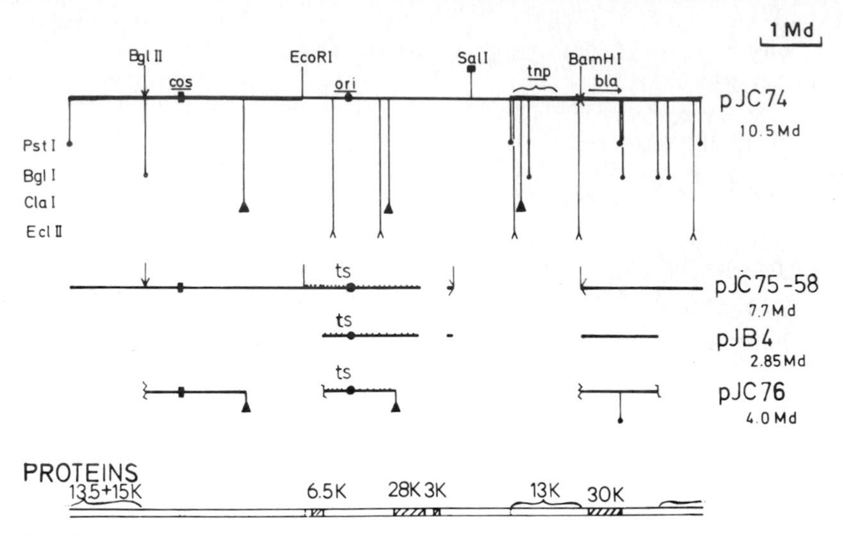

FIG. 4. Restriction map of three temperature-sensitive derivatives of pJC74. pJB4 was obtained as a spontaneous deletion from pJC75-58 during experiments to cause deletions with *Bgl*I. pJC76 was obtained by introducing a *Bgl*II-*Cla*I fragment from pJC74 into pJB4 cut with *Bam*HI and *Cla*I. The signs and indicate the ends joined to each other. The lower part of the diagram shown the positions of seven proteins coded for by pJC74. pJB4 makes only the 30,000-molecular-weight β-lactamase (Ap^R).[26] pJB4 and pJC76 have lost ColE1 immunity.

The molecular weight of the plasmid determines the maximum amount of foreign DNA that can be cloned (approximately a 31-Md cosmid), the average size of the DNA cloned (approximately a 26-Md cosmid), and the smallest size of the DNA that can be cloned (approximately a 19-Md cosmid). Hybrids as small as 10 Md have been obtained but with a low frequency, and it is not clear whether or not they were segregated by deletion from larger plasmids.

The Selection for Cosmid Hybrids. Rifampicin selection must be carried out with some care in view of the low rifampicin resistance (30 μg/ml) of strains carrying these vectors. First, a relatively long expression time ($1\frac{1}{2}$–2 hr) is needed after the transduction-adsorption step, in LB broth (1% Bacto-tryptone, 0.5% Difco yeast extract, 0.5% NaCl), before plating. The plates must not be wet and should be incubated *in the dark* for 48 hr before counting. Plates were made using 30 μg/ml rifampicin, 1.5% agar, 0.5% NaCl, 0.5% Difco yeast extract, and 1% Bacto-tryptone.

Selection for E1-immune colonies is made as follows. First, 0.1 ml of an overnight LB broth culture of CC309, a temperature-sensitive *E. coli* ColE1$^+$ in which colicin synthesis is induced at a high temperature (lethal at 42°), is spread on LB broth plates and incubated overnight at 37°. The surface growth is scraped off with a glass rod, and the plates inverted over 3 drops of chloroform (in a watch glass) for ½ hr on the bench, so as to sterilize the plates. These plates are then used for the colicin selection. Not more than 10^6 cells are added per plate. The plates are incubated at 42° so as to reduce the background of $E1^R$ mutants occurring frequently as a result of lipopolysaccharide overproduction. Colonies are picked off after 36 hr. Retesting for sensitivity to colicin E2 at 42° is used later to establish the identity of the colonies as *E. coli* and to distinguish E1 immunity (plasmid-coded) from colicin E resistance (E1 plus E2 resistance, usually a chromosomal mutation or contaminant), if required. Sensitivity to colicin E2 is tested on simple cross-streak plates on which an *E. coli* $ColE2^+$ strain has been grown, the culture scraped off, and the plate chloroformed.

Selection for ampicillin resistance (50 μg/ml) and tetracycline resistance (20 μg/ml) was carried out on LB broth plates after a 30-min (Ap^R) or a 60-min expression time (Tc^R).

Copy Number. For pJC720, and the deletions of this plasmid and pJC703, copy numbers have been estimated roughly to be similar to that of ColE1 (15 copies per cell). For the temperature-sensitive plasmids the copy number has been measured more accurately by subsequent segregation kinetics at 42°.[4] Copy numbers are 5.3 at 30°, 4.5 at 34°, and 2.1 at 37° after 12 generations at these temperatures. Tetramers and trimers of pJC75-58 have been isolated and the copy numbers estimated as 1.4 and 1.8, respectively, after 12 generations at 30°; however, the populations contain 55% (trimer) and 75% (tetramer) ampicillin-sensitive cells which have lost the plasmid. This leads to formation by the polymeric plasmids of small colonies on plates after 48 hr. This instability of vector multimers leads to a reduction in the nonhybrid background in cloning (see below).

Mobilizability. All the cosmids described here are derived from ColE1 which can be mobilized during conjugation promoted by a sex factor. pJC75-58, pJC76, and pJC77 have deletions in the region for a protein required for mobilization.[4,28] pJC75-58 and pJC76 probably still have the site at which mobilization starts and could perhaps still be complemented for transfer by a double infection with ColD and a sex factor.[32] The sexual properties of the plasmids are important in the context of the discussion of their use as EKII (German B2) safety vectors.

At present no recombinant DNA experiments may be carried out with sexually competent strains. Authorization to use cosmids of the pJC75-58, pJC76, or pJC77 type as safety vectors will largely depend on the production of a safe E. coli λ-adsorbing "incapacitated" host, such as those currently being developed in Curtiss's laboratory.[33]

[32] G. Dougan, M. Saul, G. Warren, and D. Sherratt, *Mol. Gen. Genet.* **158,** 325 (1978).
[33] R. Curtiss, III, *Annu. Rev. Microbiol.* **30,** 507 (1976).

DNA Preparation. Plasmid DNA was prepared by CsCl–ethidium bromide centrifugation of cleared lysates of strain *E. coli* C1 (*ser*$^-$) carrying the required plasmid vector. Starting with 5-liter cultures, two rounds of CsCl–ethidium bromide gradient ultracentrifugation were used to concentrate the DNA to about 500–1000 μg/ml. The DNA was finally dialyzed in 10 m*M* Tris-HCl and 0.1 m*M* EDTA, pH 7.0. The *E. coli* chromosomal DNA to be cloned was prepared by lysozyme–sodium dodecyl sulfate lysis of the cells, protease K and RNase treatment, and phenol–chloroform extraction. The detailed method varies according to the source of the DNA. It should be noted that the following criteria are important: The DNA must be of very high molecular weight, i.e., 30–50 Md or larger, and free of nucleases, and the DNA concentration should be accurately known. If the DNA is relatively free of RNA and protein, the uv absorption at 260 nm (A_{260}) can be used (1 mg/ml DNA = A_{260} 20). DNA from λ *b*515 *b*519 *c1*857 *Sam*7 was used as a control DNA for the efficiency of the packaging system using a *E. coli su* recipient for absorption and plaque titration.[5] Efficiencies of 10^7–10^8 plaque-forming units (PFU) per microgram of λ DNA should be obtained before using the packaging mix for cosmid packaging.

DNA Restriction and Ligation

Partial digests of the chromosomal DNA were made by serial dilution of the restriction enzyme (*Hin*dIII or *Bgl*II) and incubation for 30 min at 37°. The reactions were stopped by an incubation for 10 min at 70°, and the vector DNA was digested to completion. To check the molecular weight of DNA accurately in this high-molecular-weight range 0.4% agarose gels were used in an electrophoresis apparatus (H. Hözel Technik) in which no hydrostatic pressure was exerted on the gel (electrophoresis buffer—0.09 *M* Tris, 2.5 m*M* EDTA, 0.09 *M* boric acid, pH 8.4). Slab gels 3 mm thick, 13 cm wide, and 10 cm deep were run at 180 V for 3 hr. The linear vector and λ DNA were used for molecular weight markers.

The DNAs were mixed to give the indicated (Table III) ratios of vector to chromosomal DNA, over a ninefold range, at a final concentration of 300 μg/ml. If required, the DNAs were concentrated by precipitation in 70% ethanol, 100 m*M* NaCl, 0.1 m*M* EDTA, and 10 m*M* Tris, pH 7.8, for 4 hr at −20° and resuspension in 10 m*M* Tris-HCl, pH 7.8, and 0.1 m*M* EDTA after desiccation of the pellet. Resuspension of the DNA at these high DNA concentrations requires at least a ½-hour shaking, being easier if carried out with digested DNA.

The final mix for ligation (20–50 μl) was adjusted to 10 m*M* $MgCl_2$ (no more than 10 m*M* NaCl) and 10 m*M* Tris-HCl, pH 7.5, heated to 70° for 5

TABLE III

Exp.	Av. M.W. *E. coli* DNA	Vector / *E. coli*	DNA concn in ligatn: Vector μg/ml	DNA concn in ligatn: *E. coli*	DNA concn in ligatn: Total	Colonies per 2 μl DNA	Colonies per μg *E. coli* DNA	Prototrophs *pro*⁺	Prototrophs *leu*⁺	
1		3	225	75	300	$6.0 \cdot 10^4$	$4.0 \cdot 10^5$	11/4150 (0.26%)	13/4150 (0.31%)	pJC75-58 × *Bgl*II
2	$7 \cdot 10^6$	1	150	150	300	$1.1 \cdot 10^4$	$3.6 \cdot 10^4$	3/1841 (0.16%)	10/1841 (0.54%)	
3		0.33	75	225	300	$2.4 \cdot 10^2$	$5.2 \cdot 10^2$	-	-	
4		3	225	75	300	$8.6 \cdot 10^3$	$5.8 \cdot 10^4$	0/1260 (<.08%)	8/1260 (0.62%)	
5	$20 \cdot 10^6$	1	150	150	300	$5.8 \cdot 10^3$	$1.9 \cdot 10^4$	0/961 (<.1%)	2/961 (0.21%)	
6		0.33	75	225	300	$2.3 \cdot 10^2$	$5.1 \cdot 10^2$	-	-	
7	-	-	225	-	225	$4.8 \cdot 10^2$	-	0/1008 (<.1%)	0/1008 (<.1%)	
8	-	-	75	-	75	$2.4 \cdot 10^1$	-	-	-	
9		3	225	75	300	$8.7 \cdot 10^3$	$5.8 \cdot 10^4$	13/5000 (0.26%)	0/5000 (<.02%)	pJC720 × *Hin*dIII
10	$10 \cdot 10^6$	1	150	150	300	$1.3 \cdot 10^3$	$4.3 \cdot 10^3$	—	—	
11		0.33	75	225	300	$3.0 \cdot 10^2$	$6.6 \cdot 10^2$	—	—	
12		3	225	75	300	$1.1 \cdot 10^4$	$7.1 \cdot 10^4$	7/5000 (0.14%)	1/5000 (0.02%)	
13	$30 \cdot 10^6$	1	150	150	300	$1.5 \cdot 10^3$	$4.8 \cdot 10^3$	—	—	
14		0.33	75	225	300	$3.5 \cdot 10^2$	$7.7 \cdot 10^2$	—	—	
15	—	-	225	—	225	$4.0 \cdot 10^3$	—	0/1200 (<.08%)	0/1200 (<.08%)	
16	—	-	150	—	150	$1.8 \cdot 10^3$	—	—	—	
17	—	-	75	—	75	$8.1 \cdot 10^2$	—	—	—	

[a] Comparison of packaging efficiencies with cosmids pJC75-58 (7.65 Md) and pJC720 (16 Md) as a function of the average molecular weight of the foreign DNA added to the ligation mix and as a function of the ratio of vector to foreign DNA. pJC75-58 and pJC720 were cut to completion with *Bgl*II and *Hin*dIII, respectively. The foreign DNA was cut partially with the appropriate enzyme to the average molecular weight indicated in the second column, as estimated by 0.4% agarose gel electrophoresis. The ratio of vector to foreign (*E. coli*) DNA is given in the second column. As indicated in columns 4–6, the DNA concentration during ligation was kept constant at 300 μg/ml, except for the controls where vector DNA was ligated alone. Two-microliter aliquots were packaged and adsorbed to HB101. The number of cosmid-containing colonies, ampicillin-resistant (pJC75-58, 30°) or 30 μg/ml rifampicin-resistant (pJC720, 37°), is indicated in column 7, and the number of clones obtained per microgram of foreign DNA is shown in column 8. The last two columns indicate the frequency of *pro*⁺ and *leu*⁺ prototrophs among the cosmid "transductants," as tested by replica-plating the indicated number of antibiotic-resistant colonies on minimal medium. Analysis of plasmids carried by some of these hybrids indicated that the *pro*⁺ allele was carried on a 16-Md *Bgl*II fragment and the *leu*⁺ allele on a 18.5-Md *Bgl*II fragment, and that the *leu*⁺ allele had at least two *Hin*dIII sites in or close to the gene.[1] Analysis of the size of the plasmid in small, cleared lysates of isolated colonies (about 30 from each experiment) indicated that in experiment 1, 90%; experiment 2, 80%; experiment 4, 80%; experiment 5, 70%; experiment 9, 80%; and experiment 12, 70% of the plasmids were larger than the vector plasmids (average molecular weight about 26 Md). Supercoiled DNA from 5000 colonies from experiments 1 and 4 were further analyzed, as shown in Fig. 5.

min, cooled slowly at room temperature, and placed on ice for 30 min. Dithiothreitol (10 m*M*), bovine serum albumin (BSA, 100 μg/ml, filter-sterilized), 100 μ*M* ATP, and 0.2 unit of T4 DNA ligase (Boehringer) were added per 5 μg DNA. Ligation was continued for at least 8 hr at 8°. The DNA can be left in this state for several days and then frozen at −20° for several months without affecting the yields of clones subsequent to packaging.

Packaging. The *in vitro* packaging mix used was essentially that developed by Hohn,[5] in which uv irradiation of the packaging mix and addition of DNase-containing mix to the packaged mix after the initial packaging step were omitted. The packaging mix contained heat-induced *E. coli* N205 (λ *imm*434 c*Its b*2 *red*3 *Eam*4 *Sam*7) and N205 (λ *imm*434 *c*I75*ts b2 red3 Dam15Sam7*) in the following buffer: 40 m*M* Tris-HCl, pH 8.0, 10 m*M* spermidine hydrochloride, 10 m*M* putrescine hydrochloride, 0.1% mercaptoethanol, and 7% dimethyl sulfoxide. This mix had been distributed in 20-μl portions in 1.5-ml capped plastic centrifuge tubes (Eppendorf), frozen in liquid nitrogen, and stored up to 2 months at −65°. Just before use the mix was transported in liquid nitrogen to the bench and placed in ice for about 3 min; 1 μl 38 m*M* ATP was then added to the still frozen mix. A few seconds later the ligated DNA sample (1–15 μl, usually 5 μl) was added and mixed during thawing, which took place immediately. The amount of DNA added per 20 μl packaging mix was usually 1 μg, but increasing this to 4.5 μg still gave approximately the same hybrid yield per microgram of DNA. After incubation of the packaging mix at 37° (or 25°) for 30 min, DNase was added (10 μg/ml) and $MgCl_2$ (10 m*M*). When the thick pellet was again liquid (2–10 min at 37°), 0.5 ml of phage dilution buffer[5] (40 m*M* Na_2HPO_4, 20 m*M* KH_2PO_4, 80 m*M* NaCl, 20 m*M* NH_4Cl, 0.1 m*M* $CaCl_2$, 10 m*M* $MgCl_2$, 1 m*M* $MgSO_4$) and a drop of chloroform were added. After a 2-min centrifugation at 5000 *g* the supernatant was removed and used as a bacteriophage suspension for transduction of *E. coli* HB101, which had been grown to late exponential phase (A_{550} = 1.0) in L Broth containing 0.5% maltose.[5] Plating and selection were carried out as described above.

After selection for ampicillin resistance (pJC75-58) or rifampicin resistance (pJC720) colonies were replica-plated onto minimal medium lacking either proline or leucine, so as to test for complementation of these auxotrophies in HB101 by the cloned DNA.

Results

The Production of Escherichia coli Gene Banks with Cosmids pJC720 and pJC75-38[3,4]

Cosmid Transduction Frequency without Added Foreign DNA. It can be seen from the transduction frequencies obtained in experiments 7, 8,

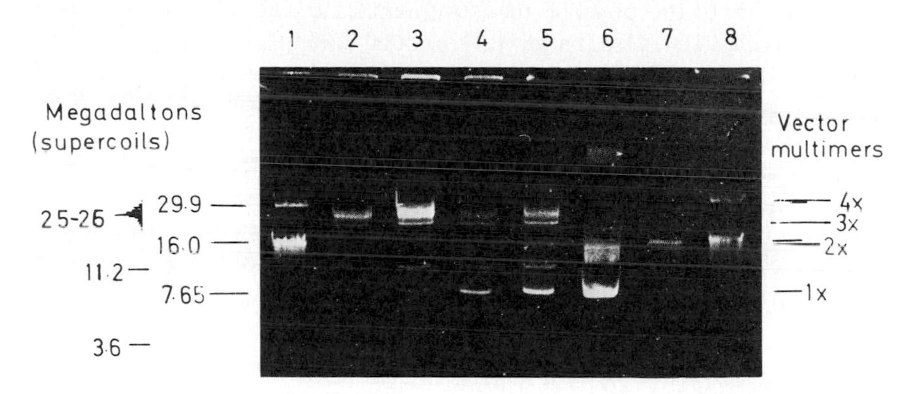

FIG. 5. Electrophoresis with 0.4% agarose gel of supercoiled DNA from gene banks produced in experiments 1 and 4 (Table III). Marker supercoil DNAs are at positions 1 and 8, pCOS10 (29.9 Md), upper band, position 6, pJC75-58 (7.65 Md), and position 7, pJC720 (16.0 Md). Positions 2 and 3 contain 100 and 400 ng, respectively, of supercoiled DNA isolated from 5000 antibiotic-resistant colonies from experiment 1 (Table III). Positions 4 and 5 contain 100 and 400 ng, respectively, of DNA from 5000 antibiotic-resistant colonies from Experiment 4 (Table III). The very faint bands at 3.6 and 11.2 Md are due to a small plasmid (and its hybrid with the vector DNA) which was present in the original foreign *E. coli* DNA cloned. Reproduced by permission of Elsevier North Holland Biomedical Press (Collins and Brüning[4]).

and 15–17 (Table III) that pJC720 gives a higher background of (nonhybrid) colonies than pJC75-58 when religated in the absence of added high-molecular-weight DNA. Other vectors, in the size range 10–12 Md and lacking the temperature-sensitive phenotype, give even higher backgrounds, sometimes reaching 10^5 colonies per microgram of vector DNA.

Effect of Added High-Molecular-Weight Foreign DNA. In experiments 1, 2, 4, 5, 9, and 12 (Table III) a clear increase in the number of transductant colonies was produced by the addition of foreign DNA of the indicated molecular weight. As discussed below, this is due to the formation of large hybrid plasmids which now make up the majority of the transductant population.

Effect of the Vector/Foreign DNA Ratio. The ratio of vector to foreign DNA is seen to have a marked effect in that the higher the ratio the higher the production of hybrids per input foreign DNA. This conforms with the hypothesis that the formation of sandwich-type molecules (a in Fig. 1) are required for packaging and the subsequent efficient production of hybrid DNA clones.

The Frequency of Hybrids among the Transduced Clones. The proportion of clones carrying large DNA fragments can be estimated either by physical measurement (Fig. 5) or by a genetic test in which one determines what percentage of the total population carries a particular gene.[34]

[34] L. Clarke and J. Carbon, *Proc. Natl. Acad. Sci. U.S.A.* **72,** 4361 (1975).

According to the physical measurements, i.e., gel electrophoresis of the supercoiled DNA from a mixture of 5000 colonies from experiments 1 and 4, the majority of the plasmids are found in the size range 25–26 Md, although in experiment 4 at least half of the DNA appears to be of the same size as the vector DNA, with small numbers of dimers and somewhat more in the trimeric form. In experiment 1, the average size of foreign DNA per hybrid is estimated to be about 26 − 7.6 = 18.4 Md.

According to Clarke and Carbon[34] the number of hybrid clones needed to give a certain probability that a particular gene is contained among the hybrids (gene bank) is

$$N = \frac{\ln (1 - P)}{\ln \{1 - [(L - X)/M]\}}$$

where N is the number of clones, P is the probability of finding the required clone, L is the average size of the fragments cloned, X is the size of the fragment (gene) screened for, and M is the size of the genome of the organism from which the cloned DNA was obtained. A group of clones having N high enough to give a P of 0.95 can be termed a gene bank, since it should contain clones carrying DNA representative of the whole donor genome. Considering $X = 0$, $M = 2.5 \times 10^9$, and $L = 18 \times 10^6$, $N = 415$; i.e., in this calculation one would expect at least 0.24% of the hybrid clones to carry any particular gene. When a large number of clones are screened, the number of specific hybrids per total clone number should approach L/M[15], i.e., $18 \times 10^6/2.5 \times 10^9 = 0.72\%$.

This criterion for a gene bank appears from the results in experiment 1 (Table III) to have been reached in this example; i.e., the genetic data support the physical data that on average the clones contain 18×10^6 daltons of foreign DNA apiece.

Preferential Exclusion of Certain Fragments. It is also apparent from these genetic analyses that another effect is being demonstrated, namely, the preferential exclusion of some fragments, depending on the restriction enzyme use for cloning, the size of the vector, or the extent of digestion of the foreign DNA. For example, in the *Hin*dIII cloning experiments the *leu* gene is almost excluded, and using partially cut *Bgl*II DNA leads to a selective reduction in the frequency of pro^+ hybrids. Analysis of the distribution of restriction endonuclease sites in the region of the *pro* and *leu* genes leads to the interpretation that large restriction fragments (>16 Md) will be eliminated from partial digests and that genes cut frequently will be absent from more complete digests and also reduced in frequency in partial digests if the neighboring fragments are of high molecular weight.[4]

Comments

The optimum efficiency of cloning is very high, e.g., in the experiment described 4×10^5 hybrid clones per microgram of foreign DNA, but efficiencies of 6×10^5 have been obtained. Thus in a single 20-μl packaging experiment approximately 10^6 hybrid clones are attainable, where the average size of the DNA insert is about 18 Md. This is considerably in excess of the number required for a gene bank from higher eukaryotes.

As demonstrated by the example above, the optimum yield can only be obtained by using the correct vector/foreign DNA ratio and with vector–foreign DNA fragments in the correct size range. The high DNA concentration for DNA ligation is also extremely important.

Complete cutting of the vector DNA is not important, as the contribution by uncut vector supercoils is insignificant.[4] One can therefore use the shortest incubation conditions possible so as to protect the cohesive ends formed from attack by exonucleases which may be contaminating the restriction enzyme used.

The use of alkaline phosphatase treatment[35] (to prevent ring closure of the vector DNA and cause the formation of hybrids with untreated DNA) is unnecessary in view of the size requirement for packaging and transduction, which essentially ensures hybrid formation. A useful application of the alkaline phosphatase treatment seems to be removal of the 5′-phosphates from the DNA fragments to be cloned. This ensures the formation of hybrids in which the cloned DNA is derived only from DNA fragments that were contiguous on the original genome. This should help remove much of the ambiguity from fine-structure mapping involving cloning of DNA adjacent to previously cloned fragments.[36]

Mention should be made of the results from earlier experiments[3] in which cosmid hybrids were constructed containing the same fragment repeated a number of times. Apart from interest in this finding as a method of gene amplification, the main observation was that the repeat units were stable and that *no* derivatives were found containing fragments in palindromic orientation (i.e., with a twofold axis of inverted symmetry). In addition, it was found that in these experiments, designed to produce repeats, nearly 10% of all hybrids carried small deletions. This points to the possibly highly unstable nature of palindromic structures (or inverted repeats?) in *E. coli,* an effect that could present serious difficulties in the cloning of large DNA fragments from higher eukaryotes.

The combinations of restriction enzymes available for cloning in con-

[35] A. Ullrich, J. Shine, J. Chirgwin, R. Pietet, E. Tischer, W. J. Rulter, and H. M. Goodman, *Science* **196,** 1313 (1977).

[36] A. Royal, A. Garapin, B. Cami, F. Perrin, J. L. Mandel, M. LeMeur, F. Brégégère, F. Gannon, J. P. LePennec, P. Chambon, and P. Kouilsky. *Nature* **279,** 125 (1979).

junction with cosmids are shown in Table I, and a list of cosmids containing single sites for these enzymes is presented in Table I. As demonstrated in the experiment below, the use of restriction enzymes with six-base-pair recognition sites can lead to the exclusion of very large fragments or very small fragments, because of the random distribution of cutting sites and the size limitations imposed by the packaging system. The small fragments may be cloneable by using partial digests but, if they are adjacent to very large fragments, they will still be eliminated. A simple way to overcome this problem in the production of complete gene banks is the use of partial digests with enzymes cutting more frequently (i.e., with four-base recognition sites or degenerate six-base sites), such as *Sau*I (for *Bam*HI or *Bgl*II vectors), *Ava*I (for *Sal*I vectors), and *Taq*I and *Hpa*II (for *Cla*I vectors). This procedure should give rise to a very random collection of fragments and to the cloning of regions representative of the whole genome, in a manner similar to the shearing and tailing procedures used in conjunction with other cloning techniques.[34] This method has been used to clone a gene of biotechnological interest from a *Klebsiella* strain, in pJC75-58, using *Sau*I partial digestion of the foreign DNA and *Bgl*II digestion of the vector. This method was successful after failures with *Hin*dIII and *Eco*RI partial and complete digests (J. Collins and H. Mayer, unpublished results).

Last, I should like to point out that the procedure for producing the packaging mix was developed entirely by Hohn.[5]

Acknowledgments

This report on attempts to optimize some of the conditions for cosmid packaging is based on original experiments conducted during a close collaboration between Barbara Hohn and myself.[3] I wish to thank her for sharing with me the belief that such a system could be developed and for her collaboration in bringing that belief to a practical and fruitful conclusion, in addition to her help in critically reading the manuscript.

[21] Transformation and Preservation of Competent Bacterial Cells by Freezing

By D. A. Morrison

The preparation of microbial cultures competent for transformation by added DNA, whether using one of the naturally transformable species or an artificial treatment to render cells permeable to DNA, is typically a multistep procedure occupying at least the greater part of one working day. In many applications of transformation where the actual transformation step is done infrequently, and the properties of the transformed cell

ISBN 0-12-181968-X

lines themselves are the principal objects of study, preparation of competent cells for each experiment presents little difficulty. However, in work where transformation is performed daily, as in studies of uptake or recombination mechanisms or where transformation is used as an analytical tool for the bioassay of DNA samples, a ready supply of competent cells of reproducible and known properties can be of great convenience, with the daily repetitive work reduced from hours to minutes. The preservation of competent or precompetent cultures by freezing has long been exploited in studies using *Bacillus subtilis*,[1] *Streptococcus pneumoniae*,[2] and *Haemophilus influenzae*.[3] This chapter presents and discusses a method that has given reasonable success, in this and other laboratories, for *Escherichia coli* cells rendered susceptible to DNA by $CaCl_2$ treatment, a procedure adapted slightly from that described by Lederberg and Cohen[4] with the addition of glycerol to allow freezing of the competent cells.

Preparation and Freezing of Competent Cells

Materials

1. 1.0 *M* $CaCl_2$
2. 1.0 *M* $MgCl_2$
3. Glycerol, reagent grade
4. Distilled water
5. L broth: 10 g Bacto-tryptone, 5 g Bacto yeast extract, 5 g NaCl, and 1 g glucose per liter, adjusted to pH 7.0 with 1 *N* NaOH[5]

(All stock solutions are sterilized in the autoclave.)

Procedure. L broth is inoculated with RR1 cells, usually from a stock of titer 10^9/ml stored at $-82°$ in broth supplemented with 10% glycerol, at an initial density of about 10^7/ml. Two 500-ml portions are incubated in 2-liter Erlenmeyer flasks with aeration at 37°. When the culture reaches $OD_{550} = 0.5$ (about 5×10^8/ml), it is combined in a 4-liter flask and swirled vigorously in a salt–ice water bath for 3 min to bring the temperature to 0°–5°. The culture is divided among five centrifuge bottles and sedimented in a Sorvall GSA rotor in the cold at 8000 rpm for 8 min. The pellets are gently resuspended, by repeated flushing with a 10-ml pipette, in a total of 250 ml of ice-cold 0.1 *M* $MgCl_2$ and assembled in one bottle. With gentle agitation this is accomplished in 5–10 min. The cells are sedimented as before. The pellet is resuspended gently in 250 ml cold 0.1 *M* $CaCl_2$, kept at 0° for 20 min, and then sedimented in the cold as before.

[1] D. Dubnau and R. Davidoff-Abelson, *J. Mol. Biol.* **56,** 209 (1971).
[2] M. S. Fox and M. K. Allen, *Proc. Natl. Acad. Sci. U.S.A.* **52,** 412 (1964).
[3] L. Nickel and S. H. Goodgal, *J. Bacteriol.* **88,** 1538 (1964).
[4] E. M. Lederberg and S. N. Cohen, *J. Bacteriol.* **119,** 1072 (1974).
[5] E. S. Lennox, *Virology* **1,** 190 (1955).

Finally, the pellet of $CaCl_2$-treated cells is resuspended in 43 ml of 0.1 *M* $CaCl_2$ mixed with 7 ml of glycerol. Resuspension of pellets is facilitated by repeated flushings with a 10-ml pipette with a wide tip.

The suspension of competent cells is distributed into cold screw-cap tubes in volumes (0.1–10 ml) determined by the projected application. This is most conveniently carried out in a cold room with a chilled pipette or sterile repeating syringe dispenser. The entire set of capped tubes is drained of water and ice briefly and placed in a bath of liquid nitrogen, or of acetone or alcohol chilled with pieces of solid CO_2. After 5 min, they are removed to a freezer for storage at −82°.

Transformation

For use, a tube of frozen cells is allowed to thaw in an ice-water bath for 10 min. Cells are then distributed, in the cold, into tubes containing DNA samples in small volumes of buffer or 0.1 *M* $CaCl_2$, kept at 0° for 30 min more, placed in a 42° bath for 2 min, and diluted into 50 volumes of broth at 37° to reduce the calcium concentration and allow resumption of growth.

Comments

1. Tubes of a batch of competent cells prepared by this method have been used without loss of transformability after 15 months at −82°.
2. When thawed and maintained on ice, cells lose transformability slowly, retaining 50% of initial activity after 3 hr.
3. The concentration of glycerol does not appear to be critical; 10 or 20% solutions give similar results.
4. Although the freezing and thawing step itself does not reduce the viability of treated cells, it often renders them more sensitive to the $CaCl_2$ and following treatments of the transformation procedure, resulting in a lower final survival. This is partially compensated for by increasing the number of cells used.
5. The $CaCl_2$ treatment procedure described by S. Kushner[6] may also be adapted for cryogenic preservation. Incorporation of 15% glycerol in all buffers, as well as in the dilution broth, preservation of washed cells by freezing in the first washing buffer (0.01 *M* morpholinopropane sulfonic acid, 0.01 *M* RbCl) before calcium treatment, and increasing the number of cells used in the final reaction volume by a factor of 10 give reproducible results similar to those described here.

[6] S. R. Kushner, *in* "Genetic Engineering" (H. W. Boyer, ed.) p. 17. North-Holland Publ., Amsterdam, 1978.

Assay

In the case of transformation involving a drug resistance marker, the products of transformation may be measured by a pour plate procedure taking advantage of diffusion-limited exposure of cells to the selective drug. The plates are designed with a drug-free zone separating the cell-containing layer from the drug agar layer, and the drug concentration is adjusted so that inhibitory levels are not reached in the cell-bearing zone until phenotypic expression is complete.

Materials

1. L broth
2. L broth containing 2% agar, maintained at 52°
3. 13 × 100 mm slip-cap tubes maintained at 49° in a heating block and containing 1.5 ml each of L agar
4. 15 × 60 mm petri dishes

Procedure. Cells are diluted appropriately in L broth, and a volume to be plated is added to L broth in a 13 × 100 mm tube to a final volume of 1.5 ml at room temperature. A 3-ml base layer of L agar is allowed to harden in each petri dish. The 1.5-ml cell sample is poured into a tube of molten agar, mixed, and poured onto the base layer. When this has hardened (about 1–2 min), a third 3-ml layer of L agar is pipetted over the cell-containing layer. Finally, for selection of drug-resistant transformants, an additional (fourth) 3-ml layer of L agar containing the drug is applied and allowed to harden. After overnight incubation at 37°, colonies appear as small spheres or disks, about $\frac{1}{2}$–2 mm in diameter, depending on the time of incubation and the total colony density on the plate. They may be counted conveniently under a dissecting microscope fitted with a square-ruled eyepiece micrometer disk.

Comments

1. The amount of drug added in the final agar layer must be chosen with care for each genetic marker and each recipient strain. The appropriate range may be determined by plating a set of identical samples of a transformed culture, within a few minutes after initial dilution into broth, on a series of plates, and incorporating a series of drug concentrations in the final agar layers. For the tetracycline resistance gene of pMB9 in RR1 recipient cells, for example, it is observed that drug levels from 15 to 60 μg/ml are satisfactory. Below 15 μg/ml, transformants are difficult or impossible to detect amid a background of small drug-sensitive colonies. Above 60 μg/ml even transformants fail to form colonies, because of drug effects on expression and/or growth.

2. The events occurring after transformation of a thawed frozen com-

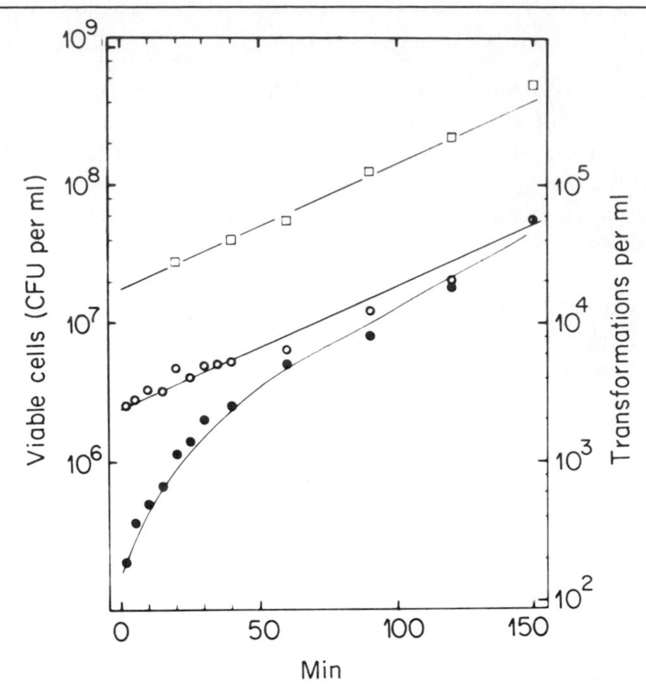

FIG. 1. Segregation and phenotypic expression of tetracycline resistance in an RR1 culture transformed with pMB9 DNA after preservation at −82°. Samples of the transformed culture taken at the indicated times after dilution into L broth were plated without drug (□), by the diffusion-limited double-overlay method (○), and by direct embedding in drug agar (●) (Morrison,[7] reproduced with permission).

petent culture of strain RR1 and dilution into broth are illustrated in Fig. 1.[7] The viable cells resume growth and division very quickly. The number of plasmid-bearing cells also begins to increase promptly, in parallel with the total viable population. There is no segregation delay, often observed in other transformation systems. This result, obtained with the diffusion-limited, double-overlay plating method described here, is compared with determinations of the number of drug-resistant cells made by plating samples directly in drug agar. It is clear that plasmid-bearing cells do not immediately express the resistance phenotype; about a 1-hr incubation in broth is required before all transformants can survive drug exposure.

3. The net result of the process described in the preceding paragraph is that, by the time of full phenotypic expression of the transformed gene, each initial transformant is represented by several descendants. The immediate double-overlay plating method allows full phenotypic expres-

[7] D. Morrison, *J. Bacteriol.* **132,** 349 (1977).

sion and also ensures that all the cells descended from a single transformant are confined to a single colony; each colony thus represents an independent transformation event if plated soon after the transformation heat pulse.

4. The plating method described combines a number of features of value for routine or high-volume work. With a standard 7×–30× dissecting microscope, the number of colonies that may be quantitated on a single small petri dish covers a very wide range, up to 50,000 with calibration of the eyepiece grid at several magnifications. Samples may be assayed immediately after transformation, as the drug plate design allows phenotypic expression after plating. The amount of agar used per assay is small. For assay of many samples, all broth and agar volumes may be dispensed into tubes or onto plates with a sterile repeating syringe, such as the Cornwall Pipetter. Selective plates need not be prepared in advance. Finally, cells from individual colonies are easily retrieved with a toothpick or Pasteur pipette.

[22] Bacterial Transformation Using Temperature-Sensitive Mutants Deficient in Peptidoglycan Synthesis

By M. Suzuki and A. A. Szalay

Bacterial transfection and transformation are essential processes in molecular cloning experiments. The frequency of transformation is determined by the uptake of DNA molecules and by the restriction system of the recipient cell. The restriction systems of different strains of *Escherichia coli* have been studied in great detail; however, very little is known about the mechanism of DNA uptake through the bacterial cell wall. Several bacterial species have developed the ability to transport DNA into the cell (Table I). In *Bacillus subtilis* and *Diplococcus pneumoniae* the uptake of DNA is nonspecific, while DNA uptake in *Haemophilus influenzae* has been demonstrated to be highly specific.[1,2,3] Recently, Sisco and Smith isolated DNA sequences from *H. influenzae* which carry the recognition signal for DNA uptake.[4] Most bacteria have not developed the ability to take up DNA ("natural" competence). To obtain competence in these bacterial species, two procedures have evolved. Mandel

[1] A Soltyk, D. Shugar, and M. Piechowski, *J. Bacteriol.* **124,** 1429 (1975).
[2] L. S. Lerman and L. J. Tolmach, *Biochem. Biophys. Acta* **26,** 68 (1957).
[3] J. J. Socca, R. L. Poland, and K. C. Zoon, *J. Bacteriol.* **118,** 369 (1974).
[4] K. L. Sisco and H. O. Smith, *Proc. Natl. Acad. Sci. U.S.A.* **76,** 972 (1979).

ISBN 0-12-181968-X

TABLE I
BACTERIAL SYSTEMS AVAILABLE FOR TRANSFECTION AND TRANSFORMATION

Host cell	Bacterial species	DNA	Transfection	Transformation	Reference
Physiologically competent cell	*Bacillus subtilis*				
	Hemophilus influenzae				
	Diplococcus pneumoniae	Phage, plasmid, bacterial	+	+	1,2
	Micrococcus luteus				
	Neicerria ghonorrhea				
	Agrobacterium tumefaciens				
Divalent cation-treated cell	*Escherichia coli*				
	Salmonella typhimurium				
	Aerobacter aerogenes	Phage, plasmid, bacterial	+	+	3,4
	Staphylococcus aureus				
	Klebsiella pneumoniae				
Frozen-thawed	*E. coli*				
	Agrobacterium tumefaciens	Phage	+	–	5,6,7,8
	Rhizobium species				
Plasmolyzed cell	*E. coli*	Phage	+	–	9,10
Phage-sensitive cell	*E. coli* with λ DNA and the helper phage	Phage	+	–	11
Spheroplast					
Osmotic shock spheroplast	*E. coli*	Phage	+	–	12
	Streptomyces kanamyceticus				
Glycine-spheroplast	*E. coli*	Phage	+	–	13
Penicillin-spheroplast	*E. coli*				
	H. influenzae	Phage	+	–	14
	H. parainfluenzae				
	E. coli with urea-treated T even phage (*S. typhimurium*. *A. aerogenes*, *Proteus vulgaris*, *Serratia marcescens*)	Phage	+	–	15,16
Lysozyme-spheroplast	*E. coli*				
	S. typhimurium	Phage	+	–	17
	A. aerogenes				

	K. pneumoniae	Phage	+	–	17
	S. marcescens				
	Proteus vulgaris				
	P. mirabilis				
	Pseudomonas aeruginosa				
Revertible spheroplast					
tsPG⁻ spheroplast	*E. coli*				
	S. typhimurium				
	A. aerogenes				
	K. pneumoniae	Phage, plasmid, bacterial	+	+	18, 19
	Providencia stuartii				
	H. parainfluenzae				
	M. luteus				

[a] Key to references:

1. N. K. Notani and J. K. Setlow, *in* "Progress in Nucleic Acid Research and Molecular Biology" Vol. 14, p. 39. Academic Press, New York, 1974.
2. J. Spizizen, B. E. Reilly, and A. H. Evans, *Annu. Rev. Microbiol.* **20,** 371 (1966).
3. A. Taketo, *Z. Naturforsch.* **30b,** 520 (1975).
4. M. Mandel and A. Higa, *J. Mol. Biol.* **53,** 159 (1970).
5. S. Y. Dityatkin, K. V. Lisovaskaya, N. N. Panzhava, and B. H. Iliashenko, *Biochim. Biophys. Acta* **281,** 319 (1972).
6. S. Hua, R. P. Mackal, B. Werninghaus, and E. A. Evans, Jr., *Virology* **46,** 192 (1971).
7. R. P. Mackal, B. Werninghaus, E. A. Evans, Jr., *Proc. Natl. Acad. Sci. U.S.A.* **51,** 1172 (1964).
8. B. I. Weinstein, R. P. Mackal, B. Werninghaus, and E. A. Evans, Jr., *Proc. Natl. Acad. Sci. U.S.A.* **62,** 420 (1969).
9. R. Benziner, H. Delius, R. Jaenisch, and P. H. Hofschneider, *Eur. J. Biochem.* **2,** 414 (1967).
10. A. Taketo and S. Kuno, *J. Biochem.* **65,** 369 (1969).
11. A. D. Kaiser and D. S. Hogness, *J. Mol. Biol.* **2,** 393 (1960).
12. A. Taketo and S. Kuno, *J. Biochem.* **65,** 361 (1969).
13. A. Taketo, *J. Biochem.* **71,** 507 (1972).
14. G. Veldhuisen and E. B. Goldberg, this Vol. 12B, p. 858.
16. D. Fraser, H. Mahler, A. Shug, and C. A. Thomas, Jr., *Proc. Natl. Acad. Sci. U.S.A.* **43,** 939 (1957).
17. G. D. Guthrie and R. L. Sinsheimer, *J. Mol. Biol.* **2,** 297 (1960).
18. M. Suzuki, manuscript in preparation.
19. M. Suzuki and R. L. Sinsheimer, manuscript in preparation.

and Higa[5] and Oishi and Cosloy[6] demonstrated transformation in $CaCl_2$-treated *E. coli* K12 cells. Later Cohen *et al.*[7] described a high frequency of transformation of calcinated *E. coli* with covalently closed circular DNA. Recently, improved methods of transformation using $CaCl_2$-treated *E. coli* cells have been described by Kushner[8] and D. Hanahan (personal communication). Using the improved method of Hanahan, transformation efficiencies up to 4×10^8 transformants/μg of pBR322 plasmid DNA have been obtained. The major disadvantage of the calcinated cell system is that many bacterial species are insensitive to calcium treatment. In addition to the calcinated cell system, a variety of spheroplast systems have been used in transfection experiments. Spheroplasts obtained by treatment with lysozyme give very high transfection frequencies up to 5×10^9 infective centers/μg of ϕx174 RFI DNA, approximately 500 times higher than that obtained with calcinated cells. However, bacterial spheroplasts cannot be converted to viable cells, and therefore, are unsuitable for transformation with plasmid DNA.

In this paper we describe a novel bacterial spheroplast transformation system which is capable of (a) high transfection frequencies, (b) high frequency of regenerating viable cells, and (c) applicable for all bacterial species containing peptidoglycan.

Mutagenesis

Conditions used to obtain temperature-sensitive peptidoglycan deficient ($tsPG^-$) mutants of *E. coli* are summarized in Table II. A comparison of six different mutagens indicated that N-methyl-N′-nitro-N-nitrosoguanidine NTG provided the highest percentage of temperature-sensitive mutants. Therefore, this mutagen was selected to generate $tsPG^-$ mutants.

Isolation of Temperature-Sensitive Cell Wall Mutants

E. coli cells grown in 10.0 ml of nutrient broth to a cell density of 2×10^8 cells/ml were harvested by centrifugation at 5000 rpm for 10 min at 4° in a Sorvall SS-34 rotor. The pellet was resuspended in 5.0 ml of 0.1 *M* potassium phosphate buffer, pH 6.0. After repeated centrifugation,

[5] M. Mandel and A. Higa, *J. Mol. Biol* **53,** 159 (1970).

[6] S. D. Cosloy and M. Oishi, *Proc. Natl. Acad. Sci. U.S.A.* **70,** 84 (1973).

[7] S. N. Cohen, A. C. Y. Chang, and L. Hsu, *Proc. Natl. Acad. Sci. U.S.A.* **69,** 2110 (1972).

[8] S. R. Kushner, Genetic Engineering, Proc. of the International Symposium on Genetic Engineering. Italy, 29–31 March (1978), p. 17. H. W. Boyer and S. Nicosia Elsevier, 1978.

TABLE II
TEMPERATURE-SENSITIVE MUTANTS OBTAINED WITH DIFFERENT MUTAGENS

Mutagen	Conditions	% of viable cells	% of ts cells
Hydroxylamine	0.8 *M* (pH 7.0), 6 hr, 37°	42.5	0.68
Ethylmethane sulfonate	0.18 *M* (pH 7.4), 2 hr, 37°	29.7	1.02
2-Aminopurine	7.5 m*M* (pH 7.2), 2 hr, 37°	38.6	0.90
Ultraviolet light	7.2 erg/mm²/sec, (pH 7.0), 0°	40.6	0.99
Sodium nitrite	1 *M* (pH 6.0), 4 min, 25°	31.2	1.07
NTG[a]	1.4 m*M* (pH 6.0), 30 min, 37°	22.0	1.30

[a] NTG, *N*-methyl-*N'*-nitro-*N*-nitrosoguanidine.

the cells were resuspended in 4.0 ml phosphate buffer and were divided into 2.0 ml samples. Nitrosoguanidine (NTG) was added to a final concentration of 200 μg/ml to both samples and the cultures were incubated at 37° for 30 min. After incubation, the cells were centrifuged for 10 min and the pellet was washed with 2 volumes of potassium phosphate buffer. The mutagen was removed by repeated washing, and the pellet was resuspended in 5.0 ml of nutrient broth containing 0.25 *M* sucrose to prevent lysis of the spheroplasts. The cells were incubated for 5 hr at 42° to stimulate spheroplast formation. The intact cell and spheroplast mixture was harvested by centrifugation at 7000 rpm for 5 min at 4° and the pellet was resuspended in 0.5 ml of nutrient broth containing 0.25 *M* sucrose. The cell suspension was incubated for 30 min at 42° and was overlayered on 7.0 ml of nutrient broth containing 0.6 *M* sucrose followed by centrifugation at 2500 rpm for 30 min at 4° (Fig. 1). Using these centrifugation conditions, two distinct bands were obtained. The upper band found in the top 2.0 ml fraction of the gradient contained mostly spheroplasts; the lower band contained predominantly cells with complete cell walls. After removal of the top 2.0 ml fraction with a wide bore pipette, the spheroplast fraction was divided into 0.5-ml fractions and each fraction was diluted fourfold by the addition of nutrient broth containing 0.25 *M* sucrose. All fractions were incubated for 12–18 hr at 30° and diluted with nutrient broth (1:10^6), and fractions (0.1 ml) were plated on nutrient agar plates and incubated at 30° for 24–48 hr. After preparing double replicas of each plate, one plate of each pair was incubated at either 30° or 42°. Usually 70% of the colonies were temperature sensitive and 50–60% of the temperature-sensitive colonies could be maintained at 42° in the presence of 0.25 *M* sucrose.

Screening for colonies competent for DNA transfection was carried out as follows. Single colonies were picked and inoculated separately in 2.0 ml of nutrient broth and grown overnight at 30° with gentle shaking.

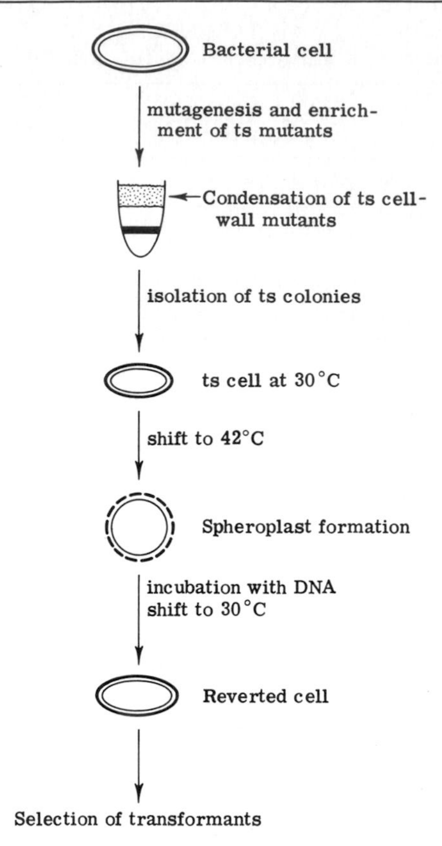

FIG. 1. Isolation and selection of temperature-sensitive spheroplasts.

Two ml of PAM (11) (see summarized laboratory procedure) containing 0.5 *M* sucrose were added to the culture and incubated at 42° for an additional 5 hr to allow spheroplast formation. Aliquots (0.25 ml) of the cell suspension were incubated with an equal volume of ϕx174 ss DNA containing 1×10^8 molecules/ml in 50 m*M* Tris HCL, pH 8.1, for 2.5 min at 42° or 7.5 min at 30°. After incubation the cells were plated on an indicator lawn to score infective centers. Approximately 500–700 mutant colonies must be screened to isolate 2–5 colonies with high transformation ability (1×10^7–1×10^9 transformants/μg ϕx174 DNA).

Reversible Spheroplast Formation in *E. coli* is Caused by a Temperature-Sensitive Mutation in Peptidoglycan Biosynthesis

As described earlier, mutagenized cells of *E. coli* maintained at 42° in the presence of 0.25 *M* sucrose could be separated by centrifugation into two distinct bands. Cells isolated from the upper band were able to survive at 42° only in the presence of 0.25 *M* sucrose. Both of these findings indicate that at the nonpermissive temperature the cell wall of the mutant cell has been altered. To determine the nature of this alteration we followed the incorporation of [^{14}C]L-alanine into *N*-acetylmuramic acid, a component of *E. coli* peptidoglycan, in temperature-sensitive mutants at both permissive (30°) and nonpermissive (42°) temperatures.[9] MM1013, a temperature-sensitive derivative of *E. coli* K12 W3350 which yields approximately 1×10^9 infective centers/μg ϕx174 DNA was chosen as a model to study peptidoglycan biosynthesis.

MM1013 cells were grown in 10 ml of nutrient broth to a cell density of 2×10^8 cells/ml. The cells were harvested by centrifugation and resuspended in 50 m*M* Tris HCl, pH 7.5, containing 8 m*M* $MgCl_2$, 10 m*M* glucose, 200 μg/ml chloramphenicol and 2.5 μCi/ml [^{14}C]L-alanine. One-half of the cell suspension was incubated at 30° and the other half was incubated at 42° for 1 hr with gentle shaking. The incubation mixture was chilled on ice and the cells were harvested by centrifugation and washed 2 times to remove unincorporated radioactive material. The soluble cellular proteins were released by boiling the cells for 3 min in 10 m*M* Tris HCl, pH 7.5 and the lysate was centrifuged at 5000 rpm for 5 min to pellet the cell wall fraction. The pellet was washed twice with 50 m*M* Tris HCl pH 7.5 containing 10 m*M* $MgCl_2$ and resuspended in 2.0 ml of the same buffer. The incorporation of [^{14}C]L-alanine into *N*-acetylmuramic acid was demonstrated by successive digestion of the cell-wall fraction with trypsin and lysozyme followed by paper chromatography. Cleavage sites for trypsin and lysozyme in the peptidoglycan structure of *E. coli* are illustrated in Fig. 2. Aliquots (200 μl) of the cell-wall fraction were digested with trypsin (500 μg/ml) at 37° for 1 hr. The reaction was terminated by heating the samples for 3.0 min at 95°. One-half of the trypsin-digested material was treated with lysozyme (100 μg/ml) at 37° for 18 hr. Cell-wall fractions treated with trypsin and trypsin followed by lysozyme were chromatographed on Whatman No. 1 filter paper in a solvent system containing *n*-butyric acid : 1.0 *M* ammonium hydroxide (1 : 0.6, v/v). The chromatogram was cut into sections and the radioactive spots were located by scintillation counting. Radioactive material obtained from cell-wall fractions digested by trypsin remained at the origin and represent the total amount of radioactivity associated with the cell-wall fraction. After lysozyme treatment, the ^{14}C-labeled *N*-acetylmuramic acid is released and moves away from the origin. Therefore, the amount of radioactivity released from the trypsin–lysozyme treated sample represents the incor-

[9] H. Matsuzawa, M. Matsuhashi, A. Oka, and Y. Sugino, *Biochem. Biophys. Res. Commun.* **36**, 682 (1969).

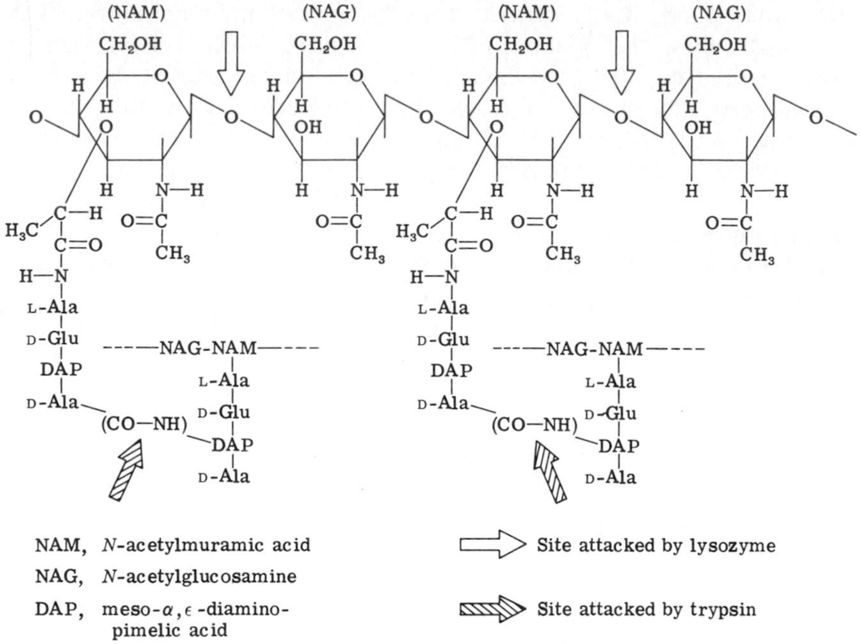

NAM, N-acetylmuramic acid

NAG, N-acetylglucosamine

DAP, meso-α,ε-diamino-pimelic acid

Site attacked by lysozyme

Site attacked by trypsin

FIG. 2. Cleavage of *Escherichia coli* peptidoglycan with lysozyme and trypsin. Lysozyme hydrolyzes the β-1,4-glycosidic bond between N-acetylmuramic acid and N-acetylglucosamine (between C-1 of NAM and C-4 of NAG). The other glycosidic linkage between C-1 of NAG and C-4 of NAM cannot be cleaved. Trypsin cleaves the peptide bond between ε-amino group of meso-α,ε-diaminopimelic acid and carboxyl group of D-alanine.

poration of [^{14}C]L-alanine into peptidoglycan. Mutants shifted to the nonpermissive temperature (42°) incorporate 80% less [^{14}C]L-alanine when compared with cells grown at the permissive temperature (30°). These results suggest that peptidoglycan biosynthesis is inhibited at the higher temperature. Comparison of DNA, RNA, and peptidoglycan synthesis in the parent strains *E. coli* K12W3350, and K12W6 with the mutant strains MM1013 and ST454 are summaried in Table III. From this table it is apparent that the ratios of DNA and RNA synthesis at permissive and nonpermissive temperatures are identical between mutant and parent strains. Significant differences in peptidoglycan biosynthesis are observed between parent and tsPG$^-$ strains at nonpermissive temperatures.

Use of tsPG$^-$ Mutants in Transfection and Transformation Experiments

Transfection of tsPG$^-$ spheroplasts of *E. coli* MM1013 was carried out as described in the screening procedure. To obtain spheroplasts, an overnight stationary culture of *E. coli* MM1013 was diluted with PAM medium

TABLE III
COMPARISON OF DNA, RNA AND PEPTIDOGLYCAN SYNTHESIS AT NONPERMISSIVE AND PERMISSIVE TEMPERATURES

	Ratio of synthesis (42°/30°)		
Strain	DNA	RNA	Peptidoglycan
E. coli K12W3350	1.95	1.90	1.33
E. coli K12W6	1.91	1.89	1.30
E. coli MM1013	1.85	1.70	0.09
E. coli ST454	—	—	0.11

containing 0.25 *M* sucrose and incubated for 5 hr at 42°. Samples (0.2 ml) of the spheroplast preparation were incubated with several phage DNAs. A standard transfection mixture contained 1×10^9 spheroplasts and 1×10^{10} DNA molecules. The spheroplasts were incubated with ϕx174 DNA and S13 DNA for 2.5 min at 42° or 7.5 min at 30°. Lambda DNA and T2 DNA incubations were carried out for 2.5 min at 42° or 17.5 min at 32°. Immediately after incubation, aliquots of the mixture were plated on indicator lawns to obtain infective centers. Table IV summarizes the transfection frequencies resulting from three independent experiments. The number of infection centers obtained from tsPG$^-$ mutants after transfection with various phage DNAs are consistently higher when compared with calcinated *E. coli* cells. Transfection frequencies from tsPG$^-$ mutants are slightly lower than those obtained from lysozyme spheroplasts.[10]

TABLE IV
TRANSFECTION FREQUENCIES[a]

	Recipient cells: *E. coli* K12W3350		
	Spheroplasts		
	Lyzozyme	tsPG$^-$	Calcinated cells
ϕX174 ss DNA	$1-2 \times 10^9$	5.5×10^8	7.5×10^5
ϕX174 RFI DNA	3.0×10^8	2.0×10^8	1.8×10^6
S13 ss DNA	5.5×10^8	1.0×10^8	2.5×10^5
S13 RFI DNA	6.0×10^7	3.0×10^7	7.5×10^5
λ DNA[b]	2.0×10^7	1.0×10^7	2.0×10^5
T2 DNA	1.0×10^5	1.0×10^5	5.0×10^1
T2-urea	1.0×10^3	3.0×10^2	non determ.
MS2 RNA	6.0×10^6	3.0×10^5	3.0×10^4

[a] Per μg DNA.
[b] 50 μg/ml protamine sulfate added.

[10] M. Suzuki and M. Azegami, *Biochim. Biophys. Acta 474,* **646** (1977).

Treatment of tsPG$^-$ spheroplasts with lysozyme or with $CaCl_2$ did not give a substantial increase in transfection frequencies.

In contrast to lysozyme spheroplasts, *E. coli* MM1013 spheroplasts maintained at 42° in 0.25 *M* sucrose can regenerate viable cells when cultured at 32°. We have obtained temperature-sensitive mutants and studied the efficiency of regeneration of viable cells in *E. coli, Agrobacterium tumifaciens, Rhizobium sp., Haemophilus influenzae,* and *Providencia stuartii.* We found that in all cases more than 95% of ts spheroplasts are capable of colony formation at 30° after 24–48 hr incubation.

The observation that these spheroplasts regenerate to viable cells provides the opportunity to use tsPG$^-$ spheroplasts of *E. coli* MM1013 in transformation experiments.

When plasmid DNA or sheared chromosomal DNA was used to transform spheroplasts, longer periods of incubation of the cells with DNA (3–4 hr) were required. We have found that the efficiency of transformation could be substantially increased when tsPG$^-$ spheroplasts were purified on nutrient broth sucrose step gradients (0.5, 1.0, 2.5 *M* sucrose) which removed DNase contamination from the preparation. Approximately 1.0×10^{10} spheroplasts were overlayered on the gradient and centrifuged at 3700 × g for 10 min. The spheroplasts formed a sharp band on the surface of the 2.5 *M* sucrose layer and were easily removed from the gradient with a large bore pipette. After dilution to a cell concentration of 5×10^8 spheroplasts/ml in PAM containing 0.25 *M* sucrose, the spheroplasts were incubated for 15 min at 42°. Approximately 1 μg of plasmid DNA or 5–10 μg of chromosomal DNA were added and the mixture was incubated for 2.5 min at 42°. Following the addition of 4 volumes of PAM containing 0.25 *M* sucrose, the incubation was continued at 30° for 4 hr. Aliquots of the cell suspension were plated on selective medium containing 0.25 *M* sucrose and transformants were scored after 48 hr. Preliminary transformation data (Suzuki, manuscript in preparation) suggest that transformation efficiency in tsPG$^-$ spheroplasts is similar to those of calcium-treated cells.

Summary

Temperature-sensitive mutations in the biosynthesis of bacterial cell-wall components lead to spheroplast formation. Temperature-sensitive mutants of *E. coli* maintained for 5 hr in the presence of 0.25 *M* sucrose at nonpermissive temperatures form spheroplasts which are competent for DNA uptake. Using a variety of bacterophage DNAs we have observed high transfection frequencies in tsPG$^-$ spheroplasts. These transfection frequencies are comparable to those obtained with lysozyme treated spheroplasts and higher than those obtained with calcinated cells.

Spheroplasts of *E. soli* (tsPG⁻) can be used for transformation with plasmid and chromosomal DNA. The unique advantage of the tsPG⁻ spheroplast system is its high efficiency (95–99%) in generating viable cells at 30°. In principle, this method can be applied to select competent cells for transformation from any bacterial species which contain peptidoglycan.

This technique will be especially useful in molecular cloning when spheroplasts isolated from certified *E. coli* strains are available.

Temperature-sensitive spheroplasts of *E. coli* are safe recipients for recombinant DNA since they require 0.25 *M* sucrose for survival and do not form colonies above 35°.

Laboratory Protocol

1. Mutagenesis
   a. Grow *E. coli* cells in 10 ml of nutrient broth (N.B.[11]) to a cell concentration of 2×10^8/ml.
   b. Pellet the cells in a sterile glass centrifuge tube.
   c. Wash the cells in 5 ml of 0.1 *M* potassium phosphate buffer, pH 6.0, and pellet.
   d. Resuspend pellet in 4 ml of 0.1 *M* potassium phosphate buffer, pH 6.0, and divide the 4 ml of washed cells into two 2-ml portions.
   e. Add 0.2 ml of nitrosoguanidine (2 mg/ml) dissolved in 0.1-*M* potassium phosphate buffer, pH 6.0.
   f. Incubate at 37° for 30 min.
   g. Centrifuge 5000 rpm for 10 min.
   h. Wash pellet with 5 ml of potassium phosphate buffer, pH 6.0.
2. Enrichment for ts mutants
   a. Resuspend pellet in 5 ml of nutrient broth containing 0.5 *M* sucrose.
   b. Incubate for 5 hr at 42°.
   c. Centrifuge at 7000 rpm for 5 min at 4°.
   d. Resuspend pellet in 0.5 ml of nutrient broth containing 0.5 *M* sucrose.
   e. Incubate at 42° for 30 min.
   f. Overlay the cells onto 7 ml of nutrient broth containing 0.6 *M* sucrose and centrifuge at 2500 rpm for 30 min at 4°.
   g. Pipet off the top 2-ml fraction with a wide bore pipet, divide into four 0.5-ml fractions, and add to each fraction 1.5 ml of nutrient broth containing 0.5 *M* sucrose.

[11] N.B., 10 g nutrient broth and 5 g NaCl/liter.

h. Incubate at 30° for 12–18 hr.
3. Selection of ts mutants
   a. Dilute each culture in nutrient broth (1:10^6) and plate 0.1 ml on nutrient broth plates.
   b. Incubate plates at 30° for 24–48 hr.
   c. Replica plate each plate twice and incubate one at 30° and the counterpart at 42°.
   d. Select colonies which grow at 30° but not at 42°.
   e. Repeat the temperature selection for all positive colonies 2–3 times.
4. Screening of ts mutants for competence of DNA transfection
   a. Inoculate 2 ml of nutrient broth with single colonies and incubate the cultures at 30° overnight with gentle shaking.
   b. Remove 0.25 ml of culture into a sterile tube containing 0.25 ml of PAM + 0.48 sucrose (PAM[12]).
   c. Incubate for 5 hr with gentle shaking at 42°.
   d. Add 0.5 ml of ϕx174 ss DNA solution containing 1 μg/ml DNA in 50 m*M* Tris HCl, pH 8.1.
   e. Incubate at 42° for 2.5 min and transfer the culture to 0–4°.
   f. Titer the infective centers.

[12] 10 g casamino acids, 10 g nutrient broth, 1 g glucose, 2 g $MgSO_4$ (per liter). G. D. Guthrie and R. L. Sinsheimer, *Biochim. Biophys. Acta* **72,** 290 (1963).

[23] *Bacillus subtilis* as a Host for Molecular Cloning

By PAUL S. LOVETT and KATHLEEN M. KEGGINS

Prior to the advent of molecular cloning technology[1,2] the two methods most commonly used to isolate quantities of cellular DNA enriched for specific genes involved the generation of specialized transducing phages that harbor specific regions of the bacterial chromosome, or the isolation of specific regions of cellular DNA on the basis of their unique physical properties or their occurrence within a subcellular organelle. The advantages of molecular cloning over these methods include its application in the isolation, in principle, of any small fragment of cellular DNA and in maintaining the selected fragment joined to a small, easily isolated replicon. The principles of molecular cloning have been developed using *Escherichia coli,* its plasmids, and its bacteriophages. In recent years the

[1] S. N. Cohen, A. C. Y. Chang, H. W. Boyer, and R. B. Helling, *Proc. Natl. Acad. Sci. U.S.A.* **70,** 3240 (1973).
[2] D. A. Jackson, R. H. Symons, and P. Berg, *Proc. Natl. Acad. Sci. U.S.A.* **69,** 2904 (1972).

ISBN 0-12-181968-X

components of the *E. coli* cloning systems have been extensively modified by genetic and biochemical manipulation to provide a variety of sophisticated phages, plasmids, and recipient bacterial strains specifically useful for molecular cloning. The current *E. coli* cloning systems have been shown to be extremely reliable for the isolation, amplification and, in many instances, genetic expression of prokaryotic and certain eukaryotic DNA fragments.

Development of easily manipulated cloning systems within nonenteric bacteria such as *Bacillus subtilis,* or eukaryotic cells such as *Saccharomyces cerevisiae,* will expand the types of DNA fragments that can be selected directly on the basis of their genetic function. Such recipient cell lines also provide a means to analyze the role of cloned genes that determine specialized biological functions. As an example, it has been estimated that more than 40 loci (operons) participate specifically in the formation of spores by *B. subtilis.*[3] Identification of fragments that harbor specific spore genes is most directly achieved by cloning the fragments into a chosen sporulation-negative mutant of *B. subtilis* and seeking complementation of the sporulation defect. Efforts to date to clone sporulation genes have used *E. coli,* a non-spore former, as the cloning recipient.[4] These experiments are technically difficult because of the absence of a direct biological selection for the cloned fragment. Similarly, identification of DNA fragments harboring genes involved in such functions as yeast cell division is most readily approached by cloning fragments directly into *S. cerevisiae.*

Plasmids occur naturally in certain strains of the related *Bacillus* species *B. pumilus* and *B. subtilis.*[5-7] The majority of these plasmids specify no known biological function and are therefore classed as cryptic. Two *Bacillus* plasmids, pPL10[8] and pPL7065,[9] determine the production of bacteriocin-like activities, and a third plasmid, pPL576,[10] reduces the ability of the host to form spores. These last-mentioned three plasmids have been transformed into *B. subtilis* 168, and each has been used as a vector for cloning DNA fragments. Unfortunately, the absence of an easily selected, plasmid-specified trait has made these experiments technically cumbersome.

The recent development of easily manipulated plasmid cloning vectors

[3] P. J. Piggot and J. G. Coote, *Bacteriol. Rev.* **40,** 908 (1976).
[4] J. Segall and R. Losick, *Cell* **11,** 751 (1977).
[5] P. S. Lovett and M. G. Bramucci, *J. Bacteriol.* **124,** 484 (1975).
[6] T. Tanaka, M. Kuroda, and K. Sakaguchi, *J. Bacteriol.* **129,** 1487 (1977).
[7] J.-C. LeHegarat and C. Anagnostopoulos, *Mol. Gen. Genet.* **157,** 167 (1977).
[8] P. S. Lovett, E. J. Duvall, and K. M. Keggins, *J. Bacteriol.* **127,** 817 (1976).
[9] P. S. Lovett, E. J. Duvall, M. G. Bramucci, and R. Taylor, *Antimicrob. Agents & Chemother.* **12,** 435 (1977).
[10] P. S. Lovett, *J. Bacteriol.* **115,** 291 (1973).

TABLE I
SOME PLASMIDS INTRODUCED INTO *Bacillus subtilis* 168 BY TRANSFORMATION

Source of plasmid	Plasmid designation	Molecular weight (Md)	Plasmid-specified trait	Reference
B. pumilus ATCC 12140	pPL10	4.4	Bacteriocin-like activity	8
B. pumilus ATCC 7065	pPL7065	4.6	Bacteriocin-like activity	9
B. pumilus NRS 576	pPL576	28.0	Oligosporogenesis	10
B. cereus	pBC16	2.8	Tet^{R}	12
S. aureus	pT127	2.9	Tet^{R}	11
S. aureus	pC194	1.8	Cm^{R}	11
S. aureus	pC221	3.0	Cm^{R}	11
S. aureus	pC223	3.0	Cm^{R}	11
S. aureus	pUB112	3.0	Cm^{R}	11
S. aureus	pUB110	2.9	Kan^{R}/Neo^{R}	14
S. aureus	pSA2100	4.7	Cm^{R} Sm^{R}	14
S. aureus	pSA0501	2.8	Sm^{R}	14
S. aureus	pSC194	4.8	Sm^{R} Cm^{R}	15, 16
S. aureus	pS194	3.0	Sm^{R}	15, 16
S. aureus	pC194	2.0	Cm^{R}	15, 16
Streptococcus sanguinis	pAM77	4.5	Ery^{R}	13

for *B. subtilis* resulted from the observation that certain small, high-copy-number, antibiotic resistance plasmids detected in *Staphylococcus aureus* could be transformed directly into *B. subtilis* where the plasmids were stably maintained and expressed the appropriate antibiotic resistance trait.[11] Since the initial observation other antibiotic resistance plasmids originating in *Bacillus*[12] and *Streptococcus*,[13] in addition to *Staphylococcus*, have been transformed into *B. subtilis* (Table I),[8–16] making available many potential plasmid cloning vectors. In this chapter we describe the use of one such plasmid, pUB110 (Table II), as a vector for cloning DNA fragments in *B. subtilis*.

Principle of the Method

Plasmids are small, circular replicons easily isolated in high purity from bacterial cells as covalently closed, circular (CCC) duplex DNA molecules.[17] Reintroduction of purified plasmids, in the CCC configura-

[11] S. D. Ehrlich, *Proc. Natl. Acad. Sci. U.S.A.* **74,** 1680 (1977).
[12] K. Bernhard, H. Schrempf, and W. Goebel, *J. Bacteriol.* **133,** 897 (1978).
[13] D. Clewell, personal communication.
[14] T. J. Gryczan, S. Contente, and D. Dubnau, *J. Bacteriol.* **134,** 318 (1978).
[15] S. Löfdahl, J.-E. Sjöström, and L. Philipson, *Gene* **3,** 149 (1978).
[16] S. Löfdahl, J.-E. Sjöström, and L. Philipson, *Gene* **3,** 161 (1978).
[17] T. F. Roth and D. R. Helinski, *Proc. Natl. Acad. Sci. U.S.A.* **58,** 650 (1967).

TABLE II
PROPERTIES OF PLASMID pUB110

Property		Reference
Molecular weight	2.9×10^6	14, 18
Sedimentation velocity	21 S	18
Plasmid-specified trait	Resistance to 5 μg/ml of neomycin sulfate	
Endonuclease sensitivity		14
*Eco*RI	One site	
*Bam*HI	One site	
*Xba*I	One site	
*Bgl*II	One site	
*Hin*dII	Two sites	
*Hpa*II	Three sites	
*Hae*III	Three sites	

tion, into appropriate plasmid-negative recipient bacteria is readily accomplished by transformation. The transformation frequency of small antibiotic resistance plasmids in the *B. subtilis* system is on the order of $0.1-10 \times 10^3$ transformants per microgram of plasmid.[14-16,18-20] Small fragments of DNA can be inserted into purified plasmids *in vitro* (see below). Plasmids containing such insertions may then be transformed into an appropriate bacterial recipient where the constructed replicon is replicated with fidelity. If the genes present on the insertion are properly expressed at the levels of transcription and translation, the cell carrying the constructed plasmid may acquire a new biological trait.

In vitro insertion of DNA fragments into plasmids can be accomplished by either of two general approaches, both of which are similar in principle.[1,2] In the technically simpler approach a vector plasmid in the CCC configuration is converted to a linear molecule by cleavage with a site-specific endonuclease such as *Hin*dIII, *Eco*RI, or *Bam*HI which fulfills the following criteria: The enzyme-sensitive site on the plasmid must occur in a region not essential to plasmid replication or to expression of a plasmid function (e.g., antibiotic resistance) that may be used as a selected trait; cleavage by the enzyme should generate short, single-stranded termini each of which is identical in base sequence, and therefore the single-stranded termini are self-complementary.[21]

Cleavage of a vector plasmid and DNA from a different source with the same site-specific endonuclease, such as *Eco*RI, generates a popula-

[18] K. M. Keggins, P. S. Lovett, and E. J. Duvall, *Proc. Natl. Acad. Sci. U.S.A.* **75,** 1423 (1978).

[19] T. J. Gryczan and D. Dubnau, *Proc. Natl. Acad. Sci. U.S.A.* **75,** 1428 (1978).

[20] S. D. Ehrlich, *Proc. Natl. Acad. Sci. U.S.A.* **75,** 1433 (1978).

[21] R. J. Roberts, *in* "Microbiology, 1978" (D. Schlessinger, ed.), p. 5. *Amer. Soc. Microbiol.,* 1978.

tion of molecules sharing the same cohesive termini. Annealing of such DNA fragments at low temperature (< 10°) facilitates the stability of the joining of the cohesive ends through hydrogen-bond formation between the complementary termini.[22] This annealing procedure favors the formation of intramolecular associations at low DNA concentrations, and the probability of intermolecular associations increases with increasing DNA concentration.[23] Thus, at moderately high DNA concentrations (e.g., 2 μg/ml) intermolecular complexes are generated. Many of these intermolecular complexes are the result of joining DNA fragments from unrelated sources, i.e., the vector plasmid and a fragment of foreign DNA. Fragments joined through hydrogen-bond formation between cohesive ends can be covalently sealed through the action of the enzyme DNA ligase.[1] Therefore, fragments of a foreign DNA molecule that are incapable of autonomous replication in a bacterium, such as *E. coli* or *B. subtilis,* can be physically joined to a replicon (plasmid) capable of autonomous replication in the chosen recipient. Transformation of an appropriate bacterial recipient with such constructed plasmids produces a cell line carrying plasmid-linked genes that the cell might not otherwise be capable of stably maintaining. If the genes on the insertion are fully expressed at the levels of transcription and translation, the cell can acquire a new biological trait. Selection of a constructed plasmid carrying a specific cloned fragment is achieved either by applying biological selection for expression of a specific cloned gene (e.g., *trpC*) or by prepurifying the desired fragment on the basis of some physical or biological characteristic unique to the fragment to be joined to the plasmid.

The second method of inserting fragments of DNA into plasmids *in vitro* is commonly referred to as the tailing method.[2,24] In this method the purified vector plasmid is linearized with a site-specific endonuclease, chosen because insertions into its site of cleavage on the plasmid do not disrupt the replicating ability of the plasmid or expression of a desired plasmid-associated trait to be used for selection. Cleavage by the endonuclease may generate either flush ends or cohesive ends. The DNA fragments to be joined to the vector plasmid can be generated by endonuclease cleavage or by mechanical shearing. In either event, the cohesive termini that will allow the plasmid vector and the foreign DNA fragments to associate are synthesized on each by using the enzyme terminal transferase. As an example, terminal transferase can be used to generate single-stranded poly(dA) tails (10 to 30 bases in length) on the free ends of the vector, and poly(dT) tails on the fragments of the foreign DNA.

[22] J. E. Mertz and R. W. Davis, *Proc. Natl. Acad. Sci. U.S.A.* **69,** 3370 (1972).

[23] A. Dugaiczyk, H. W. Boyer, and H. M. Goodman, *J. Mol. Biol.* **96,** 171 (1975).

[24] L. Clarke and J. Carbon, *Proc. Natl. Acad. Sci. U.S.A.* **72,** 4361 (1975).

Mixing of the two classes of DNA molecules permits hydrogen-bond formation to occur between the synthesized termini. Although the tailing method has not yet been tested in the *B. subtilis* cloning system, this approach has certain advantages for cloning over the use of only site-specific endonucleases that generate cohesive termini. The ability to add tails to DNA fragments generated by random mechanical shearing ensures that virtually any desired gene can be isolated on a discrete fragment, and indeed controlled shearing can result in varied sizes of DNA fragments. Plasmid transformants resulting from cloning by the tailing method all must, in principle, contain insertions, since a plasmid with poly(dA) tails cannot readily recyclize.

Materials and Reagents

Bacterial Strains

Mutant strains of *B. subtilis* 168: BR151 (*trpC2 metB10 lys-3*), BD224 (*trpC2 thr-5 recE4*), T24 (*trpE*), T22 (*trpD*), T5 (*trpC*), T12 (*trpF*), T20 (*trpB*), T4 (*trpA*).

Growth Media

Liquid media used include antibiotic medium No. 3 [penassay broth (PB) Difco] and Spizizen's minimal medium supplemented with 0.05% acid-hydrolyzed casein (MinCH). Solid media for plates include tryptose blood agar base (TBAB, Difco) and MinCH containing 0.5% acid-hydrolyzed casein solidified with 1.9% noble agar (Difco). Media containing 5 μg/ml of neomycin sulfate are indicated as TBAB-Neo, PB-Neo, etc.

Materials and Reagents for Plasmid Isolation

1. TES buffer: 0.02 *M* Tris-HCl, pH 7.5, 5 m*M* EDTA, 0.1 *M* NaCl
2. Lysozyme: 10 mg/ml of crystalline enzyme in TES
3. RNase: Pancreatic, 2 mg/ml in water, heat-shocked at 80° for 15 min
4. Pronase: 10 mg/ml in TES, predigested at 37° for 90 min
5. Sarkosyl: NL30 (ICN Pharmaceuticals, Inc.)
6. Cesium chloride: High purity (Kawecki Berylco Industries, Inc.)
7. Polyallomer tubes: $\frac{5}{8} \times 3$ inch tubes (Beckman)
8. [^{3}H]thymidine: 1 mCi/ml (New England Nuclear)
9. Deoxyadenosine: 25 mg/ml in water
10. Isoamyl alcohol
11. Fraction collector

12. Scintillation counter
13. Lysozyme buffer: 30 m*M* Tris, 50 m*M* EDTA, 50 m*M* NaCl, 25% sucrose, pH 8.0, with KOH
14. Sodium dodecyl sulfate (SDS) buffer: 2 ml of 10% SDS, 2 ml of 500 m*M* disodium EDTA, pH 8.0, 12 ml of TES buffer.
15. 5 *M* NaCl in water
16. Polyethylene glycol (PEG): 50% type 6000 in TES
17. Ethidium bromide: 10 mg/ml in TES
18. Uv light source
19. Needle (18-gauge) and syringe
20. Ti50 rotor or equivalent
21. Ultracentrifuge capable of 44,000 rpm

Materials and Reagents for Cellular DNA Isolation

1. Source of DNA: *B. subtilis* 168, *B. pumilus* NRRL B-3275, *B. pumilus* NRS576, *B. licheniformis* 9945A, *B. licheniformis* 749C
2. Lysozyme, RNase, pronase, and sarkosyl as in plasmid isolation
3. Phenol: Washed with TES buffer
4. Ethanol: 95%, cold
5. Recording spectrophotometer

Reagents for Cloning

1. Endonucleases: *Eco*RI, *Bam*HI, and *Hin*dIII, all commercially available from Bethesda Research Laboratories, Miles Research Products, and New England BioLabs
2. T4 DNA ligase: Available from New England BioLabs; and Bethesda Research Laboratories
3. *Eco*RI buffer: 0.1 *M* Tris-HCl, pH 7.5, 0.05 *M* NaCl, 0.01 *M* $MgCl_2$
4. *Hin*dIII buffer: 6 m*M* Tris-HCl, pH 7.5, 50 m*M* NaCl, 6 m*M* $MgCl_2$, 0.1 mg/ml of bovine serum albumin
5. *Bam*HI buffer: 0.1 *M* Tris-HCl, pH 7.5, 0.01 *M* $MgCl_2$
6. Dithiothreitol (DTT): 0.1 *M* in 0.1 *M* Tris-HCl, pH 7.5
7. Adenosine triphosphate (ATP): 0.01 *M* in 0.1 *M* Tris-HCl, pH 7.5

Method

Plasmid Isolation

Methods developed for isolating plasmid from *E. coli* can generally be applied, occasionally with slight modification, to the isolation of plasmids from *B. subtilis* and related species. One approach we have routinely used for isolating plasmids from *B. subtilis* and *B. pumilus* is as follows. *Ba-*

cillus subtilis 168 (pUB110) is grown to late log or early stationary phase in 200–500 ml MinCH, deoxyadenosine (250 μg/ml), nutritional supplements required by the specific bacterial strain, and 1–20 μCi of [^{3}H]thymidine per milliliter. Cells are harvested by centrifugation, washed twice with TES buffer, and resuspended to approximately $\frac{1}{10}$ the original volume in TES. Lysozyme and RNase are added to 500 and 100 μg/ml, respectively, and the suspension is incubated at 37° for 20–25 min. The lysate is diluted twofold with TES, and pronase (to 500 μg/ml) and sarkosyl (to 0.8%) are added. After incubation for approximately 30 min at 37° the lysate, which is highly viscous and generally clear, is centrifuged at 12,500 rpm in the SS34 rotor of a Sorvall RC2B centrifuge at 4° for 25 min. The resulting supernatant fraction (cleared lysate) is carefully decanted. To 5 ml of the cleared lysate is added 7.1 g of CsCl. When the salt is completely dissolved, the volume is approximately 7.1 ml. This volume is divided equally between two $\frac{5}{8}$ × 3 inch polyallomer tubes (3.5 ml per tube). Then, 1.5 ml of ethidium bromide solution (4 mg/ml in 0.1 *M* sodium phosphate buffer, pH 7) is added, the tubes are gently inverted, and paraffin oil is added to bring the tubes to volume. The tubes are centrifuged for 40 hr in a Ti50 rotor at 36,000 rpm, at 15°. The position of the plasmid peak in the gradient can be determined by fractioning the gradient into 15 portions of equal volume and counting 10-μl portions of each dried on 25-mm Whatman filter disks for radioactivity. Fractions containing the plasmid are pooled and extracted with equal volumes of isoamyl alcohol until all red color due to ethidium bromide has been visibly removed from the aqueous layer. The aqueous layer containing the plasmid is then dialyzed against 1000 volumes of TES at 4°.

A convenient alternative we are currently using to isolate relatively large quantities of unlabeled plasmid is as follows. *Bacillus subtilis* 168 (pUB110) is grown overnight (18–20 hr) in 250 ml of PB-Neo. The cells are chilled, concentrated by centrifugation, washed twice with TES, resuspended in 4 ml of lysozyme buffer, and incubated for 30 min at 37° with 500 μg/ml of lysozyme. Sixteen milliliters of SDS buffer is added, and incubation is continued at 37° for 30 min. To this solution is added 5 ml of 5 *M* NaCl. After gentle mixing the solution is placed on ice for 20 hr and then centrifuged in the SS34 rotor at 15,000 rpm for 30 min. The resulting supernatant fraction is mixed in the cold with 5 ml of 50% PEG, held on ice for 3–5 hr, and then centrifuged at 10,000 rpm at 10° for 10 min in the SS34 rotor. The resulting pellet is resuspended in 2 ml of TES containing 200 μg/ml of heat-shocked RNase and held at 65° for 30 min. Ethidium bromide is added to 500 μg/ml, and the final volume is brought to 4.0 ml with TES. Exactly 3.65 g of CsCl is then dissolved in the lysate, and the solution is transferred to a $\frac{5}{8}$ × 3 inch polyallomer tube and centrifuged at 44,000 rpm in the Ti50 rotor for 40 hr. Illumination of the tube with uv light shows two bands that fluoresce. The lower band is plasmid. This band is removed through the side of the tube with a needle and syringe. The plasmid is extracted with isoamyl alcohol and dialyzed as above.

Isolation of Bulk Cellular DNA

Strains of *B. pumilus, B. subtilis,* or *B. licheniformis* are grown at 37° in 10 ml of PB for 18–20 hr. The cells are washed twice with TES buffer, resuspended in 2.5 ml of TES, and treated sequentially with lysozyme, RNase, pronase, and sarkosyl as described for plasmid isolation. The lysate is shaken gently with an equal volume of TES-saturated phenol at room temperature for 10 min. After centrifugation at room temperature for 10 min the aqueous phase is removed with a large-bore pipette and extracted with phenol twice again. The final aqueous phase (~3 ml) is transferred to 2 volumes of cold 95% ethanol and held at 4° for 30 min. The precipitate is resuspended in 2 ml of TES and dialyzed against 1000 volumes of TES at 4° for 12–18 hr. A dilution of the DNA-containing solution is scanned in a Cary recording spectrophotometer over the range of 240–290 nm to estimate DNA concentration (OD of 1 at 260 nm ≅ 50 μg/ml of DNA) and to determine OD_{260}/OD_{280} ratios.

Cloning Method

The level of endonuclease used to digest plasmid or phenol-purified cellular DNA is determined empirically as twice the activity of a specific enzyme, such as *Eco*RI, required to convert a fixed concentration of DNA, such as pUB110, to a linear form in 30 min at 37° as monitored by electrophoresis of the intact and digested plasmid through 0.7% agarose gels.[18] pUB110 (0.5 μg) and approximately 3 μg of cellular DNA (e.g., *B. subtilis, B. pumilus,* or *B. licheniformis*) are combined in 200 μl of TES buffer and dialyzed against 1000 volumes of *Eco*RI buffer for 4 hr at 4°. The solution is then incubated at 37° with the appropriate activity of *Eco*RI to ensure complete cleavage within 30 min. After cleavage the enzyme activity is terminated by placing the digest at 65° for 15 min. Annealing is achieved by incubating the DNA at 0° for 18 hr (in an ice bath in a 4° cold room). DTT is added to 10 m*M*, and ATP to 0.1 m*M*. One unit of T4 DNA ligase is then added, and the solution is held at 4° for 8–12 hr and then moved to 15° for 10 hr. The resulting DNA preparation is dialyzed against 1000 volumes of TES buffer, and approximately 2 μg of DNA is shaken at 37° with 5×10^8 competent *B. subtilis trpC2* cells (e.g., strain BR151) in 1 ml for 1 hr. Approximately 0.9 ml of the transformation reaction mixture is added to 20 ml of PB containing 5 μg/ml of neomycin sulfate which is shaken at 37° for 20–24 hr. The remainder of the transformed cells are diluted and plated on TBAB-Neo to estimate transformation frequency. Transformation controls, including cells and DNA incubated separately, are routinely run, and each is treated as described for the transformed cells.

After overnight incubation the transformed cells in 20 ml of PB-Neo have grown to saturation ($\sim 5 \times 10^8$/ml) and there is no growth in the cell or DNA control flasks. The transformed cells are washed with Spizizen's minimal medium twice, and the culture is diluted 10-fold in MinCH (no tryptophan) containing 5 μg/ml of neomycin sulfate. This culture is shaken at 37° for 20 hr at which time the cells have grown to $2-5 \times 10^8$/ml. A loopful of the culture is streaked onto a MinCH-Neo plate, or serial dilutions of the culture are plated on MinCH-Neo and the plates are incubated at 37° overnight. A single resulting colony is purified by restreaking, and the plasmid DNA is isolated and used to transform a *trpC2* mutant of *B. subtilis* to Neo^R. Each of 50 transformants is tested for the Trp phenotype. Frequently we find that the plasmid preparations at this stage contain both a constructed *trp* plasmid and pUB110. Therefore, only 40–45 of the NeoR clones are Trp^+. One of these Trp^+ Neo^R transformant is then used as the source of the particular constructed *trp* plasmid.

Two types of recipients have been used for cloning, depending on the properties of the DNA to be cloned. If the source of the DNA to be cloned is a bacterial species sufficiently closely related to *B. subtilis* 168 such that there exists extensive homology between donor and recipient genomes, a recombination-deficient recipient is used, e.g., a recipient carrying the *recE4* mutation.[25,26] If the donor DNA shares little homology with the *B. subtilis* chromosome (e.g., DNA extracted from *B. pumilus* or *B. licheniformis*) a wild-type (Rec^+) transformation recipient is used.

Identification of the genetic constitution of cloned *trp* segments is rapidly achieved by complementation analysis. The *trp* biosynthetic genes in *B. subtilis* appear contiguous and map as a cluster in the order: *trpE D C F B A*.[27] The ability of a given plasmid to complement point mutations in each of the *B. subtilis trp* genes is determined by transforming the plasmid into six mutant derivatives of *B. subtilis* each carrying a mutation in a different *trp* gene[28] (see Materials and Reagents). Neomycin-resistant transformants are selected, and the Trp phenotype of 50–200 of the transformants is determined by patching the colonies into MinCH.

[25] D. Dubnau and C. Cirigliano, *J. Bacteriol.* **117**, 488 (1974).

[26] K. M. Keggins, E. J. Duvall, and P. S. Lovett, *J. Bacteriol.* **134**, 514 (1978).

[27] I. P. Crawford, *Bacteriol. Rev.* **39**, 87 (1975).

[28] S. O. Hoch, C. Anagnostopoulos, and I. P. Crawford, *Biochem. Biophys. Res. Commun.* **35**, 838 (1969).

Genetic and Biochemical Properties of Cloned EcoRI and BamHI trp Segments

pUB110 is sensitive to several site-specific endonucleases (Table II). Both the single *Bam*HI site and the single *Eco*RI site on pUB110 appear available for cloning. *Eco*RI-generated *trp* segments we have cloned using pUB110 as a vector have molecular weights in the range of 1.5–2.5 Md. This size is sufficient to specify two to four genes, and complementation analysis of each of the *Eco*RI *trp* plasmids shows that each complements point mutations in two, three, or four *trp* genes (Tables III and IV).

Digestion of the majority of the *Eco*RI-generated *trp* plasmids (e.g., pSL103) with *Eco*RI yields two linear fragments whose combined molecular weight (e.g., 2.3 + 2.8 Md) approximates the molecular weight of the intact plasmid (e.g., 5.0 Md). If *Eco*RI-digested pSL103 is subjected to the annealing and ligation procedure and used to transform *B. subtilis* 168 to neomycin resistance, the original vector plasmid can be removed from the constructed *trp* derivative. Occasionally *trp* plasmids constructed using *Eco*RI retain only a single *Eco*RI site. Thus digestion of one such plasmid, pSL101, with *Eco*RI generates a single linear modecule of the same molecular weight as intact pSL101. The reason for the occurrence of constructed plasmids retaining only a single *Eco*RI site is not known.

Construction of *trp* plasmids using *Bam*HI endonuclease has in our hands always yielded *trp* plasmids that retain only a single *Bam*HI site (Table V). Comparison of the inferred molecular weights of *Bam*HI-generated *trp* segments with the molecular weights of *trp* segments cloned using *Eco*RI suggests that *Bam*HI-generated fragments are larger.

Plasmid pUB110 and each of the *trp* derivatives are maintained in *B. subtilis* at approximately 50 copies per chromosome equivalent. Comparison of the Trp enzyme levels in *B. subtilis* cells carrying each of five constructed *trp* plasmids with the levels of the chromosomally specified Trp enzymes in strains harboring no plasmids generally confirms the genetic constitution of each plasmid initially deduced from complementation analysis[29] (Tables III and IV). Plasmids that complement point mutations in individual *trp* genes specify the corresponding enzyme activities. The specific activities of plasmid-specified Trp enzymes in *B. subtilis* are significantly greater than the repressed levels of the chromosomally specified enzymes and equal to or below the derepressed levels of the chromosomally specified enzymes.

Each of the constructed *trp* plasmids was initially selected for ability to complement a specific mutation in *trpC* (i.e., *trpC2*), and each of five *trp* plasmids tested contained a single *Hin*dIII-sensitive site. pUB110 is

[29] K. M. Keggins, P. S. Lovett, R. Marrero, and S. O. Hoch, *J. Bacteriol.* **139,** 1001 (1979).

TABLE III

SPECIFIC ACTIVITIES OF THE PRODUCTS OF *trp* GENES IN *Bacillus subtilis* HARBORING FOUR *Eco*RI-CONSTRUCTED *trp* PLASMIDS[a,29]

Strain	Source of cloned segment	Gene and gene product[b]					
		E, AS	*D*, PRT	*C*, InGPS	*F*, PRAI	*B*, TS-β	*A*, TS-α
Br151 d		3.49	0.87	0	52.63	17.7	4.75
Br151 r		0	0	0	0	0.51	0.29
Br151 (pSL103)	*B. pumilus*	+	+	+	+	−	−
		0	0.63	27.78	55.98	0.35	0.18
		0.15[c]					
BR151 (pSL104)	*B. pumilus*	−	±	+	+	−	−
		0	0.21	3.14	5.05	0.14	0.17
BR151 (pSL101)	*B. licheniformis*	−	−	+	+	+	+
		0	0	2.46	60.36	4.50	1.78
BR151 (pSL105)	*B. licheniformis*	−	±	+	+	−	−
		0	0.14	4.34	4.27	0.24	0.19

[a] BR151 (*trpC2*) was grown for 16 hr (stationary phase) in MinCH containing tryptophan prior to preparing extracts. BR151 r indicates cells were grown in sufficient tryptophan (20 μg/ml) to repress the appearance of the Trp enzymes. BR151 d indicates cells were grown in limiting tryptophan (0.5–1 μg/ml). BR151 harboring the various *trpC*-complementing plasmids was grown in MinCH containing no tryptophan. Specific activities refer to nanomoles of substrate disappearing or product forming per minute per milligram of protein. +, ±, and − refer to the complementing activity of each plasmid.

[b] *trp* genes *E*, *D*, *C*, *F*, *B*, and *A* specify the following enzymes: AS, anthranilate synthetase; PRT, anthranilate-5-phosphoribosylpyrophosphate phosphoribosyltransferase; InGPS, indoleglycerol phosphate synthetase; PRAI, phosphoribosylanthranilate isomerase; TS-β, tryptophan synthetase β; TS-α, tryptophan synthetase α.

[c] Specific activity detected in cell-free extracts prepared from washed cells.

TABLE IV
SPECIFIC ACTIVITIES OF THE PRODUCTS OF *trp* GENES IN BD224 (*trpC2 recE4*) AND BD224 HARBORING THE *Eco*R1-GENERATED *trp* PLASMID PSL106[a,29]

Strain	Gene and gene product					
	E, AS	*D*, PRT	*C*, InGPS	*F*, PRAI	*B*, TS-β	*A*, TS-α
BD224 d	2.90	1.32	0	26.07	13.30	4.48
BD224 r	0	0	0	0	0.42	0.11
BD224 (pSL106)	–	–	+	+	–	–
	0	0	39.71	240.9	0.58	0.33

[a] Conditions were as described in the footnote to Table III. Source of cloned segment was *B. subtilis* 168.

not cleaved by *Hin*dIII. Therefore the location of the *Hin*dIII-sensitive site in the cloned *trp* segments was deduced by insertional inactivation. Insertion of random *Hin*dIII fragments of chromosome DNA into each of four *trp* plasmids resulted in the specific loss of *trpC*-complementing activity and the elimination of detectable levels of the plasmid-specified *trpC* product, indole glycerol phosphate synthetase (Table VI). Although the *Hin*dIII insertions only inactivate *trpC*-complementing activity, the specific activity of the product of the adjacent gene, *trpF*, is reduced more than 10-fold, suggesting the insertions exert a polar effect.[29] By using the inactivation of *trpC*-complementing activity as an indication of insertion of a *Hin*dIII fragment into plasmid pSL103 we have cloned a large number of *Hin*dIII fragments from *B. subtilis* chromosome DNA, plasmids, and phage genomes.

Cloning by Insertional Inactivation of trpC

Plasmid pSL103 (0.5 μg) and the DNA to be cloned (3 μg) are combined in 250 μl of *Hin*dIII digestion buffer. *Hin*dIII is added in twofold excess, and the reaction mixture is incubated at 37° for 25 min. Inactivation of the enzyme at 65° and the procedures for annealing, ligation, and transformation are as described earlier. The transformation recipient is a highly competent *trpC* mutant of *B. subtilis* (strain BR151). Neomycin-resistant transformants are selected (on TBAB-Neo), and each is replica-plated into MinCH and MinCH plus tryptophan. Clones harboring pSL103 containing a *Hin*dIII insertion are tryptophan-requiring and are therefore easily identified. By using the ratio of DNA shown above we routinely find that 17–25% of the neomycin-resistant transformants harbor a derivative of pSL103 containing a *Hin*dIII insertion. The size of the insertions we have cloned ranges from 0.2 to 1.2 Md.

TABLE V

PROPERTIES OF SOME PLASMIDS DERIVED FROM pUB110 USING ENDONUCLEASE *Bam*HI

Plasmid	Source of cloned fragment	Molecular weight (Md)	Number of *Bam*HI sensitive sites	Complementing activity in *B. subtilis trp*							
				aroB	*E*	*D*	*C*	*F*	*B*	*A*	*his-2*
pSL121	*B. subtilis* ATCC 15841	5.2	One	0	+	+	+	+	0	0	0
pSL201	*B. pumilus* NRRL B-3275	5.1	One	0	+	+	+	+	0	0	0
pSL202	*B. licheniformis* 9945A	5.4	One	0	+	+	+	+	0	0	0

TABLE VI
EFFECT OF *Hin*dIII INSERTIONS INTO *trp* PLASMIDS ON THE GENETIC ACTIVITY OF *trpD*, *trpC*, AND *trpF*[29]

Parent plasmid[a]	Plasmid derived by *Hin*dIII insertion or removal of the insertion[b]	Source and molecular weight of insertion	Complementing activity and specific activity of gene product[c]		
			D, PRT	*C*, InGPS	*F*, PRAI
pSL103			+, 0.63	+, 27.78	+, 55.98
	pPL130, +H	*B. licheniformis* 9945A DNA, 0.4 × 10^6	+, 0.56	−, 0	+, 3.70
	pPL132, −H	−	+, 0.80	+, 27.45	+, 71.0
pSL120			+, 0.11	+, 11.79	+, 88.62
	pPL122, +H	*B. licheniformis* 9945A DNA, 0.2 × 10^6	+, 0.15	−, 0	+, 2.28
pSL105			+, 0.14	+, 4.34	+, 4.27
	pPL162, +H	*B. licheniformis* 9945A DNA, 0.2 × 10^6	+, 0.11	−, 0	+, 0.27
pSL106				+, 39.71	+, 240.9
	pPL138, +H	*B. licheniformis* 9945A DNA, 0.9 × 10^6		−, 0	+, 4.42
	pPL139, −H	−		+, 27.81	+, 171.21

[a] With one exception, properties of each parent plasmid are listed in Tables III and IV. pSL120 had as the source of its *Eco*RI cloned *trp* segment *B. subtilis* 15841.
[b] +H, an insertion; −H, removal of an insertion.
[c] Specific activities were determined in extracts of BR151 harboring the various plasmids, with one exception. pSL106 and its derivatives were introduced into BD224, and extracts of BD 224 were assayed for enzyme levels. All extracts were prepared as described in Hoch *et al.*[28]

Comments

The development of vectors useful for molecular cloning in *B. subtilis* is in the initial stages. Several plasmids have been tested for cloning and, among these, pUB110 appears to offer certain advantages including relatively high transforming activity, stable maintenance in *B. subtilis* and related bacteria, high copy number, and sensitivity to several well-characterized endonucleases. The replication of pUB110 has been shown to occur in temperature-sensitive DNA replication mutants of *B. subtilis* at temperatures that are nonpermissive for chromosome replication, indicating that this system may be useful for amplification of cloned fragments.[30] Although we have worked most extensively with pUB110, it

[30] A. G. Shibakumar and D. Dubnau, *Plasmid* **1,** 405 (1978).

seems clear that the use of other plasmids or modified forms of pUB110 may lend increasing sophistication to cloning in *B. subtilis*. For example, several *trp* derivatives of pUB110 can be used for cloning *Hin*dIII-generated DNA fragments by screening for insertional inactivation of plasmid-specified *trpC*-complementing activity. Additionally, Löfdahl, Sjöström, and Philipson[16] have demonstrated that insertion of *Eco*RI-generated DNA fragments into pSC194 inactivates plasmid-specified streptomycin resistance but plasmid-specified chloramphenicol resistance is unaffected.

The methods we have used for cloning in *B. subtilis* reflect direct extrapolation of principles established in the *E. coli* systems, and no unique problems have been detected. Of particular interest is the absence of the obvious problem of restriction of foreign DNA fragments by *B. subtilis*, since it has been reported that *B. subtilis* possesses a restriction system.[31] This restriction system may cause difficulties in cloning DNA from sources other than *Bacillus*. Should that problem be encountered, it will be necessary to use the restriction-negative mutant recently isolated.[31]

Much effort has been directed toward the development of plasmid cloning vectors and, although these approaches have been successful, it is apparent that the development of phage cloning vectors will add an additional dimension to cloning. No such vector is available, although current efforts by several laboratories should result in the development of powerful phage cloning vectors for *B. subtilis* (F. E. Young; D. Dean; J. Ito; personal communications).

At the present writing, the Office of Recombinant DNA of the National Institutes of Health has approved the system described herein for cloning DNA fragments from bacterial species closely related to *B. subtilis* (e.g., *B. licheniformis* and *B. pumilus*) and plasmids and phages that replicate in these species. The development and testing of attenuated forms of *B. subtilis* and modified plasmids will ultimately provide a system useful for cloning fragments from a broad spectrum of organisms.

[31] T. Uozumi, T. Hoshino, K. Miwa, S. Horinouchi, T. Beppu, and K. Arima, *Mol. Gen. Genet.* **152,** 65 (1977).

[24] Methods for Molecular Cloning in Eukaryotic Cells

By A. F. PURCHIO and G. C. FAREED

The transduction of new genetic information from one cell to another offers a unique opportunity to study the complex factors regulating gene expression. The discovery and purification of bacterial restriction en-

ISBN 0-12-181968-X

zymes has facilitated the isolation of specific DNA fragments spanning entire operons. These DNA fragments can be joined *in vitro* to replicons of *Escherichia coli* such as a plasmid DNA[1-4] or a bacteriophage λ genome,[5-7] and the replication and amplification of these molecules can be accomplished in *E. coli* host cells. These methods have been used to clone a wide variety of prokaryotic and eukaryotic DNA segments in *E. coli*. In addition, certain hybrid DNA molecules have been shown to complement prokaryotic cells deficient in a particular gene.[8,9]

The genomes of simian virus 40 (SV40) and polyomavirus can serve as useful vehicles for carrying new genetic material into eukaryotic cells.[10] SV40 is a member of the papovavirus group and consists of a supercoiled double-stranded DNA genome of molecular weight 3.3×10^6 enclosed in a 45-nm icosahedron with a protein capsid and no membrane coat. The virus is able to replicate in cells of its natural host (monkey cells), and this results in the release of progeny virus and cell lysis. These cells are termed permissive cells. In nonpermissive cells or cells that do not efficiently support viral multiplication (e.g., rodent cells) viral DNA synthesis is blocked. However, some of these cells may develop new phenotypic properties *in vitro* in a process termed transformation and generally carry the viral DNA in a stable, integrated form.[10,11]

Both physical and genetic maps for SV40 and polyomavirus DNAs have been throughly completed. Two functional regions exist: an early gene region spanning 0.67–0.17 SV40 map units in a counterclockwise direction with respect to the *Eco*RI site, and a late gene region from 0.67 to 0.17 map units in a clockwise direction with respect to the *Eco*RI cleavage site (Fig. 1). The early gene region is synonymous with the *A* gene which is expressed prior to the onset of viral DNA replication in infected cells. It is transcribed into at least two functional mRNA species which

[1] S. N. Cohen, A. C. Y. Chang, H. W. Boyer, and R. B. Helling, *Proc. Natl. Acad. Sci. U.S.A.* **70,** 3240 (1973).

[2] K. Timmis, F. Cabello, and S. N. Cohen, *Proc. Natl. Acad. Sci. U.S.A.* **71,** 4556 (1974).

[3] V. Hershfield, H. W. Boyer, C. Yanofsky, M. A. Lovett, and D. R. Helinski, *Proc. Natl. Acad. Sci. U.S.A.* **71,** 3455 (1974).

[4] M. Thomas, J. R. Cameron, and R. W. Davis, *Proc. Natl. Acad. Sci. U.S.A.* **71,** 4579 (1974).

[5] N. E. Murray and K. Murray, *Nature (London)* **251,** 476 (1974).

[6] J. R. Cameron, S. M. Panasenko, I. R. Lehman, and R. Davis, *Proc. Natl. Acad. Sci. U.S.A.* **72,** 3516 (1975).

[7] D. C. Tiemeier, S. M. Tilghman, and P. Leder, *Gene* **2,** 1973 (1977).

[8] C. J. Collins, D. A. Jackson, and F. A. J. DeVries, *Proc. Natl. Acad. Sci. U.S.A.* **73,** 3838 (1976).

[9] K. Struhl, J. R. Cameron, and R. W. Davis, *Proc. Natl. Acad. Sci. U.S.A.* **73,** 1471 (1976).

[10] G. C. Fareed and D. Davoli, *Annu. Rev. Biochem.* **46,** 471 (1977).

[11] J. Tooze, *in* "The Molecular Biology of Tumor Viruses" (J. Tooze, ed.), p. 391. Cold Spring Harbor Lab., Cold Spring Harbor, New York, 1973.

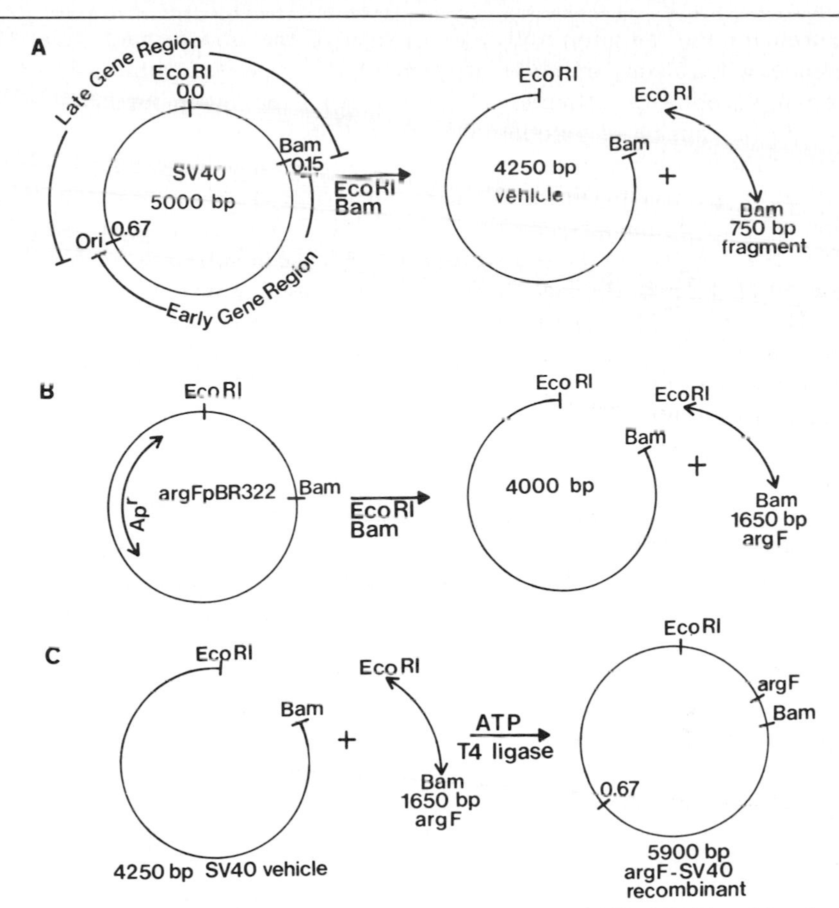

FIG. 1. (A) Schematic diagram illustrating a functional map of SV40 DNA and the cleavage sites for *Eco*RI and *Bam*HI. (B) Cleavage of pBR322*argF* with *Eco*RI and *Bam*HI generates a 4000-bp plasmid vehicle and a 1650-bp fragment. (C) The construction of the SV40-*argF* recombinant genome is depicted. bp, Base pairs; Ap^r, ampicillin resistance locus; Ori, origin for SV40 DNA replication.

sediment at 19 *S*. These messages code for big T (MW, 85,000) and small t (MW, 17,000) antigens.[12] Big T antigen may be cleaved and/or modified in some fashion to give rise to U antigen and the tumor-specific transplantation antigen. The late gene region contains the information needed to code for the viral structural proteins VP_1, VP_2, and VP_3.

In permissive cells, only the early gene region is required for viral DNA replication[13]; in nonpermissive cells, the early gene products are re-

[12] A. J. Berk and P. A. Sharp, *Proc. Natl. Acad. Sci. U.S.A.* **75,** 1274 (1978).
[13] J. Y. Chou, J. Avila, and R. G. Martin, *J. Virol.* **14,** 116 (1974).

quired for the initiation and maintenance of the transformed state.[14,15] Hence, when all or part of the late gene region is excised from SV40 DNA and replaced with a foreign DNA segment, the recombinant genome should be capable of autonomous replication in permissive cells. If the size of the recombinant genome is 70–100% of the wild-type size, mixed infection of the cells with a suitable helper virus should allow encapsidation of the recombinant into virus particles. Transfection of nonpermissive cells may lead to the transformation of these cells and the stable integration of the hybrid molecule into the host cell genome.

The foreign DNA segment may be transcribed in several ways. If it contains a functional promoter and a termination site, RNA polymerase molecules can produce transcripts of the entire DNA segment. Alternatively, transcription may initiate at the promoter for late SV40 mRNA and read through the foreign DNA segment, resulting in a hybrid RNA molecule containing both SV40 and the foreign sequences.

The first such hybrid molecule was constructed by Jackson, Symons, and Berg[16] and consisted of λ *dv-Gal* linked to the *Eco*RI cleavage site of SV40 DNA. Ganem *et al.*[17] and Nussbaum *et al.*[18] joined a segment of λ DNA to an SV40 segment containing the origin of replication and propagated the hybrid molecules in monkey cells in the presence of wild-type SV40. Goff and Berg[19] inserted a 2000-bp segment of λ DNA between the *Hpa*II and the *Bam*HI sites of SV40 using the poly(dA:dT) joint procedure and successfully propagated the hybrid molecule in monkey kidney cells by mixed infection at 41° with a *tsA 58* helper virus (a mutant of SV40 that does not make a functional T antigen at 41°). Hamer *et al.*[20] used the same method to propagate an *E. coli* suppressor tRNA gene inserted between the *Eco*RI and *Hpa*II sites of SV40. More recently[21] this hybrid molecule was used to transfect nonpermissive rat cells and was found to be carried in these cells in an episomal (nonintegrated) form.

In this chapter we illustrate the use of papovavirus vectors for molecular cloning in mammalian cells. Several other potential cloning systems which have not yet been employed for eukaryotic cells are described at the end of this chapter. Although the methods selected for isolation and joining of vector and foreign DNA fragments may vary depending on the

[14] P. Tegtmeyer, *J. Virol.* **15,** 613 (1975).

[15] J. S. Brugge and J. S. Butel, *J. Virol.* **15,** 619 (1975).

[16] D. A. Jackson, R. H. Symons, and P. Berg, *Proc. Natl. Acad. Sci. U.S.A.* **69,** 2904 (1972).

[17] D. Ganem, A. L. Nussbaum, D. Davoli, and G. C. Fareed, *Cell* **7,** 349 (1976).

[18] A. L. Nussbaum, D. Davoli, D. Ganem, and G. C. Fareed, *Proc. Natl. Acad. Sci. U.S.A.* **73,** 1068 (1976).

[19] S. P. Goff and P. Berg, *Cell* **9,** 695 (1976).

[20] D. H. Hamer, D. Davoli, C. A. Thomas, Jr., and G. C. Fareed, *J. Mol. Biol.* **112,** 155 (1977).

[21] P. Upcroft, H. Skolnik, J. A. Upcroft, D. Solomon, G. Khoury, D. H. Hamer, and G. C. Fareed, *Proc. Natl. Acad. Sci. U.S.A.* **75,** 2117 (1978).

nature of their terminal sequences (e.g., complementary cohesive restriction endonuclease sites, flush or blunt ends or complementary homopolymer tails), we have selected one example to illustrate how the SV40 genome is utilized for molecular cloning.

We describe the construction of hybrid molecules between SV40 DNA and the *argF* gene of *E. coli*. The *E. coli* K12 chromosome contains two genes coding for the enzyme ornithine transcarbamylase (OTCase): *argF*, located at 6 min near *proA/B*,[22] and *argI* located at 95 min between *pyrB* and *valS*.[23] Both genes code for an inactive protomer, three of which associate to form the active trimeric enzyme. The two enzymes catalyze the formation of citrulline from ornithine and carbamyl phosphate and cross-react immunologically. However, substantial nonhomology exists between the two genes.[24]

A 1650-bp fragment carrying the entire *argF* gene has been isolated from the specialized transducing bacteriophage ϕ80 *argF* λ *cI857* and cloned on the plasmid pBR322 between the *Bam*HI and *Eco*RI cleavage sites.[25] When amplified on this plasmid, large quantities of the *argF* gene can be obtained. The gene contains an *Eco*RI site on its left end and a *Bam*HI site on its right end; transcription initiates at a promoter near the *Eco*RI site and proceeds toward the *Bam*HI terminus. The 1650-bp fragment thus is capable of programming the synthesis of OTCase in a coupled transcription–translation cell-free system.[25]

Procedures and Results

Insertion of the argF Gene between BamHI and EcoRI Sites in the SV40 Late Gene Region

Figure 2 shows an agarose gel analysis[26,27] of the *Eco*RI and *Bam*HI cleavage products of SV40 and pBR322*argF* DNA. Cleavages of SV40 DNA with these nucleases produce 4250- and 750-bp fragments. From the pBR322*argF* DNA, 4000-bp (the plasmid portion) and a 1650-bp (the *argF* gene) fragments are generated. Our first intention was to insert the *argF* gene into SV40 at the *Eco*RI and *Bam*HI sites (Fig. 1) and use the recombinant molecule to transfect permissive monkey cells. Since the recombinant genome exceeds wild-type SV40 DNA in size, it cannot be pack-

[22] N. Glansdorff, G. Sand, and C. Verhoef, *Mutat. Res.* **4,** 742 (1967).
[23] B. Bachman, K. B. Low, and A. Taylor, *Bacteriol. Rev.* **40,** 116 (1976).
[24] D. Sens, W. Natter, and E. James, *Cell* **10,** 275 (1977).
[25] S. K. Moore, E. James, P. M. James, and G. C. Fareed, *Gene* **4,** 261 (1978).
[26] C. Aaij and P. Borst, *Biochim. Biophys. Acta* **269,** 192 (1972).
[27] P. A. Sharp, B. Sugden, and J. Sambrook, *Biochemistry* **12,** 3055 (1973).

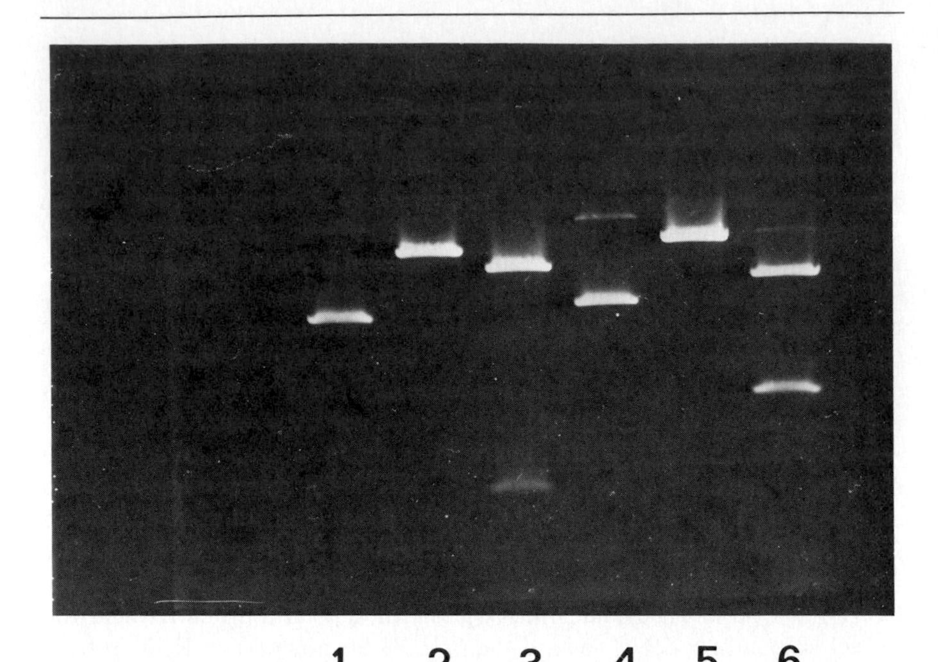

FIG. 2. Cleavage of SV40 and pBR322*argF* DNA with *Eco*RI and *Bam*HI. SV40 and pBR322*argF* were digested with *Eco*RI and *Bam*HI, and the products fractionated on a 1% agarose gel, stained with ethidium bromide (2 μg/ml H_2O, 30 min, 23°), illuminated with uv light, and photographed. (1) SV40 DNA I; (2) SV40 plus *Eco*RI; (3) SV40 plus *Eco*RI plus *Bam*HI; (4) pBR322*argF* I; (5) pBR322*argF* plus *Eco*RI; (6) pBR322*argF* plus *Eco*RI plus *Bam*HI. Each track contains 1 μg of DNA.

aged into virions. It should, however, be capable of replication in permissive cells and perhaps be retained in an episomal form.

SV40 DNA and pBR322*argF* DNA were separately cleaved with *Eco*RI and *Bam*HI, and the digestion products fractionated on preparative agarose gels. The gels were stained with ethidium bromide and illuminated with uv-long wavelength light, and the SV40 4250-bp vehicle fragment and the 1650-bp *argF* fragment were cut out with a razor blade. DNA fragments were recovered from agarose gel slices either by centrifugation of agarose DNA mixtures to equilibrium in KI gradients[28] or by the following procedure. The gel slices were dissolved in 10 volumes of 0.01 *M* Tris, pH 7.2, and 0.001 *M* EDTA saturated with KI. The solutions were chromatographed on hydroxyapatite,[29] and the columns washed with 0.05 *M* potassium phosphate, pH 6.8, until the A_{260} was zero. We

[28] N. Blin, A. V. Gabain, and H. Bujard, *FEBS Lett.* **53,** 84 (1975).

[29] D. A. Wilson and C. A. Thomas, Jr., *Biochim. Biophys. Acta* **331,** 333 (1973).

found this step to be necessary in order to obtain DNA fragments that could be efficiently joined. The DNA was eluted with 0.4 *M* potassium phosphate, pH 6.8, dialyzed against 100 volumes of STD (0.01 *M* Tris, pH 7.2, 0.001 *M* EDTA, 0.015 *M* NaCl) for 24 hr, adjusted to 0.2 *M* NaCl, and precipitated with 2.5 volumes of 95% ethanol. The samples were stored at −20° until further use.

In order to determine the optimal joining time 2 μg of *argF* 1650-bp fragment was mixed with 4 μg of SV40 vehicle (molar ratio or 1650-bp fragment/4250-bp fragment ratio = 1.4) in a total volume of 150 μl containing 0.1 M Tris, pH 7.4, 50 m*M* NaCl, 7 m*M* $MgCl_2$, 6.4 m*M* β-mercaptoethanol, and 80 μ*M* ATP. Then, 0.084 unit of bacteriophage T4 polynucleotide ligase was added and the reaction incubated at 12°. Samples were removed after 2, 4, and 8 hr and analyzed on agarose gels (Fig. 3). As can be seen, the products of the reaction consisted of *argF* dimers, SV40 dimers, and SV40-*argF* monomers and higher oligomers.

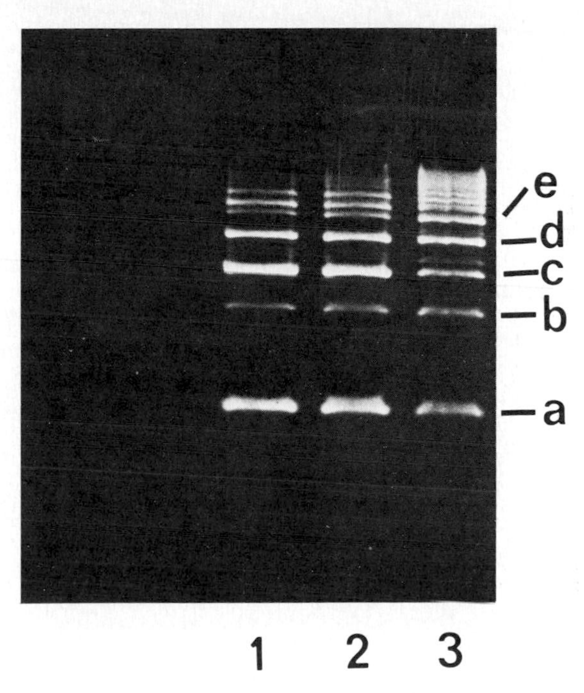

FIG. 3. Time course for ligation of SV40 (*Eco*RI, *Bam*) vector to *argF* DNA. SV40 (*Eco*RI, *Bam*) vector (4 μg) was mixed with 2 μg of *argF* DNA and ligated in a total volume of 150 μl as described in the text. Samples of the reaction were removed at the indicated times, fractionated in agarose gels (1%), stained with ethidium bromide, illuminated with uv light, and photographed. (1) Two-hour ligation; (2) 4-hr ligation; (3) 8-hr ligation. (a) Position of the *argF* 1650-bp monomer; (b) 1650-bp dimers; (c) SV40 (*Eco*RI, *Bam*HI) vector linears; (d) SV40-*argF* linear recombinant; (e) SV40 vector dimers.

Notice that with time there is an accumulation of high-molecular-weight products.

From Fig. 3 we determined that a 4- to 6-hr joining time was optimal for the production of SV40-*argF* monomers. The joining reaction mixture was scaled up, and 75 μg of 1650-bp *argF* DNA was joined with 150 μg of [^{3}H]thymidine-labeled SV40 4250-bp vehicle fragment in 1.5 ml (three separate 500-μl reactions). We found that about one-third of the ligase concentration used in the reaction shown in Fig. 3 (0.072 unit/500 μl) was sufficient. After 6 hr at 12°, a sample of the reaction was electrophoresed on an analytical agarose gel, and the remainder of the material was precipitated with 3 volumes of enthanol at −20°. When the joining reaction had proceeded to a satisfactory level, the SV40-*argF* linears were purified

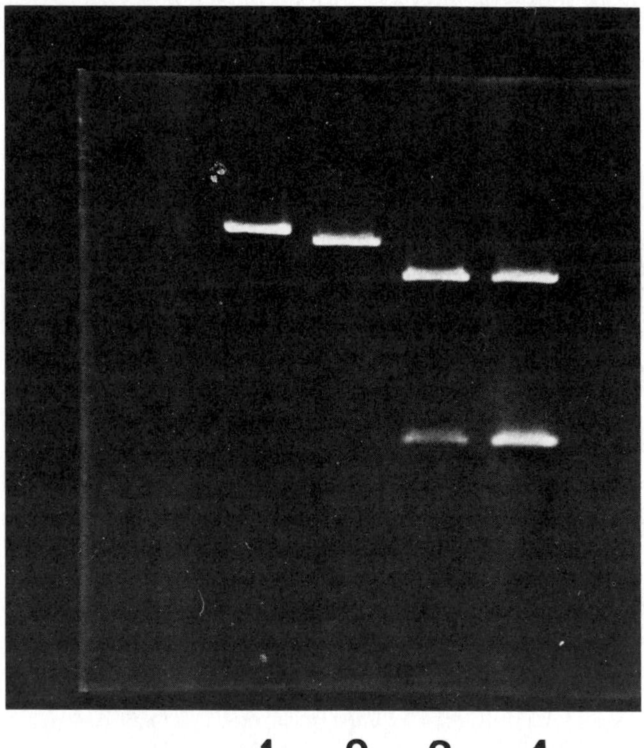

FIG. 4. Analysis of SV40 (*Eco*RI, *Bam*)-*argF* linear recombinant. SV40 (*Eco*RI, *Bam*HI) and *argF* DNA were ligated, and the recombinant band purified from the reaction products after fractionation on agarose gels (1%) as described in the text. (1) Purified SV40 (*Eco*RI, *Bam*HI)-*argF* linear recombinant; (2) pBR322*argF* linear DNA (5650 bp); (3) SV40 (*Eco*RI, *Bam*HI)-*argF* linear molecules digested with *Eco*RI and *Bam*HI; (4) SV40 (*Eco*RI, *Bam*HI) 4250-bp vector and *argF* marker DNAs.

from a preparative agarose gel as described above. The final yield was about 12 μg of recombinant DNA after dialysis.

Figure 4 shows an agarose gel analysis of the purified linear recombinant. It is evident that the cleavage with *Eco*RI and *Bam*HI gives two DNA fragments which coelectrophorese with 4250- and 1650-bp markers. The linear recombinant was circularized in a 500 μl ligation reaction containing 0.5 μg DNA and 0.026 unit of T4 ligase for 48 hr at 12°. Additional ligase (0.036 unit) and ATP (80 μM) were added after 24 hr. Figure 5

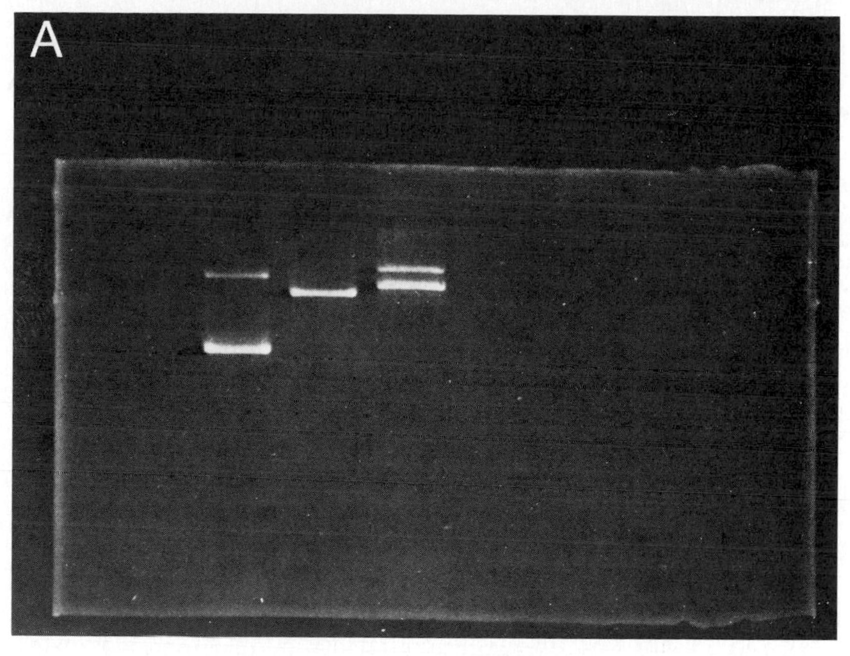

FIG. 5. Circularization of SV40 (*Eco*RI, *Bam*HI)-*argF* linear recombinant. SV40 (*Eco*RI, *Bam*HI)-*argF* DNA was circularized at a concentration of 1 μg/ml as described in the text and the products of the reaction analyzed as follows. (A) Agarose gel (13). (1) pBR322*argF* form I (supercoil) and form II (open circles); (2) SV40 (*Eco*RI, *Bam*HI)-*argF* linears; (3) SV40 (*Eco*RI, *Bam*HI)-*argF* molecules after circularization. The majority of the DNA (75%) is uncircularized and migrates in the position of the SV40 (*Eco*RI, *Bam*HI)-*argF* linears shown in (2). The circularized material migrates just behind this band. The band near the top of the gel in (3) is due to a small amount of SV40-*argF* dimers formed during the reaction. (B) CsCl–ethidium bromide equilibrium centrifugation. A sample of the circularization reaction (6000 cpm) described above was subjected to dye density centrifugation in CsCl–ethidium bromide. The gradients were collected by bottom puncture and counted. Centrifugation is from right to left. The arrow indicates the position of SV40 linear molecules run in a parallel gradient.

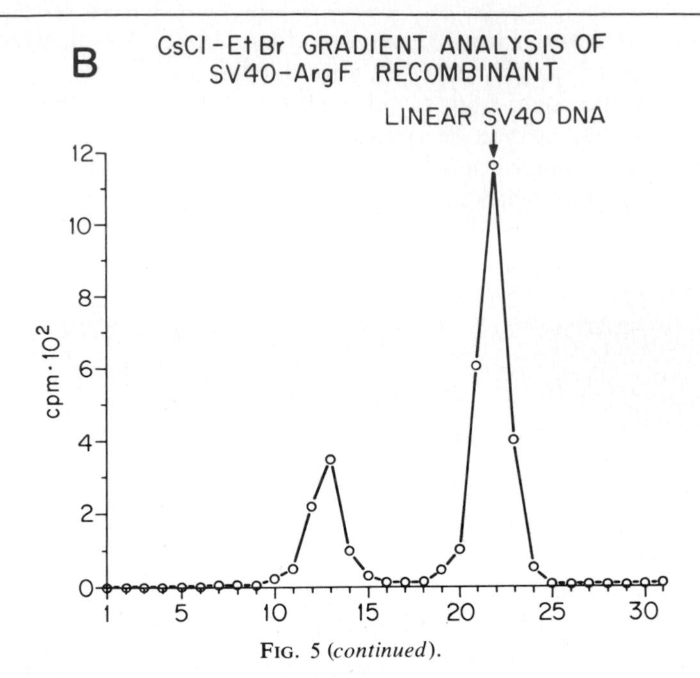

FIG. 5 (*continued*).

shows that about 25% of the DNA was circularized under these conditions as analyzed by electrophoresis through agarose gels or CsCl–ethidium bromide isopycnic centrifugation.

Propagation of the Recombinant Genome in the Absence of a Helper Virus

A 2-μg sample of the SV40-*argF* recombinant (circularized as described above) was used without further purification to transfect a 75 cm² flask of rhesus monkey kidney cells (LLC-MK2 cells). Transfections were performed in the presence of 1 mg/ml DEAE-dextran.[20] The cells were expanded and tested for T antigen by the indirect immunofluorescence method. They were approximately 5% T-antigen-positive 3 days after transfection.

Supercoiled DNA was extracted by the Hirt procedure,[30] digested with *Eco*RI and *Bam*HI, fractionated on agarose gels, transferred to nitrocellulose strips,[31] and hybridized with nick-translated ^{32}P-labeled pBR322-*argF* DNA.[32,33] Figure 6 shows that, when supercoiled DNA was

[30] B. Hirt, *J. Mol. Biol.* **26,** 365 (1967).
[31] E. M. Southern, *J. Mol. Biol.* **98,** 503 (1975).
[32] T. Maniatis, S. G. Kee, A. Efstratidis, and F. C. Kafatos, *Cell* **8,** 163 (1976).
[33] P. W. J. Rigby, M. Dieckmann, C. Rhodes, and P. Berg, *J. Mol. Biol.* **113,** 237 (1977).

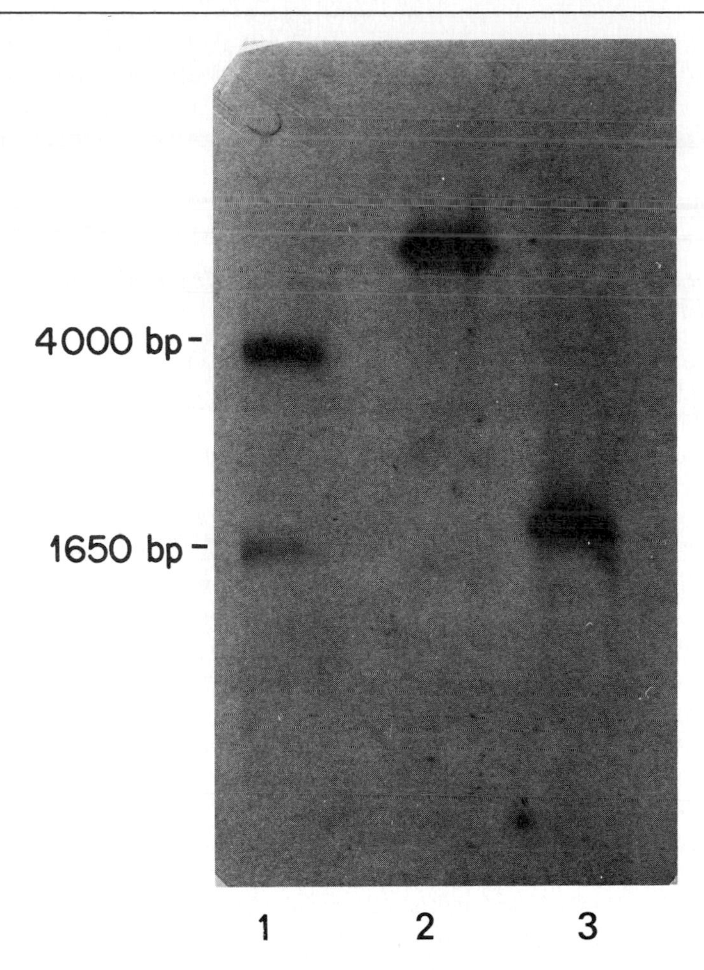

FIG. 6. Analysis of superhelical DNA from CV1-P cells transfected with the SV40 (*Eco*RI, *Bam*HI)-*argF* circularized recombinant. A 75-cm² flask of CV1-P cells was transfected with 2 μg of DNA from the circularization reaction shown in Fig. 5. The cells were expanded, and supercoiled DNA was prepared by the method of Hirt.[26] This DNA was cleaved with *Eco*RI and *Bam*HI, fractionated on an agarose gel, transferred to a nitrocellulose filter,[17] and hybridized to ^{32}P-labeled pBR322*argF* DNA. (1) pBR322*argF* DNA cleaved with *Eco*RI plus *Bam*HI (100 pg); (2) superhelical DNA from transfected CV1-P cells, cleaved with *Eco*RI; (3) superhelical DNA from transfected CV1-P cells cleaved with *Eco*RI plus *Bam*HI.

isolated from these cells and cleaved with *Eco*RI, the DNA that hybridized with the labeled probe migrated more slowly than the 4000-bp pBR322 marker, as expected for a linear molecule of 5900 bp. When the DNA was cleaved with *Eco*RI and *Bam*HI and hybridized with the probe, a single band which comigrated with the 1650-bp *argF* marker was observed, indicating that the recombinant SV40-*argF* molecule had replicated and had retained these restriction sites. Within weeks, however, the cells were all T-antigen-negative, and no *argF* DNA sequences could be detected. The procedure was repeated using African green monkey kidney cell lines (BSC-1, CV-1P, and TC-7 cells), and the same results were obtained.

Propagation of an SV40-argF recombinant DNA by Lytic Infection with a Helper Virus

We next focused our attention on the construction of hybrid molecules that could be propagated lytically in monkey cells by coinfection with a *tsA* helper. To do this, the *argF* gene had to be inserted into SV40 in such a way that the recombinant could be encapsulated. We therefore purified a 3325-bp vector from a *Hae*II (0.83 map unit) and *Bam*HI (0.15 map unit) digest of SV40 DNA and a 2900-bp vector from a *Hpa*II (0.725 map unit) and *Bam*HI digest of SV40. Both vectors contained the origin of replication of SV40 DNA and the entire early gene region. Our plan was then to link the *argF* gene region to the SV40 vehicles through the *Bam* site and infect cells with the linear recombinant along with circular DNA from SV40. In this approach, one depends upon cellular enzymes to circularize the recombinant.

In order to limit the products of the ligation reaction and increase our yields of recombinant, we treated our DNA samples with alkaline phosphatase to inhibit ligation at undesirable sites. pBR322*argF* (1 μg) was cleaved to the linear form with *Eco*RI, ethanol-precipitated, dissolved in 1 ml of 0.01 *M* Tris, pH 8.0, and treated with 48 μg of bacterial alkaline phosphatase (BAP, Worthington, 35.6 u/mg) for 1.75 hr at 37°. The reaction mixture was extracted three times with phenol, ethanol-precipitated twice, and cleaved with *Bam*HI; then the 1650-bp *argF* fragment was purified as described above. As shown in Fig. 7, this fragment was only able to form dimers in our ligation reaction, since it can only be joined at the *Bam*HI site; the addition of polynucleotide kinase to this reaction (0.03 unit/25 μl, P-L Biochemicals) resulted in the appearance of higher-molecular-weight ligation products.

SV40 DNA was separately cleaved with *Hpa*II and *Hae*II and treated with BAP as described above (the DNA from the *Hae*II digest was treated with BAP at 65° for 30 min), phenol-extracted three times, precipitated twice with ethanol, and digested with *Bam*HI; then the 3325-bp (*Hae*II, *Bam*HI) and 2900-bp (*Hpa*II, *Bam*HI) fragments were purified as described above. The SV40-*argF* recombinants were constructed by ligating the SV40 vectors in the presence of a fivefold molar excess of 1650-bp *argF* which had been treated with BAP at its *Eco*RI site. Figures

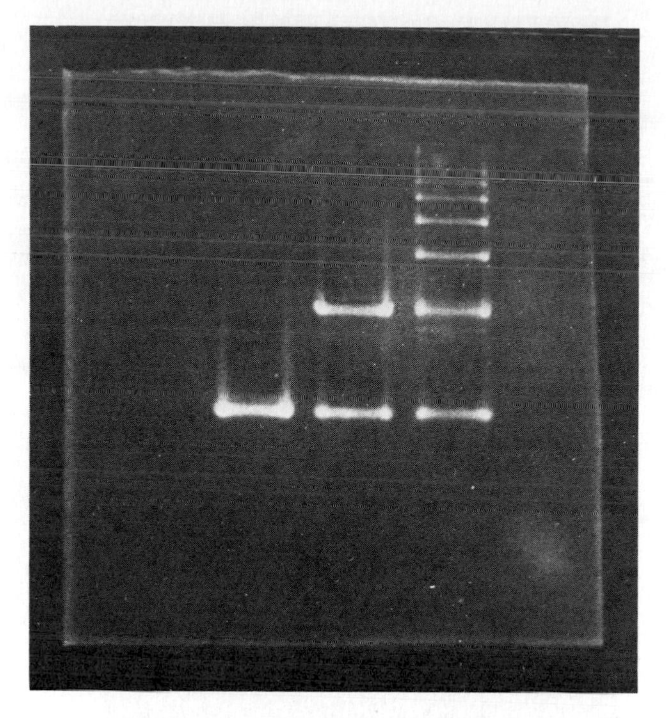

FIG. 7. Ligation of *argF* DNA after treatment with alkaline phosphatase. pBR322*argF* DNA was cleaved with *Eco*RI, treated with alkaline phosphatase, and cleaved with *Bam*HI; the 1650-bp *argF* fragment was purified as described in the text. (1) *argF* DNA, unligated; (2) *argF* DNA, ligated; (3) *argF* DNA, ligated in the presence of polynucleotide kinase.

8A and B show the results of such a ligation. Five species of DNA are observed: unligated SV40 vehicle, unligated *argF* insert, SV40 vehicle dimers, *argF* dimers, and SV40-*argF* recombinant linears. Note that in Fig. 8a the SV40 vehicle (3325 bp) migrates just behind the *argF* dimers, while in Fig. 8b the *Hpa*II, *Bam*HI SV40 vehicle (2900 bp) migrates faster than the *argF* dimers.

The DNA from the joining reactions shown in Fig. 8a and b was used for transfection without further purification. DNA (0.75 μg containing 0.15 μg of recombinant) was mixed with 0.15 μg of *tsA 58* DNA and used to transfect a 60-cm^2 dish of CV1-P cells in the presence of 1 mg/ml DEAE-dextran as described.[34] The cells were overlaid with agar, and plaques were detected by staining with neutral red as described by Mertz and Berg.[34] Positive plaques were detected by a modification of the Viller-

[34] J. E. Mertz and P. Berg, *Virology* **62,** 112 (1974).

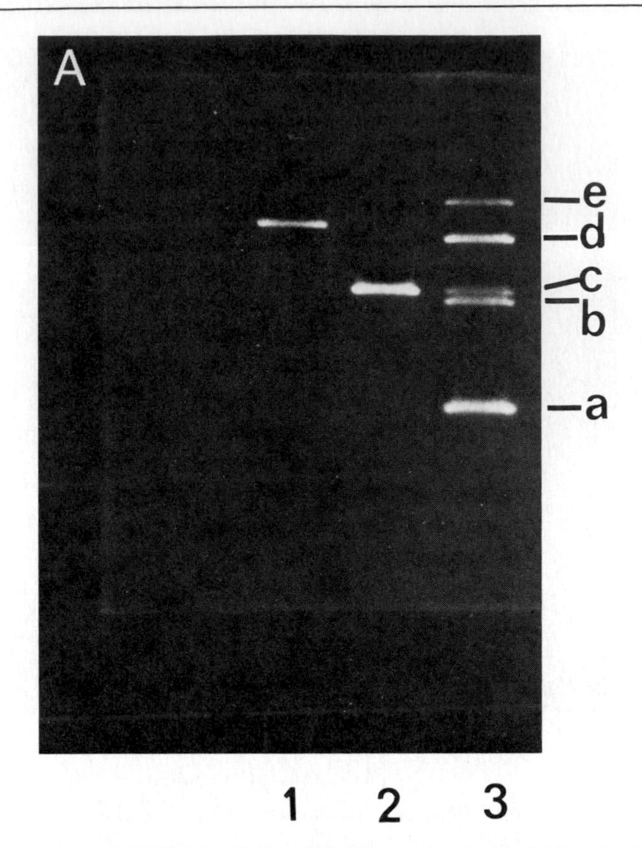

FIG. 8A. Ligation of SV40 (*Hae*II, *Bam*HI) DNA treated with BAP at its *Hae*II end to *argF* DNA treated with BAP at its *Eco*RI end. SV40 DNA was cleaved with *Hae*II, treated with BAP and cleaved with *Bam*HI; the 3325 bp vector was purified on agarose gels as described in text. This DNA was ligated to a fivefold molar excess of *argF* DNA which had been similarly treated with BAP at its *Eco*RI end, and a sample of the DNA from this reaction analyzed on an agarose gel (1%). (1) pBR322*argF* linear DNA (5650 bp); (2) SV40 (*Hae*II, *Bam*HI) vector DNA (3325 bp); (3) SV40 (*Hae*II, *Bam*HI) ligated to *argF* DNA. (a) The position of the *argF* 1650-bp monomers; (b) *argF* 1650-bp dimers; (c) SV40 (*Hae*II, *Bam*HI) vector monomers; (d) SV40 (*Hae*II, *Bam*HI)-*argF* linear recombinant; (e) SV40 (*Hae*II, *Bam*HI) vector dimers.

real and Berg procedure.[35] Each plaque was aspirated and used to infect a 25-cm^2 flask of CV-1P cells. When the cells had completely lysed, 50 μl of each lysate was applied to a Millipore filter disk under vacuum. The disks were treated as described,[35] hybridized with ^{32}P-labeled, nick-translated pBR322*argF* DNA, and autoradiographed. Figure 9 shows the detected of positive plaques from cells infected with SV40 (*Hae*II, *Bam*HI)-*argF* re-

[35] L. P. Villarreal and P. Berg, Science **196,** 183 (1977).

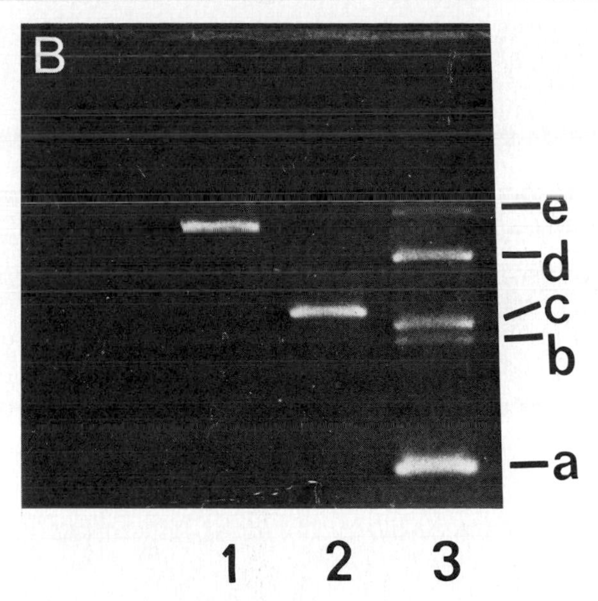

FIG. 8B. Ligation of SV40 (*Hpa*II, *Bam*HI) DNA treated with BAP at its *Hpa*II end to *argF* DNA treated with BAP at its *Eco*RI end. SV40 DNA was cleaved with *Hpa*II, treated with BAP, and cleaved with *Bam*HI; the 2900-bp vector was purified as described in the text. This DNA was ligated to a fivefold molar excess of the 1650-bp *argF* DNA which had been similarly treated with BAP at its *Eco*RI end, and a sample of the DNA from this reaction analyzed on an agarose gel. (1) pBR322*argF* linear DNA (5650 bp); (2) SV40 (*Hae*II, *Bam*HI) vector DNA (3325 bp); (3) SV40 (*Hpa*II, *Bam*HI) ligated to *argF* DNA. (a) The position of the *argF* 1650-bp monomers; (b) SV40 (*Hpa*II, *Bam*HI) vector monomers; (c) *argF* dimers; (d) SV40 (*Hpa*II, *Bam*HI)-*argF* linear recombinant; (3) SV40 (*Hpa*II, *Bam*HI) vector dimers.

combinant. While this procedure involves expansion of virus stocks in addition to the one described by Villareal and Berg, we found it more desirable since we did not have to match the autoradiograph with both the filter disk and the agar from the culture dish to locate the appropriate plaques. Positive virus stocks from cells transfected with the SV40 (*Hpa*II, *Bam*HI)-*argF* recombinant were obtained in a similar fashion.

Since we relied on cellular enzymes to circularize the linear recombinants, we suspected that nucleotide sequences may be lost from either end. In order to examine clones that had retained a significant portion of *argF* gene, we prepared superhelical DNA from cells infected with positive virus stocks, cleaved it with *Hin*dIII and *Bam*HI, fractionated it on agarose gels, transferred it to nitrocellulose strips,[31] and hybridized it to ^{32}P-labeled nick-translated pBR322*argF*. *Hin*dIII cleaves the 1650-bp fragment about 150 bases in from the *Eco*RI site.[25] Since the *Eco*RI site on the recombinant would be very likely lost upon circularization, the *Hin*dIII,

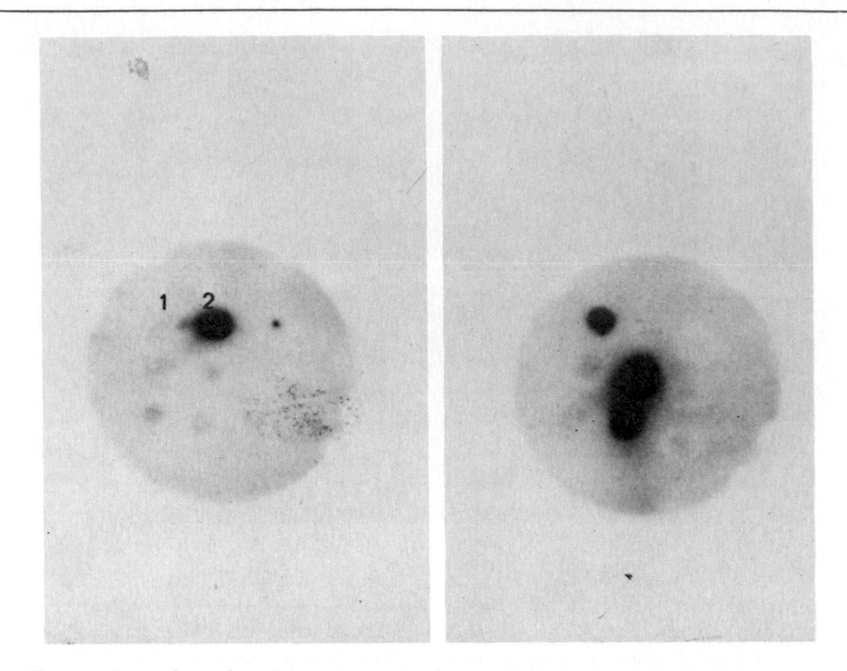

FIG. 9. Detection of positive clones by filter hybridization. Individual plaques from CV1-P cells transfected with the SV40 (*Hae*II, *Bam*HI)-*argF* recombinant were aspirated and used to infect a 25 cm^2 flask of CV1-P cells. The cells were allowed to lyse after 8–10 days in a total volume of 5 ml. Fifty microliters of the cellular lysates from each plaque was applied to a Millipore filter and dried under vacuum. The filter was processed as described,[30] hybridized to ^{32}P-labeled pBR322*argF* DNA, and autoradiographed. Each filter contained nine separate lysate samples. Position 1 contains a control lysate from cells infected with SV40; position 2 contains the same lysate to which 100 pg of denatured pBR322*argF* DNA has been added. The remaining positions contain lysates from cells infected with SV40 (*Hae*II, *Bam*HI)-*argF* plaques.

*Bam*HI fragment is then the largest intact *argF* DNA segment that could be cleaved out of the SV40-*argF* hybrid.

Figure 10 shows the results of such an analysis on several of our clones. Clones 3, 4, and 5 yield DNA fragments which coelectrophorese with a *Hin*dIII digest of the 1650-bp *argF* gene. Clones 1 and 2 produce fragments larger than the 1650-bp fragments and probably arose as a result of the loss of the *Hin*dIII site on the *argF* gene and consist of *argF* sequences joined to SV40 sequences starting from 0.67 (a *Hin*dIII site on SV40 DNA). Clones 3 and 4 were obtained from cells transfected with the SV40 (*Hae*II, *Bam*HI)-*argF* hybrid, while clone 5 was from cells transfected with the SV40 (*Hpa*II, *Bam*HI)-*argF* hybrid.

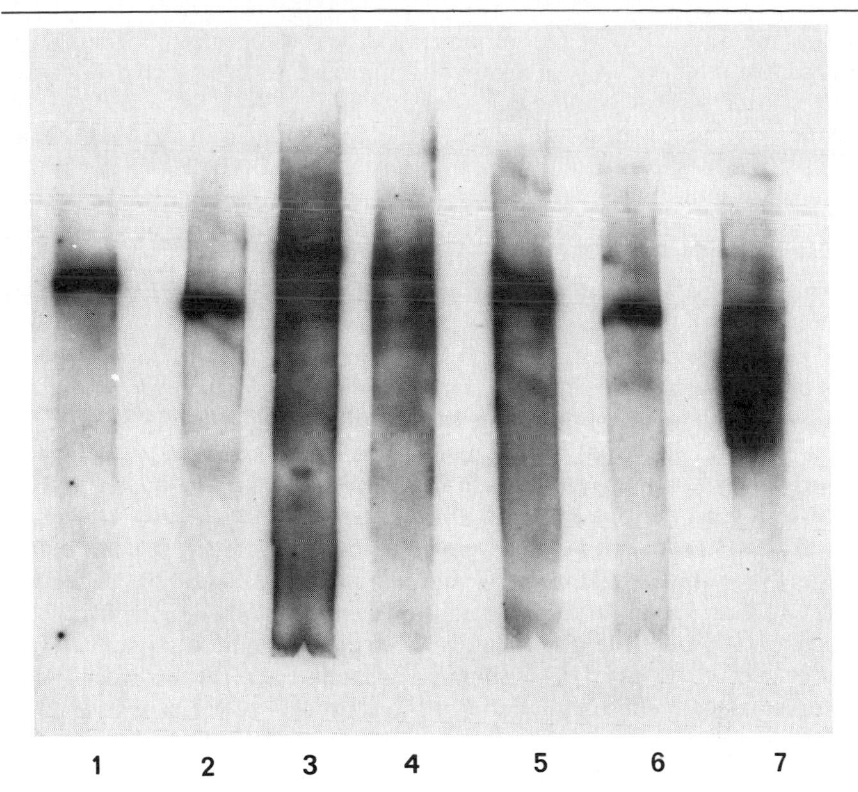

FIG. 10. Analyses of superhelical DNA from CV1-P cells infected with recombinant virus stocks. CV1-P cells were infected with either SV40 (*Hae*II, *Bam*HI)-*argF* or SV40 (*Hpa*II, *Bam*HI)-*argF* virus stock. Superhelical DNA was purified by the method of Hirt (26) 64 hr postinfection. The DNA samples were digested with *Hin*dIII plus *Bam*HI, fractionated on a 1% agarose gel, transferred to a nitrocellulose filter, and hybridized with ^{32}P-labeled pBR322*argF* DNA. (1) Two nanograms of *argF* DNA; (2) 2 ng of *argF* DNA cleaved with *Hin*dIII; (3–6) SV40 (*Hae*II, *Bam*HI)-*argF* clones 1–4 (5 μg); (7) SV40 (*Hpa*II, *Bam*HI)-*argF* clone 5 (5 μg).

Discussion

Techniques are currently available for propagating foreign DNA segments in mammalian cells using SV40 and other papovaviruses as cloning vectors. The foreign DNA may be inserted into the late region of SV40 (1) between two appropriate restriction endonuclease sites,[17,18,20] (2) by the poly(dA:dT) joining procedure,[19] or (3) by joining at one endonuclease site and allowing for circularization *in vivo*. If the size of the inserted DNA segment is less than or equal to the fragment excised from the late gene region, one may coinfect permissive cells with a *tsA* helper viral

DNA and hybrid and helper genomes will be packaged into virions at the restrictive temperature. An analogous approach to this involves insertion of the foreign DNA into the early gene region of SV40 DNA and the subsequent cloning with a late gene *ts* mutant helper (*tsB* or *C*). Alternatively, one may construct hybrids larger than wild-type SV40 and propagate these molecules as free-replicating episomes in appropriate permissive or semipermissive host cells. A limitation of this approach at present is the difficulty in the selection of cloned cells lines that retain viability, express the SV40 T antigen, and continue to replicate nonrearranged recombinant viral genomes.

We attempted to propagate an SV40-*argF* hybrid containing approximately 1100 bp more than wild-type SV40 DNA by insertion of the 1650-bp *argF* segment between the *Eco*RI and *Bam*HI sites on SV40 genomes. A significant fraction of transfected cells (5%) became T-antigen-positive and were presumably replicating the SV40-*argF* hybrid. However, this episome was lost from these cells after a few weeks.

We have propagated in mammalian cells the *argF* gene from *E. coli* by joining it to the *Bam*HI of SV40 and allowing circularization to occur *in vivo*. Joining at one restriction endonuclease site and allowing for *in vivo* circularization has the disadvantage of losing nucleotide sequences from either end of the linear recombinant, and one thus has to screen hybrid virus stocks for those containing a large portion of DNA insert. Three clones that we have examined have retained a large portion of the gene (at least 90%), and we have detected polyadenylated *argF*-specific RNA from cells infected with these virus stocks. We suspect that the polyadenylated mRNA may be due to the proximity of the *Bam*HI site to the termination site for late SV40 mRNA species (0.17 map unit).

Concluding Remarks

Several methods are currently available to introduce new genetic information into cells. The thymidine kinase (TK) gene has been stably introduced into a line of thymidine kinase-deficient (TK^-) mouse cells by direct transfection with DNA segments from herpes simplex virus (HSV).[36–39] Clonal isolates of cells with a TK^+ phenotype were shown to contain the HSV TK gene in an integrated form. It may soon be possible to use the HSV TK gene fragment as a carrier or vector for the transduction of additional genes into TK^- host cells. These investigators have more recently shown that the high-molecular-weight DNA from a variety

[36] M. Wigler, S. Silverstein, L.-S. Lee, A. Pellicer, Y.-C. Cheng, and R. Axel, *Cell* **11,** 223 (1977).

[37] N. J. Maitland and J. K. McDougall, *Cell* **11,** 233 (1977).

[38] S. Bacchetti and F. Graham, *Proc. Natl. Acad. Sci. U.S.A.* **74,** 1590 (1977).

[39] A. Pellicer, M. Wigler, R. Axel, and S. Silverstein, *Cell* **14,** 133 (1978).

of higher eukaryotes can also serve as donors for stable biochemical transformation of TK^- mouse cells.[40,41] Given an appropriate selection procedure, this approach should be feasible for other eukaryotic genes provided a suitable somatic cell mutant for that gene is available.[42]

The only eukaryotic cloning vectors of practical significance to date, however, are papovaviruses. SV40 has proven to be a useful vector for the propagation of new DNA in monkey cells. It can also be used to transduce foreign DNA into nonpermissive hosts, such as rodent cells.[21] Another potentially useful cloning vector is the DNA of polyomavirus. Polyomavirus is a member of the papovavirus group and, like SV40, contains a superhelical genome of about 3×10^6 daltons which has been extensively mapped.[10,11] Mouse cells are permissive hosts for the virus, although it is oncogenic for newborn laboratory mice.[11] Polyomavirus, like SV40, is able to transform cells of other species, such as those from rats and hamsters.[11] While polyoma DNA has not received as much attention as SV40 as a cloning vehicle, it would be a useful agent to use for the introduction of DNA into mouse cells, as well as the cells of other rodents. Recombinant molecules could be propagated "helper-free" or by coinfected with the *tsA* mutant of polyoma. Since several TK^- mouse cell lines exist, a logical development in the near future will be to construct a bivalent cloning vector in which a functional TK gene is inserted into the late gene region of polyomavirus.

Finally, the DNA from adenoviruses could serve as a vector for introducing foreign DNA into avian, murine, or simian cells. The adenoviruses from these animal species contain linear DNA genomes of 2.5×10^7 daltons, which are able to replicate in and transform a variety of cell types.[11] Certain regions of the adenovirus genome are nonessential for replication and can be replaced by foreign DNA segments. Such naturally occurring DNAs are the human adeno–SV40 hybrids that contain a DNA genome consisting of both adenovirus and SV40 sequences.[10,43] While the DNA of an adenovirus is almost an order of magnitude larger than that of SV40 and therefore more difficult to manipulate, improvements in DNA transfection and genetic complementation techniques for larger-sized DNA viruses should yield additional cloning vehicles.

Acknowledgments

We are indebted to Dr. Eric James from the University of Kentucky School of Medicine for invaluable contributions to the research project described here and to the National Cancer Institute, DHEW, for its support (Grant No. CA 20794).

[40] M. Wigler, A. Pellicer, S. Silverstein, and R. Axel, *Cell* **14,** 725 (1978).

[41] A. C. Minson, P. Wildy, A. Buchan, and C. Darby, *Cell* **13,** 581 (1978).

[42] F. E. Farber, J. L. Melnick, and J. S. Butel, *Biochim. Biophys. Acta* **390,** 298 (1975).

[43] A. M. Lewis, M. J. Levin, W. H. Wiese, C. S. Crumpacker, and P. H. Henry, *Proc. Natl. Acad. Sci. U.S.A.* **63,** 1128 (1969).

Section V

Screening and Selection of Cloned DNA

[25] Colony Hybridization

By MICHAEL GRUNSTEIN and JOHN WALLIS

Colony hybridization allows bacterial clones to be rapidly screened by RNA–DNA or DNA–DNA hybridization for those containing specific DNA sequences.[1] Bacterial colonies are grown on nitrocellulose filters lying on agar petri plates. Reference sets of colonies are made by replica plating and stored at 2°–4°. The cells on the filter are lysed, and their DNA is denatured *in situ* without dislodging the remnants of the cells from their initial site. Radiolabeled RNA or DNA probe is then hybridized with the filters. After washing away unhybridized probe, autoradiography reveals which of the DNA prints on the filter and therefore which colonies in the reference set contain the DNA sequence in question (Fig. 1). This procedure has been modified by many investigators. It is now also possible to screen phage plaques in a similar manner, with the added advantages of lambda phage cloning, mainly the ability to screen larger numbers of plaques per plate.[2] It is possible to combine the advantages of lambda and plasmid cloning by the partial lysis of bacteria lysogenic for a thermoinducible phage. This allows a large number of colonies to be screened and obviates the need for replica plating a reference set.[3] The colony hybridization procedure has also been adapted for the screening of yeast transformants[4] and animal cell SV40-induced plaques[5] and has also been used in conjunction with specific antibodies to probe for gene activity (i.e., protein synthesis) in phage lambda plaques[6] and bacterial clones.[7]

Procedure

Formation of the Filter and Reference Set of Colonies

Nitrocellulose filters (obtained from Millipore, HA, 0.45-μm pores) are cut into circles 8.2 cm in diameter. Each filter is oriented with a pencil mark and placed on an agar plate, care being taken to avoid trapping air

[1] M. Grunstein and D. S. Hogness, *Proc. Natl. Acad. Sci. U.S.A.* **72,** 3961 (1975).
[2] W. D. Benton and R. W. Davis, *Science* **196,** 180 (1977).
[3] B. Cami and P. Kourilsky, *Nucleic Acids Res.* **5,** 2381 (1978).
[4] A. Hinner, J. B. Hicks, and G. R. Fink, *Proc. Natl. Acad. Sci. U.S.A.* **75,** 1929 (1978).
[5] L. P. Villarreal and P. Berg, *Science* **196,** 183 (1977).
[6] B. Sanzey, O. Mercereau, T. Ternynck, and P. Kourilsky, *Proc. Natl. Acad. Sci. U.S.A.* **73,** 3394 (1976).
[7] L. Villa-Komaroff, A. Efstradiatis, S. Broome, P. Lomedico, R. Tizard, S. P. Naber, W. L. Chick, and W. Gilbert, *Proc. Natl. Acad. Sci. U.S.A.* **75,** 3727 (1978).

ISBN 0-12-181968-X

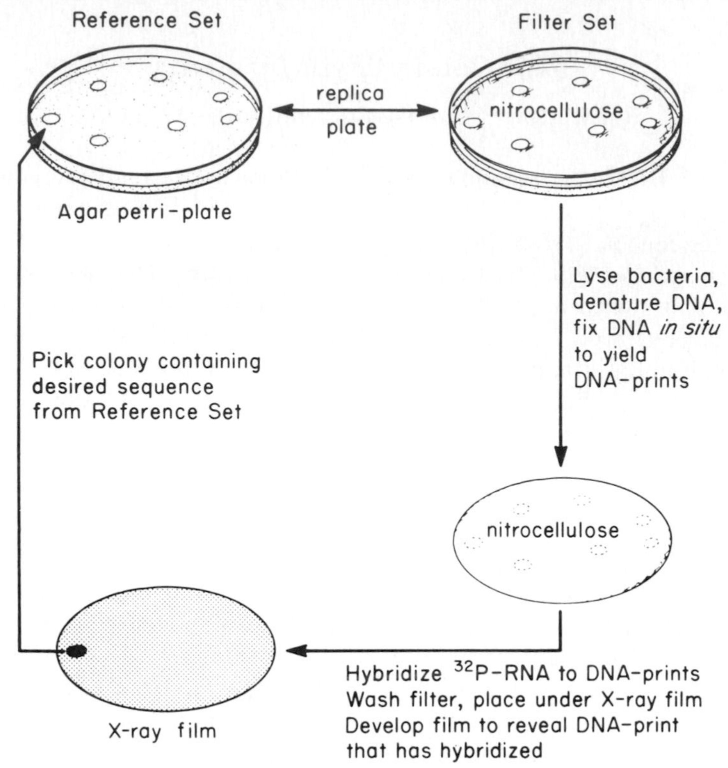

FIG. 1. Colony hybridization rationale.

bubbles beneath the filter. After the filter has absorbed moisture uniformly from the plate (approximately 30 sec) bacteria may be transferred to the filter surface with sterile toothpicks or by standard spreading. Spreading should be performed immediately after delivering 0.2 ml of cells onto the filter surface. The filter plate is incubated at 37° until the colonies are approximately 1.0 mm in diameter. A reference set of colonies is then made by lightly pressing the colonies onto velvet backed by a cylindrical steel or polyvinyl chloride block, and the velvet is pressed onto an additional agar petri plate with the same orientation as the filter set.[8] The reference set of bacteria is allowed to form colonies at 37° and then is stored in an airtight container at 2°–4°C. If the colonies contain colicinogenic plasmids which may be amplified in the presence of chloramphenicol,[9] the following procedure is employed. When the colonies on the filter are approximately 1 mm in diameter, the filter is transferred to

[8] W. Hayes, *in* "The Genetics of Bacteria and Their Viruses," p. 187. Wiley, New York, 1968.

[9] V. Hershfield, H. W. Boyer, C. Yanofsky, M. A. Levett, and D. R. Helinski, *Proc. Natl. Acad. Sci. U.S.A.* **71,** 3455 (1974).

an agar petri plate containing chloramphenicol (170 μg/ml) (L. Simpson, unpublished). These plates are poured after chloramphenicol has been completely dissolved in agar at 60°. The filter is left on the chloramphenicol-containing petri plates for 15–20 hr at 37°.

Cell Lysis and DNA Denaturation

All subsequent steps are performed at room temperature unless otherwise indicated. To prevent movement of the bacteria or the DNA from the initial site during cell lysis and denaturation, it is necessary to apply the first solution through the underside of the filter, allowing it to diffuse into the colonies. The filter is removed from the petri plate and placed, colonies uppermost, on four sheets of Whatman 3 MM absorbent filter paper soaked in 0.5 *N* NaOH. After 5 min the filter is transferred to a filter block and sleeve made to accommodate 8.2-cm filters (Fig. 2), and suction is ap-

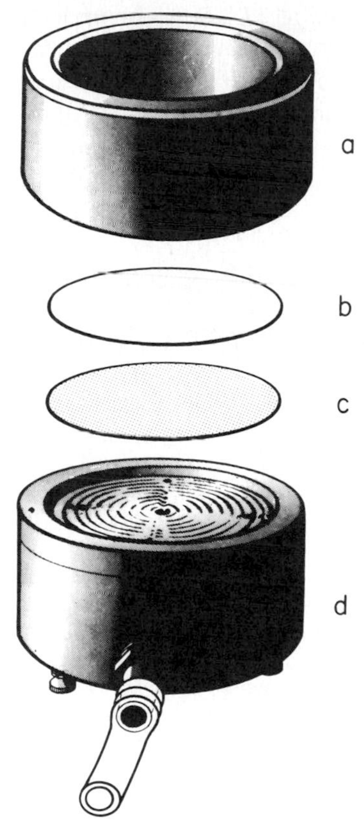

FIG. 2. Filtration block. (a) Stainless-steel collar. (b) Nitrocellulose filter (8.2 cm). (c) Stainless-steel perforated disk. (d) Polyvinyl chloride base. For exact measurements refer to Grunstein and Hogness.[1]

plied by means of a water aspirator until the colony residues appear dry (1–2 min). At this point there is little danger of movement from the colony site, and the remaining solutions are applied to the upper side of the filter while moderate to strong suction is applied. Then, 12C ml 1 *M* Tris-HCl, pH 7.6, is poured through the filter in 30-ml aliquots. This is followed by 120 ml 1.5 *M* NaCl 0.5 *M* Tris-HCl, pH 7.6, and then by 120 ml chloroform, both applied in 30-ml aliquots. The filter is removed and allowed to dry on a sheet of absorbent paper at 60° for 15 min followed by baking at 80° *in vacuo* for 2 hr.

Hybridization and Autoradiography

Isotopically Labeled RNA. The filter is moistened with 1.0 ml hybridization buffer (5× SSC,[9a] 50% formamide) containing the isotopically labeled RNA and incubated at 37° for 16 hr under mineral oil or between sheets of Saran Wrap (Dow Chemical) to prevent evaporation. The filter is then washed twice in 50 ml 2× SSC, 30 min each, and then in 50 ml 2× SSC and 20 μg/ml pancreatic RNase for 30 min.[10] These washes take place on a gently rotating shaker platform. The filter is blotted to remove excess moisture and then covered with Saran Wrap and autoradiographed under Kodak X-ray film (XR-1) and a Dupont intensifying screen (Chronex Quanta II) at −77°.

Isotopically Labeled DNA. After cell lysis and DNA denaturation the filter is soaked in 3× SSC, 0.02% polyvinylpyrrolidone, 0.02% Ficoll, 0.02% BSA at 60° for 5 hr[11] and then dried at 60° for 15 min. The isotopically labeled DNA in 0.9 ml 0.01 *M* Tris-HCl, pH 7.5, and 0.001 *M* EDTA is heated in a boiling water bath for 5 min. Then 0.1 ml 20× SSC is

[9a] The following abbreviations are used in this chapter. SSC, standard saline citrate (0.15 *M* NaCl, 0.015 *M* $Na_3C_6H_5O_7 \cdot 2H_2O$); G + C content, mole percent of deoxyguanylic acid (G) and deoxycytidylic acid (C); BSA, bovine serum albumin; EDTA, (ethylenedinitrilo)tetraacetic acid; kb, kilobases—1000 bases or base pairs in single or double-stranded DNA, respectively; Dm, a segment of *Drosophila melanogaster* DNA; cDm and pDm, hybrid plasmids consisting of a Dm segment inserted into ColE1 and pSC101 DNAs, respectively; cRNA, RNA complementary to DNA; C3Sp2, a hybrid plasmid consisting of the plasmid Δ395 (5.6 kb), referred to here as C3, and the histone gene segment Sp2 from the sea urchin *S. purpuratus*. L. Kedes, A. C. Y. Chang, D. Houseman, and S. N. Cohen, *Nature* (*London*) **255,** 533 (1975).

[10] The RNase step has been omitted in some hybridization experiments utilizing ^{34}P-labeled cRNA (2 × 10^7 dpm/μg) as the probe. Instead, the final wash was extended to 90 min in 2× SSC. Although it is possible to obtain specific positive responses without the ribonuclease step, the filter radiation background may on occasion be greater and uneven.

[11] D. T. Denhardt, *Biochem. Biophys. Res. Commun.* **23,** 641 (1966).

added to complete the hybridization solution; the filter is moistened with the hybridization solution and incubated at 55° for 16 hr under mineral oil.

After incubation, the filter is washed three times (60 min each) in 100 ml 2× SSC at 55°. The filter is blotted dry, covered with Saran Wrap, and autoradiographed at −77° under Kodak X-ray film (XR-1) and a Dupont intensifying screen (Chronex Quanta II).

If a colony screen is performed, it is sometimes advantageous to have the DNA prints on the filter set easily visible. This helps in identifying the colony in question on the reference set when a large number of colonies is screened. The DNA print may easily be seen by staining the filter in 0.25% methyl green in water for 1 min and destaining under running water for 3–5 min.

Comments

Amount of Radioactive Probe on the Filter Set

The amount of radioactively labeled RNA or DNA necessary to obtain a strong autoradiographic signal depends on several variables, including the isotope used, the sequence complexity of the probe, the number of plasmid copies per cell, the temperature of hybridization, and the hybridization buffer. An adequate autoradiographic response is an optical density of 0.5–1.0 on the x-ray film, measured by scanning through the middle of the exposed positive signal with a Joyce-Loebl densitometer. A response of this intensity was obtained in the following examples.

1. The RNA probe was *Drosophila* 28 S rRNA (4000 bases) labeled in cell culture with ^{32}P to a specific activity of 2.0×10^5 dpm/μg. Hybridization was carried out with 10,000 dpm/cm^2 of filter in 5× SSC and 50% formamide for 16 hr at 37°. The *Escherichia coli* DNA print scored contained the (pSC101) plasmid hybrid pDm103 (Dm103, 17 kb) which contained a single ribosomal DNA repeat and was present in approximately four copies per cell.[12] Exposure time was 24 hr using Kodak XR-1 x-ray film. If a Dupont Quanta II intensifying screen is used and autoradiography is allowed to take place at −77°, the exposure time may be reduced 10- to 12-fold.

2. cRNA was synthesized *in vitro* using *E. coli* RNA polymerase and ColE1 plasmid DNA (6 kb) as the template[13] and ^{32}P-labeled UTP. Spe-

[12] D. M. Glover, R. L. White, D. J. Finnegan, and D. S. Hogness, *Cell* **5**, 149 (1975).

[13] P. Wensink, D. J. Finnegan, J. E. Donelson, and D. S. Hogness, *Cell* **3**, 315 (1974).

cific activity of the cRNA was 3.0×10^7 dpm/μg. The DNA prints were prepared from cells containing ColE1 plasmids grown without chloramphenicol amplification. Using 30,000 dpm/cm^2 of filter and the same hybridization conditions as in example 1, a positive response was obtained after 4 hr of autoradiography (without the use of an intensifying screen).[1]

3. The probe was a plasmid containing a sea urchin histone gene fragment (C3Sp2) nick-translated with DNA polymerase[14] and ^{32}P-dTTP to a specific activity of 1.8×10^7 dpm/μg. The *E. coli* DNA print scored contained the plasmid C3Sp2 amplified with chloramphenicol (170 μg/ml) as described in the text. Hybridization was carried out with 510 dpm/cm^2 of filter in 5× SSC 50% formamide for 16 hr at 37°. Autoradiography was for 16 hr at −77° with a Dupont Quanta II intensifying screen.

Colony Hybridization: DNA Driven

Note that the specific activity of the labeled probe was not mentioned as a factor contributing to the density of signal obtained by autoradiography. Colony hybridization is relatively insensitive to the specific activity of the probe over a wide range tested (2.5×10^4–2×10^7 dpm/μg). This argues for the presence of a vast excess of DNA in each DNA print compared to the amount of RNA added.

Is there a vast excess of DNA? Each pDm103 DNA print is calculated to contain 2 ng pDm103 DNA.[1] However, as much as 30 ng RNA (750 dpm) has been added per filter, and yet the extent of hybridization observed (0.1–0.2% RNA per 1-mm diameter DNA print) is similar when 0.038 ng (750 dpm) RNA is added. Evidently, despite the apparent conditions of DNA excess, only a small percentage of the RNA has hybridized in 16 hr in 5× SSC and 50% formamide at 37°. This may be explained by (1) assuming that not all the RNA on the filter is available for hybridization. Perhaps because of slow diffusion of RNA through the nitrocellulose membrane, only the radioactive RNA layered directly on top of the DNA print hybridizes readily. Alternatively (2) all the RNA is equally available for hybridization, but the rate is such that we observe only a fraction of the hybrid that will eventually form at infinite time. This latter alternative would predict that having more DNA per DNA print would increase the hybridization rate and the extent of hybridization, since DNA appears to leach from the filters during lengthy incubations.

Chloramphenicol causes the amplification of colicinogenic plasmids in the bacterial cell. In liquid culture this causes the number of plasmids per cell to increase from 20–30 copies to several thousand copies per *E. coli* cell in the presence of the drug.[9] It is found that, by transferring the filter

[14] P. W. J. Rigby, M. Dieckmann, C. Rhodes, and P. Berg, *J. Mol. Biol.* **113**, 237 (1977).

to an agar petri plate containing chloramphenicol (170 μg/ml), the density of the autoradiographic signal may be increased as much as 7- to 10-fold. A likely explanation is that the hybridization reaction occurs in a DNA excess and that its rate is dependent on the concentration of plasmid DNA in the DNA print.

Hybridization Solution and Temperature

Colony hybridization is unlike nucleic acid hybridization in solution or on nitrocellulose (DNA) filters where the only components are the nucleic acid and nitrocellulose. Most of the DNA in the DNA print is not tightly bound to the nitrocellulose filter even after fixation. For example, if the DNA print (colony residue) is removed with a scalpel after the filter is dried, there is very little hybridization of labeled nucleic acid probe with the position of the DNA print.[15] Therefore, the thermal stability of the nucleic acid hybridization is dependent on the stability of the binding of the DNA print to the filter, at least as far as the autoradiographic assay for colony hybridization is concerned. This suggests that the optimal temperature for colony hybridization is not necessarily the same as the optimal temperature for standard nucleic acid hybridization and merits independent investigation.

Nick-translated C3Sp2 DNA labeled with ^{32}P was renatured with DNA prints containing the same DNA sequences. The optimal temperature of the reassociation (T_{opt}) with the DNA prints was determined in 5× SSC and 50% formamide and also in 2× SSC (Fig. 3). The T_{opt} was 30° in the formamide buffer. The T_{opt} (60°) in 2× SSC produced an almost twofold greater autoradiographic response than the formamide T_{opt} (perhaps because of the leaching of DNA from the colonies in formamide). These results suggest that the T_{opt} for colony hybridization in the formamide hybridization solution is approximately 7°–10° lower than that expected for standard DNA reassociation of sequences having a similar (45–50%) G + C content.[16] In 2× SSC, the colony hybridization T_{opt} is approximately 15° lower than the expected value.[16] It was found that, while a reassociation temperature of 60° in 2× SSC gave the strongest autoradiographic response, the filters became brittle and difficult to handle at this temperature; 55° proved to be a more convenient temperature.

[15] To further illustrate this point, colonies may be grown on filter paper other than nitrocellulose, which binds specifically to single-stranded DNA. For example, Whatman No. 1 and No. 540 cellulose have been used (P. Wensink, unpublished) to obtain specific colony hybridization. In our hands, however, the filter radioactive noise or background is somewhat greater with these filters than with nitrocellulose.

[16] B. L. McConaughy, C. D. Laird, and B. J. McCarthy, *Biochemistry* **8,** 3289 (1969).

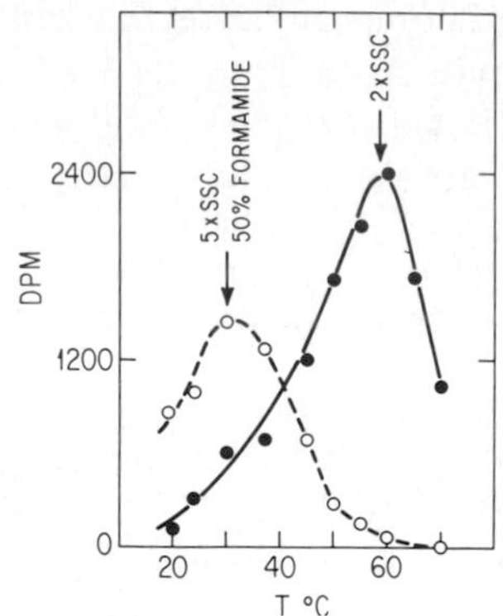

FIG. 3. Optimal temperature of DNA–DNA reassociation. For a given temperature three filters (diameter, 13 mm) containing nine colonies per filter were each hybridized with 20 μl hybridization buffer containing 1640 dpm of nick-translated ^{32}P-labeled C3Sp2 DNA with a specific activity of 10^7 dpm/μg. Hybridization took place for 19 hr. The filters were washed as described in the text, and the three filters were pooled and radioactivity measured by scintillation. Values for background radiation using HB101 bacterial colonies, lacking the plasmid C3Sp2, on the filter were subtracted to give the results shown.

We also measured the T_{opt} for C3Sp2 cRNA hybridization with bacterial colonies containing C3Sp2 DNA (data not shown). The T_{opt} in 5× SSC and 50% formamide was 37°–42°. The T_{opt} in 2× SSC was 60°.

Background Radiation on the Filter Set

There are two types of background radiation in colony hybridizations. In the first the nitrocellulose binds a contaminant of the radioactive precursor used to prepare ^{32}P-labeled cRNA. This binding is less intense in the spot of a DNA print. The DNA print containing the plasmid DNA of interest shows a strong hybridization signal, so three regions are observed on the autoradiograph: some DNA prints showing no hybridization signal, others showing a positive signal, and background filter radiation of intermediate intensity (Fig. 4). The degree of background filter radiation caused by this contaminant varies extensively with the batch of radioactive ribonucleoside triphosphates used. The contaminant chromatographs with RNA on Sephadex G50, coprecipitates in ethanol, and is not easily removed by dialysis. However, it may be completely removed by electrophoresis of the ^{32}P-labeled RNA preparation through agarose in the following manner.[17]

A 10-ml, sawed-off plastic pipette is filled with 4 ml 1% agarose in TA buffer (0.4 *M* Tris, 0.02 *M* CH_3COONa, 0.002 *M* EDTA, 0.2% sodium dodecyl sulfate, pH 7.8). A short length of dialysis tubing, previously

[17] M. Grunstein and P. Schedl, *J. Mol. Biol.* **104,** 323 (1976).

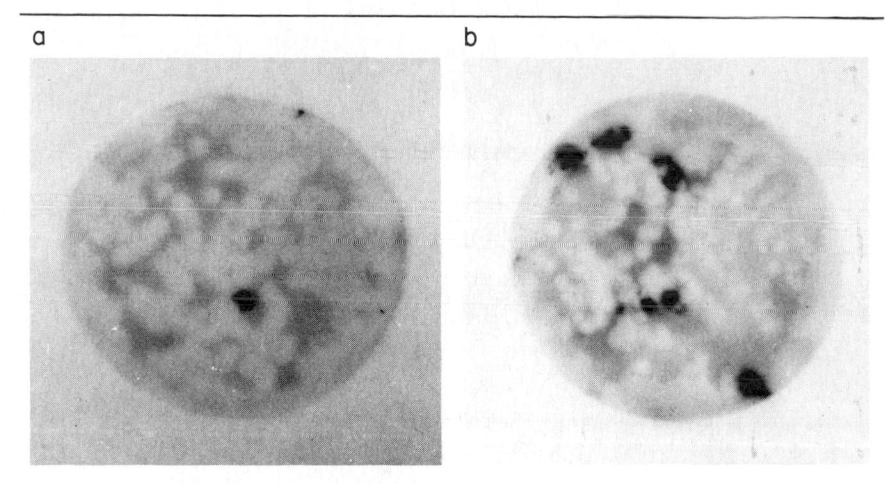

FIG. 4. Filter background radiation. In this case the filter radiation is intermediate between that of the colonies lacking the plasmid ColE1 (HB 101) and those carrying the plasmid (HB101/ColE1). (a) A mixture of ColE1$^+$ and ColE1$^-$ bacteria (HB101) in a 1:100 ratio was spread on a 47-mm filter (area, 17.3 cm^2) to obtain a total of 1–2 × 10^2 colonies per filter; 5 × 10^5 dpm ^{32}P-cRNA (2 × 10^7 dpm/μg) in 250 μl was applied to the filter. Exposure time was 4 hr. Amplification of the autoradiographic signal by the use of either chloramphenicol or an intensifying screen was omitted in this experiment. (b) A similar experiment in which ColEl$^+$ and ColE1$^-$ bacteria were spread in a 1:10 ratio.

boiled three times in 0.1 *M* EDTA and washed in distilled water, is knotted at one end, filled with buffer, and fit securely over the tapered end of the pipette. It is important to exclude air bubbles from the dialysis bag. The labeled RNA (in 0.03% bromphenol blue, 10% glycerol, 2 m*M* EDTA) is electrophoresed (4 mA, 14 hr) into the dialysis tubing containing 0.5 ml TA buffer until the dye has completely entered the dialysis bag. The contaminant does not pass through the gel. After electrophoresis, during which the dye dialyzes from the bag, the sample is shaken with an equal volume of water-saturated phenol and centrifuged for 2 min in an Eppendorf microcentrifuge (Model 3200). 0.1 volume 3 *M* NaCl, 50 μg *E. coli* tRNA (as carrier), and 2 volumes cold ethanol are then added to the aqueous phase. After 15 min at −70°, the precipitated RNA is pelleted by centrifugation. The RNA pellet is washed twice by adding 70% ethanol to the tube (without disturbing the pellet), followed by centrifugation. The pellet is dried under vacuum after removal of the last ethanol wash and resuspended in hybridization buffer.

A second type of background radiation is observed in which the filter is relatively free of background radiation, while *E. coli* DNA prints that do not contain any plasmid display a weak positive response ranging from $\frac{1}{30}$ to $\frac{1}{10}$ of the normal positive response. This occurs when the probe is

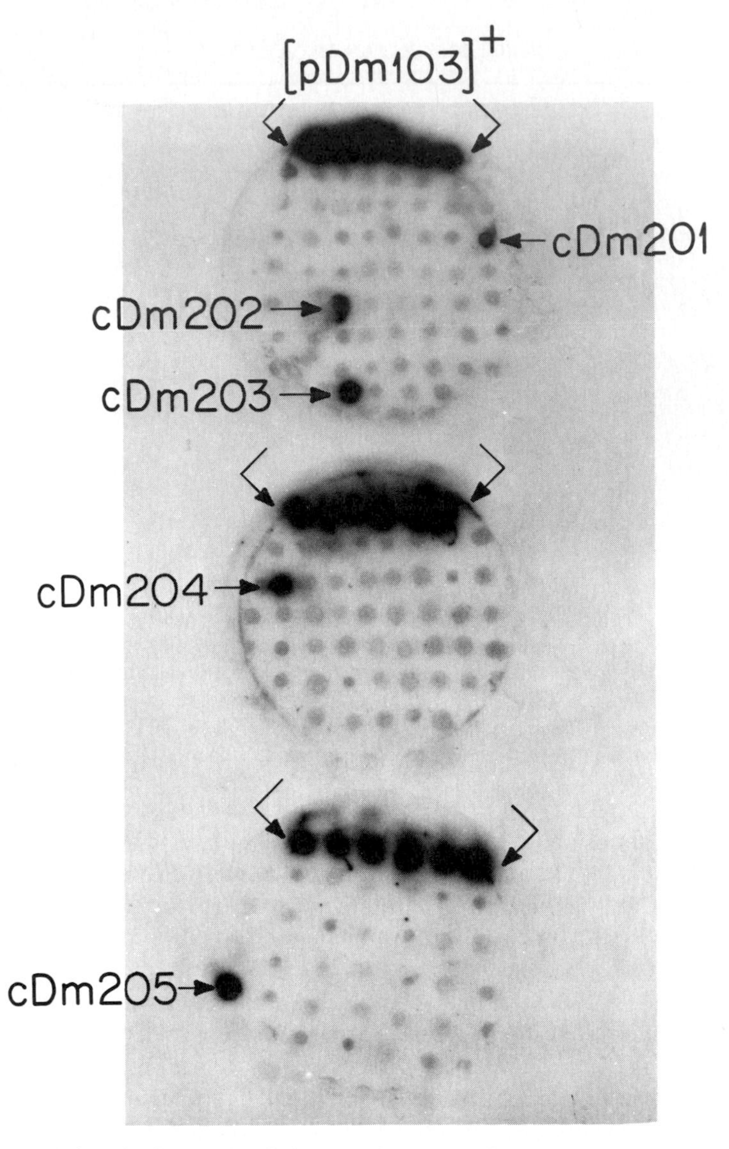

FIG. 5. Colony background radiation. In this case, the filter background is very low; however, the HB101 colonies, lacking the plasmids to be screened, show a slightly positive autoradiographic response. In this experiment hybrids containing ColE1 plasmid DNA joined to random segments of *Drosophila melanogaster* DNA were screened by ^{32}P-cRNA transcribed from pDm103 (rDNA cloned in the tetracycline-resistant plasmid pSC101).[1] Three hundred independent transformants were transferred to six 47-mm filters, each of which contained six control colonies of HB101/pDm103 at the top of the pattern. Then, 5×10^5 dpm of pDm103 ^{32}P-cRNA (2×10^7 dpm/μg) was applied per filter for the colony hybridization. The autoradiograph shown here resulted from three of the six filters after a 5-hr exposure and shows the five rDNA hybrids (cDm201-205) identified by this screening procedure. The experiments in Figs. 4 and 5 are presented to show the types of radiation background that may be obtained and were performed as described in a previous publication.[1]

cRNA prepared from partially purified plasmid DNA and probably results from contamination of the template with *E. coli* sequences. It may be reduced by repurification of the plasmid by CsCl–ethidium bromide centrifugation.[18] An example of this type of response is shown in Fig. 5.

Under conditions where a strong response is expected because of the availability of sufficient purified radioactive RNA or DNA probe, neither background source constitutes a problem. However, background radiation may be a problem if only a small fraction of the nucleic acid probe contains the sequence of interest, and in this case it is important that contaminants that may bind to either the filter or the DNA print be removed.

Acknowledgments

The authors are grateful to Dr. D. S. Hogness, in whose laboratory the procedure described was originally developed. This research was supported in part by USPHS Grant No. 5 RO1 GM 23674.

[18] R. Radloff, W. Bauer, and J. Vinograd, *Proc. Natl. Acad. Sci. U.S.A.* **59,** 838 (1968).

[26] A Sensitive and Rapid Method for Recombinant Phage Screening

By SAVIO L. C. WOO

The use of bacteriophage lambda as a DNA vector for cloning unique-sequence eukaryotic genes from genomic DNA is a very attractive method, since large numbers of recombinant phage plaques can be readily generated using minute quantities of target DNA. A major obstacle to successful cloning, however, has been one's inability to screen large numbers of recombinant bacteriophages for the presence of a particular gene. Since Grunstein and Hogness[1] first reported the *in situ* hybridization procedure for screening recombinant bacterial colonies grown and lysed on nitrocellulose filters, Kramer *et al.*[2] have reported a similar procedure for the screening of recombinant bacteriophages. Individual phage plaques were handpicked and allowed to infect bacterial hosts grown on nitrocellulose filters. The phage DNAs were subsequently denatured and fixed on

[1] M. Grunstein and D. S. Hogness, *Proc. Natl. Acad. Sci. U.S.A.* **72,** 3961 (1975).
[2] R. A. Kramer, J. R. Cameron, and R. W. Davis, *Cell* **8,** 227 (1976).

ISBN 0-12-181968-X

the filters by treatment with alkali. The filters were allowed to hybridize with ^{32}P-labeled RNA from yeast ribosomes, and several recombinant bacteriophages containing yeast ribosomal genes were identified on the filters upon subsequent radioautography. This is a very sensitive method, because the DNAs from entire complements of bacteriophage plaques are fixed onto the nitrocellulose filters which are subsequently employed for *in situ* hybridization. Although the method is a sensitive one for recombinant phage screening, it requires handpicking of individual phage plaques and makes screening of large numbers of bacteriophages impractical. Nevertheless, using a direct application of this procedure, Tilghman *et al.*[3] and Tonegawa *et al.*[4] have identified several recombinant bacteriophage clones containing the mouse β-globin gene and the immunoglobulin light-chain gene, respectively, after screening thousands of recombinant bacteriophages. The recombinant phages in these reports were generated by transfecting bacterial hosts with DNA fractions several hundredfold enriched for these DNA sequences.

Bacteriophage cloning of unique-sequence eukaryotic genes was revolutionized subsequently by the development of an extremely rapid and direct screening method. Benton and Davis[5] have found that a single phage plaque contains sufficient DNA for *in situ* hybridization and that the DNA can be fixed onto a nitrocellulose filter by making direct contact between the phage plaques and the filter. Since no handpicking of individual phage plaques is required, it is entirely feasible to screen hundreds of thousands of recombinant phages in a relatively short period of time using this procedure. Although this new screening method is undoubtedly much more rapid than the original *in situ* screening method of Kramer *et al.*,[2] it is not as sensitive a method. This is because only limited quantities of phage DNAs can be transferred to the nitrocellulose filters upon direct contact with the phage plaques. Although the transferred phage DNAs are sufficient for the subsequent *in situ* hybridization step, radioautographic signals generated from positive clones are generally much weaker than those obtained with the original procedure. Weak positive signals necessitate prolonged radioautography, which in turn increases background and causes the appearance of false positive signals. This difficulty has recently been resolved by a modified method of the Benton and Davis procedure, which implements an amplification step prior to *in situ* hybridization and is detailed in this chapter.

[3] S. M. Tilghman, D. C. Tiemeier, F. Polsky, M. H. Edgell, J. G. Seidman, A. Leder, L. W. Enquist, B. Norman, and P. Leder, *Proc. Natl. Acad. Sci. U.S.A.* **74,** 4406 (1977).

[4] S. Tonegawa, C. Brack, N. Hozumi, and R. Schuller, *Proc. Natl. Acad. Sci. U.S.A.* **74,** 3518 (1977).

[5] W. D. Benton and R. W. Davis, *Science* **196,** 180 (1977).

Principle of the Method

One way to improve the sensitivity of the Benton and Davis screening method is to have more phage DNA fixed onto the nitrocellulose filter. This can be best accomplished by implementing an amplification step prior to *in situ* hybridization. The nitrocellulose filter is presoaked in a suspension of host bacteria and dried prior to the transfer of phage DNA and intact phage particles directly from the phage plaques. The filter is transferred to a fresh nutrient agar plate and incubated at 37°. Bacteria will grow on the filter and establish a lawn, while intact phage particles transferred onto the filter will infect the growing bacterial cells and form plaques. The filter, containing entire complements of individual phage plaques, is then employed subsequently for *in situ* hybridization. Thus, strong positive signals can be obtained with no background after only a few hours of radioautography, greatly enhancing the sensitivity of the Benton and Davis procedure.

Materials and Reagents

Escherichia coli LE392/Thy A and the certified EK2 bacteriophage vector λ gtWES · λ B[6] were kindly provided by Philip Leder of the National Institutes of Health. The phage DNA vector was digested with *Eco*RI, and the λ B fragment was separated from λ gtWES DNA by preparative agarose gel electrophoresis. The restriction endonuclease *Eco*RI was purchased from Bethesda Research Laboratories, and *E. coli* DNA polymerase I was obtained from Boehringer Mannheim Company. Nitrocellulose filters precut to circles 8.6 cm in diameter were obtained from Millipore Corporation.

Methods

Transfection

The transfection is carried out by the procedure of Tiemeier *et al.*[7] Briefly, the cells are grown at 37° with vigorous shaking in L broth supplemented with 50 μg/ml of thymidine to an A_{600} of 0.6. The cell suspension is chilled and centrifuged in the cold for 5 min at 2500 *g*. The cell pellet is resuspended in 1 volume of an ice-cold solution of 25 m*M* Tris-HCl, pH 7.6, containing 10 m*M* NaCl, and the swollen cells are pelleted by centrifugation. The cells in the pellet are gently resuspended on ice in 0.5 original volume of 25 m*M* Tris-HCl, pH 7.6, containing 10 m*M* NaCl and

[6] P. Leder, D. Tiemeier, and L. Enquist, *Science* **196,** 175 (1977).

[7] D. C. Tiemeier, S. M. Tilghman, and P. Leder, *Gene* **2,** 173 (1977).

50 m*M* $CaCl_2$. After 20 min, the cells are pelleted by centrifugation at 800 *g* for 5 min and gently resuspended in 0.1 original volume of the $CaCl_2$ solution. The final cell suspension is aliquoted into 0.4-ml portions in disposable glass test tubes. To each tube of cells is added 2–5 μl of a solution containing 40–100 μg/ml of recombinant DNA or 0.05–5 μg/ml of intact phage vector DNA. The tubes are allowed to stand on ice for 30 min, transferred to a 42° water bath for 1 min, and chilled. The contents of the tubes are immediately mixed with 4 ml of 0.7% agar in T broth and plated on 1% agar in L broth. Both the T soft agar and L hard agar are supplemented with 10 m*M* $MgSO_4$ and 50 μg/ml of thymidine. The plates are incubated at 37° overnight to allow the formation of phage plaques. The plates are subsequently stored in a refrigerator for at least 2 hr in order to harden the agar prior to screening.

Amplification

An overnight culture of LE392/Thy A in T broth containing 10 m*M* $MgSO_4$ and 50 μg/ml of thymidine is diluted 10-fold with 180 ml of culture medium at room temperature in a 15-cm-diameter petri dish. Nitrocellulose filters, precut to circles 8.6 cm in diameter, are marked with a ballpoint pen, antoclaved for 10 min, and submerged in the diluted cell suspension. The filters, now coated with cells, are blotted on autoclaved Whatman 3 MM papers. The wet nitrocellulose filters are transferred to a second Whatman 3 MM paper and allowed to air-dry for 30–60 min. The entire procedure is carried out sterilely in a decontaminated Bio-Gard hood. The nitrocellulose filters are then layered onto the phage-containing agar plates and stored in a refrigerator for 1–30 min to allow the transfer of phage particles. The petri dishes are marked according to the filters which are gently peeled off the phage-containing soft agar. They are transferred onto fresh L agar plates supplemented with 10 m*M* $MgSO_4$ and 50 μg/ml of thymidine, with the side that has come into contact with phage plaques facing the air. The fresh petri dishes, containing the nitrocellulose filters, are then incubated at 37° overnight. During this time, the bacterial cell coating will grow to establish a lawn on the nitrocellulose filters, and phage particles transferred to the nitrocellulose filters from the original plates will infect the cells and form individual plaques. If the nitrocellulose filters have been properly dried after coating with the bacterial cell suspension, replicate patterns of phage plaques on the original plates are routinely obtained. The nitrocellulose filters, containing entire complements of individual phage plaques, are then employed for *in situ* hybridization.

In Situ Hybridization

In situ hybridization is carried out by a modification of the procedure of Grunstein and Hogness.[1] The nitrocellulose filters are lifted off the agar plates and allowed to rest for 15 min at room temperature on Whatman 3 MM filters saturated with 0.5 *N* NaOH and 1.5 *M* NaCl. The bacteria and phage particles are lysed, and the denatured DNAs are fixed onto the nitrocellulose during this treatment. The filters are neutralized by transferring successively to Whatman 3 MM papers saturated with 1 *M* Tris-HCl, pH 7.4, and 0.5 *M* Tris-HCl, pH 7.4, containing 1.5 *M* NaCl. They are then blotted on Whatman 3 MM papers for drying. The air-dried filters are baked in a 68° oven for 2 hr and coated at 68° overnight with a solution containing 6× SSC and 0.2% each of bovine serum albumin, Ficoll and polyvinylpyrrolidone according to the procedure of Denhardt.[8] The coated filters are transferred to fresh petri dishes to which small volumes of the Denhardt solution containing the ^{32}P-labeled hybridization probes and 0.5% sodium dodecyl sulfate (SDS) are added. Hybridization

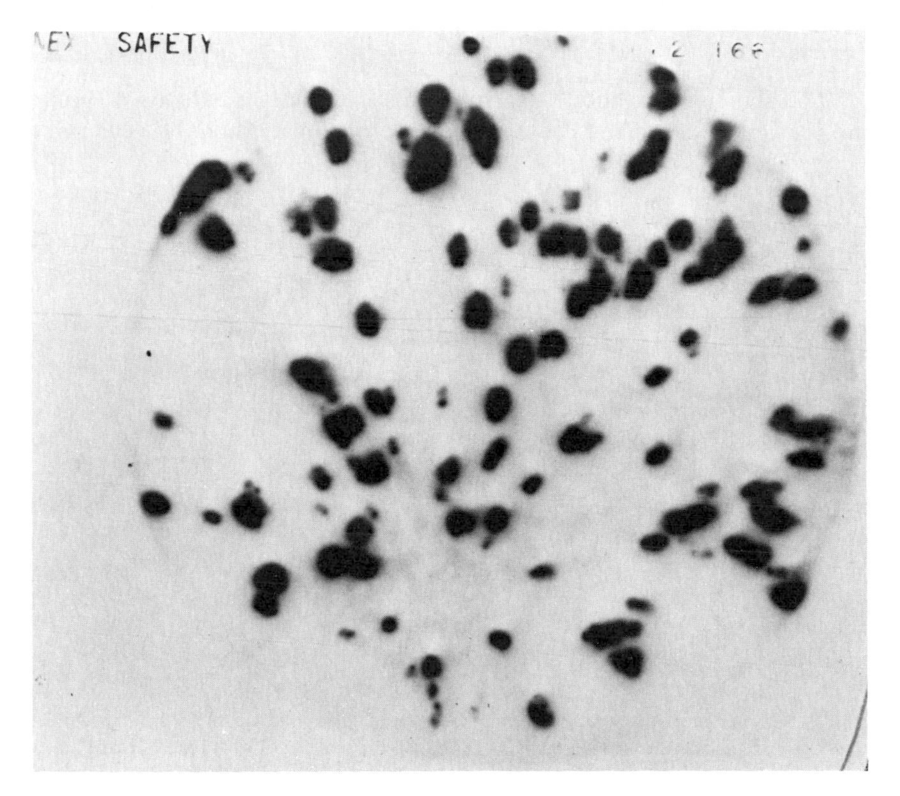

FIG. 1. Screening of λ gtWES · λ B phage plaques by *in situ* hybridization using ^{32}P-labeled λ B DNA as the probe. Phage particles from individual plaques on the agar plates were transferred to a nitrocellulose filter and amplified as described in the text. Hybridization was carried out at 68° for 12 hr, and radioautography took place at −20° for 4 hr.

[8] D. T. Denhardt, *Biochem. Biophys. Res. Commun.* **23,** 641 (1966).

[9] T. Maniatis, A. Jeffrey, and D. G. Kleid, *Proc. Natl. Acad. Sci. U.S.A.* **72,** 1184 (1975).

is carried out at 68° overnight with 10^6 cpm of hybridization probe per filter. The nitrocellulose filters are then washed at 68° for a total of 4 hr with three changes of 1× SSC containing 0.5% SDS. They are then blotted on Whatman 3 MM papers at room temperature and subjected to radioautography at −20° using Dupont Cronex 4 x-ray films in the presence of intensifying screens. Using as hybridization probe the λ B DNA fragment after labeling to a specific activity of 2×10^8 cpm/μg by nick translation according to the procedure of Maniatis *et al.*,[9] strong hybridization signals were obtained on the x-ray films with parental λ gtWES · λ B phage plaques after only a few hours of radioautography (Fig. 1). Hybridization signals of this intensity were obtained regardless of the original plaque sizes, and every phage plaque was detected in this experiment.

Discussion

The screening method described here has the advantages of both of the previously described bacteriophage screening methods. It is as rapid a

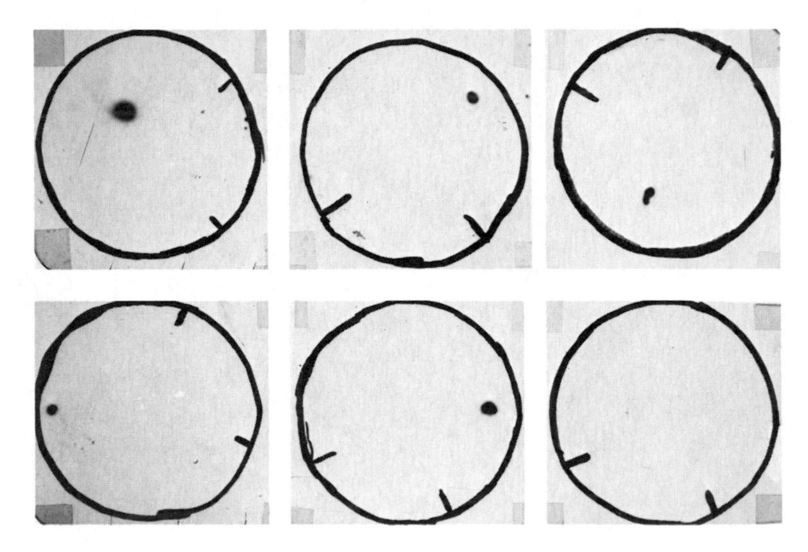

FIG. 2. Screening of recombinant phages containing genomic chick ovalbumin DNA sequences by *in situ* hybridization using as probe ^{32}P-labeled pOV230, a recombinant plasmid containing a full-length ovalbumin DNA synthesized from ovalbumin mRNA.[10] In this experiment, there were approximately 500 individual phage plaques per plate. The procedures for phage transfer and amplification were as described in the text. Radioautography was carried out for 10 hr at −20°. Radioautograms of the filters that yielded hybridization signals are presented. A filter that contained no signal is also shown at the lower right-hand corner for comparison purposes. These signals can readily be detected after only 1 hr of radioautography.

method as that described by Benton and Davis[5] because the extra day needed for the amplification step can be saved with the short radioautography time. Yet it is as sensitive a method as that described by Kramer *et al.*,[2] because the DNAs from entire complements of individual phage plaques are employed for *in situ* hybridization. Thus it should be a sensitive, rapid, and reliable method for recombinant phage screening.

A critical step in this screening procedure is the transfer of phage particles from the original plates to the bacteria-coated nitrocellulose filters. The transfer will not be quantitative if the filter is not dried thoroughly. The instantaneous wetting of the filter when layered onto the phage-containing agar is indicative of this problem. When this occurs, there is significant diffusion of phage particles and the plaques formed on the filter will not be a replicate of those on the original plate. If the filter has been dried thoroughly, it will slowly become wet when layered onto the agar, at a rate similar to that of an untreated nitrocellulose filter. Second, the density of the phage plaques on the original plate should be less than confluent in order to ensure the formation of replicate plaques on the filter. The successful production of replicate phage plaques on the filter is imperative in the application of this procedure for recombinant phage screening. With the use of this procedure, recombinant phages containing various *Eco*RI fragments of the unique-sequence chick ovalbumin gene have been identified (Fig. 2).[10] Analyses of these cloned DNA fragments have revealed that the structural ovalbumin gene sequences are separated into eight portions by seven intervening DNA sequences.[11–13]

Acknowledgments

The author is indebted to Ms. Sarah Meeks whose excellence in laboratory skills has transformed a mere idea into reality. Part of the work described in this article was supported by the Baylor Center for Population Research and Reproductive Biology. The author is an Investigator of the Howard Hughes Medical Institute.

[10] S. L. C. Woo, A. Dugaiczyk, M.-J. Tsai, E. C. Lai, J. F. Catterall, and B. W. O'Malley, *Proc. Natl. Acad. Sci. U.S.A.* **75,** 3688 (1978).

[11] A. Dugaiczyk, S. L. C. Woo, E. C. Lai, M. L. Mace, L. A. McReynolds, and B. W. O'Malley, *Nature* **274,** 328 (1978).

[12] E. C. Lai, S. L. C. Woo, A. Dugaiczyk, and B. W. O'Malley, *Cell* **16,** 201 (1979).

[13] A. Dugaiczyk, S. L. C. Woo, D. A. Colbert, E. C. Lai, M. L. Mace, and B. W. O'Malley, *Proc. Natl. Acad. Sci. U.S.A.* **76,** 2253 (1979).

[27] Selection of Specific Clones from Colony Banks by Suppression or Complementation Tests

By LOUISE CLARKE and JOHN CARBON

A wide variety of *Escherichia coli* strains containing mutations in well-characterized genes involved in amino acid, protein, and nucleotide biosynthesis are readily available. Fulfillment of a particular metabolic requirement in these strains can often be achieved by relatively low levels of expression of the needed gene product. Thus the ability of a cloned segment of DNA to suppress or complement *E. coli* auxotrophic mutations *in vivo* provides a simple way to identify cloned segments of *E. coli* DNA[1,2] and is a sensitive assay for the identification and meaningful expression of segments of yeast DNA[3-6] cloned in *E. coli*.

Rapid and convenient screening procedures have been developed to test recombinant DNA plasmids for the ability to complement bacterial mutations. A rapid screening method is one of several requirements that must be met in order to establish a generally useful hybrid plasmid colony collection. Other requirements include a method of generating DNA segments to be cloned that leaves desired gene systems intact on at least a portion of the cleaved segments, and the use of an efficient cloning procedure that yields sufficient transformant clones containing hybrid plasmids to be representative of the entire genome of the organism under study. The experimental system employed must at least permit the isolation of a large number of *E. coli* genes on hybrid plasmids to show convincingly that the cloning and complementation procedures are adequate.

This chapter describes the establishment in *E. coli* of both *E. coli* and yeast hybrid plasmid colony collections and the selection of specific plasmid-bearing clones from these banks by suppression or complementation tests.

Bacterial Strains

All bacterial strains used in this study are derivatives of *E. coli* K12. Our collection of colonies containing hybrid ColE1 DNA (*E. coli*)

[1] L. Clarke and J. Carbon, *Proc. Natl. Acad. Sci. U.S.A.* **72,** 4361 (1975).
[2] L. Clarke and J. Carbon, *Cell* **9,** 91 (1976).
[3] B. Ratzkin and J. Carbon, *Proc. Natl. Acad. Sci. U.S.A.* **74,** 487 (1977).
[4] K. Struhl, J. R. Cameron, and R. W. Davis, *Proc. Natl. Acad. Sci. U.S.A.* **73,** 1471 (1976).
[5] L. Clarke and J. Carbon, *J. Mol. Biol.* **120,** 517 (1978).

ISBN 0-12-181968-X

plasmids is established in strain JA200 (F^+/Δ*trpE5 recA1 thr leuB6 lacY*).[2] *Escherichia coli* strains harboring the F plasmid, along with the nontransmissible plasmid ColE1 or some of its hybrid derivatives, transfer both plasmids with high efficiency to an F^- recipient.[2,7] Thus F-mediated transfer of hybrid plasmids to appropriate auxotrophs provides a convenient screening technique which is described below. (It should be noted that certain hybrid vectors, such as pBR313 and pBR322, which contain only a portion of ColE1, do not transfer readily in the presence of F.[8]) The *recA* marker is introduced into the host strain, JA200, to avoid recombination of ColE1 DNA (*E. coli*) plasmids at homologous regions of the *E. coli* chromosome.

The colony bank composed of clones harboring ColE1 DNA (*Saccharomyces cerevesiae*) plasmids is established in strain JA221 (*hsdM*$^+$ *hsdR*$^-$ Δ*trpE5 recA1 leuB6 lacY*).[5] This host is *hsdR*$^-$ to prevent restriction of the foreign (*S. cerevesiae*) DNA but is *hsdM*$^+$ to permit proper methylation of the cloned hybrid plasmid DNA, so that further complementation tests can be carried out with the plasmids in a wide variety of auxotrophic *hsdR*$^+$ strains. The cloning of yeast or other eukaryotic DNAs in *E. coli* strains bearing F or other conjugative plasmids is not permitted by the National Institutes of Health (NIH) guidelines for recombinant DNA research.

DNA

Covalently closed circular plasmid DNA from *E. coli* strain JC411 ColE1[9] and hybrid plasmid DNA from various transformants are purified as described by other workers[10,11] and elsewhere in this volume. High-molecular-weight *E. coli* DNA (9–20 × 10^7 daltons) and yeast DNA are purified by the procedures of Saito and Miura[12] and Smith and Halvorson,[13] respectively. The yeast DNA is placed on a 5–20% neutral sucrose gradient and spun at 4° in a Spinco SW27 rotor for 17 hr at 26,500 rpm to remove small pieces of DNA. Only fractions containing DNA pieces larger than 2 × 10^7 daltons (the bottom 3 ml of each tube) are pooled and used for establishing the colony collection. For the subse-

[6] J. Carbon, B. Ratzkin, L. Clarke, and D. Richardson, *Brookhaven Symp. Biol.* **29,** 277 (1977).
[7] P. Fredericq and M. Betz-Bareau, *C. R. Seances Soc. Biol. Ses Fil.* **147,** 2043 (1953).
[8] I. G. Young and M. I. Poulis, *Gene* **4,** 175 (1978).
[9] D. B. Clewell and D. R. Helinski, *Proc. Natl. Acad. Sci. U.S.A.* **62,** 1159 (1969).
[10] D. B. Clewell, *J. Bacteriol.* **110,** 667 (1972).
[11] P. Guerry, D. J. LeBlanc, and S. Falkow, *J. Bacteriol.* **116,** 1064 (1973).
[12] H. Saito and K. Miura, *Biochim. Biophys. Acta* **72,** 619 (1963).
[13] D. Smith and H. O. Halvorson, this series, Vol. 12, p. 538.

quent cloning manipulations described below, it is necessary to begin with DNA as large and as free of nicks as possible.

Construction of ColE1 $(dT)_n \cdot (dA)_n$ DNA (*Escherichia coli* or *Saccharomyces cerevesiae*) Annealed Circles

In constructing hybrid plasmid banks, we routinely use the poly(dA · dT) "connector" method[14,15] for joining DNA segments to the plasmid vector. This procedure gives a high yield of recombinant DNA circles *in vitro* and a relatively high number of transformants bearing hybrid plasmids *in vivo* and can be used with randomly sheared DNA samples, ensuring that any gene remains intact on at least a portion of the DNA pieces.

Plasmid ColE1 is used as a vector because it it small (4.2×10^6 daltons),[16] amplifiable,[17] transfers readily in the presence of F, and seems to maintain stably large inserts of foreign DNA. Construction of the 3'-$(dT)_n$-tailed vector, ColE1 DNA L_{RI}(exo)-$(dT)_{150}$, is as follows: ColE1 covalently closed circular DNA is digested to completion with *Eco*RI restriction endonuclease[18] in a reaction mixture containing 10 m*M* Tris-HCl pH 7.5, 50 m*M* NaCl, 5 m*M* $MgCl_2$, 100 μg/ml of gelatin, and 100 μg/ml of DNA. The reaction is allowed to proceed for 30 min at 37° and then is immediately chilled to 0° in an ice bath. One-half volume of 200 m*M* potassium glycine, pH 9.4, and 12 m*M* $MgCl_2$ is added to the reaction mixture with 400 units/ml of λ 5'-exonuclease,[19] and the mixture is allowed to incubate on ice for 30 min. NaEDTA (10 m*M*) is added to stop the reaction, and the DNA L_{RI}(exo) is immediately phenol-extracted and precipitated in 67% ethanol. It has been previously determined that the λ exonuclease treatment removes approximately 25 nucleotides from the 5'-phosphoryl termini of the DNA L_{RI}. This treatment facilitates the subsequent terminal transferase reaction.[14,15]

Following the method of Lobban and Kaiser,[14] poly(dT) extensions are added to the 3'-hydroxyl termini of ColE1 DNA L_{RI}(exo) in a reaction containing 100 m*M* potassium cacodylate, pH 7.0, 8 m*M* $MgCl_2$, 4 m*M* 2-mercaptoethanol, 7.5 m*M* KH_2PO_4, 0.5 m*M* $CoCl_2$, 150 μg/ml of bovine serum albumin, 0.1 m*M* ^{3}H-dTTP, 100 μg/ml of DNA, and 55 μg/ml

[14] P. Lobban and D. Kaiser, *J. Mol. Biol.* **78,** 453 (1973).

[15] D. A. Jackson, R. H. Symons, and P. Berg, *Proc. Natl. Acad. Sci. U.S.A.* **69,** 2904 (1972).

[16] M. Bazarol and D. R. Helinski, *J. Mol. Biol.* **36,** 185 (1968).

[17] V. Hershfield, H. W. Boyer, C. Yanofsky, M. A. Lovett, and D. R. Helinski, *Proc. Natl. Acad. Sci. U.S.A.* **71,** 3455 (1974).

[18] P. J. Greene, M. C. Betlach, H. M. Goodman, and H. W. Boyer, *in Methods Mol. Biol.* **7,** 87 (1974).

[19] J. W. Little, I. R. Lehman, and A. D. Kaiser, *J. Biol. Chem.* **242,** 672 (1967).

of calf thymus deoxynucleotidyl terminal transferase.[20] The reaction is incubated for 25 min at 37° and terminated by the addition of NaEDTA (10 m*M*). Polythymidylylated ColE1 DNA L_{RI}(exo)-$(dT)_n$ is phenol-extracted and ethanol-precipitated. Under the above conditions approximately 300 residues of dT are added per molecule DNA, or 150 residues per end. Small preliminary time trials should be carried out to determine the exact conditions necessary to add approximately 75–150 residues per 3′-end. The cacodylate buffer must be freshly prepared. Some batches of commercial cacodylic acid contain impurities that strongly inhibit the tailing reaction.

The *E. coli* or yeast DNA to be cloned is sheared to an approximate size of 8–12 × 10^6 daltons (12–18 kbp) by hydrodynamic shearing. The DNA is suspended at 100 μg/ml in 0.01 *M* Tris-HCl, pH 7.5, 0.01 *M* NaCl, and 0.001 *M* Na_2EDTA (STE), and fragmented by high-speed stirring in a stainless-steel cup (capacity, 1 ml) at 0° for 45 min using a setting of 4.5 (approximately 5400 rpm) on a Tri-R Stir-R motor (Model S63C) fitted with a Virtis shaft and microhomogenizer blades. The fragmented DNA (average molecular weight, 8.4 × 10^6 daltons) is treated with λ 5′-exonuclease as described above, and $poly(dA)_{150}$ extensions are added to the 3′-ends of the DNA in a reaction mixture similar to that described above, omitting the $CoCl_2$ and using 0.1 m*M* ^3H-dATP, 100 μg/ml of sheared, λ exonuclease-treated *E. coli* or yeast DNA, and 180 μg/ml of terminal transferase. The reaction is incubated at 37° for 45 min, terminated by the addition of NaEDTA (10 m*M*), and the polydeoxyadenylylated DNA L_{sh}(exo)-$(dA)_{150}$ is phenol-extracted and ethanol-precipitated.

In our experience, terminal transferase reactions that proceed much more rapidly than those described above indicate that the DNA is probably extensively nicked. Tails are thus added to nicks, which leads to branched hybrid circle formation and ultimately to poor transformation efficiency.

Equimolar concentrations of ColE1 DNA L_{RI}(exo)-$(dT)_{150}$ and *E. coli* or yeast DNA L_{sh}(exo)-$(dA)_{150}$ are mixed at a total concentration of 5 μg/ml in 10 m*M* Tris-HCl, pH 8.0, 100 m*M* NaCl, and 1 m*M* NaEDTA and allowed to anneal for 16 hr at 37°. After annealing, DNA samples are prepared for electron microscopy by the aqueous method of Davis *et al.*,[21] and molecules are scored according to structure. The annealed mixtures should contain between 15 and 25% large hybrid circular DNA molecules. DNA preparations that contain very few circular molecules or that contain a large number of branched structures (tailed circles) yield few, if any, transformants harboring hybrid plasmids. Annealed DNA prepara-

[20] K. Kato, J. M. Gonçalves, G. E. Houts, and F. J. Bollum, *J. Biol. Chem.* **242,** 2780 (1967).
[21] R. W. Davis, M. Simon, and N. Davidson, this series, Vol. 21, p. 413.

tions are concentrated by ethanol precipitation and stored at 4° in 10 m*M* Tris-HCl, pH 8.0, 10 m*M* NaCl, and 1 m*M* NaEDTA.

Establishment of Hybrid Plasmid Colony Banks

Approximately 12 μg of annealed ColE1 $(dT)_n \cdot (dA)_n$ DNA (*E. coli*) or 25 μg of ColE1 $(dT)_n \cdot (dA)_n$ DNA (*S. cerevesiae*) is used to transform strains JA200 (F^+/*recA1*) or JA221 (*hsdR*$^-$ *hsdM*$^+$), respectively, according to a modification of the method of Mandel and Higa[22] described by Wensink *et al.*,[23] except that after exposure to DNA cells are diluted 10-fold into L broth (10 g/liter tryptone, 10 g/liter NaCl, 5 g/liter yeast extract) and incubated with shaking for 30–90 min at 37°. The cells are washed once and resuspended in $\frac{1}{3}$ volume L broth. In individual tubes, 0.25 ml cells are mixed with partially purified colicin E1[24] and 0.75 ml L broth. Colicin E1 purified to the ammonium sulfate step of reference 24 is satisfactory. The exact amount of colicin E1 to be added to the tubes must be determined for each new colicin preparation by plating dilutions of colicin with JA200 or JA221 cells, and with cells bearing a ColE1-derived plasmid. After incubation for 20 min at room temperature, 4 ml of warm, 0.8% L agar is added to each tube, and the contents are plated onto 1.5% L agar plates. After incubating the plates for 24 hr at 37°, approximately 2000 colonies for the *E. coli* collection and 6000 colonies for the yeast collection are transferred to 1.5% L agar plates (48 colonies per plate) freshly prepared with an overlay of colicin E1 in 5 ml of 1.5% L agar.

A separate control culture of JA200 or JA221 is simultaneously treated in an identical manner to the transformed culture described above, except that cells are not exposed to DNA. From the number of colicin E1-resistant cells in this mock culture, it can be estimated that about 70–90% of the colicin-resistant cells in the transformed culture contain ColE1 plasmids and that the remaining clones are colicin-tolerant but contain no plasmid. When random clones are picked from the collections and screened for the presence of plasmids,[25] at least 70% of the colonies should contain hybrid plasmids 15–30 kbp in total size.

The colonies in the collections are arranged in a grid pattern of 48 per plate such that they can easily be transferred via a wooden block of 48 needles to standard 96-well MicroTest II dishes (Falcon Plastics), with each well containing 0.2 ml of L broth, colicin E1, and 8% dimethyl sulfoxide (DMSO).[26] The colonies can therefore be individually maintained,

[22] M. Mandel and A. Higa, *J. Mol. Biol.* **53,** 159 (1970).
[23] P. C. Wensink, D. J. Finnegan, J. E. Donelson, and D. S. Hogness, *Cell* **3,** 315 (1974).
[24] S. A. Schwartz and D. R. Helinski, *J. Biol. Chem.* **246,** 6318 (1971).
[25] J. A. Meyers, D. Sanchez, L. P. Elwell, and S. Falkow, *J. Bacteriol.* **127,** 1529 (1976).
[26] J. Roth, this series, Vol. 17, p. 3.

permanently stored at −80°, and used repeatedly by inoculating fresh plates.

We have determined the transformant colony bank size needed to obtain a plasmid collection representing 90–99% of the *E. coli* or yeast genome as follows.[2] Given a preparation of cell DNA fragmented to a size such that each fragment represents a fraction (f) of the total genome, the probability (p) that a given unique DNA sequence is present in a collection of N transformant colonies is given by the expression

$$P = 1 - (1-f)^N$$

or

$$N = \ln (1-P)/\ln (1-f)$$

A sample calculation for *E. coli* (genome size, 2.7×10^9 daltons) for $P = 0.99$ is

$$N = \frac{\ln (1\text{-}0.99)}{\ln [1 - (8.5 \times 10^6/2.7 \times 10^9)]} = 1437$$

Thus, using a preparation of DNA randomly sheared to an average size of 8.5×10^6 daltons for the construction of annealed hybrid circular DNA, a colony bank of only about 1400 transformants for *E. coli* or 5400 transformants for yeast is adequate to give a probability of 99% that any *E. coli* or yeast gene will be on a hybrid plasmid in one of the clones.

The transformation frequency of annealed hybrid DNA is on the order of 1000 transformants per microgram of DNA. The transformant clones in each experiment above that are not picked and stored as individual cultures may be scraped from the plates, pooled, stored in 8% DMSO at −80°, and later used to prepare mixed hybrid plasmid DNA stocks.

Screening Colony Banks

Selection for Phenotypic Reversion of Genetic Loci in Bank Host Strains

Suppression or complementation of a marker in the host bacterial strains can easily be tested by simply plating the colony collections onto the appropriate selective plates. In this way, we have identified in our ColE1 DNA (yeast) colony bank, for example, a number of clones containing different hybrid plasmids which complement or suppress the *leuB6* mutation in JA221.[3,6] These pYe*leu* plasmids fall into two groups: (1) a plasmid, pYe(*leu2*)10, capable of complementing any mutation in *leuB*, the structural gene for β-isopropylmalate dehydrogenase, and (2) a group of plasmids from different regions of the yeast genome which specifically

TABLE I
Escherichia coli GENES IDENTIFIED ON ColE1 HYBRID PLASMIDS IN THE COLONY BANK

Approximate map location (nearest minute)	*E. coli* gene(s) identified on plasmid[a]	Plasmid number, pLC No.[b]
0/100	*thr*	17-22
1	*araC*	24-41
2	*leuA, ftsI*	26-6
	ace, nadC, guaC	37-40
4	*dnaE*	26-43
		34-20
5	*gpt*	4-34
6	*gpt, proAB*	44-11
	proAB	7-19
		25-25
		32-46
		40-18
	proA	28-33
8	*lac*	20-30
10	*acrA*	4-44
		34-42
11	*dnaZ*	5-1
		5-2
		6-2
		10-24
		10-26
		30-3
		30-4
		30-23
	plsA	24-31
		28-31
		28-32
12	*purE*	8-25
		15-15
		33-18
17	*bio, uvrB*	25-23
21	*pyrD*	31-38
		43-25
22	*fabA*	29-15
25	*purB*	7-23
25–29	*flaKLM*	11-15
		24-46
		35-44
		36-11
27	*trpE*	4-6
		5-23
		29-41
		32-12
		32-27
		41-15

TABLE I—*Continued*

Approximate map location (nearest minute)	*E. coli* gene(s) identified on plasmid[a]	Plasmid number, pLC No.[b]
38	*xthA*	10-4
		26-8
40	*fadD*	1-21
		15-17
		15-32
		30-32
42	*flaGH*, *cheB*	21-2
		24-15
	flaGH, *cheBA*, *mot*	1-28
	cheA, *mot*, *flaI*	27-20
		38-14
		38-36
42	*uvrC*, *flaD*	7-18
		13-12
43	*flaD, hag, flaN*	24-16
		26-7
	flaN, *flaBCOE*, *flaAPQR*	41-7
44	*his*	14-29
		26-21
	gnd	33-30
48	*nrdA*, *glpT*, *ftsB*	19-24
	glpT, *ftsB*	3-46
		42-17
	glpT	8-12
		8-24
		8-29
		14-12
	ftsB	1-41
		6-16
50	*purF*	28-44
	fabB	26-23
		33-1
		39-16
54	*guaAB*	32-25
		34-10
55	*nadB*, *pss*	34-44
		34-46
58	*recA*	17-38
		24-27
		30-20
	recA srl	17-43
		18-42
		21-33
		22-40
		24-32

(*Continued*)

TABLE I—*Continued*

Approximate map location (nearest minute)	*E. coli* gene(s) identified on plasmid[a]	Plasmid number, pLC No.[b]
68	*argG*	7-27
		7-34
		8-7
		16-36
		21-11
		28-20
72	*aroE*, *rrnD*	16-1
72	*aroE, rrnD,* $tRNA_1^{Ile}$, $tRNA_{1b}^{Ala}$	12-24
		16-6
		22-11
73	*trpS*	13-42
	aroB	29-47
77	*plsB*	2-43
		3-22
		9-28
79	*xyl*	10-15
		32-9
	xyl glyS	1-3
		42-22
80	*gpsA*	32-10
83	*rrnC*	21-9
	trpT, rrnC, ilvE, $tRNA^{Glu}$, $tRNA^{Asp}$	22-36
	rho ilv	30-17
83	*ilv*	21-35
		22-3
		22-31
		26-3
		27-15
		30-15
	ilv, *cya*, *rep*	44-7
	cya	23-3
		29-5
		36-14
		41-4
		43-44
85	*rrnA*	5-31
		16-11
	rrnA, $tRNA_1^{Ile}$, $tRNA_{1b}^{Ala}$	19-3
	glnA	28-35
		28-36
		41-35
86	*rha*	3-24
		5-5
87	*tpi pfkA*	16-4
	pfkA	30-43

TABLE I—*Continued*

Approximate map location (nearest minute)	*E. coli* gene(s) identified on plasmid[a]	Plasmid number, pLC No.[b]
88	*argH*	20-10
		41-13
	rrnB rplLK, $tRNA^{Glu}$	34-34
89	*purD*, *purH*	8-22
90	*pgi*	37-5
91	*uvrA*	33-42
		38-48
	uvrA, *dnaB*, *lexA*	44-14
	dnaB	11-9
93	*psd*	8-47
		44-13
	purA	14-14
99	*dnaC*	4-39
		8-9
		25-8
		28-5
		30-24
		31-39
99	*deoD*	1-13
		22-4
		39-40
100/0	*serB*, *trpR*	32-33
	trpR, *thr*	35-1
	thr	17-22
?	*fabF*	28-43
?	*ftsE*	6-37
		19-48
		31-16
		31-32
?	*ftsI*	4-14
?	*ftsM*	4-16
?	*hpt*	37-20
?	ribRNA, $tRNA_1^{Ile}$, $tRNA_{1b}^{Ala}$, $tRNA^{Asp}$	7-21
?	ribRNA, $tRNA^{Glu}$	23-30
?	ribRNA	24-26
		27-40
		28-21
		43-34

[a] The gene assignments listed here have been made in numerous laboratories and communicated to the authors by mail. In some cases, they are tentative and await confirmation. Since the average size of the *E. coli* DNA inserts in these plasmids is large (about 12 kbp or 0.3–0.4 min on the *E. coli* gene map), the genes listed here are not necessarily the only ones carried by the plasmids; closely adjacent genes can also be present.

[b] The colonies in the bank are numbered as follows: the first number is the plate number

(*Continued*)

suppress the *leuB6* mutation in strain JA221. The nature of this suppression has not been investigated, nor is the exact nature of the *leuB6* mutation known, although it is most likely a missense mutation. Plasmid pYe(*leu2*)10, however, has now been shown to carry the yeast *leu2* (β-isopropylmalate dehydrogenase) gene and can complement both *leuB* bacterial mutations and *leu2* yeast mutations.[27]

Screening by F-Mediated Transfer to Suitable F^- Recipients

Escherichia coli strains containing the fertility plasmid F and the nontransmissible plasmid ColE1 transfer both plasmids very efficiently to an F^- recipient.[2,7] Therefore, collections of colonies bearing hybrid plasmids constructed from *E. coli* DNA plus ColE1 can be rapidly screened for numerous markers by replica mating in the following way. The entire collection, for example, of about 2000 ColE1 DNA (*E. coli*) hybrid plasmid-bearing clones, which has been stored at −80° in microtiter dishes, is thawed and patched onto 44 L agar plates using a wooden block of 48 needles as described above. The plates are incubated overnight to allow colony patches to grow up, and the colonies from each plate are transferred to a sterile velvet pad. From a single velvet up to 5 or 6 fresh L plates can be inoculated. The process is repeated for each plate, giving five or six copies of the entire collection, one set for each F^- recipient to be tested. The freshly inoculated master plates are grown at 37° for about 5 hr. Colonies from a set of 44 plates are then transferred, using velvet, to minimal plates spread with 0.2 ml of a fresh overnight culture of a particular F^- auxotroph or recipient of interest. The minimal plates should counterselect the bank host strain (e.g., by the omission of tryptophan for JA200) and select by complementation for transfer of the appropriate plasmid to the auxotroph. Upon incubation, recipient cells that have received complementing plasmids grow into fairly heavy patches on the selective plates. Occasionally, a single or a few colonies appear on selective plates instead of a solid patch. These may be either revertants of the recipient strain or, if the recipient is *recA*$^+$, clones that have received a wild-type allele through rare *Hfr* formation in the donor.

The replica-mating technique has been used by us and by many other investigators to identify genes on hybrid plasmids in the ColE1 DNA (*E.*

[27] A. Hinnen, J. B. Hicks, and G. R. Fink, *Proc. Natl. Acad. Sci. U.S.A.* **75**, 1929 (1978).

[b] (*Continued*)
(1 through 44); the second number (after the hyphen) designates the number of the colony on the plate (1 through 48). The 6 × 8 colony grid is numbered in sequence from left to right in each row, starting with the upper left-handed corner (the plate is oriented with the agar surface up and the reference mark toward the top).

coli) colony bank.[2] Table I lists the hybrid plasmids tentatively identified to date and some of the markers carried by these plasmids. The probability of finding any particular *E. coli* gene on a plasmid in the collection is quite high (about 80%).

F-mediated transfer is a rapid and convenient way to screen colony banks for particular hybrid plasmids and to transfer plasmids to other strains for testing. It should be noted, however, that current versions of the National Institutes of Health guidelines forbid its use in any system other than for hybrid plasmids constructed from DNA from *E. coli* or other class I bacteria that normally exchange genetic information with *E. coli*. To identify by complementation tests certain yeast DNA segments cloned in *E. coli,* we have used the following method of screening.

Transformation of Escherichia coli Auxotrophic Mutants with Mixed Hybrid Plasmid DNA Preparations

From pools of up to 50,000 transformant colonies, supercoiled hybrid plasmid DNA is prepared from chloramphenicol-treated cells by banding in CsCl.[10,11] The source of the transformant colonies is a stock of pooled clones stored in L broth and 8% DMSO at $-80°$, originally scraped from the selection plates in transformations with annealed ColE1 $(dT)_n \cdot (dA)_n$ DNA (*E. coli* or yeast). Thus the stock is composed of transformants that have been grown up on plates as individual clones. When a mixed plasmid DNA preparation is made, a fresh culture is inoculated from the stock to a relatively high density ($A_{600} = 0.3$), so that the cells only go through a few doublings before the addition of chloramphenicol. Thus the loss of certain plasmids due to segregation or to slower growth rate of a clone is minimized. A very large number of pooled transformants in the stock, many times the number needed to cover the genome of the organism under study, is desirable, since certain plasmids might be preferentially lost during pooled growth of transformants.

An *E. coli* auxotrophic mutant is then transformed with an amount of the supercoiled hybrid plasmid DNA pool calculated to give an extremely large number of transformants to colicin E1 resistance, and the auxotrophic recipient is plated directly onto media selective for the desired complementation. We use enough ColE1 DNA (yeast) (10–20 μg), for example, to yield at least 10^5 total transformants to colicin E1 resistance per selection, even though about 5400 independent transformants are sufficient for a 99% probability of covering the entire yeast genome. In this way we have identified by complementation segments of yeast DNA cloned in *E. coli* that carry the *arg4, his3, trp7,* and *trp5* yeast gene systems.[3,5,6,28]

[28] A. Walz, B. Ratzkin, and J. Carbon, *Proc. Natl. Acad. Sci. U.S.A.* **75,** 6172 (1978).

Complementation of auxotrophic mutations and subsequent selection is a powerful tool that has permitted identification of many cloned DNA segments of *E. coli, yeast, and Neurospora*.[29] Selective pressure introduced by the expression of a gene on a cloned segment of DNA can result in the isolation of plasmids with significant alterations in cloned DNA structure, however.[5,28] Similar alterations would be selected for even if some unknown gene on the cloned DNA produced a product harmful to the bacterial host. Clearly any foreign DNA segment cloned in *E. coli* must be directly compared to the analogous segment from the foreign genome before meaningful structure–function studies are attempted.

[29] D. Vapnek, J. A. Hautala, J. W. Jacobson, and N. H. Giles, *Proc. Natl. Acad. Sci. U.S.A.* **74,** 3508 (1977).

[28] Cloning of Yeast Genes Coding for Glycolytic Enzymes[1]

By MICHAEL J. HOLLAND, JANICE P. HOLLAND, and KIMBERLY A. JACKSON

Regulation of the glycolytic pathway at the enzyme level is well established, however, little is known about regulation of the genes that code for these enzymes. Large changes in the specific activity of yeast glycolytic enzymes do occur with changes in the growth environment of the cells.[2,3] The kinetics of yeast glycolytic enzyme synthesis after a shift from growth on acetate to glucose as carbon source suggest that induction of these enzymes is coordinately regulated,[3] however, the mechanism of induction is unknown. One approach to understanding coordinate regulation of glycolytic genes is to isolate and characterize structural genes that encode glycolytic enzymes. Most of the glycolytic enzymes of yeast are present at high intracellular concentration,[4] as are the mRNAs that encode them.[5,6] mRNAs coding for enolase, phosphoglycerate kinase, and glyceraldehyde-3-phosphate dehydrogenase have been isolated by preparative polyacrylamide gel electrophoresis of total cellular poly(A)-containing mRNA.[6] The purity of the isolated mRNAs ranges from 15 to 40%. In this chapter we describe a rapid method for isolating *Escherichia*

[1] This investigation was supported by grants from the National Institutes of Health and the National Foundation March of Dimes.

[2] F. A. Hommes, *Arch. Biochem. Biophys.* **114,** 231 (1966).

[3] P. K. Maitra and Z. Lobo, *J. Biol. Chem.* **246,** 475 (1971).

[4] B. Hess, A. Boiteux, and J. Kruger, *Adv. Enzyme Regul.* **7,** 149 (1968).

[5] M. J. Holland, G. L. Hager, and W. J. Rutter, *Biochemistry* **16,** 8 (1977).

[6] M. J. Holland and J. P. Holland, *Biochemistry* **17,** 4900 (1978).

ISBN 0-12-181968-X

coli transformants containing hybrid plasmid DNA composed of a bacterial plasmid vector and a segment of yeast genomic DNA containing a glycolytic enzyme structural gene. Radioactively labeled complementary DNA (cDNA) synthesized from partially purified mRNA is used as a hybridization probe for selection of the transformant. Isolation of a yeast glyceraldehyde-3-phosphate dehydrogenase structural gene is reported to illustrate the cloning procedure.

Cloning Method

Principle. The first step of the procedure involves determining the molecular weight of a restriction endonuclease cleavage fragment of yeast genomic DNA containing the structural gene of interest. Restriction endonuclease-cleaved genomic DNA is electrophoresed on an agarose slab gel, transferred to nitrocellulose paper, and hybridized with ^{32}P-labeled cDNA synthesized from partially purified mRNA. Since the concentration of each unique genomic cleavage fragment is identical, the amount of hybrid formed (visualized by autoradiography) is proportional to the relative concentration of each component of the hybridization probe.[7] The molecular weight of the cleavage fragment complementary to the most abundant species in the cDNA is determined.

The second step of the procedure involves ligation of total restricted genomic DNA to a suitable cloning vector and transformation of competent *E. coli* with the ligation mixture. Total supercoiled hybrid plasmid DNA is then isolated from the shotgun collection of transformants. An aliquot of this DNA is electrophoresed on an agarose slab gel, transferred to nitrocellulose paper, and hybridized with the ^{32}P-labeled cDNA. Since the relative concentration of genomic DNA sequences in the hybrid plasmid DNA preparation is dependent on the efficiency of ligation and transformation of each cleavage fragment, as well as on the number of hybrid plasmid DNA copies per transformant,[8] significant hybridization of supercoiled hybrid plasmids with minor components of the cDNA probe is routinely observed at this step. It is therefore necessary to identify hybrids formed between the cDNA probe and supercoiled hybrid plasmid with the molecular weight expected for the vector DNA plus the cleavage fragment identified in the first step of the procedure.

The final step of the procedure involves electrophoresis of a second aliquot of the total hybrid plasmid DNA preparation on an agarose gel,

[7] This assumption is valid if the extent of hybridization is below that necessary to saturate genomic DNA sequences complementary to the cDNA probe.

[8] We observe greater than 10-fold differences in the number of hybrid plasmid DNA copies per transformant. These differences are related in an unknown way to the segment of yeast DNA contained in each hybrid plasmid.

slicing the appropriate molecular weight region from the gel, elution of the hybrid plasmid from the gel slice, and transformation of competent *E. coli* with the eluted hybrid plasmid supercoils. The transformant containing plasmid DNA sequences complementary to the most abundant sequences in the probe is then selected by hybridization of lysed colonies, immobilized on filter paper, with ^{32}P-labeled cDNA.

Reagents

1. STE buffer: 10 m*M* Tris, pH 7.4, 0.1 *M* NaCl, 1 m*M* EDTA
2. Tris–acetate electrophoresis buffer: 40 m*M* Tris, 20 m*M* sodium acetate, 2 m*M* EDTA, pH 8.1
3. 18× SSC: 2.7 *M* NaCl, 0.27 *M* sodium citrate, pH 7.0
4. Triton lytic buffer: 0.1% Triton X-100, 60 m*M* EDTA, 50 m*M* Tris, pH 7.8

Bacterial Strains and Media

1. *Escherichia coli* K12 strain RR101 (*hsdS1 proA2 leu-6 ara-14 galK2 lacY1 xyl-5 mtl-1 str-20 thi-1 supE44* $F^- \lambda^-$), a *recA*$^+$ derivative of strain HB101,[9] was used as host for transformations (supplied by R. Rodriguez).
2. The bacterial vector was pSF2124, a ColE1 ampicillin-resistant recombinant plasmid,[10] isolated from strain C600 (Thr^- Leu^- Thi^-) (donated by S. Falkow).
3. L broth: 2% tryptone (Difco), 1% yeast extract (Difco), 1% NaCl, 0.5% glucose.
4. L broth plates: 1.5% Bacto-agar (Difco). 2× L broth: 4% tryptone (Difco), 2% yeast extract (Difco), 1% NaCl, 0.5% glucose.
5. Minimal medium M-9: 0.7% Na_2HPO_4, 0.3% KH_2PO_4, 0.1% NaCl, 0.5% NH_4Cl, 0.4% casamino acids (Difco), 0.1 m*M* $CaCl_2$, and 1 m*M* $MgSO_4$ supplemented with 0.5% glucose, 10 μg/ml thiamine, and 100 μg/ml each of proline and leucine.

Materials. Restriction endonuclease EcoRI was purchased from Bethesda Research Laboratories, ampicillin from Bristol Laboratories, T4 DNA ligase from New England Biolabs, and lysozyme from Miles Laboratories. Nitrocellulose membrane filters (0.45 μm) were obtained from Schleicher and Schuell. [α-^{32}P]dCTP (2000–3000 Ci/mmol) was purchased from Amersham Corporation.

Containment. Escherichia coli K12 strain RR101 and the plasmid vector pSF2124 consitute an EK1 host–vector system. The experiments described here were carried out in a P2 level containment facility in accordance with the National Institutes of Health guidelines for recombinant DNA research.

[9] H. W. Boyer and D. Roulland-Dussoix, *J. Mol. Biol.* **41,** 459 (1969).
[10] M. So, R. Gill, and S. Falkow, *Mol. Gen. Genet.* **142,** 239 (1975).

Preparation of Glyceraldehyde-3-Phosphate Dehydrogenase cDNA

cDNA is synthesized according to Efstratiades *et al.*[11] with reverse transcriptase isolated from avian myeloblastosis virus. Purified reverse transcriptase was supplied by the Office of Program Resources and Logistics, Viral Oncology Program, National Cancer Institute. Reaction mixtures (50 μl) contain 50 m*M* Tris buffer, pH 8.3, 60 m*M* NaCl, 6 m*M* $MgCl_2$, 1 m*M* dithiothreitol, 100 μg/ml of actinomycin D, 20 of μg/ml oligo(dT) (12-18), 0.3 m*M* dATP, 0.3 m*M* dGTP, and 0.3 m*M* dTTP, 100 μCi of [α-^{32}P]dCTP (2000–3000 Ci/mmol), 0.4 μg of reverse transcriptase, and 20 μg/ml of partially purified glyceraldehyde-3-phosphate dehydrogenase mRNA. Glyceraldehyde-3-phosphate dehydrogenase mRNA was isolated from total poly(A)-containing mRNA by preparative acrylamide gel electrophoresis in 98% formamide as described by Holland and Holland.[6] The estimated purity of this mRNA is 33%. Reverse transcriptase reaction mixtures are incubated for 45 min at 37°, adjusted to 1 ml with buffer containing 0.1 *M* NaCl, 10 m*M* Tris, pH 7.4, and 1 m*M* EDTA (STE) with 10 μg/ml depurinated salmon sperm DNA and extracted with an equal volume of STE-saturated phenol. After centrifugation at 5000 *g* for 5 min, the aqueous phase is extracted with 2 volumes of ether. The cDNA is chromatographed on a 25 × 0.9 cm Sephadex G-50 column, equilibrated in STE buffer, to remove unincorporated [α-^{32}P]dCTP. Fractions containing the cDNA are pooled, adjusted to 0.6 *N* NaOH, and incubated for 1 hr at 37° to hydrolyze the mRNA template. After neutralization with 2 *N* HCl, the cDNA is precipitated with 5 μg/ml of depurinated salmon sperm DNA carrier by the addition of 2 volumes of ethanol. The precipitated cDNA is collected by centrifugation at 10,000 *g* for 30 min at 4°. The pellet is rinsed with 70% ethanol, recentrifuged, dried under nitrogen, suspended in 100 μl of 0.1 m*M* EDTA, and stored on ice at 4°. Approximately 40×10^6 cpm of cDNA is recovered (sp. act., 1×10^9 cpm/μg).

Identification of EcoRI Cleavage Fragments of Yeast Genomic DNA that Hybridize with the cDNA Probe

Yeast cellular DNA[12] (5 μg) is digested with 15 units of *Eco*RI in a reaction mixture (50 μl) containing 100 m*M* Tris, pH 7.4, 50 m*M* NaCl, 5 m*M* $MgCl_2$, and 2 m*M* 2-mercaptoethanol at 37° for 1 hr. The amount of restriction endonuclease necessary for limit digestion is established by titrating the enzyme for conversion of pSF2124 supercoiled DNA, which

[11] A. Efstratiades, T. Maniatis, F. C. Kafatos, A. Jeffrey, and J. N. Vournakis, *Cell* **4**, 367 (1975).

[12] M. J. Holland, G. L. Hager, and W. J. Rutter, *Biochemistry* **16**, 16 (1977).

contains a single *Eco*RI cleavage site, to linear molecules. A two- to threefold excess of that amount of *Eco*RI is then employed. The reaction is terminated by the addition of 0.2 volume of a solution containing 25% glycerol, 0.125 *M* EDTA, pH 7.8, and 2% sarkosyl. Digested DNA is loaded directly onto a 1% agarose vertical slab gel (30 × 14 × 0.3 cm), and electrophoresis is carried out at 40 V for 14 hr in Tris–acetate buffer, pH 8.1, containing 0.5 μg/ml of ethidium bromide. Figure 1[13,14] is a photograph of the slab gel after electrophoresis. Figure 1A and B shows mixtures of marker DNA fragments and Fig. 1C shows the distribution of *Eco*RI cleavage fragments obtained from the yeast DNA.

The agarose gel is transferred to a plastic container containing 300 ml of 1 *N* NaOH and allowed to stand at room temperature for 20 min. Neutralization is brought about by the addition of 210 ml of 1 *M* Tris-HCl, pH 7.4, and 300 ml of 1 *N* HCl, and incubation is allowed to continue for 1 hr at room temperature. Following removal of the denaturation solution, the gel is equilibrated in 300 ml of 6× SSC for 20 min. Transfer of DNA from the gel to nitrocellulose paper is as described by Ketner and Kelly.[15] The agarose gel, reduced to a thickness of less than 0.05 mm, is removed from the nitrocellulose membrane which is then soaked in 200 ml of 2× SSC for 15 min at room temperature, blotted with filter paper, and baked *in vacuo* for 2 hr at 80°. The filter is prepared for hybridization by a modification of the technique outlined by Jeffreys and Flavell.[16] The nitrocellulose filter is incubated at 65° in a Pyrex container covered with a Mylar sheet sealed with autoclave tape in (1) 300 ml of 3× SSC for 30 min, (2) 300 ml of 3× SSC containing 0.02% polyvinylpyrrolidine, 0.02% Ficoll, and 0.02% bovine serum albumin[17] for 3 hr, and (3) the solution described in step 2 supplemented with 0.1% sodium dodecyl sulfate and 10 μg/ml of depurinated salmon sperm DNA for 1 hr. The nitrocellulose filter is then blotted with filter paper and air-dried.

Hybridization is carried out for 24 hr at 65° under mineral oil in the Pyrex container described above. One milliliter of hybridization solution (identical to that used in step 3 described above) containing $1-1.5 \times 10^5$ cpm of partially purified glyceraldehyde-3-phosphate dehydrogenase cDNA is used for a 40-cm² nitrocellulose filter. After hybridization the filter is washed three times with 200 ml of chloroform at room temperature to remove the oil, followed by five 15-min washes of 300 ml of 6× SSC at 65°. Autoradiography is carried out with Kodak X-OMat R film

[13] The molecular weights of the *Hin*dIII cleavage fragments of adenovirus 5 were provided by Dr. T. Shenk and Ms. M. Fitzgerald.

[14] J. H. Cramer, F. W. Farrelly, and R. H. Rownd, *Mol. Gen. Genet.* **148,** 233 (1976).

[15] G. Ketner and T. J. Kelly, Jr., *Proc. Natl. Acad. Sci. U.S.A.* **73,** 1102 (1976).

[16] A. J. Jeffreys and R. A. Flavell, *Cell* **12,** 429 (1977).

[17] D. T. Denhardt, *Biochem. Biophys. Res. Commun.* **23,** 641 (1966).

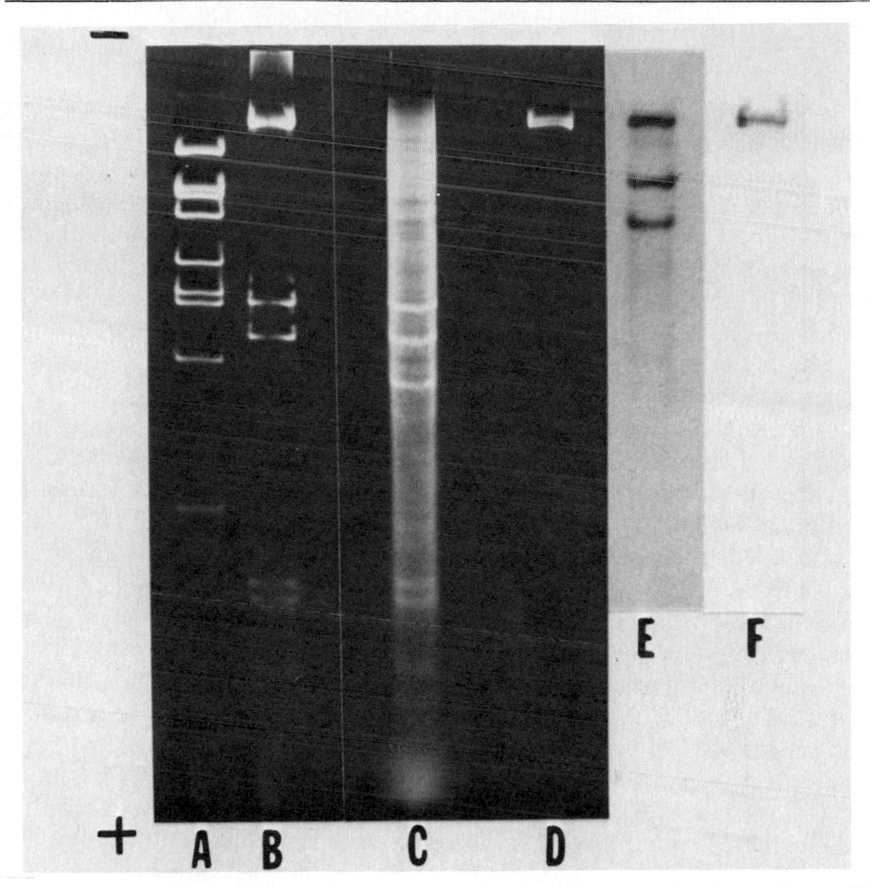

FIG. 1. Identification of *Eco*RI cleavage fragments of yeast genomic DNA and p*gap*492 DNA that are complementary to a glyceraldehyde-3-phosphate dehydrogenase cDNA hybridization probe. DNA was visualized by ethidium bromide fluorescence after electrophoresis on a 1% agarose slab gel. (A) *Hin*dIII cleavage fragments of adenovirus-5 DNA. The molecular weights (kilobase pairs) of these marker fragments[13] are 7.7, 5.6, 5.2, 4.6, 3.2, 2.8, 2.65, 2.1, and 1.2. (B) *Eco*RI cleavage fragments of p*rib*20, a hybrid plasmid isolated in our laboratory which contains a portion of a yeast ribosomal gene repeat. The molecular weights (kilobase pairs) of these marker fragments[14] are 11.5, 2.8, 2.4, 0.57, 0.49, 0.3, and 0.23. (C) *Eco*RI-digested yeast genomic DNA. (D) *Eco*RI-digested p*gap*492 DNA. (E) An autoradiogram of hybrids formed between glyceraldehyde-3-phosphate dehydrogenase cDNA and *Eco*RI-digested yeast DNA (C) immobilized on a nitrocellulose filter. (F) An autoradiogram of hybrids formed between glyceraldehyde 3-phosphate dehydrogenase cDNA and *Eco*RI-digested p*gap*492 DNA (D) immobilized on a nitrocellulose filter.

and a Dupont Lightning-Plus intensifying screen for 15 hr. The results of the hybridization are shown in Fig. 1E. Three major hybrids are formed with genomic DNA cleavage fragments of molecular weight >11, 5.5, and 4.3 kb.

Construction of a Collection of Transformants Containing EcoRI-Digested Yeast DNA

One microgram of pSF2124 DNA and 5 μg of yeast DNA are digested with *Eco*RI as described above. The digested DNAs are extracted with STE-saturated phenol, ether-extracted, adjusted to 0.2 *M* sodium acetate, and precipitated with 2 volumes of ethanol at −20°. After centrifugation at 10,000 *g* for 30 min the DNA precipitates are suspended in 10 μl each of 10 m*M* Tris, pH 7.4, and 1 m*M* EDTA. The DNAs are mixed and ligated in a 50-μl reaction containing 50 m*M* NaCl, 50 m*M* Tris, pH 7.4, 0.5 m*M* EDTA, 10 m*M* $MgCl_2$, 0.1 m*M* ATP, 10 m*M* dithiotheitol, and 1 Weiss unit of T4 DNA ligase for 18 hr at 12°. This ligation mixture is used to transform competent *E. coli* strain RR101.

Escherichia coli strain RR101 is grown in M-9 minimal medium supplemented with 0.5% glucose, 10 μg/ml thiamine, and 100 μg/ml each of proline and leucine. Cells are harvested at OD_{660} = 0.4–0.5 by centrifugation at 4000 *g* for 5 min. The cells are washed with cold 10 m*M* NaCl, centrifuged, suspended in 30 m*M* $CaCl_2$ (the volume of 30 m*M* $CaCl_2$ is equal to the volume of culture harvested), and allowed to stand on ice for 20 min. Following centrifugation at 4000 *g* for 5 min the cells are resuspended in 30 m*M* $CaCl_2$ (the volume of 30 m*M* $CaCl_2$ is equal to 5% of the volume of culture harvested). Transformation reactions contain the ligation mixture described above (50 μl), 25 μl of 6 mg/ml bovine serum albumin in 0.125 *M* $CaCl_2$, 25 μl of H_2O, and 200 μl of the $CaCl_2$-treated cell suspension. Transformation reaction mixtures are allowed to stand on ice for 60 min with occassional swirling, followed by a 60-sec incubation at 42°. Three milliliters of 2× L broth medium is added, and the cells are incubated for 1 hr in a shaker bath at 37°. The number of transformants is determined by titering on L broth agar plates containing 20 μg/ml of ampicillin. Approximately 8000 transformants are derived from the ligation mixture described above.

Isolation of Hybrid Plasmid DNA

The total mixture of transformants isolated above are used to innoculate a 1-liter L broth culture containing 10 μg/ml of ampicillin. The culture is harvested in stationary phase [OD_{660} (undiluted) = 2.0–2.1] by centrifugation at 2000 *g* for 10 min. Hybrid plasmid DNA is isolated from the cells by the method of Herschfield *et al.*[18] The cell pellet is suspended in 10 ml of 25% sucrose (w/v) in 50 m*M* Tris, pH 7.8. Four milliliters of 0.25 *M* EDTA, pH 7.8, is added, and the suspension is allowed to stand

[18] V. Herschfield, H. W. Boyer, C. Yanofsky, M. A. Lovett, and D. R. Helinski, *Proc. Natl. Acad. Sci. U.S.A.* **71**, 3455 (1974).

on ice for 5 min. Two milliliters of a 5 mg/ml lysozyme solution in 0.25 *M* Tris, pH 7.8, is added, and the suspension is allowed to stand on ice for 15 min. Finally, 16 ml of Triton lytic buffer is added. After 15 min on ice the lysate is centrifuged at 35,000 *g* for 40 min, and the supernatant is extracted twice with equal volumes of STE-saturated phenol. Nucleic acid is precipitated with 2 volumes of ethanol at −20°. After centrifugation the precipitate is suspended in 25 ml of 50 m*M* Tris, pH 8.0, and 1 m*M* EDTA, to which 26.2 g of CsCl and 1.5 ml of 10 mg/ml ethidium bromide in H_2O are added. The hybrid plasmid DNA is banded by centrifugation for 36 hr at 38,000 rpm in a Beckman Ti60 rotor. Supercoiled DNA is visualized in the centrifuge tube with a long-wavelength uv lamp and removed by puncturing the centrifuge tube from the side with a syringe. Ethidium bromide is removed from the DNA by chromatography on a 2-ml Dowex 50W-X8 column equilibrated with 10 m*M* Tris, pH 8.0. Following dialysis against 2 liters of 10 m*M* Tris, pH 7.4, and 1 m*M* EDTA to remove CsCl, the DNA solution is adjusted to 0.2 *M* sodium acetate, pH 5.4, and precipitated with 2 volumes of ethanol at −20°. Approximately 40 μg of hybrid plasmid DNA is recovered and suspended at a final concentration of 0.4 μg/μl in 10 m*M* Tris, pH 7.4, 1 m*M* EDTA.

Hybridization of Hybrid Plasmid DNA Supercoils with cDNA

Ten micrograms of hybrid plasmid DNA is adjusted to 5% glycerol, 30 m*M* EDTA, and 0.04% sarkosyl and loaded on a 0.85-cm lane of a 0.8% agarose slab gel (30 × 14 × 0.3 cm). Electrophoresis is in Tris–acetate buffer containing 0.5 μg/ml of ethidium bromide for 20–24 hr at 40 V. A series of supercoiled DNA markers are electrophoresed in a parallel lane. Following electrophoresis the gel is irradiated for 10 min with a short-wavelength uv lamp in order to nick the supercoiled DNA. DNA is then transferred, after denaturation, to nitrocellulose paper and hybridized with ^{32}P-labeled cDNA as described above. Figure 2 illustrates the results obtained after hybridization of the supercoiled hybrid plasmids, immobilized on nitrocellulose paper, with the cDNA. Three hybrids are detected corresponding in molecular weight to the expected hybrid plasmids containing each of the *Eco*RI fragments identified in Fig. 1. The 22-kb hybrid plasmid was cloned as described below.

A second 10-μg aliquot of hybrid plasmid DNA isolated as described above is electrophoresed on a 0.8% agarose slab gel in Tris–acetate buffer without ethidium bromide. Following electrophoresis, the gel is stained in a solution of 0.5 μg/ml ethidium bromide for 20 min. The gel is then transferred to a glass plate and illuminated from above with a long-wavelength uv lamp. By utilizing the mobilities of the supercoiled plasmid DNA markers, the mobility of the 22-kb hybrid plasmid is calculated, and

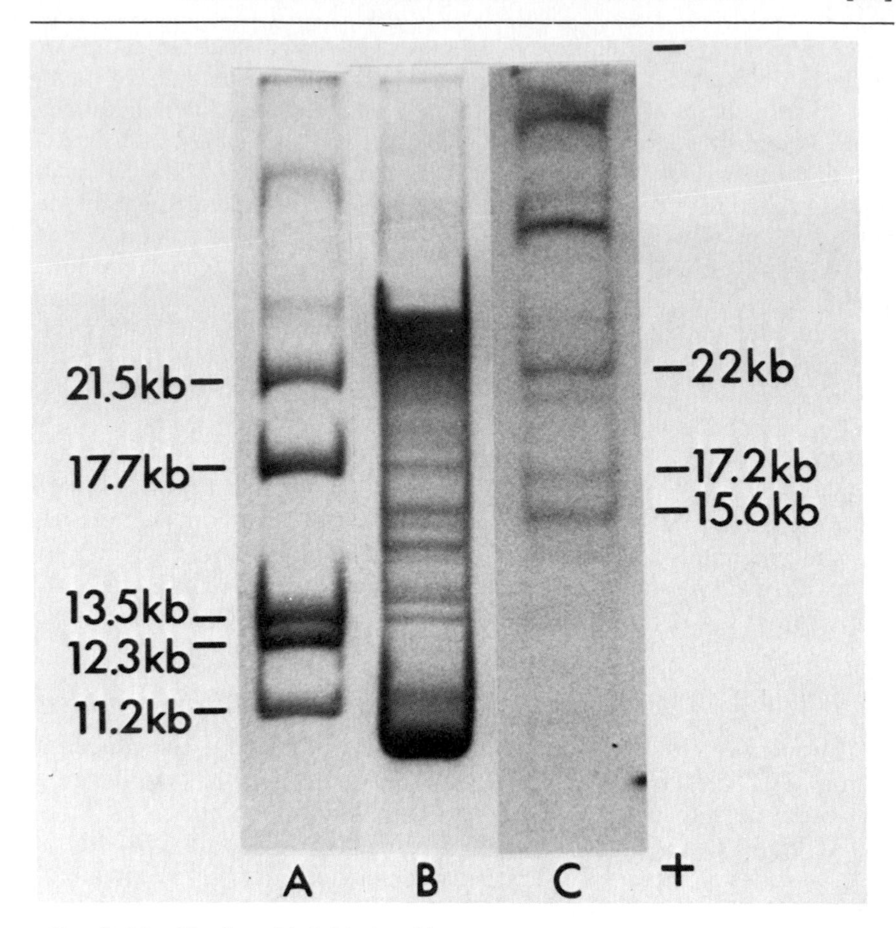

FIG. 2. Identification of hybrid plasmid supercoiled DNAs containing *Eco*RI cleavage fragments of yeast DNA complementary to glyceraldehyde-3-phosphate dehydrogenase cDNA. DNA was visualized by ethidium bromide fluorescence after electrophoresis on a 0.8% agarose slab gel. (A) Known molecular weight supercoiled DNA markers. (B) Total hybrid plasmid DNA supercoils isolated from a shotgun collection of transformants containing *Eco*RI cleavage fragments of yeast genomic DNA. (C) An autoradiogram of hybrids formed between glyceraldehyde-3-phosphate dehydrogenase cDNA and the hybrid plasmid DNA shown in (B) after transfer to a nitrocellulose filter.

the section of the agarose gel containing the 22-kb supercoil is minced in 0.2 ml of buffer containing 10 m*M* Tris, pH 8.0, 1 *M* NaCl, and 1 m*M* EDTA by 20 passages through a 20-gauge syringe needle. The suspension is incubated for 16–18 hr at 50° to elute the plasmid DNA from the gel. Following a 20,000 *g* centrifugation for 30 min, the supernatant is adjusted to 10 μg/ml with yeast rRNA carrier and extracted with an equal volume of STE-saturated phenol. After an ether extraction, plasmid DNA is precipitated with 2 volumes of ethanol at −20°. Plasmid DNA is collected by centrifugation and suspended in 50 μl 10 m*M* Tris, pH 7.4, 1 m*M* EDTA. The DNA suspension is used to transform *E. coli* strain RR101 as described above.

Colony Hybridization

Transformants (200–400) are spread on 15-cm-diameter L broth agar plates containing 20 μg/ml of ampicillin. The agar plates are dried for 2 days at 37° in an upright position before use. Colony filter hybridization is then carried out as described by Beckman *et al.*[19] Colonies are transferred to sterile Whatman 540 filter disks (12.5 cm diameter) by pressing the filter paper to the surface of the agar plate. After transfer, the plate is reincubated for 8–12 hr at 37° to allow the colonies to regrow. This plate now becomes the master for the recovery of transformants. Colonies adsorbed to the filter disk are lysed by dropwise addition of 0.5 *N* NaOH. Sufficient 0.5 *N* NaOH is added to just moisten the filter paper. Addition of excess NaOH results in streaking and subsequent difficulty in identifying colonies that form hybrids with the cDNA probe. Lysis and denaturation of plasmid DNA is allowed to occur for 15 min at room temperature. Excess NaOH is then removed by aspiration, and the filter is neutralized by dropwise addition of 0.2 *M* Tris, pH 7.5, and 0.6 *M* NaCl. The pH of the filter is monitored with pH indicator paper. The filter is then dried at 70° for 2 hr. The DNA is then fixed to the filter by irradiation for 10 min with a short-wavelength uv lamp at a distance of 10 cm. The filter is preincubated for 4 hr in a solution containing 6× SSC, 0.02% Ficoll, 0.02% bovine serum albumin, and 0.02% polyvinylpyrrolidine[17] at 65°. Following incubation the filters are blotted to remove excess liquid and air-dried. Hybridization is carried out in 2 ml of the same solution described for the preincubation containing 1×10^6 cpm of ^{32}P-labeled cDNA and 10 μg/ml of depurinated salmon sperm DNA for 16–20 hr at 65° under mineral oil. After hybridization the filter is washed as described above for the nitrocellulose filters. Colony hybridization is monitored by autoradiography as described above. Figure 3 illustrates the results of the colony hybridization. Hybrid plasmid DNA was isolated from six positive transformants and analyzed by *Eco*RI digestion. All six contained the vector plus a 12-kb fragment of yeast DNA. One of these hybrid plasmids, designated p*gap*492, was digested with *Eco*RI and electrophoresed on a 1% agarose gel in parallel with *Eco*RI-digested yeast DNA. The results of hybridizing the digested DNA, after transfer to nitrocellulose paper, with the cDNA probe are shown in Fig. 1E and F. These data demonstrate that the transformants identified by colony hybridization contain the high-molecular-

[19] J. S. Beckman, P. F. Johnson, and J. Abelson, *Science* **196**, 205 (1977).

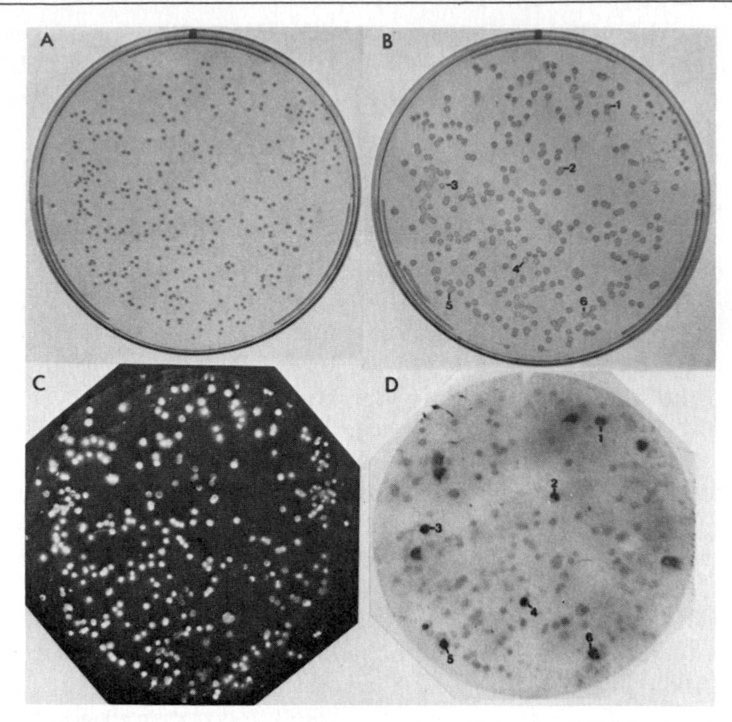

FIG. 3. Identification of an *E. coli* transformant containing yeast genomic DNA complementary to a glyceraldehyde-3-phosphate dehydrogenase cDNA. Approximately 350 colonies were spread on an L broth agar plate containing 20 μg/ml ampicillin. (A) Colonies before transfer to Whatman 540 filter paper. (B) Colonies regrown after transfer to Whatman 540 filter paper. (C) Colonies (visualized with a long-wavelength uv lamp) immobilized on the Whatman 540 filter. (D) Autoradiogram of the filter after hybridization with the glyceraldehyde-3-phosphate dehydrogenase probe. The positions of six colonies that contain p*gap*492 hybrid plasmid DNA are indicated in (B) and (D).

weight *Eco*RI cleavage fragment of yeast genomic DNA that is complementary to the cDNA probe.

Conclusion

The cloning procedure described in this chapter is suitable for cloning fragments of yeast genomic DNA that are complementary to partially purified hybridization probes. Microgram quantities of genomic DNA are needed, and it is not necessary to fractionate restriction endonuclease cleavage fragments prior to ligation and transformation. Isolation of a known molecular weight supercoiled hybrid plasmid DNA from a mixture of hybrid plasmid DNAs minimizes the number of false positive transformants isolated on the basis of hybridization with minor components of the partially purified probe. Small quantities of supercoiled hybrid plasmid DNA are required, since the efficiency of transformation of competent *E. coli* strain RR101 with supercoiled DNA is 500–2000 transformants per nanogram of DNA.

The glyceraldehyde-3-phosphate dehydrogenase structural gene reported here has been extensively characterized by restriction endonuclease and transcriptional mapping analysis. A portion of the nucleotide sequence of the isolated structural gene has been determined and shown to encode a portion of the known amino acid sequence of yeast glyceraldehyde-3-phosphate dehydrogenase. These results will be reported elsewhere.

[29] Hybridization with Synthetic Oligonucleotides

By J. W. SZOSTAK, J. I. STILES, B.-K. TYE, P. CHIU, F. SHERMAN, and RAY WU

Recent advances in chemical[1] and enzymatic[2] synthesis of oligonucleotides have greatly increased the availability of these compounds. Oligonucleotides 10 to 20 bases long are potentially useful as hybridization probes for the detection of unique genes in Southern blot filter hybridization experiments[3] and for the screening of colony or bacteriophage banks for particular sequences. The amino acid sequence of many interesting proteins has been determined. From this information it is possible to deduce a partial nucleotide sequence for the corresponding mRNA or gene. The degeneracy of the genetic code results in ambiguity at the second base of some codons and at the third base of most codons. This effect can be minimized by selecting a region of the protein sequence consisting predominantly of unique codons (Met and Trp) and the other least ambiguous codons (Asp, Asn, Cys, His, Phe, Tyr, Glu, Gln, and Lys). In these cases, the uncertainty is between A and G or T and C, and the effect of possible mismatches is minimized by selecting G for A or G ambiguity and T for T or C ambiguity in the oligonucleotide.[4] This results in either correct base pairing or a G = T mismatch. This type of mismatch is expected to cause less destabilization of the helix than any other mismatch.[5]

In this chapter we describe procedures for the use of synthetic oligo-

[1] R. Wu, C. P. Bahl, and S. A. Narang, *Prog. Nucleic Acid Res. Mol. Biol.* **21,** 101 (1978).
[2] S. Gillam, R. Jahnke, and M. Smith, *J. Biol. Chem.* **253,** 2532 (1978).
[3] E. Southern, *J. Mol. Biol.* **98,** 503 (1975).
[4] R. Wu, *Nature (London), New Biol.* **236,** 198 (1972).
[5] O. C. Uhlenbeck, F. H. Martin, and P. Doty, *J. Mol. Biol.* **57,** 217 (1971).

ISBN 0-12-181968-X

nucleotides for Southern blot experiments and gene bank screening, and demonstrate the effect of various mismatches on the efficiency of hybridization.

Sensitivity Versus Specificity

To use synthetic oligonucleotide probes for hybridization they must first be end-labeled and then annealed with single-stranded DNA bound to a nitrocellulose filter. The temperature should be 15°–20° below the estimated T_m of the hybrid; in practice the conditions of the hybridization reaction must be carefully optimized in order to achieve high sensitivity and specificity. The specificity of the probe is determined by its length (and therefore the number of times its complementary sequence occurs in the DNA being probed), and by the stringency of the reaction conditions. If the conditions are insufficiently stringent, the probe will hybridize with many closely related sequences. However, the efficiency of the hybridization reaction declines as the reaction conditions are made more stringent. A balance must therefore be found between the opposing requirements of sensitivity and specificity.

Hybridization with a 12-nucleotide-long fragment (12-mer), with one mismatch, is sufficiently sensitive for the detection of correct binding to a restriction digest of λ DNA. We have tested the effects of several different mismatches and find that errors near the middle of the sequence are more critical than errors near either end.

An oligonucleotide 13–15 nucleotides in length is sufficient for the detection of a unique gene in total yeast DNA by Southern blot analysis,[3] and a 15-mer can be used in the screening of a yeast DNA bank cloned in a λ vector. Conditions must be optimized and the specificity determined by Southern blot experiments before plaque hybridization is performed. However, analysis of mammalian DNA by Southern blots may not be possible because of the greater complexity of this DNA, although plaques could still be detected. The use of single-stranded DNA phages such as certain M13 derivatives as cloning vectors may considerably enhance the usefulness of oligonucleotide hybridization in gene bank screening because of the greater sensitivity attainable with these vectors.

Materials

Oligonucleotides: d(A-G-C-A-C-C-T-T-T-C-T-T-A-G-C), complementary to bases 24–39 of the yeast iso-1-cytochrome *c* mRNA, was synthesized in our laboratory[6] by the phosphotriester method. d(G-A-G-C-

[6] J. W. Szostak, J. I. Stiles, C. P. Bahl, and R. Wu, *Nature (London)* **265**, 61 (1977).

G-G-A-T-A-A-C-A-A-T-T) and d(C-C-G-C-T-C-A-C-A-A-T-T, complementary to the *Escherichia coli lac* operator, were a gift from S. A. Narang.

5'-End-labeling was carried out with T4 polynucleotide kinase (New England Biolabs) and [γ-^{32}P]ATP (New England Nuclear) at a specific activity of 1000–3000 Ci/mmol.[7]

DNA: Yeast DNA was prepared as described[8] from the following strains: D311-3A (a *his1 trp2 lys2*), used as a standard *CYC1* (wild type) strain; B-955 (a *CYC1-91-A his1 trp2 lys2*) which has CAA in place of GAA[9] and thus lacks the *Eco*RI site at bases 3–9; B-2185 (a *CYC1-183-AD his1 trp2 lys2*) which has a deletion encompassing amino acids 8–12[10], the binding site of the 15-mer; ϕ80 *dlac O*^{c} mutants RV1, RV10, RV17, RV51, RV116, and RV120 which were a gift from J. R. Sadler.[11]

Restriction enzymes: Purchased from New England Biolabs.

Hybridization: DNA fragments were transferred to nitrocellulose paper (Schleicher and Schuell) by the blotting procedure of Southern[3] or the plaque transfer procedure of Benton and Davis.[12] All hybridizations were carried out in 2× SSC, with 0.2% polyvinylpyrrolidone, Ficoll, and bovine albumin (Sigma) in sealed plastic bags.

Hybridization of a 12-mer to the λ Endolysin Gene

The amino acid sequence of the λ endolysin gene is known[13] and contains a region with low-nucleotide-sequence ambiguity. We[14] synthesized a 12-mer complementary to this sequence, with four possible ambiguities of the G = T type. Hybridization to exonuclease III-digested λ DNA was observed, and primer extension and DNA sequencing work showed that hybridization with the correct site had occurred. It was not possible, however, to determine how many mismatched base pairs, if any, were actually present.

[7] A. Maxam, and W. Gilbert, *Proc. Natl. Acad. Sci. U.S.A.* **74,** 560 (1977).

[8] D. R. Cryer, R. Eccleshall, and J. Marmur, *Methods Cell Biol.* **12,** 39 (1975).

[9] F. Sherman and J. W. Stewart, *in* "The Biochemistry of Gene Expression in Higher Organisms" (J. K. Pollak and J. W. Lee, eds.), p. 56. Australian and New Zealand Book Co., LTD, Sydney, Australia, 1973.

[10] J. W. Stewart and F. Sherman, *in* "Molecular and Environmental Aspects of Mutagenesis" (L. Pradash *et al.*, eds.), p. 102. Thomas, Springfield, Illinois, 1974.

[11] A. Jobe, J. R. Sadler, and S. Bourgeois, *J. Mol. Biol.* **85,** 231 (1974).

[12] W. D. Benton and R. W. Davis, *Science* **196,** 205 (1977).

[13] M. Imada and A. Tsugita, *Nature (London), New Biol.* **233,** 230 (1971).

[14] R. Wu, C. D. Tu, and R. Padmanabhan, *Biochem. Biophys. Res. Commun.* **55,** 1092 (1973).

Hybridization of Oligonucleotide Probes to the *Escherichia coli lac* Operator

The effect of type and position of mismatch on oligonucleotide hybridization was tested. A-C, C-T, G-T, and G-A mismatches near the middle or ends of the complementary sequence were studied. The mismatches occurred when hybridizing DNA from a series of ϕ80 d*lac* O^c mutants to synthetic single-stranded *lac* operator probes carrying the unmutated sequence. Hybridizations were carried out at different temperatures to determine the stability of the hybrids. The position and base change of each mutation[15] are shown along with the sequence of the probes.

1. A synthetic 15-mer was hybridized with two mutants:

```
5'  G-A-G-C-G-G-A-T-A-A-C-A-A-T-T  3'     15-mer probe
3'—C-T-C-G-C-C-T-A-T-T-G-T-T-A-A—5'      lac operator sequence
    ↓ ↓
    A C                                   Single base change
  RV10 17                                 Mutant strain
```

DNAs, isolated from the ϕ80 d*lac* O^c mutants and from the wild type RV80, were digested with restriction enzyme *Eco*RI and analyzed after agarose gel electrophoresis by Southern blotting.[3]

Hybridization occurred specifically to one fragment size. At 31°, RV17 DNA hybridized with the 15-mer as well as RV80, but RV10 was significantly weaker. This means that a G-A mismatch on the end of the complementary sequence is more destabilizing than an A-C mismatch in the penultimate position. A mismatch on the end would presumably be less destabilizing than one further in. This advantage is probably nulled by the destabilization of the neighboring base pair by the purine to purine mismatch.[16] Thus, the relative stabilities of these hybrids is determined by mismatch type rather than position.

2. A synthetic 12-mer was hybridized to *Hin*dIII digested DNA from several of the ϕ80 d*lac* O^c mutants, shown below:

```
      1   3   5   7   9   11
5'   C-C-G-C-T-C-A-C-A-A-T-T  3'      12-mer probe
3'—G-G-C-G-A-G-T-G-T-T-A-A—5'         lac operator sequence
       ↓ ↓ ↓ ↓ ↓     ↓
       A T T G T     A                Single base change
  RV120   51  10     1                Mutant strain
       116  17
```

DNA from ϕ80 (lacking the d*lac*) as well as the wild type RV80 were

[15] W. Gilbert, J. Gralla, J. Majors, and A. Maxam, *in* "Protein-Ligand Interactions" (H. Sund and G. Blauer, eds.), p. 193, de Gruyter, Berlin, 1975.

[16] J. B. Dodgson and R. D. Wells, *Biochem.* **16**, 2367 (1977).

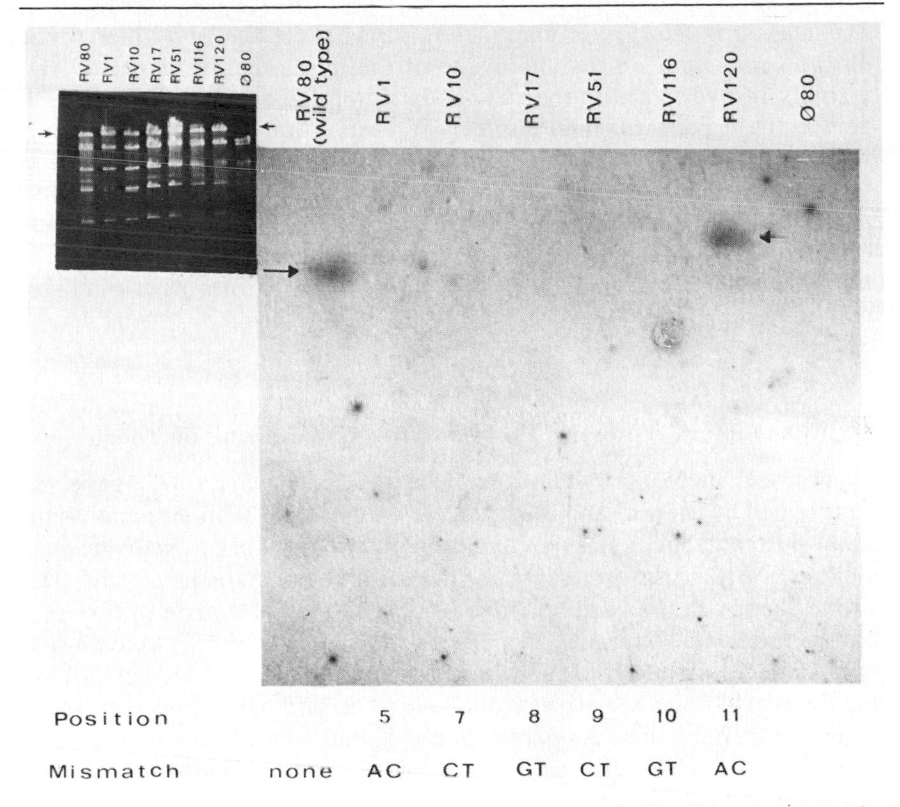

FIG. 1. Hybridization at 30° of a synthetic 12-mer probe to the *E. coli lac* operator region in ϕ80 d*lac* DNA. Insert shows the ethidium bromide stained gel of *Hin*dIII digested ϕ80 d*lac* DNA. The arrows in the insert indicate the *Hin*dIII fragment containing the *lac* operator sequence, and the two arrows in the figure show the location of the hybridized bands in RV80 and RV120. The filter was prehybridized in 5 ml of 10X Denhardt's solution made with 2XSSC for 14 hr at 65°. ^{32}P-labeled probe was added to a concentration of 3×10^5 cpm/ml. Hybridization was carried out at 30° for 13 hr. The hybridized filter was washed at 23° in 2XSSC for 2 hr, with 4 changes of buffer, before using to expose x-ray film.

included as controls. The results at 30° are shown in Fig. 1. The synthetic 12-mer probe hybridized to RV120 DNA as strongly as to RV80 DNA. At a more stringent temperature (32°) this signal was weaker than RV80, and at 37°, only the wild type hybridized. The mismatch with RV120 DNA is C-A which is presumably less stable than a G-T mismatch, such as found with RV116 and RV17. Hybridizations with the 15-mer, however, indicated that an A-C mismatch is relatively stable at 31°. Thus, the determining factor for stable hybridization to the 12-mer is position rather than the nature of the mismatched base pair. The importance of position is especially striking when comparing RV116 and 120, in which the mismatches are G-T at position 3 and C-A at position 2, respectively. One explanation is that in an oligonucleotide as short as the 12-mer probe, all the nucleotides on the short side of the mismatch are unpaired. The stability of hydridization then depends simply on a sufficient number of perfect base pairings. The number of base pairs before the mismatch minus one (due to the destabilizing effect of the mismatch[16]) can be termed the effective probe size. With RV116 the effective probe size is 8 and with RV120 it is 9. In the conditions of the experiment, a minimum effective probe size of 9 (with 4 G-C pairs) seems to be required for stable hybridization. This finding is in general agreement with other published results.[4]

Hybridization of Synthetic Probes to the *CYC1* Gene of Yeast

The sequence of the first 44 nucleotides of the yeast *CYC1* gene was determined by Stewart and Sherman[10] by amino acid sequence analysis of frameshift mutants. A 15-mer complementary to part of this sequence was synthesized for use as a hybridization probe by Szostak *et al.*[6] This 15-mer has been used in two ways to detect the *CYC1* gene in Southern blot experiments. First, the 15-mer was used to make cDNA complementary to the *CYC1* mRNA as previously described.[6] The labeled cDNA is a highly specific and sensitive probe, but it is difficult to prepare and is available in only small amounts. Second, end-labeled 15-mer was used directly as a hybridization probe. While this probe is easily prepared and is sufficiently specific, the efficiency of the hybridization is low and a large excess of labeled probe must be used.

Yeast DNA from a wild-type strain has an *Eco*RI site at bases 3–9 of the CYC1 gene (amino acids 2–3 of the iso-1-cytochrome *c*), whereas DNA from a strain carrying the *CYC1-91-A* mutation does not have this *Eco*R1 site. Since the strains are isogenic, we expect that the *CYC1* *Eco*R1 fragment will be larger in the mutant than in the wild type, while all other *Eco*R1 fragments will be of the same size. DNA from the two strains was digested with excess *Eco*R1, electrophoresed on a 1% agarose gel, and transferred to nitrocellulose paper. The filter was annealed with *CYC1* cDNA (50–150 bases long) at 65°, washed, and autoradiographed (Fig. 2). Hybridization was observed with DNA fragments of 4.45 Md from the wild-type strain, and 6.2 Md in the mutant. The size shift conclusively identified these fragments as carrying the *CYC1* gene. The 6.2-Md fragment was the fusion product of the 4.45- and 1.75-Md fragments from the wild-type strain.

In a separate experiment, hybridization of end-labeled 15-mer alone was observed with the 1.75-Md fragment. This fragment did not appear with DNA from a strain carrying the *CYC1-183-Ad* deletion which exactly

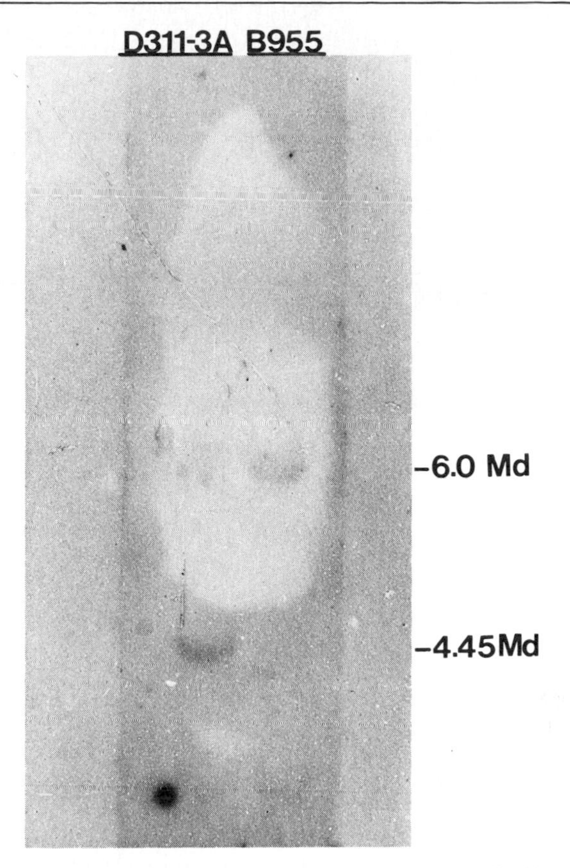

FIG. 2. Hybridization of a polymerase-extended 15-mer (*CYC1* cDNA) with yeast DNA from the wild type (D311-3A) and from a mutant lacking the *Eco*RI site (B-955). Loss of the *Eco*RI site results in the fusion of two adjacent fragments to give the one larger fragment (6.2 Md) seen in the mutant.

covered the 15-mer binding site. The 1.75-Md fragment must have been on the 3′-end of the 6.2-Md fragment and was the fragment carrying the structural gene for iso-1-cytochrome *c*.

Hybridization of labeled 15-mer to *Eco*R1-, *Bam*HI-, and *Eco*R1 plus *Bam*HI-digested wild-type yeast DNA is illustrated in Fig. 3. In each case there is a major and a minor band. The major band in the *Eco*RI digest is the 1.75-Md fragment carrying the *CYC1* gene. The minor band is probably due to a DNA fragment carrying a closely related sequence that also binds the 15-mer.

Hybridization was repeated at several temperatures between 25° and 50°. The optimum for sensitivity and specificity was in the 42°–45° range.

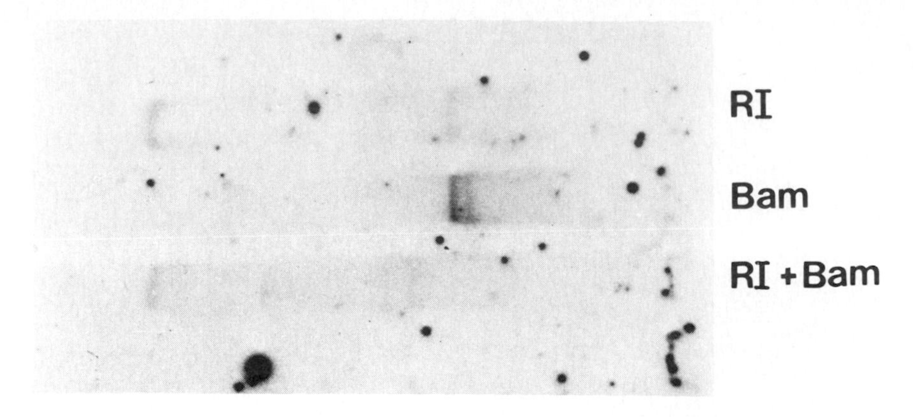

FIG. 3. Hybridization of a labeled 15-mer to restriction enzyme digests of yeast DNA. The lower band in the *Eco*RI digest is at 1.75 Md and is due to hybridization with the *CYC1* gene.

The efficiency of the hybridization declined rapidly above 45°, and at lower temperatures many additional bands were seen. Only one minor band remained at 44°.

We have used the labeled 15-mer to screen a λ phage bank of *Eco*R1 fragments of total yeast DNA. With the conditions optimized as described above, and using the plaque transfer procedure,[12] we observed a clear pattern of hybridization with individual plaques; however, the sensitivity was greatly enhanced by spotting the plaques to a grid before proceeding with the plaque transfer. In addition to the regular pattern, plaques prepared in this manner gave stronger signals and were larger in size, making them more distinct from the background (Fig. 4). Analysis of these clones has shown both the *CYC1* gene and the cloned fragment responsible for the minor band seen in Fig. 3.

Hybridization of a synthetic 13-mer with restriction enzyme digests of total yeast DNA has been reported by Montgomery *et al.*[17] They found hybridization with seven different fragments, including the *CYC1* gene. The correct band was identified with the *Eco*R1 site mutant described above. A *CYC1* clone was also identified by hybridization with a 13-mer. DNA of the correct size was isolated from a gel and used to prepare a phage λ bank. The plaques were transferred to a grid and screened by hybridization with labeled 13-mer. Hybridization with a clone carrying the *CYC1* gene was detected.

[17] D. L. Montgomery, B. D. Hall, S. Gillam, and M. Smith, *Cell* **14,** 673 (1978).

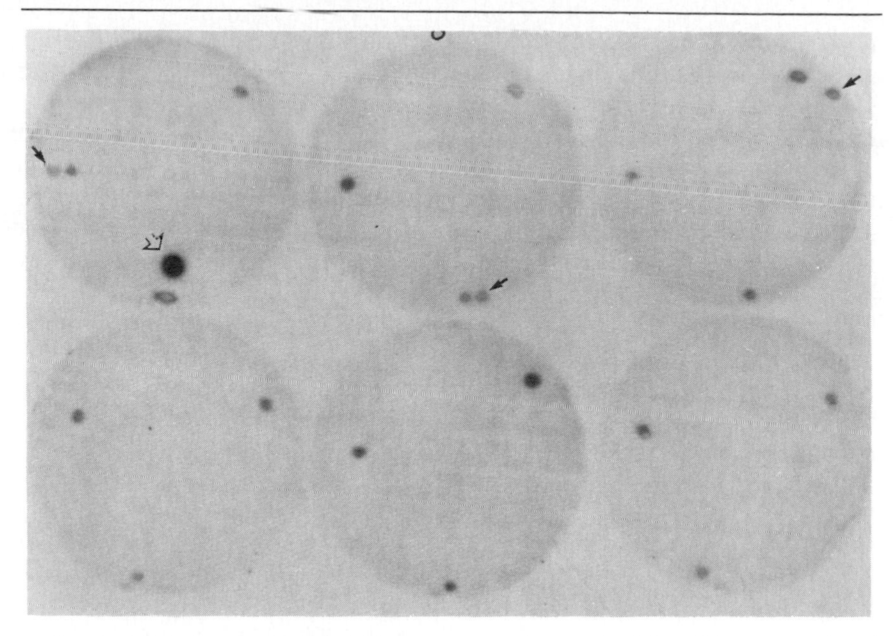

FIG. 4. Hybridization of labeled 15-mer with a λ phage bank of *Eco*RI fragments of total yeast DNA. Plaques transferred to a grid so that there were approximately 200 plaques per plate. The three common spots on all filters are a previously isolated clone containing the minor band seen in the *Eco*RI track in Fig. 3. The solid arrows indicate positive plaques and the open arrow nonspecific binding.

Problems with Filter Hybridizations

A common technical problem encountered in filter hybridization is a heavy background of dark spots on the film, which can obscure the desired bands. This problem becomes severe with the very high levels of radioactive probe used for oligonucleotide hybridization. The background can be minimized by using the lowest possible concentration of probe (usually 10^5–10^6 cpm/ml is sufficient) and by performing the hybridization in 0.2% polyvinylpyrrolidone, Ficoll, and bovine serum albumin. It is also important that the filter be thoroughly wetted before the labeled probe is added. Nitrocellulose filters wet poorly in high salt; the best procedure is to wet the filters in distilled water. They are then sealed in plastic bags and incubated briefly in hybridization buffer at 65°, after which the labeled probe is added (R. Rothstein, personal communication).

Cloning Eukaryotic Genes

The yeast genome contains 2×10^7 base pairs of DNA. A sequence of 12 nucleotides is likely to be unique. Since neither a 13-mer (seven bands) nor a 15-mer (two bands) hybridized with a single site, the hybridization conditions were probably not sufficiently stringent to eliminate binding to closely related sequences. The complexity of mammalian DNA requires a 15- to 16-nucleotide-long sequence to be unique, and probably an 18- to 20-nucleotide sequence is more desirable to avoid unwanted hybridization with related sequences. This would be difficult to derive from the amino acid sequence of a protein. An alternative may be to screen a phage bank with two shorter probes 11–14 nucleotides in length and to search for clones to which both short probes bind.

Recently the phage M13 has been modified so as to be a useful cloning vector.[18] This phage produces plaques containing up to 10^9 phage particles. Plaque hybridization with this system is approximately 100 times more efficient than with λ plaques (R. Rothstein, personal communication). This property may greatly enhance the usefulness of synthetic oligonucleotides in screening phage banks. Probes short enough to be derived from amino acid sequences may be sufficiently long to have the sensitivity required for M13 plaque hybridization.

Acknowledgments

This work was supported in part by NSF Research Grant 77-20313 awarded to R. W., NSF Grant PCM78-02341 and NIH Grant A1 14980-01 awarded to B. K. T., NIH grant GM12702 awarded to F. S., and NIH postdoctoral Fellowship GM05441-02 to J. I. S. and in part by the U.S. Department of Energy at the University of Rochester, Department of Radiation Biology and Biophysics. This paper has been designated Report No. UR-3490-1581.

[18] B. Gronenborg and J. Messing, *Nature* (*London*) **272,** 375 (1978).

[30] *In Situ* Immunoassays for Translation Products

By DAVID ANDERSON, LUCILLE SHAPIRO, and A. M. SKALKA

One approach to the isolation of specific recombinants from a clone bank has been immunological detection of cloned gene translation products produced in *Escherichia coli*. Immunological selection provides an approach when it is not possible to use selection techniques that depend

[1] A. Skalka and L. Shapiro, *Gene* **1,** 65 (1976).

[2] B. Sanzey, O. Mercereau, T. Ternynck, and P. Kourilsky, *Proc. Natl. Acad. Sci. U.S.A.* **73,** 3394 (1976).

ISBN 0-12-181968-X

on nucleic acid hybridization or functional expression of the cloned gene. The first procedures developed for immunological selection were *in situ* immunoassays.[1,2] Recombinant phage or *E. coli* containing recombinant plasmids are grown on agar plates containing antisera produced against the protein encoded in the desired gene. Phage or *E. coli* expressing antigenic determinants of the protein are detected by observation of an immunoprecipitin reaction or an amplification of the immunoprecipitin reactions in the agar. Recently, other procedures have been developed that utilize radioiodinated antibody or radioiodinated protein A from *Staphylococcus aureus* and have the theoretical ability to detect one antigen molecule per *E. coli* cell.[3,4] However, the value of the *in situ* immunoassay remains, because of its simplicity. It is not necessary to iodinate protein or purify antibody from antisera or $F(ab)_2$ fragments of antibody. Furthermore, the *in situ* immunoassay itself is quite sensitive. Using *E. coli* β-galactosidase as a model antigen, it can be detected at the uninduced level which is in the range of 10–20 molecules per cell.[5]

Materials and Reagents

Reagents

1. N-Z amine A: Humko Sheffield Chemical, Norwich, New York
2. Salt free lysozyme: Worthington Biochemical Corporation, Freehold, New Jersey
3. Agarose: Seakem powder, Marine Colloids, P. O. Box 308, Rockland, Maine, or Sigma Chemical Company, Saint Louis, Missouri
4. Bacto-agar, Bacto-tryptone, and casamino acids: Difco Laboratories, Detroit, Michigan
5. 5-Bromo-4-chloro-3-indolyl-β,D-galactoside (XG): Bachem, Marina Del Ray, California
6. Isopropyl-β-thiogalactoside (IPTG): Sigma Chemical Company, Saint Louis, Missouri
7. Sarkosyl NL97: Ciba-Geigy Corporation, Greensboro, North Carolina
8. Rabbit anti-β-galactosidase serum: Gift of Dr. A. Fowler, University of California, Los Angeles

[3] H. Erlich, S. Cohen, and H. McDevitt, *Cell* **13,** 681 (1978).

[4] S. Broome and W. Gilbert, *Proc. Natl. Acad. Sci. U.S.A.* **75,** 2764 (1978).

[5] J. Miller, "Experiments in Molecular Genetics," p. 398. Cold Spring Harbor Lab., Cold Spring Harbor, New York, 1972.

Solutions

1. Tryptone broth: 1% Bacto-tryptone, 0.5% NaCl, 0.001 *M* $MgCl_2$
2. Soft tryptone agar: Tryptone broth with 0.6% Bacto-agar
3. Tryptone bottom agar: Tryptone broth with 1.5% agar
4. N-Z bottom agar: 1% N-Z amine A, 0.5% NaCl, 0.4% casamino acids, 10 μg/ml thymine, 1.5% Bacto-agar
5. N-Z bottom agarose: 1% N-Z amine A, 0.5% NaCl, 0.4% casamino acids, 10 μg/ml thymine, 1.3% agarose
6. Water-soft agar or agarose: 0.6% agar or agarose in H_2O
7. Tris-EDTA soft agarose: 0.6% agarose, 0.1 *M* Tris-Cl, 0.01 *M* ethylenediaminetetraacetic acid, pH 8.0
8. Lysozyme solution: 5 mg/ml salt-free lysozyme in 0.1 *M* Tris-Cl, 0.01 *M* EDTA
9. PBS: 8.0 g NaCl, 0.2 g KCl, 1.15 g $Na_2HPO_4 \cdot 2H_2O$, 0.2 g KH_2PO_4 per liter

Phage and bacterial strains. Escherichia coli S90C [F^- ara^- Δ(*lac pro*) *strA*] was used as host strain for the phages λ *plac5* (i^- z^+ y^-) *c*1857 *sRI*λ*3*^{0} *sRI*λ*2*^{0} *sRI*λ*1*^{0}[6] (hereafter called simply λ *plac*) and λ *c*1857 (thereafter called λ) in the plaque assay. The same phage strains were used to construct S90C lysogens. *Escherichia coli* strains W3110 (*lac* i^+ z^+ y^+), and its lysogenic derivative (λ *cI857*) were also employed.

Method

General Comments

Agar or agarose plates for growth of cells were poured thin to conserve antiserum and enhance detection of immunoprecipitates. For small-scale tests, 2–3 ml of agar was used in 35-mm petri dishes. In large-scale screening experiments with the colony assay, 12 ml of agar was put in 100-mm-diameter dishes. Antiserum (prewarmed to 45°) was added to molten agar or agarose medium at 45° immediately before pouring.

Antisera Preparations

Whole blood obtained from immunized rabbits was allowed to clot at

[6] A. Rambach and P. Tiollais, *Proc. Natl. Acad. Sci. U.S.A.* **71,** 3927 (1974).

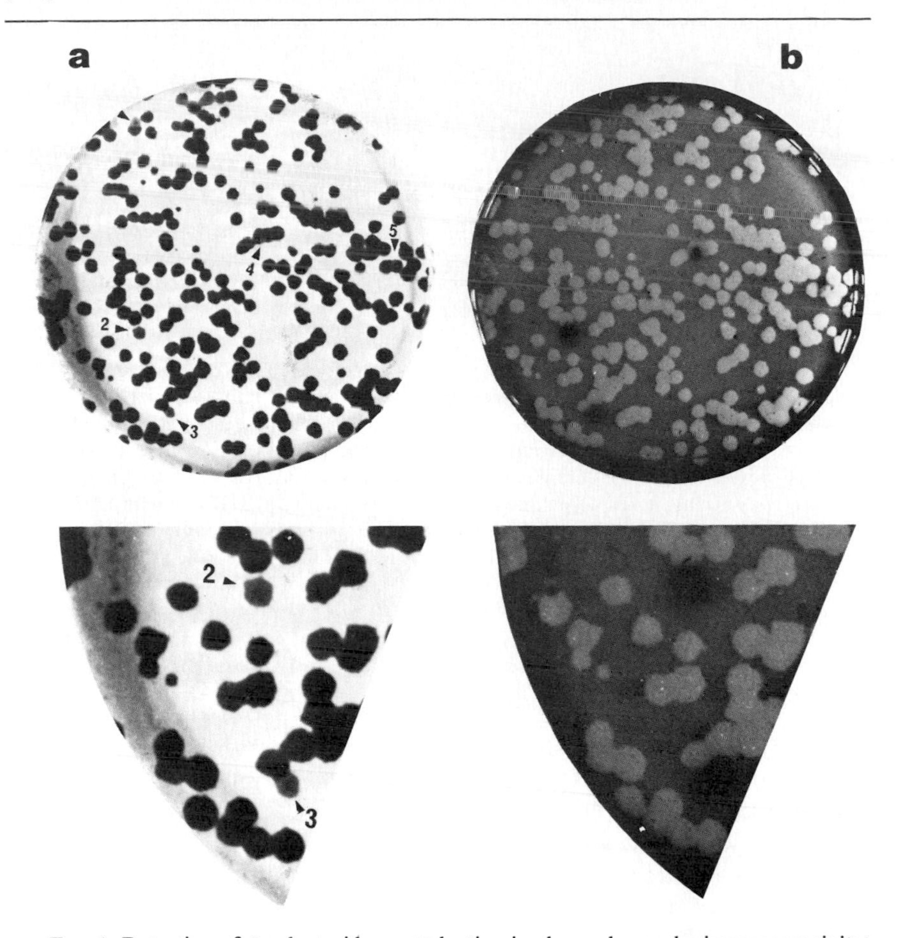

FIG. 1. Detection of β-galactosidase production in phage plaques by immunoprecipitate formation and subsequent confirmation by color development in the presence of the XG indicator substrate. The immunodetection plate (a), containing a mixture of λ *plac* and λ in a ratio of 1:60, was prepared as described in the text with anti-β-galactosidase antibody in the soft agar overlay. The plate was photographed against a background so that clear plaques appear black. Turbid (gray) plaques containing the immunoprecipitate are indicated with numbered arrows. The 35-mm immunodetection plate was subsequently overlaid with 0.5 ml of soft agar containing 80 μg/ml of indicator substrate XG and then incubated at 37° for 90 min. (b) The blue color, which appears dark in the photograph, indicates β-galactosidase activity.

20° and then placed at 4° overnight. The clot was discarded, and the serum was filtered through 0.45-μm sterile nitrocellulose membranes.

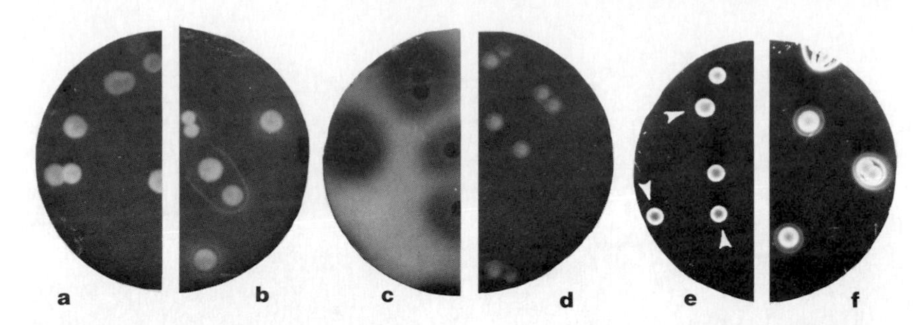

FIG. 2. Immunoprecipitates surrounding bacterial colonies lysed by prophage induction. Colonies of *E. coli* lysogenic for λ were grown at 30° on N-Z bottom agar plates with or without antibody, incubated at 42° for 4 hr, and then, in some cases, overlayed with 0.6% agar containing XG indicator substrate or antibody. Photography was against a black background to reveal clearly the colonies and precipitin rings. In the case of XG staining photography was against a white background. (a) *Escherichia coli* S90C (λ) with antibody in overlay; (b) *E. coli* S90C (λ *plac*) with antibody in overlay; (c) *E. coli* S90C (λ *plac*) with XG in overlay; (d) *E. coli* S90C (λ *plac*) with antibody in bottom agar; (e) *E. coli* W3110 (λ) with antibody in overlay; (f) *E. coli* W3110 (λ) with 1×10^{-3} *M* IPTG in bottom agar and antibody in overlay.

Immunoassay for Recombinant Phage λ Plaques

Escherichia coli host (S90C was used in the original experiments to test the method, but any host appropriate for the vector could be used) is grown to stationary phase, centrifuged, and resuspended in ½ volume of 0.02 *M* MgCl. One volume of phage solution is mixed with 2 volumes of resuspended bacteria, incubated at 37° for 20 min, and then mixed with 25 volumes of molten tryptone soft agar (45°). An appropriate amount (see below) of antiserum (prewarmed at 45°) is then added before pouring and spreading (0.4 ml overlay per 35-mm dish, 2.8 ml overlay per 100-mm dish) on thin tryptone bottom agar petri dishes. It is important for the agar overlay to be kept completely molten until distributed on the bottom agar. After incubation at 37° for 16 hr, antigen–antibody complexes can be detected by turbidity within appropriate plaques. For weak antigen–antibody interactions, subsequent incubation at 4° may enhance the observation of turbidity.

Figure 1 provides an example of the immunoplaque assay. λ *plac* phage containing the β-galactosidase *Z* gene are identified in a mixture with λ phage as confirmed by XG staining.

[7] K. Borck, J. D. Beggs, W. J. Brammar, A. S. Hopkins, and N. E. Murray., *Mol. Gen. Genet.* **146,** 199 (1976).

Immunoassay with Bacterial Colonies

Lysis by Induction of λ Prophage. At least two possibilities exist for lysis by induction of λ prophage. Recombinant phage made with vectors that retain the λ *att* site[7] can be used to make lysogens, or plasmid recombinants can be grown in cells that are already lysogens. However, the Na-

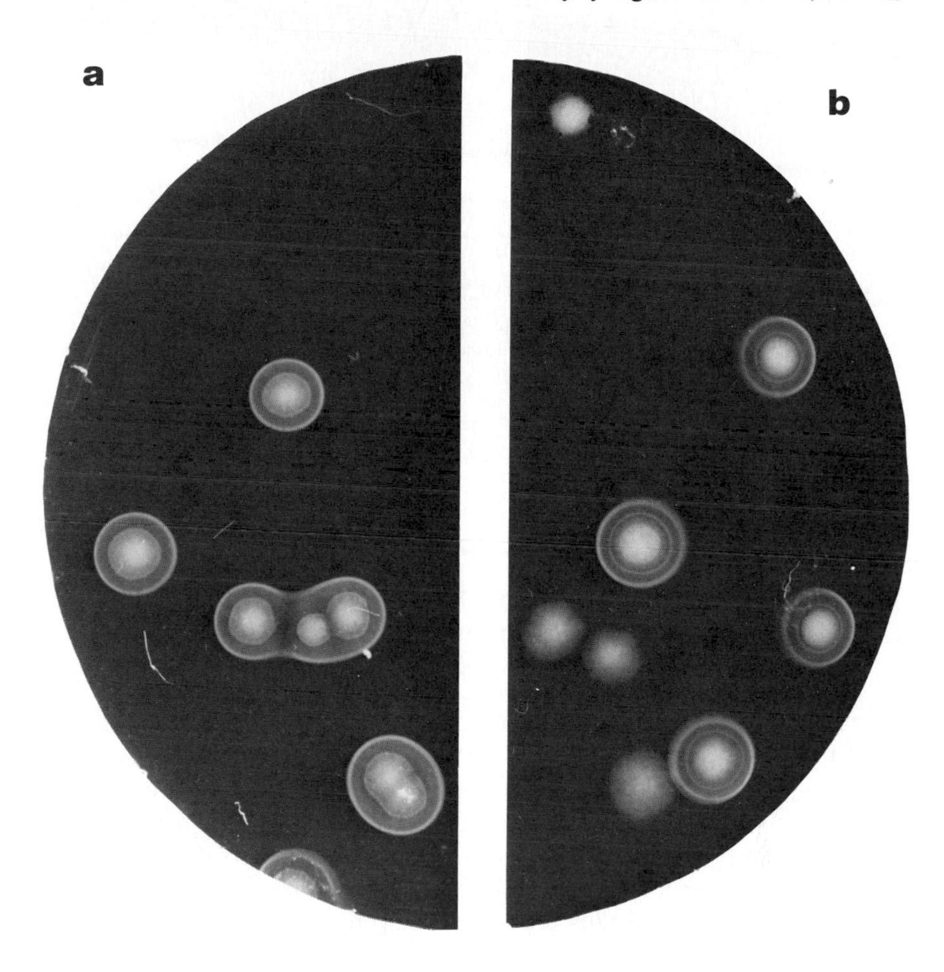

FIG. 3. Immunoprecipitates surrounding bacterial colonies lysed by biochemical procedures. Colonies were grown on N-Z bottom agarose plates containing antibody and 1×10^{-3} *M* IPTG. Colonies were treated as described in the text but without the chloroform vapor treatment. (a) *Escherichia coli* W3110 (a *lac*$^+$ strain)—all colonies showed an immunoprecipitation reaction. (b) A mixture of *E. coli* W3110 (a *lac*$^+$ strain) and S90C (a *lac*$^-$ strain)—only about 50% of the colonies showed an immunoprecipitation reaction.

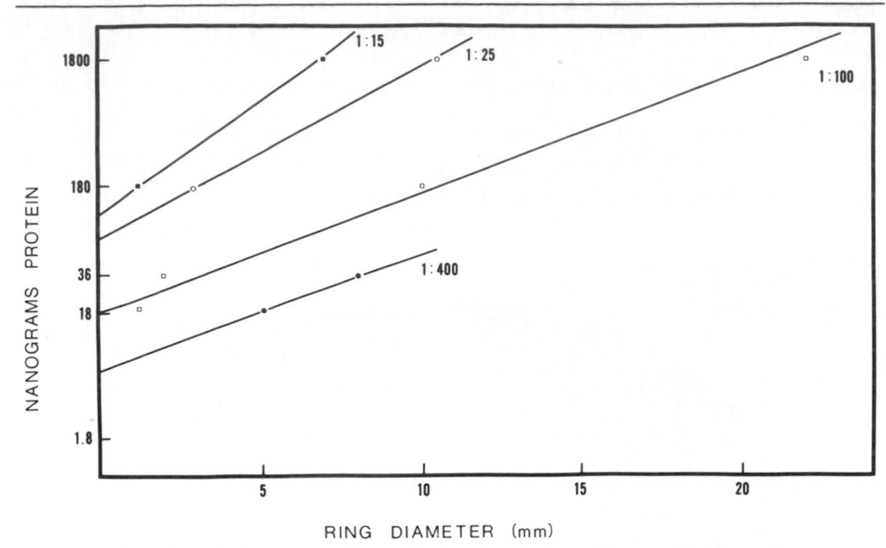

FIG. 4. Results of ring test to determine best antiserum dilution. Rabbit antiserum prepared using commercial glyceraldehyde-3-phosphate dehydrogenase was diluted in N-Z bottom agar as indicated for each line of the graph and poured in petri dishes as in the immunoassay. Five 2-mm holes were punched per plate, and five levels of glyceraldehyde-3-phosphate dehydrogenase (1.8 μg, 180 ng, 36 ng, 18 ng, 1.8 ng) were added to the holes in each plate. The plates were incubated at 37° overnight, and immunoprecipitin ring diameters measured.

tional Institutes of Health guidelines for recombinant DNA research currently disallow these types of experiments for many types of recombinants. We have not attempted superinfection[4] of colonies with λ *vir* as means of lysis.

Colonies of *E. coli* lysogenic for λ are grown for 24–48 hr at 30° on petri dishes containing thin layers of N-Z bottom agar. The plates are then incubated at 42° for 4 hr to permit induction of the heat-inducible λ prophages. Next, the colonies are embedded in an overlay (0.5 ml/35 mm dish) of molten soft water–agar containing an appropriate amount of antiserum. After incubation at 37° for 10–16 hr, immunoprecipitin rings can be detected around colonies producing antigen. Alternatively, antiserum is added to the bottom agar at the time the plates are poured. In this case the soft agar overlay is unnecessary.

If the overlay technique is used, the plates must be well dried before use, otherwise some colonies may lift off into the overlay solution.

Lysis by Biochemical Procedures. Cells containing recombinant plasmids are replica-plated onto plates containing N-Z bottom agarose and antiserum (Bacto agar or Noble agar could not be used because an unknown constituent precipitated the lysozyme added in a subsequent step) and allowed to grow overnight at 37°. Media should also contain the appropriate selective antibiotic to force retention of recombinant plasmids. The plates must be dried before use to prevent colonies from floating off in the subsequent (lysis mix) overlay. Plates are generally left open and inverted at 37° for 1 hr after solidifying.

After colony growth, the plates are placed open and inverted in air saturated with chloroform vapor for 10 min. Cells on 35-mm plates are then lysed with an overlay mix of 0.9 ml Tris-EDTA soft agarose and 0.1 ml lysozyme solution (4.5 ml Tris-EDTA soft agarose and 0.5 ml lysozyme solution for 100-mm plates). The overlay must be applied gently so the cells remain attached to the bottom agar. The overlayed plates are incubated at 37° for 1 hr, and then 0.1 ml of 2% sarkosyl NL97 solution is added to the surface of 35-mm plates (0.5 ml/100-mm plate). Incubation is continued at 37° overnight to allow lysis and immunoprecipitation of antigen.

An alternative procedure is to omit the antisera from the bottom agarose and add it in a second overlay after cell lysis. After incubation with sarkosyl for 1 hr at 37°, sarkosyl is removed from the surface by two washings with distilled water. The plates are dried open and inverted for 1 hr at 37°. The antiserum is then added in 0.6% molten agarose in water using the same volume as the previous overlay.

Examples of lysis by phage induction and biochemical lysis are shown in Figs. 2 and 3, respectively. It is noteworthy that immunoprecipitation can be detected in W3110 (λ) *lac*$^+$ *E. coli* in the absence of IPTG inducer (Fig. 2e).

Determination of Proper Antiserum Dilution

In order to detect a precipitive reaction, approximate equivalence of antibody and antigen concentrations are required. The dilution of antiserum required to give equivalence at low antigen concentrations is determined via a ring test. N-Z bottom agarose plates are poured containing several antiserum dilutions. Holes (2 mm in diameter) are punched in the agarose of each plate, and 2 μl of antigen dissolved in PBS is added to each at several concentrations over a broad range. Plates are incubated at 37° overnight. At each antigen dilution, the immunoprecipitin ring diameter is directly proportional to the log of the amount of antigen. An example of the results of this test are shown in Fig. 4 for antiserum prepared against yeast glyceraldehyde-3-phosphate dehydrogenase (GAPDH). The theoretical limit of sensitivity at each antiserum dilution is where the line intersects the *y* axis. In this case, an antiserum dilution of approximately 1:100 was chosen, since it allowed a wide range of detection with less than 18 ng of antigen as the lower limit.

Comments

When observing and photographing immunoassay plates, it is necessary to use a black background and indirect lighting. Weak antigen–antibody interactions are more readily detected after incubation at 4°. Also, if the biochemical lysis is used, removal of some of the sarkosyl by rinsing the plates with several changes of PBS may improve detection of weak antigen–antibody interations.

It is important to include control plates in immunoassay tests which have preimmune antisera. Occasionally cells can produce nonspecific precipitation reactions which can be mistaken as a positive response.

The *in situ* immunoassay type of procedure should be most useful for selecting recombinant genes from organisms whose transcription and translation signals are recognized in *E. coli*. Theoretically, any cloned gene could be detected immunologically by forcing expression by read-through from a phage or plasmid promoter into cloned genes in phase. However, in those cases, it may be better to use the more sensitive immunoassays that utilize autoradiography.[3,4]

[31] Selection of Specific Clones from Colony Banks by Screening with Radioactive Antibody

By LOUISE CLARKE, RONALD HITZEMAN, and JOHN CARBON

In the process of identifying within a colony collection those clones containing plasmids that carry a specific segment of foreign DNA, it is important to have available screening methods that do not depend upon functional expression of a eukaryotic protein or on complementation of a specific bacterial mutation. It is not expected that all eukaryotic genes will be accurately expressed in bacteria, and indeed many eukaryotic genes may contain inserts within the structural gene sequences. In addition, the total number of eukaryotic genes for which complementation assays could eventually be developed is relatively small. Therefore, we have developed methods for the screening of colony collections (constructed as

ISBN 0-12-181968-X

described in this volume [27]) using radioactively labeled antibodies directed against purified eukaryotic proteins of interest. Such a technique should be specific, sensitive, and effective, even if only a small quantity of antigen, or perhaps a fragment of an antigen that still retains antigenic determinants, is produced. The method does not rely upon complementation or enzymatic activity and, since it avoids selective pressure for functional expression, the structural alterations within the cloned DNA segment resulting in efficient expression in the bacterial cell would not be expected to occur.[1,2]

This chapter describes both a plastic well method and an agar plate–paper disk method for screening colony banks with radioactive antibody.

Antibody Preparation

Antiserum made to the desired purified protein whose gene is being sought in a collection of colonies containing cloned foreign DNA is produced by standard procedures. Ten to 20 ml of blood are kept at room temperature for 2 hr, and the clot is loosened with a glass rod. After storage at 4° for 16 hr, the clot is centrifuged down at 10,000 *g* for 15 min. The antiserum supernate is then made 50% in $(NH_4)_2SO_4$ (31.3 g/100 ml of antiserum) while gently stirring at 4°. The precipitate is collected by centrifugation at 9000 *g* for 15 min at 4° and redissolved at its original volume in 10 m*M* potassium phosphate, pH 6.8. The 50% $(NH_4)_2SO_4$ precipitation is repeated, and the precipitate is dissolved at the original volume in phosphate-buffered saline (PBS, 10 m*M* sodium phosphate, pH 7.0, 0.14 *M* NaCl). This fraction is dialyzed against PBS and used for covalent attachment to CNBr-activated paper as described below.

A portion (1–3 ml) of the above is further purified by affinity chromatography as described by Shapiro *et al.*[3] to obtain purified antibody for well coating and for labeling with $Na^{125}I$ to a specific activity of 10^6–10^7 cpm/μg by the lactoperoxidase method of David and Reisfeld.[4] Ten micrograms of this purified preparation is enough for labeling disks for the screening of about 40 agar plates, or for screening a colony collection of 6000 clones by the plastic well method (see below). All antibody prepara-

[1] J. Carbon, B. Ratzkin, L. Clarke, and D. Richardson, *Brookhaven Symp. Biol.* **29,** 277 (1977).

[2] L. Clarke and J. Carbon, *J. Mol. Biol.* **120,** 517 (1978).

[3] D. J. Shapiro, J. M. Taylor, G. S. McKnight, R. Palacios, C. Gonzalez, M. L. Kiely, and R. T. Schimke, *J. Biol. Chem.* **249,** 3665 (1974).

[4] G. S. David and R. A. Reisfeld, *Biochemistry* **13,** 1014 (1974).

tions are stored in small aliquots at −20° to avoid repeated freeze-thawing.

Plastic Well Screening Method

The plastic well antibody screening technique is a modification[5] of the method of Ling and Overby[6] and is based on the observation of Catt *et al.*[7] that antibodies bind very tightly to plastic surfaces and remain competent to bind antigen. The bound antigen (if multivalent) is further able to bind radioactively labeled antibody with a "sandwich" effect.

Our method[8] involves coating wells of plastic microtiter dishes (96 wells/dish) with purified antibody (rabbit) directed against a specific antigen. The wells are then thoroughly washed, and crude extracts of plasmid-bearing clones, or known amounts of antigen in control extracts, are added to the wells and incubated. If the appropriate antigen is contained within one or more of the extracts, it will bind to the antibody coating the well and be retained. After washing, the wells are further incubated with the same antibody labeled with ^{125}I to high specific activity (10^6–10^7 cpm/μg). The labeled antibody binds to antigen already bound to the cold antibody coating the plate. Upon subsequent washing and counting of wells, the clone(s) bearing plasmids of interest are identified. A similar plastic well assay has also been described by Erlich *et al.*[9]

Preparation of Crude Cell Extracts

Escherichia coli colonies carrying hybrid plasmids containing foreign DNA segments are transferred to S agar plates (48–50 colonies per plate) and are grown up overnight as individual clones. S agar is 32 g/liter tryptone, 5 g/liter NaCl, 20 g/liter yeast extract (omitted if cloned foreign DNA is from yeast), 15 g/liter Difco agar, and 0.2 g/liter NaOH. The combined colonies from each plate are scraped with a sterile spatula into tubes containing 2 ml 0.05 *M* Tris-HCl, pH 7.5, and 0.5 mg egg-white lysozyme (Sigma). The tubes are incubated for 30 min at room temperature and 1 hr at 0° and then frozen at −80° and thawed a total of three times. Finally, extracts are incubated at 37° for 10 min and stored at −80°. An entire colony collection of many clones is thus reduced to a relatively small number of pools for assaying.

[5] P. P. Hung and S. G. Lee, personal communication.

[6] C. M. Ling and L. R. Overby, *J. Immunol.* **109,** 834 (1972).

[7] K. J. Catt, G. W. Tregear, H. G. Burgess, and C. Skermer, *Clin. Chim. Acta* **27,** 267 (1970).

[8] J. Carbon, L. Clarke, C. Chinault, B. Ratzkin, and A. Walz, *in* "Biochemistry and Genetics of Yeasts" (M. Bacila, B. L. Horecker, and A. O. M. Stoppani, eds.), p. 428. Academic Press, New York, 1978.

[9] H. A. Erlich, S. N. Cohen, and H. O. McDevitt, *Cell* **13,** 681 (1978).

Plastic Well Assay

Wells of flexible, plastic microtiter dishes (No. 220-29, Cooke Engineering Company, San Mateo, California) are coated by the addition to each well of 150 μl purified antibody (7 μg/ml) in 0.01 *M* Tris, pH 9.2. Coating proceeds overnight at room temperature, and the wells are thoroughly washed with water and allowed to dry. Cell extracts, prepared by lysozyme treatment and repeated freezing and thawing, are added to the coated wells (150 μl) and incubated at 37° for 4–6 hr with gentle shaking. The wells are again thoroughly washed with water, filled with 150 μl ^{125}I-antibody in 0.01 *M* Tris, pH 7.5, 0.14 *M* NaCl, and 25% fetal calf serum (approximately 10^5 cpm/well), and incubated overnight at room temperature. After washing thoroughly in 0.8% saline, the microtiter dishes are dried, and individual wells are cut out and counted in a scintillation counter. Optimal sensitivity of the assay is obtained by using antibody purified by affinity chromatography.

In reconstruction experiments using known amounts of a specific antigen in buffer or in control cell extract, less than a nanogram of antigen can be detected per well. Such an experiment is shown in Fig. 1, where known amounts of yeast hexokinase in buffered calf serum were incubated in antihexokinase-coated wells. After washing and incubation with ^{125}I-antibody (5×10^6 cpm/μg), as little as 0.1 ng hexokinase is easily de-

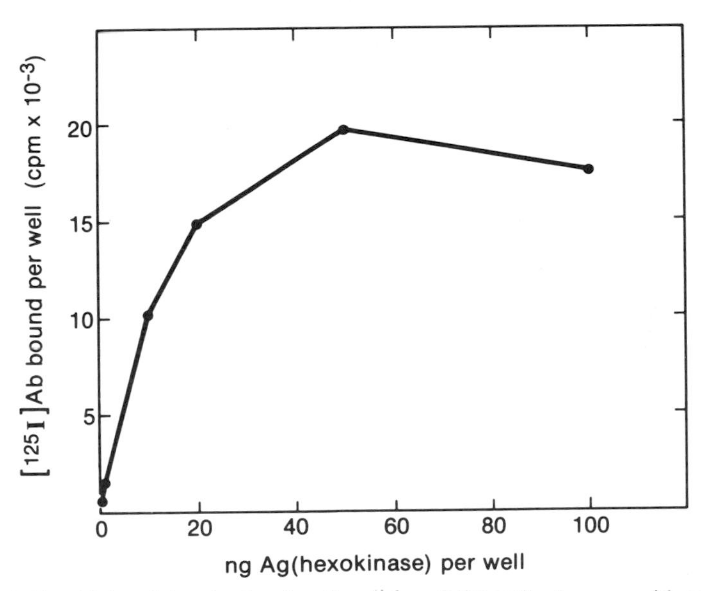

FIG. 1. Sensitivity of the plastic microtiter dish well ^{125}I-antibody assay with yeast hexokinase as antigen.[8] At least 100 pg of hexokinase per well is detectable by this procedure. Similar results are obtained with other antigen–antibody systems.

tected per well (approximately 1100 cpm/well over a background of 600 cpm/well).

Using the above method, we have screened a collection of 6240 clones, each containing a hybrid plasmid constructed of ColE1 DNA and randomly sheared yeast DNA, for colonies harboring plasmids carrying the yeast hexokinase genes.[8] In order to facilitate the screening of such a large number of colonies, crude extracts were prepared from pools of 48 clones each (about 130 such pooled extracts were screened). Using this procedure, we have detected eight such pools from the collection that respond to antihexokinase. The 48 colonies in each of four of the hexokinase-positive pools were grown on a petri plate in an 8×6 grid array, and extracts of pools of horizontal and vertical rows (14 extracts for each original pool) were prepared. In every case, one horizontal pool and one vertical pool per plate responded in the radioactive antibody assay, pinpointing the single active clone out of the 48 in the original pool. Thus, only 144 assays are necessary to identify a single positive clone in the collection of 6240 colonies.

Colony Screening on Agar Plates

Recently two methods have been published for radioimmunological screening to detect specific translation products produced by individual colonies or phage plaques on agar plates.[9,10] The first method, developed by Erlich *et al.*,[9] involves the covalent attachment of $F(ab)'_2$ antibody fragments (Fc fragments are removed by pepsin digestion and affinity chromatography) to diazobenzyloxymethyl-cellulose paper,[11] followed by contact of the paper with lysed colonies or bacterial plaques on agar plates. The papers are then incubated with antibody that has not had the Fc portion removed by pepsin digestion. The location of the antigen is visualized by incubation with ^{125}I-labeled *Staphylococcus aureus* protein A, which binds to the Fc ends of antibodies attached to antigen on the paper, and by autoradiography. The second method, that of Broome and Gilbert,[10] involves the coating of plastic disks with antibody, followed by exposure of the plastic to antigen present in colonies lysed on agar plates. The locations of antigen binding are then detected by incubation with ^{125}I-labeled antibody, followed by autoradiography.

Our laboratory has developed a third method involving the covalent attachment of antibody to CNBr-activated paper, followed by incubation of the paper with lysed colonies on agar plates. Colonies releasing antigen that can bind to the Ab on the paper are then detected by incubation with ^{125}I-labeled antibody, followed by autoradiography. The method can eas-

[10] S. Broome and W. Gilbert, *Proc. Natl. Acad. Sci. U.S.A.* **75,** 2746 (1978).

[11] J. C. Alwine, D. J. Kemp, and G. R. Stark, *Proc. Natl. Acad. Sci. U.S.A.* **74,** 5350 (1977).

ily detect the presence of 50–100 pg of antigen (yeast hexokinase or 3-phosphoglycerokinase) when added to a lysed colony on a plate. Screening of our collection of 6240 *E. coli* colonies containing hybrid yeast DNA ColE1 plasmids with labeled hexokinase antibody has revealed 15 positive clones.

Preparation of CNBr-Activated Paper

This paper is prepared for the covalent attachment of protein by a modification of the method used by Axén and Ernback[12] to produce CNBr-activated agarose or cellulose powder for the covalent attachment of enzymes. Twenty Whatman No. 40 (ashless), 9-cm-diameter paper disks are washed in a 1-liter beaker with 400 ml 0.1 *M* $NaHCO_3$ by occasional swirling for 15 min. A similar wash with H_2O then follows. The disks are placed in a 2-liter beaker containing 12 g CNBr (all work with this compound must be done in a fume hood) dissolved in 480 ml H_2O at 23°–25°C. The reaction is started by the addition of 2 *N* NaOH to obtain a pH of about 11. The pH should be maintained near 11 by the occasional addition of 2 *N* NaOH while swirling the beaker vigorously by hand for 8 min. (Stirring with a stirring bar is not satisfactory, and no more than 20 disks should be treated during a single preparation.) The reaction is stopped by decanting the CNBr solution, and the disks are washed with two 500-ml portions of 0.1 *M* $NaHCO_3$. Five filters at a time are then washed by suction using 250 ml 0.1 *M* $NaHCO_3$, 250 ml H_2O, 200 ml 50% acetone in H_2O, and 50 ml acetone. The disks are vacuum-dried and stored over desiccant at 4°. The treated disks are very stable to storage for at least several months.

Binding of Antibody to Treated Paper Disks

Thirty to 60 μl of the $(NH_4)_2SO_4$ fraction of antiserum is mixed with 3 ml 20 m*M* sodium phosphate, pH 6.8, in a glass petri dish. More antiserum can be used if the antibody titer is low. A single CNBr-activated paper disk is then put in each dish so that no air bubbles are trapped between the paper and the glass. These petri dishes are stacked level, sealed with Parafilm, and left at 4° for at least 18 hr. The paper disks are then removed and put in a solution containing 0.1 *M* glycine, 0.5% bovine serum albumin, 0.14 *M* NaCl, and 10 m*M* sodium phosphate at pH 7.0 (5 ml/disk) for 6 hr at 37°. The disks can be used at this point, left overnight at 4°, or frozen at −20° for later use. Immediately before use, they are soaked with a solution (5 ml/disk) containing 5% calf serum in PBS (CS/PBS) for 30 min, followed by a suction wash of each disk with 25 ml

[12] R. Axén and S. Ernback, *Eur. J. Biochem.* **18**, 351 (1971).

of CS/PBS. Finally, they are kept in a solution (4 ml/disk) of CS/PBS until use.

Colony Lysis on Agar Plates and Application of Paper Disks

Colonies from collections of hybrid plasmid-bearing transformants are grown from 1 to 2 days on S agar plates. The plates are inverted for 30 min over a 7-cm 3 MM Whatman paper disk containing 1 ml chloroform. One hour after removing the plates, a top agar layer is applied to each plate, containing 1.5 ml top agar (1 g tryptone, 0.8 g Difco purified agar, 0.8 g NaCl/100 ml, pH 6.8), 100 μl 5 mg/ml sodium dodecyl sulfate, and 100 μl 12.5 mg/ml lysozyme in PBS (added immediately before plating). The plates are spotted with 1-μl controls of 10 ng, 1 ng, and 0.1 ng purified antigen in CS/PBS. The orientation of the paper on the plate can later be identified by these spots and a stab at their location on the plates. The paper disks containing covalently bound antiserum are blotted on paper towels and applied to the plates. Care must be taken to remove air bubbles between the paper and the agar. The papers are left in contact at room temperature for 5–9 hr. They are then removed four at a time and stirred gently for 1–2 hr with 100 ml CS/PBS in a 1-liter beaker. Each disk is then washed by suction with 200 ml PBS, and four disks are again put into each beaker with 50 ml CS/PBS.

^{125}I-Labeled Antibody Interaction with the Disks

Ten micrograms of antibody purified by affinity chromatography is prelabeled with 0.5 mCi Na^{125}I to a specific activity of 5–10 × 10^6 cpm/μg using the lactoperoxidase method.[4] The CS/PBS is removed from the four disks in each beaker and replaced with 4 × 10^6 cpm ^{125}I-labeled antibody in 10 ml of a solution containing 25% calf serum, 10 m*M* Tris, pH 7.5, and 0.14 *M* NaCl. The beakers are covered and swirled gently for 5–7 hr at room temperature. The ^{125}I-containing solution is then replaced by 50 ml of CS/PBS. After 30 min the disks are each washed by suction using 200 ml PBS and finally soaked for 15 min in a pan containing PBS. After drying with the labeled side up, the disks are exposed for 2 days at −70° to Kodak XRP1 x-ray film in the presence of a DuPont Cronex Lighting-Plus intensifying screen.

[32] Immunological Detection and Characterization of Products Translated from Cloned DNA Fragments

By HENRY A. ERLICH, STANLEY N. COHEN, and HUGH O. MCDEVITT

The ability to detect the translation products of cloned eukaryotic DNA fragments in bacterial cells represents an important advance in recombinant DNA technology. This capability makes possible the identification of clones that have acquired plasmids or bacteriophage containing a specific eukaryotic DNA sequence for which no nucleic acid hybridization probe is available. It also permits analysis of the conditions that would allow the transcription and translation of a previously identified eukaryotic DNA sequence.

In some cases, the biological activity of a gene product encoded by the cloned DNA fragments can be used for the direct phenotypic selection of clones in which the specific DNA sequence is expressed as a functional protein.[1-4] Such detection systems, however, require a homologous bacterial function that can be inactivated in the host strain and, consequently, are usually limited to use with eukaryotic proteins that have a bacterial counterpart.

The most general assay for screening bacterial clones for gene products encoded by cloned foreign DNA is one based on antibody-mediated recognition of antigenic determinants. (This approach requires that the determinants be located on polypeptides rather than on carbohydrate side chains.) Such an immunoassay is capable of detecting incompletely translated products as well as proteins that have no easily detectable or selectable function. Furthermore, proteins unable to function in a bacterial cell environment would be expected to retain their immunologically reactive sites.

We have developed a simple and sensitive solid-phase indirect radioimmunoassay for the detection of antigen in liquid cultures or *in situ* in bacterial colonies and phage plaques.[5] Moreover, this radioimmunoassay is capable of identifying antigen in polyacrylamide gels, making possible the electrophoretic characterization of translation products of foreign gene fragments cloned into plasmid or phage vectors. The assay, which

[1] K. Struhl, J. R. Cameron, and R. Davis, *Proc. Natl. Acad. Sci., U.S.A.* **73,** 1471 (1976).
[2] G. Ratzkin and J. Carbon, *Proc. Natl. Acad. Sci. U.S.A.* **74,** 487 (1977).
[3] D. Vapnek, J. A. Hautala, J. W. Jacobson, N. H. Giles, and S. R. Kushner, *Proc. Natl. Acad. Sci. U.S.A.* **74,** 3508 (1977).
[4] A. C. Y. Chang, J. H. Nunberg, R. J. Kaufman, H. A. Erlich, R. T. Schimke, and S. N. Cohen, *Nature (London)* **275,** 617 (1978).
[5] H. A. Erlich, S. N. Cohen, and H. O. McDevitt, *Cell* **13,** 681 (1978).

ISBN 0-12-181968-X

can be carried out using either derivatized cellulose filters or polyvinyl sheets as a solid-phase matrix, has been used to identify and to characterize electrophoretically mouse dihydrofolate reductase (DHFR) antigen in *Escherichia coli* cells containing a chimeric plasmid coding for the mouse DHFR.[4] The presence of histone H2B antigen in bacteria with a chimeric plasmid coding for sea urchin histone H2B has also been detected by this technique (unpublished experiments).

Alternative immunoassays based on enzyme-coupled detection of protein in λ plaques,[6] or on immunoprecipitation in top-layer agar for λ plaques and bacterial colonies,[7] are relatively insensitive. The immunoassay reported by Broome and Gilbert[8] utilizes the same principle as the solid-phase radioimmunoassay described here, except that it requires purification and labeling of antibody.

Principles of the Method

The detection of antigen utilizes an antibody "sandwich" method in which the antigen is bound to an insoluble matrix by $F(ab)_2'$ IgG fragments coupled to the solid phase. The bound antigen is then detected by a second layer of antibody, and the bound antibody is in turn detected by ^{125}I-labeled protein A from *Staphylococcus aureus*. This indirect sandwich radioimmunoassay is based on the ability of protein A to discriminate between the $F(ab)_2'$ fragment and the intact IgG. The Fc portion of the IgG, which is specifically recognized by protein A,[9] has been removed by pepsin digestion in the preparation of the $F(ab)_2'$ fragment. Since the radiolabeled detection reagent is protein A, not the second layer of antibody, this indirect sandwich immunoassay does not require purification and labeling of specific antibody. This feature is important for screening with low-titer, low-affinity antiserum (e.g., antihistone antiserum), since iodination of some antibodies reduces antigen-binding activity. Furthermore, the same preparation of ^{125}I-labeled protein A can be used in conjunction with different antisera and with the same antiserum over a range of antibody concentrations.

The sandwich immunoassay, which requires the recognition of independent antigenic sites on the same molecule, results in a unique capacity for identifying fused or hybrid proteins. Hybrid peptides, such as those observed in recent experiments in which eukaryotic DNA fragments have

[6] B. Sanzey, O. Mercereau, T. Ternynok, and P. Kourilsky, *Proc. Natl. Acad. Sci. U.S.A.* **73,** (1976).

[7] A. Skalka and L. Shapiro, *Gene* **1,** 65 (1976).

[8] S. Broome and W. Gilbert, *Proc. Natl. Acad. Sci. U.S.A.* **75,** 2746 (1978).

[9] A. Forsgven and J. Sjöquist, *J. Immunol.* **97,** 822 (1966).

been inserted into prokaryotic structural genes,[10] can be detected with solid-phase $F(ab)'_2$ fragments directed against antigenic determinants on one protein and intact IgG-specific for the other protein.

Two different materials, polyvinyl chloride (PVC) and a chemically derivatized cellulose filter, can be used as the solid phase, depending on the nature of the immunoassay. The wells of a PVC microliter plate, coated with $F(ab)'_2$ fragments, are used as the insoluble matrix for quantitative determinations of antigen in bacteria growing in liquid culture; quantitation is obtained by cutting out the flexible wells and measuring the bound ^{125}I radioactivity. For the *in situ* assay of bacterial colonies or phage plaques, PVC disks, coated with $F(ab)'_2$ fragments or cellulose filters to which $F(ab)'_2$ fragments have been covalently coupled, are overlaid on the plate colonies or phage plaques. Antigen transferred from the plaques or lysed bacterial colonies to the solid-phase $F(ab)'_2$ fragment is detected by subsequent incubations of the filter or disk with antibody and ^{125}I-labeled protein A, followed by autoradiography. For the identification of antigen in polyacrylamide gels, the $F(ab)'_2$-coupled cellulose filter is overlaid on the gel following electrophoresis, and the antigen transferred from gel to the filter is detected as described above.

The advantages of using PVC as the supporting matrix are the relative ease of coupling $F(ab)'_2$ fragments and the low level of background radioactivity binding to the matrix. The main advantages of using the derivatized filter paper is that antigen in bacterial colonies, phage plaques, and bands in polyacrylamide gels is transferred with significantly greater resolution to the filter, because of its absorbing properties, then to PVC.

Preparation of Reagents

Iodination of Protein A

Protein A (Pharmacia) was labeled with ^{125}I using chloramine T in a modification of the original method of Hunter and Greenwood.[11] Twenty microliters of a 1 mg/ml protein A solution in 0.5 *M* phosphate buffer, pH 7.0, was mixed with 20 μl of neutralized $Na^{125}I$ (New England Nuclear) and 10 μl of a 1 mg/ml solution of chloramine T in 0.5 *M* phosphate buffer. After 1 min the reaction was stopped by adding 50 μl of 1 mg/ml tyrosine in 0.5 *M* phosphate buffer, and 100 μl of 1% bovine serum albumin (BSA) was added as carrier protein. The mixture was applied immediately to a 10-ml column of Sephadex G-25 and eluted with 1%

[10] L. Villa-Komaroff, A. Efstratiadis, S. Broome, P. Lomedico, R. Tizard, S. P. Naber, W. L. Chick, and W. Gilbert, *Proc. Natl. Acad. Sci. U.S.A.* **75,** 3727 (1978).

[11] W. M. Hunter and F. C. Greenwood, *Nature (London)* **194,** 495 (1962).

BSA. The peak fractions were pooled and stored at 4° in the presence of 0.5% sodium NaN_3. A fresh batch of reagent was labeled every 6–8 weeks. The specific activity of the labeled reagent, assuming no protein loss, varied between 15 and 30 μCi/μg.

Preparation of F(ab)$'_2$ Fragment of Rabbit Anti-β-Galactosidase Serum

The F(ab)$'_2$ fragment was derived from rabbit anti-*E. coli* β-galactosidase serum according to a modification of the method of Madsen and Rodkey.[12] One milliliter of serum was incubated with an equal volume of saturated $(NH_4)_2SO_4$. The pellet resulting from two precipitation steps with 50% $(NH_4)_2SO_4$ was resuspended in 1 ml of 0.1 *M* sodium acetate (pH 7.0) and dialyzed overnight against the acetate buffer (pH 7.0). The protein solution was adjusted to a pH of 4.3 with 1 *M* HCl and incubated with 100 μl of pepsin (Worthington Biochemical Company) at 4 mg/ml in 0.1 *M* acetate buffer, pH 4.3, for 8 hr at 37°. After dialysis against 0.05 *M* phosphate-buffered saline, pH 7.4 (PBS), the protein solution was applied to a Sephadex G-100 (Pharmacia) column and then passed over a 3-ml affinity column of protein A coupled to cyanogen bromide activated Sepharose 4B to remove any residual immunoglobulin molecules bearing Fc determinants. Passage over the protein A–Sepharose column alone proved sufficient for obtaining clean preparations of F(ab)$'_2$ fragments.

Preparation of F(ab)$'_2$-Coupled Cellulose Filters

Whatman 540 cellulose paper is converted to aminobenzyloxymethyl (ABM) paper by treatment with 1-[(*m*-nitrobenzyloxy)methyl]pyridinium chloride (BDH Chemicals, distributed by Gallard-Schlesinger Chemical Manufacturing Corporation) according to the method of Alwine *et al.*[13] A sheet (14 × 25 cm) of Whatman 540 paper is incubated at 60° with a 10-ml solution containing 0.8 g of *m*-nitrobenzyloxymethyl pyridinium chloride and 0.25 g of sodium acetate for 10 min with constant rubbing to achieve even spreading of the solution in the paper. The paper is dried in an oven at 60° for 10 min and then at 135° for 30 min. It is then washed twice with about 200 ml of water for 20 min with shaking, dried in a 60° oven, and washed twice with about 200 ml of benzene for 20 min. The paper is air-dried and can be stored in a dessicator in the NO_2 form. For conversion to the NH_2 (ABM) paper, the paper is incubated at 60° in 150 ml containing 30 g of sodium hyposulfite with shaking for 30 min. It is then washed twice with water for 20 min, with 30% acetic acid for 20 min, and

[12] L. H. Madsen and L. S. Rodkey, *J. Immunol. Methods* **9,** 355 (1976).
[13] J. Alwine, D. J. Kemp, and G. R. Stark, *Proc. Natl. Acad. Sci. U.S.A.* **74,** 5350 (1977).

once again with water. The ABM paper is stored dessicated at 4°; it is stable for up to 3 weeks.

Circles are cut out from the ABM paper and converted to the reactive diazobenzyloxymethyl (DBM)-cellulose paper by treatment with a solution containing 40 ml of water, 80 ml of 1.8 *M* HCl, and 3.2 ml of fresh solution of $NaNO_2$ (10 mg/ml) for 30 min at 4°. After 30 min, excess HNO_2 is destroyed by adding solid urea. The papers are then washed four times with cold water for 5 minutes and then washed twice with 25 m*M* phosphate buffer at pH 6.5. The paper turns bright yellow during the washing process. It is then incubated with the $F(ab)'_2$ fragment (0.05 mg/ml in 25 m*M* phosphate buffer, pH 6.5) overnight at 4°. The coupling capacity of the activated cellulose filters is about 0.2 $\mu g/cm^2$. The unreacted diazonium groups are inactivated and nonspecific binding sites blocked by incubating the filters with 1 *M* glycine and 0.5% BSA for 3 hr at 37°.

Method

The method's application is illustrated here using *E. coli* β-galactosidase as a model protein. Using rabbit anti-*E. coli* β-galactosidase antiserum, the presence of specific antigen was detected in bacterial cultures, in bacterial colonies, and in phage plaques. The sequence of binding reactions in the indirect sandwich radioimmunoassay is represented diagrammatically in Fig. 1. The initial reaction is the coupling of $F(ab)'_2$ fragments derived from pepsin digestion of the rabbit serum immunoglobulin to the insoluble matrix.

The $F(ab)'_2$ fragments are adsorbed onto the surface of the wells of PVC microliter plates or PVC disks by incubating the $F(ab)'_2$ preparation (0.01–0.05 mg/ml in PBS) in the wells or with the disk for 1 hr at room temperature. The binding capacity of the wells is about 50 ng of protein/25 μl, so that the unbound material in the $F(ab)'_2$ preparation can be recovered, stored at −20°, and reutilized in later experiments. After adsorption of the $F(ab)'_2$ fragments, the wells or disks are washed and incubated for an additional hour with PBS containing 5% fetal calf serum (FCS) and 0.1% BSA (wash buffer) to block residual binding sites on the PVC.

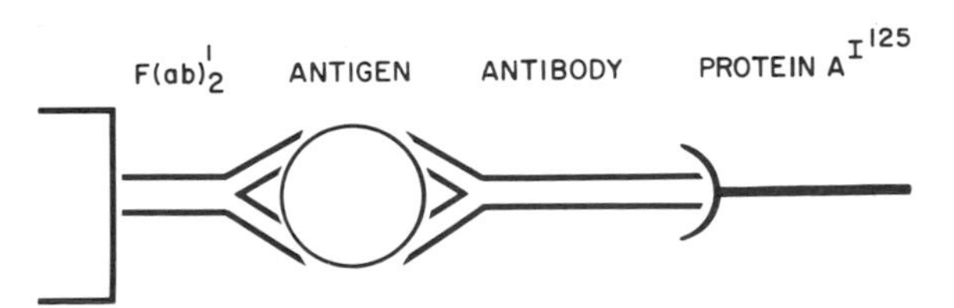

FIG. 1. Schematic diagram of the indirect radioimmunoassay.

Plate Assay

The microtiter plate wells are coated with $F(ab)'_2$ fragments (75 μl of a solution 0.025 mg/ml in PBS), the unbound protein removed by washing with wash buffer, and the wells incubated with 75 μl of wash buffer for 1 hr at room temperature. The antigen (50 μl diluted in wash buffer) is then incubated in the wells for 2–3 hr. Next, the rabbit anti-β-galactosidase antiserum (50 μl, diluted $\frac{1}{1000}$ in wash buffer) is incubated in the wells for 2–3 hr. Specific antibody interacts with the antigen which is bound via the $F(ab)'_2$ fragment to the solid phase. In the final step, ^{125}I-labeled protein A (50 μl, diluted in wash buffer to a final concentration of 0.1–0.3 μg/ml and to a final activity of about 100,000 cpm/well) which specifically recognizes the immunoglobulin Fc portion, is incubated in the well for 2–3 hr. After each step, unbound reagents are removed by washing. ^{125}I-labeled protein A molecules bound to the wells are measured by cutting and counting the wells in a gamma counter.

Purified antigen is assayed at various concentrations to generate a standard curve, and the cell lysate to be tested is assayed over a range of dilutions. Three sets of negative controls involving the substitution of wash buffer for the $F(ab)'_2$ preparation, of wash buffer for the antigen preparation, and of nonimmune rabbit serum for the specific rabbit antiserum are necessary for the determination of nonspecific background binding of ^{125}I-labeled protein A. Figure 2 shows the results of the β-galactosidase immunoassay for a culture of *E. coli* with a deletion for the

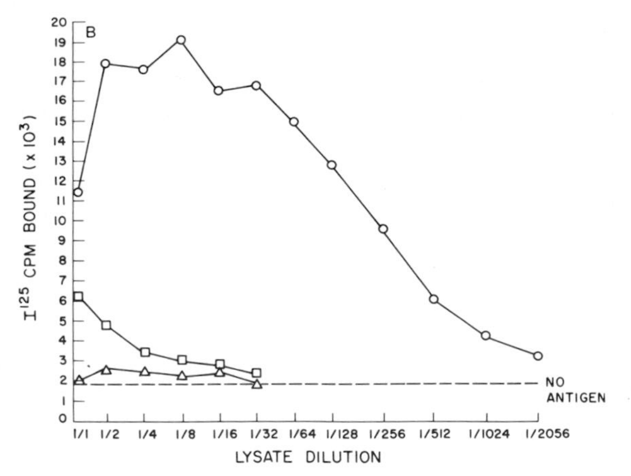

FIG. 2. Comparison of *lac*$^+$ and *lac* ∇ lysates in the indirect radioimmunoassay. Lysates (50 μl) from MC1000 [*lac*(Δ)] and D7001 (*lac*$^+$ cultured in the absence (□) or presence (○) of 10^{-3} *M* IPTG were added to the $F(ab)'_2$-coated wells of a microtiter plate. After a 2-hr incubation, 50 μl of antiserum (a 1:1000 dilution in 1% BSA) was added to the well. Two hours later ^{125}I-labeled protein A (50 μl) was incubated in the well (100,000 cpm/well) overnight at 4°. All samples were tested in duplicate.

β-galactosidase structural gene (*lac*∇) for a *lac*$^+$ culture grown in the presence of 10^{-3} *M* isopropyl-β-thiogalactoside (IPTG), an inducer of β-galactosidase, and for an uninduced *lac*$^+$ culture. All *E. coli* strains were grown to saturation in L broth; cultures were lysed by the addition of chloroform (2 drops/ml) and sodium dodecyl sulfate (SDS) to a final concentration of 0.1%, followed by vortexing. Uninduced *lac*$^+$ cultures, estimated to contain 10–20 β-galactosidase molecules per cell[14] are detected easily and reliably when the lysate from about 5×10^7 cells is assayed. In experiments with purified antigen, the detection threshold for β-galactosidase was shown to be less than 100 pg or 1×10^8 molecules. When cultured in the presence of IPTG, 1 *lac*$^+$ cell per 10^3 *lac*$^\Delta$ cells can be detected.[5] The titration curve for the radioactivity bound to the induced *lac*$^+$ lysate indicates that about 500 times as much immunologically reactive material is present as in the uninduced *lac*$^+$ lysate. The quantitation of β-galactosidase antigen by the indirect radioimmunoassay is in general agreement with measurement of enzyme activity for these same cultures.

The chloroform–SDS lysis method was developed to optimize the rapid mass screening of microcultures. If the number of clones is small, however, concentrated sonicated extracts of bacterial cells can be used, thus avoiding the problem of detergent inhibition and allowing the extract from more than 10^9 cells to be measured in a single well. Under these conditions, the sensitivity of the assay is enhanced by at least an order of magnitude, permitting the detection of clones producing as little as one molecule per cell or the identification of 1 *lac*$^+$ in 10^4 *lac*$^\nabla$ cells.

In Situ Immunoassay for Phage Plaque and Bacterial Colonies

By adsorbing the $F(ab)_2'$ fragments onto the surface of PVC disks or by covalently coupling the $F(ab)_2'$ fragments to a chemically activated cellulose filter, the indirect radioimmunoassay can be adapted, using autoradiography, to the *in situ* analysis of bacterial colonies and of phage plaques. After the $F(ab)_2'$ fragments have been reacted with the diazonium groups on the activated cellulose filter, the unreacted diazonium groups are inactivated by treatment with 1 *M* glycine and 0.5% BSA to ensure that antigen transfer to the filter from the plate is mediated by the solid-phase $F(ab)_2'$ fragment. The filter is then applied directly to the phage plate or the plate containing lysed bacterial colonies for 1–3 hr at 4°. The filter is then incubated with antiserum for 2–3 hr, subsequently with ^{125}I-labeled protein A for 2–3 hr, and finally washed and analyzed by

[14] J. H. Miller, "Experiments in Molecular Genetics." Cold Spring Harbor Lab., Cold Spring Harbor, New York, 1972.

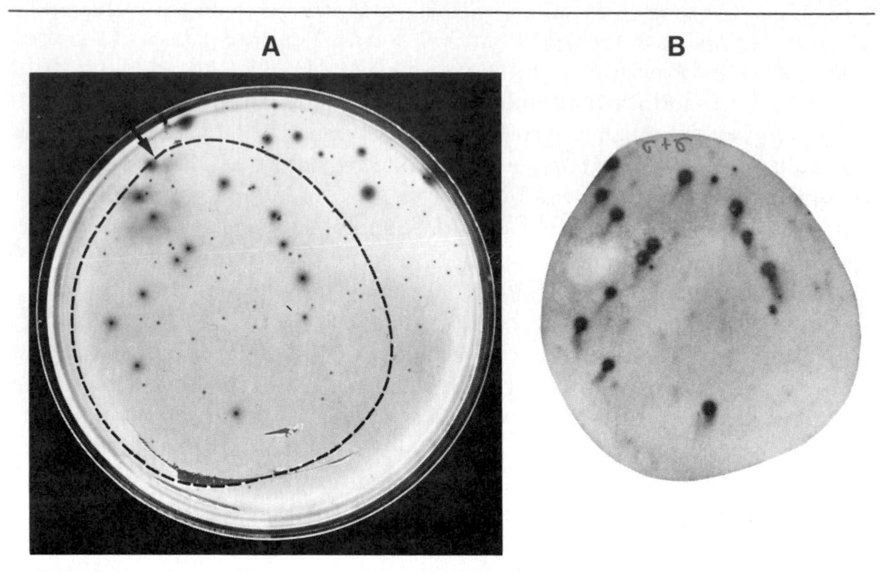

FIG. 3. *In situ* radioimmunossay for β-galactosidase in λ C1857 p5 *lac*∇(*att*) SR2 plaques. First, 0.1 ml of a L-broth dilution of λ C1857 p5 *lac*∇(*att*) *SR2* mixed with 0.1 ml of a dilution of λ C1857 was plated with 0.1 ml of a saturated MC1000 (lac^{∇}) culture in 2 ml of H top agar containing 1 mg of X-Gal on YT plates. The plates were incubated at 37° for 5 hr and then overlaid with a $F(ab)'_2$-coated cellulose filter for 1 hr. The filter was incubated with antiserum (1:1000) for 3 hr and then with a ^{125}I-labeled protein A solution for 3 hr. The washed and dried filter was overlaid with Kodak NS-T2 x-ray film for 6 hr. (A) YT plate with X-Gal stain in top layer agar. The dotted outline indicates the area overlaid by the cellulose filter. (B) Autoradiogram of the treated filter.

autoradiography. The use of the *in situ* immunoassay of β-galactosidase for phage plaques is illustrated in Fig. 3.

A mixture of λ p5 *lac* and wild-type λ was plated with a culture of lac^{∇} *E. coli* on YT[14] plates containing the β-galactosidase stain X-Gal (5-bromo-4-chloro-3-indoyl-β-D-galactoside). Plaques resulting from cells infected with λ p5 *lac* were identified by the enzymatic stain. The β-galactosidase antigen present in phage plaques after a 5-hr incubation at 37° was assayed by placing the $F(ab)'_2$-coupled filter on the plate for 2 hr at 4° followed by sequential incubations of the filter with antiserum and ^{125}I-labeled protein and autoradiography of the filter. All plaques, identified by the X-Gal stain, are clearly detected as discrete dark spots on x-ray film. The lower limit of sensitivity of the *in situ* phage plaque immunoassay determined with a λ carrying a *lac* allele with a low-efficiency promoter mutation, was shown to be less than 1×10^7 molecules per plaque.[5]

The lysis of bacterial colonies for the *in situ* immunoassay can be achieved by exposing the open petri dish for 5 min to chloroform vapors in a desiccator and then overlaying the plate with 1.5 ml of top agar containing lysozyme (0.25 mg) and SDS (0.025%) (Fig. 4). Exposure to chloroform vapor alone is usually sufficient to release antigen from cells in bacterial colonies, although colony lysis is incomplete. Heat induction of λ CI857 prophage and direct application of a λ vir suspension to individual colonies represent alternative methods for colony lysis.[8] $F(ab)'_2$-coated PVC disks can be substituted for the $F(ab)'_2$-coupled filters for the *in situ* immunoassay.

In Situ Identification of Antigen in Gels

After electrophoresis, gels were placed in a bath of PBS–0.5% Triton X-100 at room temperature for 30–60 min, in order to remove SDS, and then laid on a pile of moist filters in a trough containing PBS. The upper surface of the gel was then gently blotted to remove pools of buffer.

Filters were marked for identification and pressed gently into the appropriate part of the gel. After all filters were in place, they were overlaid with dry pieces of Whatman 3 MM paper and a weight to maintain pressure. PBS was added to the trough as needed. After 2–6 hr of transfer, the filters were washed with buffer and incubated with antiserum and ^{125}I-labeled protein A, as described above. The resolution of bands transferred to strips of $F(ab)'_2$-coated PVC proved inadequate; we have used $F(ab)'_2$-coupled filters exclusively for the identification of antigen in gels.

The method of *in situ* antigen identification is illustrated in Fig. 5 by the detection of mouse DHFR in bacterial and mouse cell extracts.

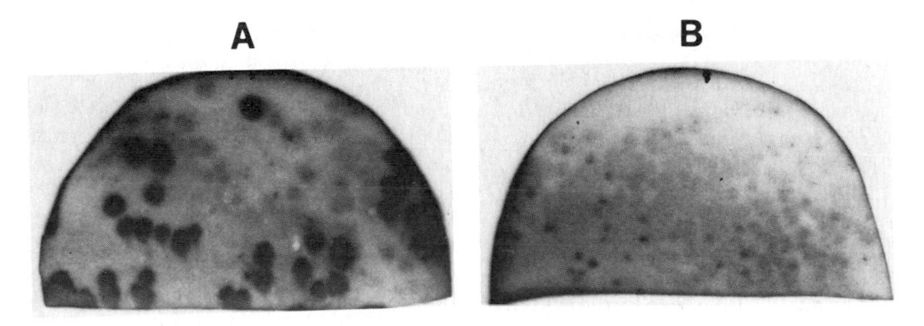

FIG. 4. *In situ* radioimmunoassay for β-galactosidase in induced and in uninduced colonies of D7001 (*lac*$^+$). A mixture of MC1000 (*lac*$^\nabla$) and D7001 (lac$^+$) cultures was plated on YT plates with IPTG (A) or without IPTG (B) and grown up to microcolonies (about 0.5 mm in diameter) at about 1000 colonies per plate. The plates were exposed to chloroform vapors for 5 min and then overlaid with 1.5 ml of H top agar containing lysozyme (0.25 mg) and SDS (0.025%). The $F(ab)'_2$ fragment-coated filter was applied directly to the plate surface for 2 hr and treated with antiserum and ^{125}I-labeled protein A as described in Fig. 3. The filters were placed on Kodak NS-72 for autoradiography. (A) An x-ray film exposed for 18 hr. (B) An x-ray film exposed for 24 hr.

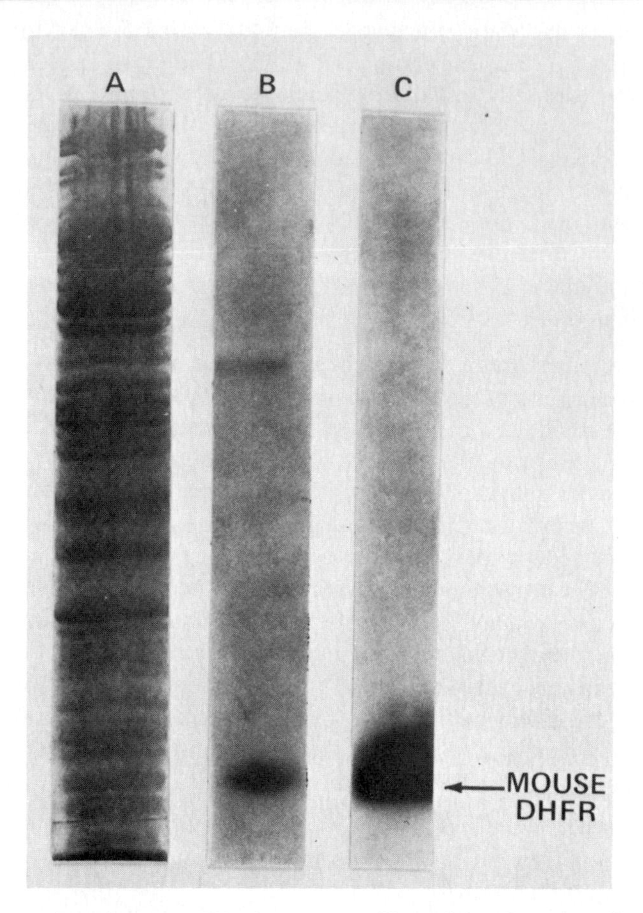

FIG. 5. Detection of mouse DHFR in extracts of bacterial and mouse cells. Extracts of bacterial cells containing the chimeric plasmid pDHFR7 that codes for mouse DHFR[4] and of mouse cells were electrophoresed in an 11% SDS–polyacrylamide slab gel. (A) Coomassie blue stain of proteins (ca. 200 μg/lane) in the bacterial extract. (B) Autoradiogram of the anti-mouse DHFR D(ab)$'_2$-coupled filter, following transfer from a gel containing the bacterial extract; the filter was incubated with rabbit anti-mouse DHFR antiserum (diluted $\frac{1}{300}$ in 1% BSA) overnight and then with ^{125}I-labeled protein A. (C) Autoradiogram of the filter, following transfer from a gel containing the mouse cell extract. The x-ray films shown here were exposed for 36 hr.

Comments

For quantitation of antigen, all reagents [F(ab)$'_2$, antiserum, and ^{125}I-labeled protein A] must be present at saturating levels. In Fig. 2, for example, β-galactosidase antigen in the induced *lac*$^+$ culture is not limiting until the lysate has been diluted about 50-fold. The background level of nonspecific radioactive protein-A binding, however, rises slightly as the concentration of the detecting reagents, particularly the antiserum, is increased. The concentration of these reagents should therefore be varied according to the estimated amount of antigen to be detected. Here, the background binding, revealed by ^{125}I-labeled protein A binding to a negative control containing wash buffer and no antigen, is due to the nonspecific binding of the second layer of antibody to the solid phase. Another potential source of background binding results from our observation that many rabbit sera contain antibodies directed against bacterial proteins. These antibodies must be removed from the serum by absorption, or the specific antibodies directed against the eukaryotic gene product must be purified by affinity chromatography. The simplest, most general approach is removal by absorption. We have passed rabbit antiserum over Sepharose columns to which bacterial extracts had been covalently coupled. Sonicated extracts were prepared from a 20-ml culture of *E. coli* containing the plasmid vector without the eukaryotic DNA insert. The extract was dialyzed overnight against coupling buffer (0.1 *M* Na_2CO_3, 0.3 *M* NaCl, pH 8.5) and incubated overnight at 4° with 1.3 g (about 5 ml) of activated cyanogen bromide–Sepharose 4B (Pharmacia). The gel was washed with 1 *M* ethanolamine, pH 9.0, and then, in alternating cycles, with 0.1 *M* acetate buffer, pH 4.0, and 0.1 *M* borate buffer, pH 8.0, according to the manufacturer's instructions.

Rabbit antiserum (2.0 ml diluted 1:100 in 1% BSA) was passed over a 3-ml column followed by 6 ml of 1% BSA. The absorbed serum was therefore diluted 1:400. The column was washed with 1 *M* propionic acid and stored in PBS in the presence of 0.1% NaN_3.

A recent modification of this method utilizes solid-phase protein A and bound antibody instead of solid-phase $F(ab)'_2$ fragments to bind antigen. The radiolabeled detecting reagent, ^{125}I-labeled protein, fails to bind to the first antibody, since it is already bound by the solid-phase protein A. This modification has the advantages that no preparation of $F(ab)'_2$ fragments from antiserum is required and that the capacity for binding antigen is higher than with solid-phase $F(ab)'_2$ fragments. The background level of nonspecific ^{125}I-labeled protein A binding, however, is significantly higher.

Acknowledgments

This work was supported by grants from the National Institutes of Health, National Science Foundation, and the American Cancer Society. H. A. E. is an F. M. Kirby Fellow of the Cancer Research Institute.

[33] R-Looping and Structural Gene Identification of Recombinant DNA*

MICHAEL ROSBASH, DALE BLANK, KAREN FAHRNER, LYNNA HEREFORD, ROBERT RICCIARDI, BRYAN ROBERTS, STEPHANIE RUBY, and JOHN WOOLFORD

The advent of recombinant DNA technology has made possible studies on the organization and expression of eukaryotic genes, which were previously restricted to prokaryotic and viral genes. In large measure, these studies exploit and are dependent upon the availability—through recombinant DNA molecules—of large amounts of individual eukaryotic DNA sequences. They are also dependent upon an accurate definition of the structural gene sequence or sequences present in any recombinant DNA molecule of interest. In other words, a recombinant DNA molecule must first be positively identified as containing the structural gene sequence or sequences of interest before more detailed studies on or with it can be pursued. It is often advisable as well to define and localize all other structural gene sequences contained in the same recombinant molecule.

In some cases, these preliminary characterizations are fairly straightforward. This is particularly true when the complementary RNA sequences are abundant in the tissue or cells of interest. Since rRNA or an abundant mRNA species can often be extensively purified by size fractionation alone, complementarity between this RNA and a recombinant DNA molecule can usually be detected by hybridization of the RNA and DNA followed by the protection of one labeled species from single-strand-specific nuclease digestion. In the case of abundant mRNAs, complementarity can also be detected by hybrid-arrested translation.[1,2]

The vast majority of eukaryotic mRNAs are, however, not abundant; most mRNA species are relatively rare in that each one constitutes approximately 0.01% of the total mRNA or less. Even a moderately abundant mRNA, such as a ribosomal protein mRNA of yeast, is present in the cell at a "concentration" of approximately 0.1% of the total mRNA.[3] The detection of complementarity between low-abundance mRNAs and re-

[1] B. M. Paterson, B. E. Roberts, and E. L. Kuff, *Proc. Natl. Acad. Sci. U.S.A.* **74,** 4370 (1977).

[2] N. D. Hastie and W. A. Held, *Proc. Natl. Acad. Sci. U.S.A.* **75,** 1217 (1978).

[3] L. M. Hereford and M. Rosbash, *Cell* **10,** 463 (1977).

* The first publication of this work, and the appropriate original reference, is Woolford, J. L. and Rosbash, M., *Nucleic Acids Res.*, **6,** 2483 (1979).

ISBN 0-12-181968-X

combinant DNA requires a sensitive assay for individual low-abundance mRNAs and the ability to achieve a high signal/noise ratio during or after hybridization. The assay must be able to detect and recognize nanogram or even subnanogram quantities of individual mRNAs, while the hybridization must be able to purify or at least enrich substantially the complementary mRNA.

A traditional approach to this problem is the fixation of DNA to a solid support,[4] hybridization of RNA with the DNA matrix, elution of the RNA, and assay of the hybridized RNA by a sensitive means with which it can be detected and identified. While this method is of clear merit, it has certain disadvantages, the most outstanding of which are the fixation of the DNA and the relatively long times normally required to achieve a substantial amount of hybridization. In order to obviate these and other more minor disadvantages, we have developed a new procedure for the hybridization of RNA with recombinant DNA and the separation of the hybridized from the unhybridized RNA. R loops are formed during hybridization in liquid, after which the R loops are separated from unhybridized RNA by column chromatography, providing an effective separation between hybridized and unhybridized RNA. The hybridized RNA is then assayed by standard procedures.

Principle of the Method

In high concentrations of formamide, RNA–DNA duplexes have a higher thermal denaturation temperature (T_m) than the corresponding DNA–DNA duplexes.[5–7] Although it is presently not known what physical interactions are responsible for this difference in thermal stability, it is the cause of the formation and stability of R loops. When DNA and complementary RNA are incubated under these conditions, R loops readily form. For the formation of R loops with recombinant DNA, our general procedure consists of incubating sufficient quantities of DNA with limited amounts of mRNA such that the rate of the reaction is DNA-driven. At sufficiently long times and under appropriate conditions, all the complementary RNA is in R loops, while the unhybridized RNA population is missing those species complementary to the recombinant DNA molecule(s) used in the annealing reaction.

The rate of R-loop formation and therefore the time required for the reaction to go to effective completion (~ 10 times the $C_0t_{1/2}$) is a sensitive function of the incubation temperature (T_i).[6] At high concentrations of

[4] D. Gillespie and S. Spiegelman, *J. Mol. Biol.* **12,** 824 (1965).
[5] R. L. White and S. S. Hogness, *Cell* **10,** 177 (1977).
[6] M. Thomas, R. L. White, and R. W. Davis, *Proc. Natl. Acad. Sci. U.S.A.* **73,** 2294 (1976).
[7] J. Casey and N. Davidson, *Nucleic Acids Res.* **4,** 1539 (1977).

formamide and salt, the maximal rate of R-loop formation appears to occur at a T_i equal to the irreversible melting temperature (T_{ss}, the temperature at which the DNA is irreversibly denatured to single strands) of the DNA sequence participating in R-loop formation. This is presumably because the melting of the participating portion of the DNA duplex is rate-limiting at temperatures below the T_{ss} of that sequence, while the small difference in temperature ($\leq 10°-15°$) between T_i and the irreversible melting temperature of the specific RNA–DNA duplex formed is rate-limiting at temperatures above the T_{ss} of the corresponding DNA–DNA duplex.[6] The efficient formation of R loops appears therefore to require determination of the T_{ss} of every DNA sequence that participates in R-loop formation. In principle, this requires a precise definition of the RNA complementary regions of the recombinant DNA molecules as well as their T_{ss}.[6]

In order to obviate these difficult and tedious measurements, we have developed a simple agarose gel assay for the measurement of T_{ss} values of DNA restriction fragments. For a restriction fragment lying entirely within the eukaryotic portion of a recombinant DNA molecule, this assay results in an accurate measurement of the T_{ss} value of this eukaryotic sequence. Since the mRNA complementary sequences are not defined within this fragment, this T_{ss} value is a *maximum* value for the T_{ss} of any and all mRNA complementary regions of this restriction fragment. Consequently, incubation at such a temperature *must* be at or above the T_{ss} of all DNA sequences within the plasmid that will form R loops with eukaryotic RNA. Incubation over a range of temperatures, starting at such a temperature and decreasing to well below this value, is very likely to be at the T_{ss} of the mRNA complementary DNA sequences at some time.

In fact, we have simplified this procedure even more by adopting the following general strategy with yeast DNA-containing plasmids. T_{ss} values of restriction fragments from the eukaryotic portions of a number of recombinant plasmids (chosen at random) were measured. In addition, a variety of specific plasmids were similarly investigated—plasmids known to contain sequences complementary to abundant mRNAs. All the restriction fragments examined had T_{ss} values between 45° and 55° (in 0.56 M Na^+ and 70% formamide as described in the following section). Therefore, we concluded that most yeast sequences, including mRNA-coding sequences, had T_{ss} values between 55° and 45°. It is likely that the lower value corresponds roughly to the T_{ss} of DNA with a GC content of approximately 40%, the GC content of *Saccharomyces cerevisiae* DNA. While this relationship, appropriately correct for differences in [Na^+], is somewhat different than previously published comparisons between T_{ss} and GC content,[6] it should be viewed as an approximation because of the large contribution GC-rich clusters make to the T_{ss} of a molecule as compared to their contribution to the overall GC content and T_m. Indeed, when linearized by digestion with a restriction enzyme with a single site in the recombinant DNA molecule, all recombinant molecules have a T_{ss} of 58–59°, identical to the T_{ss} of pMB9. As previously suggested, this high T_{ss} value is likely to be due to the pressure of one or more GC-rich clusters within the pMB9 sequence.[8]

These T_{ss} measurements, along with previously available data from the literature, provide a rationale with which one can establish a fairly general scheme for the formation of R loops under conditions of DNA excess and without measuring the T_{ss} of every individual eukaryotic gene or insert. Recombinant DNA, linearized with a single restriction enzyme cut, is incubated with RNA between 55° and 45°. Since the initial incubation temperature is high relative to the T_{ss} of most yeast DNA fragments, it is likely that most of the mRNA hybridization occurs with single-stranded yeast DNA. Although the yeast DNA is denatured at this temperature, the pMB9 DNA is below its T_{ss} so that the two complementary DNA strands of the hybrid plasmid are held in register. At the end of the incubation, when the reaction temperature is lowered well below the T_{ss} of the yeast DNA (i.e., to room temperature), any unhybridized yeast DNA rapidly renatures, forming bonafide R loops. A similar notion has been previously presented for the formation of R loops after single-stranded DNA–RNA hybridization in high formamide concentrations.[8]

By decreasing linearly the incubation temperature between these two values over a 4-hr period, most yeast DNA sequences should be within 4° of their T_{ss} for approximately 2 hr. At the sequence concentrations used in this procedure [3 μg/100 μl of recombinant DNA of maximum sequence complexity 20 kilobases (kb)] and under conditions of DNA excess, this is sufficiently past the R-looping $C_0t_{1/2}$ such that most or all of the complementary RNA is found in R loops at the end of the incubation. This is consistent with previously published data on the rate of R-loop formation as well as our experience in forming R loops with pMB9 yeast plasmids and yeast RNA.[6,8]

The next requirement is a method by which one can rapidly and efficiently separate R loops from unhybridized RNA. One can take advantage of the significant fraction of every recombinant DNA molecule that has no sequence homology with eukaryotic mRNA. This is a consequence of the fact that each molecule contains the vector sequences (in the case of the experiments presented here, this consists of 5.5 kb of pMB9 DNA) and the fact that only a fraction of eukaryotic DNA is complementary to mRNA. An RNA molecule in an R loop therefore acquires many of the hydrodynamic properties of double-stranded DNA, some of which differ

[8] D. S. Holmes, R. H. Cohn, H. H. Kedes, and N. Davidson, *Biochemistry* **16**, 1504 (1977).

markedly from single-stranded RNA. In particular, double-stranded DNA, has a relatively extended conformation in high concentrations of sodium ion while RNA has a relatively compact structure.[9]

We have exploited this difference in the hydrodynamic properties of RNA and DNA to separate R loops from unhybridized RNA—and therefore to separate hybridized from nonhybridized RNA. This has been simply accomplished by gel filtration chromatography on agarose A-150m. At high concentrations of NaCl, all DNA of molecular weight greater than that of pMB9—and therefore all R loops formed on pMB9 containing recombinant DNA molecules—is excluded, while all RNA is included. The agarose column also serves to exchange the buffer of the excluded molecules into high-ionic-strength aqueous buffer. Since the RNA is not assayed until after the column chromatography, the R loops that form must remain intact during several hours at room temperature in high-ionic-strength aqueous buffers, consistent with previously published data.[6] It is possible that some partial strand displacement occurs in the aqueous buffer, but most of the R loops must remain at least somewhat intact, since most or all of the complementary RNA is found in the excluded volume.

Materials and Reagents

Materials

Isolation and Purification of DNA. The plasmid vector pMB9 or recombinant DNA molecules containing insertions of yeast DNA constructed by Petes *et al.*[10] were isolated from *Escherichia coli* K12 strain HB101 by standard procedures.[10] After dialysis of the CsCl, DNA was extracted twice with redistilled phenol and twice with $CHCl_3$–isoamyl alcohol (24:1). DNA was stored in TE (0.01 *M* Tris, 0.001 *M* EDTA, pH 8.3) at 4°. Linearized plasmid pPW311 DNA, a hybrid of pMB9 and sheared *Drosophila melanogaster* DNA, constructed with AT tailing like the p(Y) plasmids, was a kind gift of P. Gergen and P. Wensink.

Preparation of RNA. Total RNA and poly(A) RNA were isolated from yeast as described previously.[3]

Reagents

Formamide (99%, Matheson, Colemen and Bell) was deionized as described by Maniatis *et al.*[11] and stored at −80°. Piperazine-*N*,-*N'*-bis-

[9] F. W. Studier, *J. Mol. Biol.* **11,** 373 (1965).

[10] T. D. Petes, J. R. Broach, P. C. Wensink, L. M. Hereford, G. R. Fink, and D. Botstein, *Gene* **4,** 37 (1978).

[11] T. Maniatis, A. Jeffrey and J. H. van de Sande, *Biochemistry* **14,** 3787 (1975).

2-ethanesulfonic acid (PIPES) was purchased from the Sigma Chemical Company. Wheat germ was kindly supplied by General Mills, Vallejo, California. Radioactive amino acids were purchased from Amersham Corporation and from New England Nuclear. Wheat germ tRNA (Sigma Chemical Company) was further purified by phenol extraction and DEAE-cellulose column chromatography as described in Dudock *et al.*[12] Restriction endonucleases were purchased from New England Biolabs.

Method

Determination of T_{ss} Values

As described above, it is convenient to determine the T_{ss} of the eukaryotic portion of the plasmid(s) of interest or to estimate the range of T_{ss} values found in a number of inserts from the same organism. This is done as follows.

Plasmid DNA (1–100 μg) is digested to completion with a restriction enzyme or enzymes under standard conditions. After digestion, the DNA is generally (but not necessarily) extracted with phenol and $CHCl_3$–isoamyl alcohol as described in Materials and Reagents. The DNA is precipitated with ethanol with or without the presence of tRNA carrier. (The tRNA runs off the agarose gel so it does not interfere with the assay.) The DNA pellet is washed twice with 70% ethanol, dried *in vacuo,* and resuspended in R-loop buffer (70% formamide, 0.4 *M* NaCl, 0.1 *M* PIPES, pH 7.8, a final $[Na^+] \cong 0.56$ *M* including the contribution from the PIPES).[13] The amount of DNA used in each T_{ss} determination is not critical and is a function of the number of DNA bands (and their molecular weight) that one wishes to visualize at each temperature. In general, approximately 2 μg divided among 20 samples (0.1 μg per temperature point) is quite adequate. For such a protocol the 2 μg DNA is resuspended in 10 μl H_2O to which is added 70 μl deionized formamide plus 20 μl 5× salt buffer (2 *M* NaCl, 0.5 *M* PIPES, pH 7.8, 0.005 *M* EDTA). The composition of this R-looping buffer should be identical—and if possible composed of the identical solutions—as the buffer used for R-loop formation; i.e., if R loops are formed under different salt or buffer conditions, the same conditions should apply for T_{ss} determinations. The DNA is then distributed, 5 μl into 18 or 19 siliconized glass tubes (10 × 75 mm). These tubes are kept on ice. One tube is retained as an untreated control. The others are incubated, one at a time, at increasing temperatures spanning a reasonable range of T_{ss} values. (The difference in the T_{ss} between this procedure

[12] B. S. Dudock, G. Katz, E. K. Taylor, and R. W. Holley, *Proc. Natl. Acad. Sci. U.S.A.* **62,** 941 (1969).

[13] More recently, we have used the same $[Na^+]$, but the PIPES at pH 6.8.

and one in which the 5-μl samples are heated in sealed capillaries is $\leq 1°$). Each tube is placed for 3 min at a certain temperature in a water bath. At the end of this time, 20 μl of ice-cold quench buffer (80% TE, ~20% glycerol plus bromphenol blue plus xylene cyanol F.F.) is added as the tube is removed from the water bath, the liquid rapidly mixed, and the tube immediately placed in a dry ice–ethanol bath. The temperature of the water bath is then raised a fixed increment and, when it reaches the correct temperature, the next tube is then incubated for 3 min. After the penultimate tube is frozen, the temperature is generally raised well above any possible T_{ss} value (e.g., 75% in the buffer cited above) and the final tube incubated and frozen.

These samples are then applied to a 1% agarose gel and subjected to electrophoresis at 3 V/cm for 16 hr under standard conditions.[14] They have been stored at $-20°$ for 24 hr prior to electrophoresis; the single-stranded DNA may be stable indefinitely in the frozen state. Two precautions are generally taken: (1) The samples are melted one at a time and loaded as soon as the 25 μl is melted and homogeneous. (2) The samples are generally melted and loaded in reverse order, i.e., the highest temperature first; this is to ensure that any reannealing will occur maximally in the sample heated to the highest temperature and therefore be highly visible in this sample in which all restriction fragments should be completely single-stranded.

The irreversible melting profile of pMB9 is shown in Fig. 1. Although not an exact copy of the protocol described above, the major features of the experiment are the same. The T_{ss} of pMB9 under these conditions is $>58°$ and $\leq 60°$. It should be noted that, when the DNA is relatively free of single-stranded nicks, the two complementary strands often separate from each other during electrophoresis. In Fig. 2, the T_{ss} of several restriction fragments from two recombinant DNA molecules is determined.

In this particular experiment, the R-loop buffer contained 80% formamide which reduced the T_{ss} approximately 5° as compared to 70% formamide (unpublished results). pMB9 melts under these conditions at approximately 54°, equal to the T_{ss} determined earlier in 70% formamide. All the lower-molecular-weight restriction fragments have T_{ss} values between 40° and 48°, equivalent to between 45° and 53° in 70% formamide, consistent with additional experiments not described here. Since these recombinant plasmids were originally constructed with terminal transferase and AT tails, each has one or two fragments with a high T_{ss} as a result of the persistence of yeast pMB9-containing fragments after digestion with *Bam*HI and *Eco*RI. This then is the very simple technique used to determine T_{ss} values.

[14] P. A. Sharp, B. Sugden, and J. Sambrook, *Biochemistry* **12,** 3055 (1973).

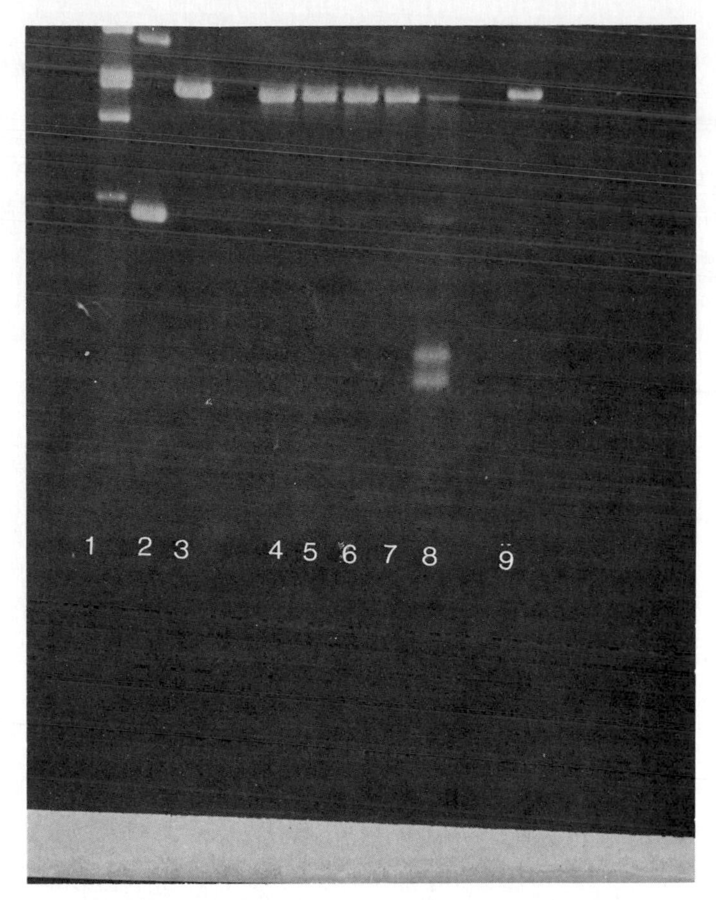

FIG. 1. T_{ss} determination of *Eco*RI-cut pMB9. One microgram of pMB9 was digested with *Eco*RI to completion, ethanol- precipitated, washed, and resuspended in 50 μl 1× R-loop buffer as described in the text. Samples were distributed, incubated, and processed as described in the text. (1) *Eco*RI-cut λ DNA, molecular weight standard; (2) pMB9 DNA, no restriction enzyme digestion; (3) *Eco*RI-cut pMB9, layered directly on gel, no ethanol precipitation; (4) T = 52°; (5) T = 54°; (6) T = 56°; (7) T = 58°; (8) T = 60°; (9) unheated.

R-Loop Formation and Hybridized RNA Separation and Assay

Our general laboratory protocol is as follows. Recombinant DNA is linearized by digestion to completion with a restriction enzyme which has only one site in the molecule. After digestion, the DNA is extracted with phenol and $CHCl_3$ as described in Materials and Reagents. Three micrograms of plasmid DNA (of maximum sequence complexity of 20 kb) is mixed with 5 μg total RNA in a final volume of 100 μl containing 70% (v/v) deionized formamide, 0.1 M PIPES, pH 7.8, 0.01 M Na_2EDTA, and

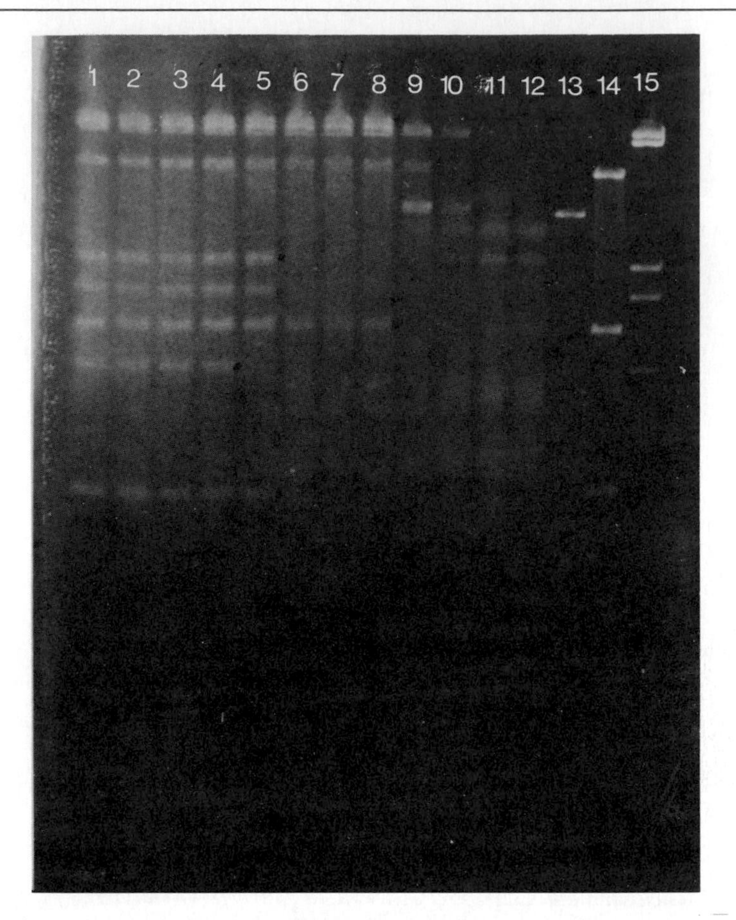

FIG. 2. T_{ss} determination of *Bam*HI-*Eco*RI-cut plasmids. Four micrograms of pY11-10 and 4 μg pY9-78, both digested to completion with restriction enzymes *Eco*RI and *Bam*HI, were precipitated separately with ethanol and resuspended each in 100 μl of 80% formamide–R loop buffer. Five microliters of each was added to siliconized 10× 75 tubes. To samples 9–12 (in lanes 9–12) 2 μl pMB9, digested with *Eco*RI, also in 80% formamide–R-loop buffer, was added in addition. Samples were processed as described in the text. (1) Unheated; (2) 36°; (3) 38°; (4) 40°; (5) 42°; (6) 44°; (7) 46°; (8) 48°; (9) 50°; (10) 53°; (11) 54°; (12) 70°; (13) pMB9 alone, unheated; (14) pY11-10 alone, unheated; (15) pY-9-78 alone, unheated.

0.4 NaCl (a final [Na^+] of ~0.56 *M* including the contribution from the PIPES.[8] Recently, we have switched to a pH of 6.8, closer to the pK of PIPES but with the same [Na^+]. Incubations are performed by linearly decreasing the temperature from 55° to 45° over a 4-hr period using a Lauda constant-temperature bath equipped with a Neslab TP-2 temperature programmer. Alternatively, the temperature is decreased ~3.3° every hour for 3 hr and the incubation continued for an additional hour. Upon completion of the incubation, 200 μl column buffer (0.8 *M* NaCl, 0.01 *M* Tris, pH 7.5, 0.001 *M* EDTA) is added to the 100 μl reaction mixture and applied to a 1 × 25 cm agarose A-150m 100–200 mesh column (Bio-Rad Laboratories) equilibrated with column buffer. Fractions of 0.25 ml are collected at a flow rate of 3 ml/hr at room temperature. The fractions containing the excluded and included volumes can be identified by any of a number of means. Most important is a careful pooling of a narrow cut of the excluded volume so that no included RNA contaminates the excluded DNA and R loops. Ethidium bromide staining is convenient for this purpose. However, the preferred procedure at the present time in our laboratory is to add a small amount (~2000 cpm, $>10^4$ cpm/μg) of ^{3}H-labeled plasmid DNA to the incubation mixture just prior to the column chromatography. Five to ten percent of each fraction is then counted, which locates accurately the excluded volume. This internal standard has reduced markedly the frequency with which the excluded volume contains detectable amounts of nonhybridized RNA. After pooling the excluded fractions, the subsequent two or three fractions are passed over, after which a large number of fractions are pooled to collect nucleic acid which is partly and completely included. In fact, the pooling and assay of the included fractions is almost always superfluous and is shown here in Fig. 3 for purposes of describing the method. The usual practice is to assay exclusively the excluded fractions. The excluded and included fractions are each ethanol-precipitated overnight at −20° by the addition of 10 μg purified wheat germ tRNA and 2.5 volumes of ethanol. The precipitated nucleic acid is washed twice with 70% ethanol, resuspended in sterile distilled water, heated to 100° for 1 min to melt all duplexes, and frozen in a dry ice–ethanol bath.

RNA samples, including those purified as described above, are translated in a wheat germ extract containing [^{35}S]methionine or [^{3}H]leucine and [^{3}H]lysine as described by Paterson *et al.*[15] Often, no stimulation of protein synthesis above endogenous levels is seen with RNA from the excluded fractions of the column, since very small quantities of individual mRNAs are present. The products of the cell-free translation are analyzed by electrophoresis on any number of gel systems. In the absence of any prior knowledge about the products, the radioactive peptides are analyzed by electrophoresis on SDS slab gels containing a linear 10–15% gradient of polyacrylamide[16] stabilized by a 5–25% (w/v) sucrose gradient. In general, approximately 10% of the included fraction translation product is applied to a lane in order to permit a side-by-side comparison of ex-

[15] B. M. Paterson, B. E. Roberts, and E. L. Kuff, *Proc. Natl. Acad. Sci. U.S.A.* **74,** 4370 (1977).

[16] U. K. Laemmli, *Nature (London), New Biol.* **227,** 680 (1970).

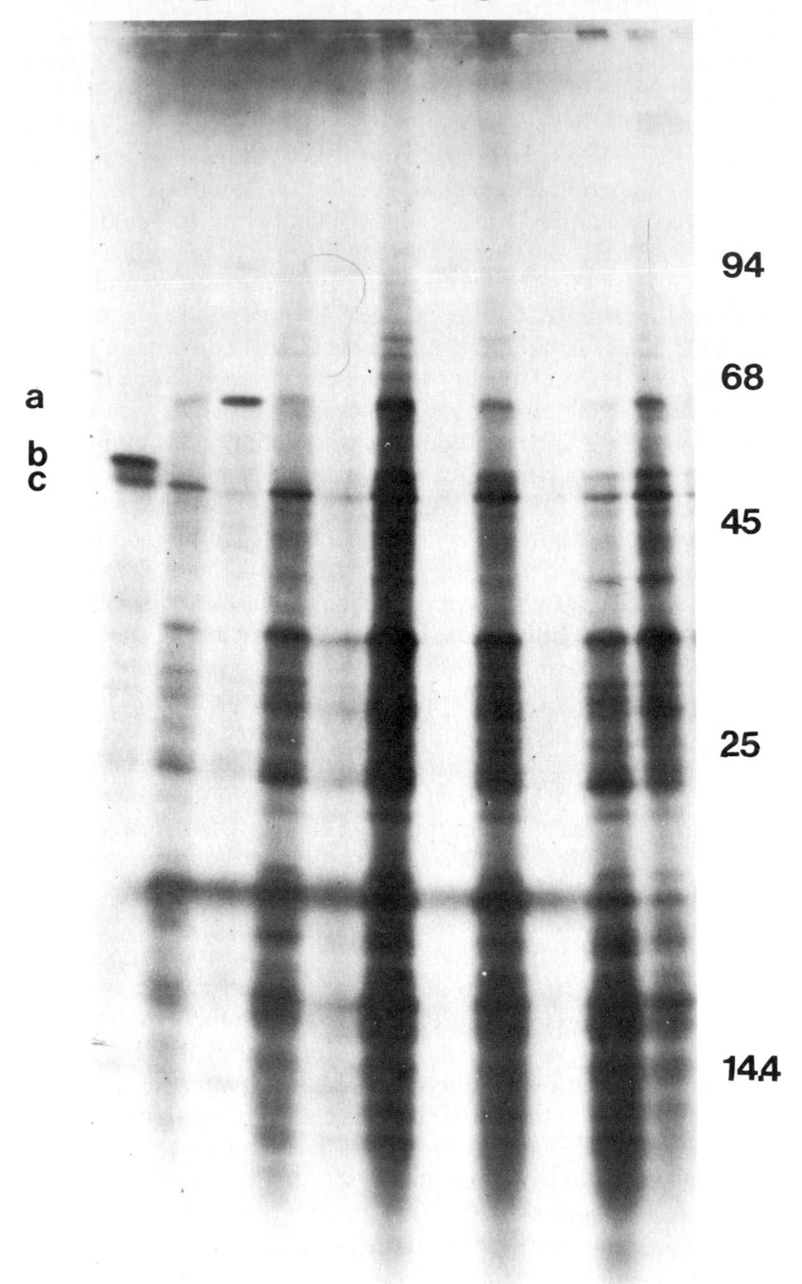

FIG. 3. Polypeptides synthesized *in vitro* from purified yeast RNA molecules. The supercoiled recombinant DNA molecules pY3-83 and pY11-10 were linearized by digestion with restriction endonuclease *Bam*HI, which cuts each of them once in the vector. Seven micro-

cluded and included RNA. Otherwise, the large number of protein species in the vicinity of the appropriate band or bands generally makes the included lane overexposed and uninterpretable after the time required to develop the excluded lane. Protein gels are almost always (in the case of ^{3}H-labeled proteins, always) subjected to fluorography.[17]

Figure 3 shows the results of applying this method to two recombinant DNA molecules (pY3-83 and pY11 10) containing yeast nuclear DNA inserted into the plasmid vector pMB9. These molecules were chosen for the purpose of examining the efficacy of this method, because results from colony hybridization suggested that they might contain sequences complementary to major yeast mRNAs.

pY3-83 DNA hybridizes with mRNA that codes for two polypeptides of molecular weight 51,000 and 48,000 daltons (lane 1, peptides b and c). Electron microscope examination of the R loops suggests that two mRNAs are encoded by pY3-83. pY11-10 DNA codes for a single polypeptide of approximately 63,000 daltons (lane 3, peptide a). In both cases, these are the only visible polypeptide products. The presence of biologically active mRNAs in the included volume (lanes 2, 4, 6, and 8) implies that most or all of the mRNA survivies the entire procedure relatively intact. No translationally active yeast mRNA is found in the excluded volume of the column when pMB9 DNA (lane 5), *Drosophila* DNA (lane 7), or no DNA (lane 9) is incubated with yeast RNA. By comparing lanes 4 and 11, it can be seen that hybridization with pY11-10 DNA removes most of the mRNA coding for peptide a from the included volume. A similar conclusion is reached for peptide b by comparing lanes 2 and 11. Peptide c is difficult to visualize in the translation of total RNA, because it migrates very close to a major polypeptide of slightly lower molecular weight.

There is on occasion some background in the translation products of the excluded volume. This is manifest by the appearance of a set of bands from the excluded volume RNA equivalent to the pattern obtained from

[17] W. H. Bonner and R. A. Laskey, *Eur. J. Biochem.* **46**, 83 (1974).

grams of total yeast RNA were hybridized with recombinant plasmid DNA, chromatographed, and assayed by cell-free translation as described in the text. The resultant ^{35}S-labeled polypeptide products were electrophoresed on sodium dodecyl sulfate 10–15% gradient polyacrylamide gel, which was embedded with PPO, dried, and fluorographed for 1 week at −80° using Kodak XR-5 x-ray film. Molecular weights, in thousands of daltons, of marker proteins run on the same gel are shown on the right. (1) Hybridization with pY3-83 DNA, excluded volume, major products are polypeptide bands b and c; (2) hybridization with pY3-83 DNA, included volume; (3) hybridization with pY11-10 DNA, excluded volume, major product is polypeptide a; (4) hybridization with pY11-10 DNA, included volume; (5) hybridization with pMB9 DNA, excluded volume; (6) hybridization with pMB9 DNA, included volume; (7) hybridization with a *Drosophila melanogaster* plasmid DNA, pPW311, excluded volume; (8) hybridization with pPW 311 DNA, included volume; (9) hybridization with no DNA, excluded volume; (10) hybridization with no DNA, included volume; (11) translation products of total yeast RNA.

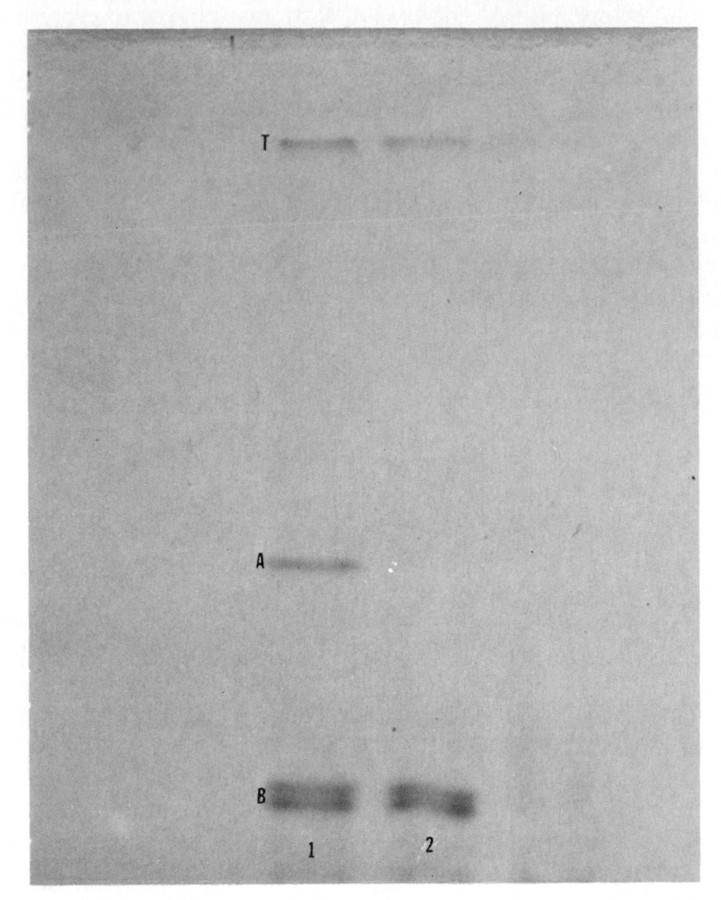

FIG. 4. R looping of histone plasmids. Three micrograms of plasmid were nicked as described in the text (see Comments) and annealed to 3 μg poly(A)$^+$ RNA as described in the text. The nucleic acid in the excluded volume was added to a 25-μl wheat germ reaction mix and translated in the presence of [^{3}H]leucine and [^{3}H]lysine. Ten microliters of products were examined on a 15% acrylamide SDS slab gel. The gel was subjected to fluorography and exposed for 1 week at $-80°$. (1) Plasmid TRT-1; (2) plasmid TRT-2. T, top of gel; A, larger protein; B, histone doublet.

the translation of unfractionated yeast RNA (e.g., lane 5). The seriousness of this problem is somewhat exaggerated by the fact that only ~10% of the included RNA translation product is applied to the gel, as compared to all of the excluded volume RNA translation product. As described above, the inclusion of an internal DNA marker reduces substantially the frequency of such a pattern, which makes us suspect that it is largely due to slight inaccuracies in pooling the excluded volume, resulting in the presence of a small fraction of included mRNA. However, other possibilities such as aggregation are not excluded.

When no discrete bands are seen in the translation products of the excluded RNA, it may be that the complementary mRNAs are not abundant. In this case, and whenever such a circumstance is suspected, the annealing reaction is performed with 3–5 μg poly(A)$^+$ RNA instead of 5 μg total RNA. This increases the available mRNA by 20- to 50-fold. All mRNA species, except for abundant species present at >1% mRNA each, are still in DNA excess. (The lack of excess DNA is of no significance unless the included volume is assayed for the disappearance of a particular mRNA species). An example of such a hybridization is shown in Fig. 4. In this experiment, two yeast plasmids known to code for mRNAs for yeast histones H_2A and H_2B were hybridized with 3 μg poly(A)$^+$ RNA. (In this experiment, the plasmid DNA was not linearized by restriction enzyme digestion but was converted to open circles as described in the following section.) From other kinds of data, these two mRNAs are suspected to be moderately abundant, each being present at about 0.1% of the total mRNA. Both plasmids hybridize with two mRNAs that code for the protein doublet B. This doublet has been shown to consist of histones H_2A and H_2B. Interestingly, the procedure revealed a

FIG. 5. Two-dimensional gel. Ten microliters of product from the translation of excluded volume RNA after hybridization with TRT-2 (Fig. 4, lane 2) was analyzed on two-dimensional Triton gels.[18] The gel was subjected to fluorography and exposed for 3 weeks at −80°.

third protein (A) of higher molecular weight. More extensive studies suggest that this third mRNA is present at a lower intracellular concentration than the histone mRNAs and encoded by one of the histone plasmids. The success in revealing such a low-abundance protein suggests that this technique, even in the absence of any other modifications, may be sensitive enough to visualize complementarity to low-abundance mRNAs.

Figure 5 is a two-dimensional gel profile of the proteins shown in Fig. 4, lane 2.[18] This gel system, originally designed for histones, clearly shows two spots. These spots, far removed from the diagonal because of the affinity of these proteins for the detergent Triton included in the second dimension, comigrate with marker histones H_2A and H_2B. The absence of other radioactive spots is striking and demonstrates the degree of purification afforded by this procedure.

Comments

1. RNA is included, while DNA $\geqslant$ 5.5 kb (pMB9) is entirely excluded. Even silk fibroin mRNA (16,000 nucleotides in length) is partially included and separated by a few fractions from the excluded DNA. rRNA and all mRNAs of reasonable length are well separated from the excluded DNA. DNA of 3 kb is almost entirely excluded (~1 fraction different from 17 kb DNA) and can be used in this experiment as well as larger DNA.

2. Preliminary experiments in this laboratory indicate that recombinant phage (using Charon 4) will also function under these conditions. The use of this vector has not been extensively examined, but specific mRNAs are moved into the excluded volume of the agarose column. The only change in protocol is the use of 10 μg phage DNA per 100 μl incubation solution to compensate for the increased sequence complexity of the λ phage as compared to that of plasmids.

3. Up to 20 μg clean plasmid DNA has been added to a 25-μl wheat germ reaction with no effect on incorporation. Therefore we often use four plasmids in a single hybridization reaction and column (3 μg/100 μl for each plasmid or a 12 μg/100 μl concentration of total plasmid). If the appropriate protein is found, these four plasmids are then examined individually to determine which one contains the appropriate sequence.

4. If the temperature is raised above the T_{ss} of pMB9 prior to incubation for R-loop formation, the same reaction products should result, since the conditions for R-loop formation on yeast DNA should be adequate for the renaturation of pMB9 DNA.

5. The *Bam*HI restriction fragments of adenovirus fractionate the appropriate adenovirus mRNAs into the excluded volume of the agarose

[18] S. Spiker, *Nature* (*London*) **259**, 418 (1976).

column. The *Bam*HI B fragment, which contains the leader sequences,[19] also fractionates all of the appropriate adenovirus mRNAs into the excluded volume of the agarose column. Since this DNA contains approximately 215 base pairs homology with the late adenovirus mRNAs[19,20] the minimum required length of homology must be less than this value. With regions of homology of less than 100 bases, it is likely that successful fractionation will depend on the precise characteristics of the sequence as well as a more careful definition of the hybridization and chromatography conditions. Furthermore, with short regions of homology, strand displacement in aqueous buffer might go to completion.

6. It is not known what degree of sequence mismatch this procedure will tolerate. Because 70–80% formamide increases the T_{SS} of RNA–DNA approximately 10° above the T_{ss} of DNA–DNA,[6] it is expected that a region of at least 90% homology is required. This is probably a minimum value, since strand displacement in aqueous buffer might take place with such mismatched sequences. In any case, it is almost certain that comparable hybridization procedures with single-stranded DNA are more tolerant of mismatching than R-loop formation, making this method a more advantageous approach for the identification of genes by mRNA hybridization.

7. Recently we have switched to a procedure in which the supercoils are converted to open circles by mild hydrolysis rather than linearization with a restriction enzyme. (This was the procedure used in the experiment shown in Figs. 4 and 5.) Three micrograms of plasmid DNA is resuspended in 20 μl 5× PIPES buffer (2 *M* NaCl, 0.5 *M* PIPES, pH 6.7–6.8, 0.05 *M* EDTA), sealed in a capillary, and heated to 75° for 3 hr. This converts approximately 50% of the supercoils to open circles (somewhat more than 50% for larger plasmids and somewhat less for pMB9). At the end of 3 hr, this 20 μl is added to 10 μl RNA in H_2O to which is added 70 μl formamide. The R-looping buffer is thus constituted, and incubation at 55° begins.

8. All plasmids containing yeast or *Drosophila* DNA segments were propagated under P2 EK1 containment, in compliance with the National Institutes of Health guidelines for recombinant DNA research.

Acknowledgments

M. Rosbash is grateful to Ray White for helpful discussions and suggestions. This work was supported by Grants HD 08887 and GM 23549 from the National Institutes of Health. M. Rosbash and B. Roberts were supported by Research Career Development Awards from the NIH, L. Hereford by a fellowship from the Medical Foundation, and J. L. Woolford by a fellowship from the NIH, and R. Ricciardi by a fellowship from the American Cancer Society.

[19] S. M. Berget, C. Moore, and P. A. Sharp, *Proc. Natl. Acad. Sci. U.S.A.* **74,** 3171 (1977).
[20] L. T. Chow, R. E. Gelinas, T. R. Broker, and R. J. Roberts, *Cell* **12,** 1 (1977).

Section VI

Detection and Analysis of Expression of Cloned Genes

[34] Maximizing Gene Expression on a Plasmid Using Recombination *in Vitro*

By THOMAS M. ROBERTS and GAIL D. LAUER

Some proteins of biochemical interest cannot be isolated in large quantities from their natural sources. Over the last few years a variety of methods utilizing recombinant DNA technology have been applied to this problem. In the simplest case, the gene coding for the protein in question, along with its promoter, is inserted into a high-copy-number *Escherichia coli* plasmid present in about 30–50 copies per cell. If the gene–promoter combination functions as well when located on the plasmid as it did when chromosomally located, roughly a 30- to 50-fold increase in the amount of the desired protein per transformed cell occurs.[1] For some proteins such as the *E. coli lacZ* gene product, the procedure works extremely well (*lacZ* protein can represent up to 30% of the total soluble protein of the cell), but the strategy is not always successful. At times the difficulty may lie in the promoter of the gene. For instance, if the gene is not of *E. coli* origin, its promoter may not function properly in *E. coli*. Some promoters function normally in *E. coli* but are too weak to direct sufficient transcription of their genes. A logical solution to such problems is the fusion *in vitro* of the gene of interest and a highly efficient *E. coli* promoter, such as the promoter of the *lac* operon. Carried on a high-copy-number *E. coli* plasmid, such fusions can direct the synthesis of large amounts of protein when present in appropriate *E. coli* hosts.[2-4]

Even the fusion of a strong promoter with a gene ensures only frequent transcription of that gene. High levels of protein production from a cloned gene require both frequent transcription of the gene and efficient translation of the resulting mRNA. Efficient translation of an mRNA, in turn, is dependent on a large number of factors including stability of the message and its ability to bind ribosomes. Since factors affecting the posttranscriptional fate of an mRNA are poorly understood at the molecular level, it is difficult to design a promoter–gene fusion in order to optimize the translational utilization of the message. However, recently the problem has been attacked on two fronts:

1. For reasons not yet clearly understood, the amount of gene product produced in transformed cells by plasmids carrying a given gene–

[1] D. R. Helinski, V. Hershfield, D. Figurski, and R. J. Meyer, *in* "Recombinant Molecules: Impact on Science and Society" (R. F. Beers, Jr. and E. G. Basset, eds.), p. 151. Raven, New York, 1977.

[2] K. Backman, M. Ptashne, and W. Gilbert, *Proc. Natl. Acad. Sci. U.S.A.* **73,** 4174 (1976).

[3] K. Backman and M. Ptashne, *Cell* **13,** 65 (1978).

[4] J. Hedgpeth, M. Ballivet, and H. Eisen, *Mol. Gen. Genet.* **163,** 197 (1978).

ISBN 0-12-181968-X

promoter fusion is dependent on the separation between the promoter and the gene. In cases of fusion of a restriction fragment bearing the *E. coli lac* promoter and either the *cI* gene or the *cro* gene of λ, protein production has been found to vary by as much as 20-fold with changes in gene–promoter separation of as little as three base pairs.[4,5] Since the cause(s) of these differences is unknown, the best course in optimizing the protein output of a given type of gene–promoter fusion is, at present, merely to vary the separation at random.

2. Efficient translation of an mRNA is thought to require, among other things, that the message bear a ribosome-binding site. Current theory holds that a ribosome-binding site on an *E. coli* message includes the AUG or GUG signaling the starting point of translation and a so-called Shine and Dalgarno (SD) sequence, a group of from three to nine bases in the leader of the message which are complementary to the bases at the 3′-end of the 16 S rRNA.[6,7] Each *E. coli* gene examined to date has been found to have associated with it a region coding for an SD sequence. Since the 3′-end of the 16 S rRNA is conserved in prokaryotic evolution, genes from other prokaryotes, too, may be preceded by sequences which, when transcribed, can function as efficient SD sequences in *E. coli*. However, eukaryotic genes are not preceded by regions coding for suitable *E. coli* SD sequences. Hence in the case of eukaryotic genes it is necessary to do more than simply fuse an *E. coli* promoter with the gene in order to ensure its efficient translation in *E. coli*. This problem can be overcome in theory by fusing with the gene in question a DNA restriction fragment containing not only an *E. coli* promoter but also the region downstream from the promoter coding for an *E. coli* SD sequence. In practice no eukaryotic gene has been activated in this manner. However, it has been shown that the DNA fragment bearing the promoter and region coding for SD of the *E. coli lacZ* gene can be fused to the 5′-end of the λ *cI* gene to form the coding sequence for a very effective ribosome-binding site.[3]

In this chapter, we present a method, utilizing a combination of restriction endonuclease cleavage and digestion with *E. coli* exonuclease III (Exo III) and S1 nuclease, which allows one to position a restriction fragment bearing the promoter of the *lacZ* gene of *E. coli* and the region coding for the SD sequence of the *lacZ* gene at virtually any distance in front of any cloned gene. This technique has been used to produce *E. coli* strains which can provide up to 200,000 monomers of *cro* protein per transformed cell.[5] It should be useful in constructing plasmids that direct

[5] T. M. Roberts, R. Kacich, and M. Ptashne, *Proc. Natl. Acad. Sci. U.S.A.* **76,** 760 (1978).
[6] J. Shine and L. Dalgarno, *Nature* (*London*) **254,** 34 (1975).
[7] J. A. Steitz, *in* "Biological Regulation and Control" (R. Goldberger, ed.), p. 349. Plenum, New York, 1979.

production of large amounts of various *E. coli* proteins and, in theory, of proteins from other prokaryotic and even eukaryotic cells.

Experimental Plan

The protocol consists of three steps:

1. Cloning the gene in question: This is the most variable part of the project. In general each gene is best dealt with in a slightly different manner. In each case, however, two basic design features are maintained. The gene is cloned on a high-copy-number plasmid such as pMB9 or pBR322, and the cloning is engineered in such a manner that the resulting plasmid contains, just upstream from the gene, a unique site for cleavage by a restriction endonuclease.

2. Removing various amounts of the DNA between the unique restriction site and the beginning of the gene: DNA from the plasmid constructed in step 1 is opened by cleavage at the unique restriction site upstream from the gene. The resulting linear plasmid DNA is digested for varying lengths of time with *E. coli* Exo III followed by digestion with S1 nuclease to remove single-stranded tails. In this manner some or all of the DNA between the site of restriction used in opening the plasmid and the ATG signaling the starting point of translation of the gene can be removed.

3. Inserting a restriction fragment carrying the promoter of the *E. coli lac* operon: In theory, at this point, a flush-ended, promoter-bearing DNA restriction fragment could be ligated to the nuclease-treated plasmid DNA generated in step 2 using blunt-end ligation mediated by T4 ligase. Since the blunt-ended fragment could be ligated to the plasmid backbone in either of two orientations, the resulting plasmid would contain the promoter fragment pointing in the correct orientation (toward the gene in question) only 50% of the time. In practice the plasmid constructed in step 1 is usually engineered to contain a single *Eco*RI site (though as we will see below other restriction endonuclease sites can be substituted for the *Eco*RI site) upstream from the unique restriction site used in opening the plasmid in step 2. Thus, if in step 1 a unique *Bam*HI site had been placed 50 base pairs upstream from the ATG signaling the starting point of translation for the gene, care would have been taken to ensure the presence of a single *Eco*RI site greater than 50 base pairs upstream from the *Bam*HI site. Then, to insert the promoter fragments, the DNA generated in step 2 is cut with *Eco*RI, and a version of the *lac* promoter-containing fragment bearing an *Eco*RI sticky end upstream from the promoter and a flush (*Alu*I-generated) end downstream from the promoter is ligated into placc

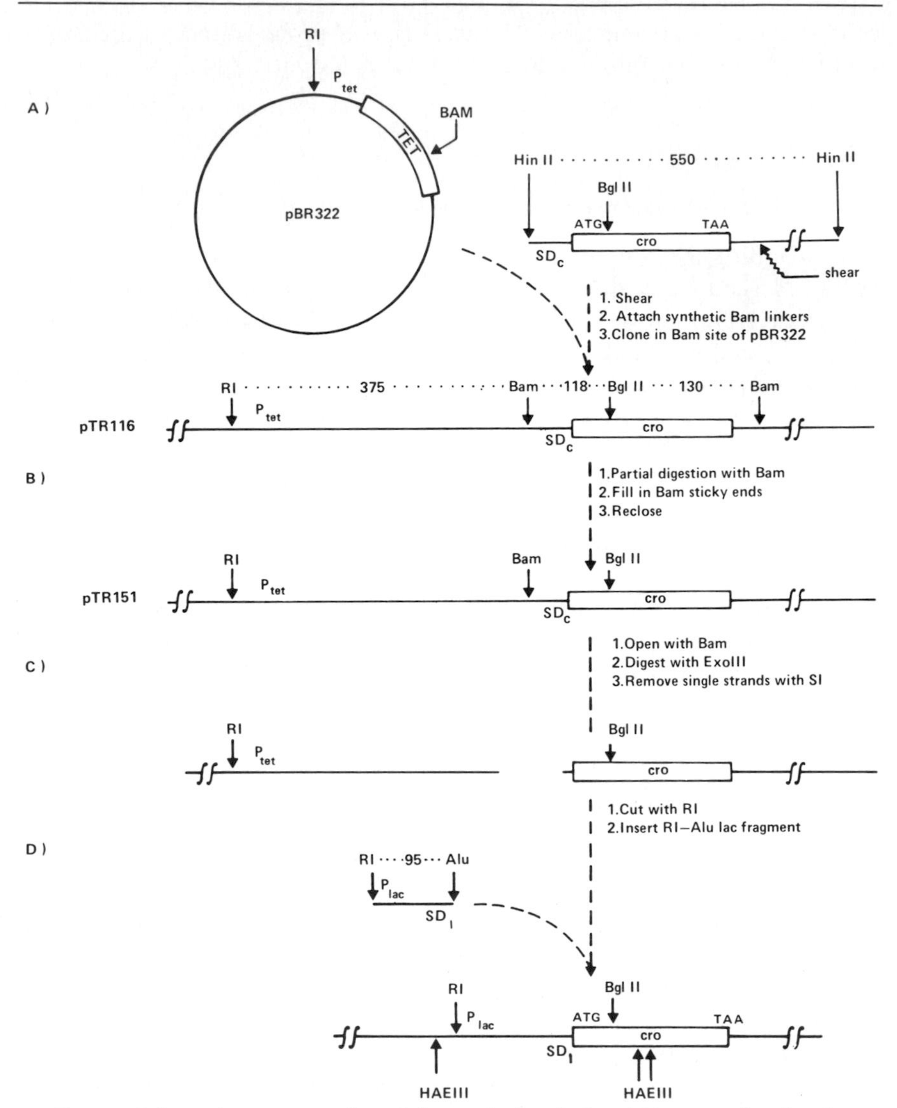

FIG. 1. Schematic representation of the methods of plasmid construction for the *cro* gene of phage λ. The approximate locations of several restriction endonuclease cleavage sites are shown for the plasmid pBR322, for a DNA fragment bearing the *cro* gene of phage λ, and for a DNA fragment bearing the promoter of the *lacZ* gene (see Backman and Ptashne,[3] for the source of this fragment). The location of the *tet* and *lacZ* gene promoters are indicated, as are the extents of the *tet* and *cro* genes. SD_c and SD_l indicate the SD sequences of the *cro* and *lacZ* genes, respectively. AUG and UAA, respectively, are the start and stop signals for translation of the *cro* protein. Distances are indicated in base pairs. (A) The fragment bearing the *cro* gene is shortened by shearing to remove certain λ control elements near the 3′-end of the gene, and the smaller fragment is inserted into the *Bam*HI site in pBR322 using *Bam*HI linkers. (B) The *Bam*HI site near the carboxyl terminus of the *cro*

in the plasmid backbone. Finally, the plasmid DNA is transformed into an appropriate *E. coli* host.

Details of the protocol are presented below. A specific example of the protocol as applied to the *cro* gene of phage λ is shown in Fig. 1.

Materials and Reagents

Strains

Escherichia coli strain MM294 (endoI r_k^- m_k^+ *Bl* pro^-) was used as the host for all plasmids. Cells were normally grown in tryptone broth supplemented with 0.1% yeast extract. Cells used for plasmid DNA isolation were grown in M-9 salts[8] supplemented with 0.5% glucose, 0.3% casamino acids, and 0.01% thiamine.

Enzymes

All restriction enzymes were obtained from New England BioLabs. T4 ligase was purified by the method of Panet *et al.*[9] through the phosphocellulose step. S1 nuclease was purified by the method of Vogt[10] through the DEAE-cellulose chromatographic step. *Escherichia coli* DNA polymerase I and Exo III (the generous gift of W. McClure) were purified by the procedure of Jovin *et al.*[11] through step 7.

Buffers

*Pol*I reactions were carried out in 50 m*M* Tris-HCl, pH 7.8, 5 m*M* $MgCl_2$, 10 m*M* β-mercaptoethanol, 50 μg/ml bovine serum albumin, and 2 μ*M* of each deoxynucleoside triphosphate.[2] T4 ligase was used in 0.2 *M* Tris-HCl, pH 7.6, 0.01 *M* $MgCl_2$, 0.05 *M* β-mercaptoethanol, 0.001 *M* ATP. Restriction enzymes requiring low-salt conditions were used in 6.6

[8] J. H. Miller, ed., "Experiments in Molecular Genetics." Cold Spring Harbor Lab., Cold Spring Harbor, New York, 1972.

[9] A. Panet, J. H. van de Sande, P. C. Loewen, H. G. Khorana, A. J. Raae, J. R. Lillehaug, and K. Kleppe, *Biochemistry* **12,** 5045 (1973).

[10] V. M. Vogt, *Eur. J. Biochem.* **33,** 192 (1973).

[11] T. M. Jovin, P. T. Englund, and L. L. Bertsch, *J. Biol. Chem.* **244,** 2996 (1969).

gene is eliminated. (C) The plasmid is opened at the *Bam*HI site and varying amounts of DNA removed by Exo III and S1. (D) The partially resected plasmid is cut at the *Eco*RI site, and the *lac* promoter (bearing the *UV5* mutation rendering it CAP-independent) inserted by sticky-end ligation at its *Eco*RI end and by blunt-end ligation to the resected plasmid DNA at its *Alu*I end. In this process the *Eco*RI fragment carrying the P_{tet} was eliminated. The efficiency of steps C and D is fairly high—about 200–400 plasmids result from 1 μg of pTR151 input.

m*M* Tris-HCl, pH 7.4, 6.6 m*M* $MgCl_2$, and 6.6 m*M* β-mercaptoethanol.[3] High-salt-requiring restriction enzymes were used in the above buffer plus 50 m*M* NaCl. Exo III reactions were carried out in high-salt restriction buffer, and S1 reactions were carried out in high-salt restriction buffer plus 0.05 *M* NaOAc–HOAc, pH 4.0, 0.15 *M* NaCl, and 0.006 *M* $ZnSO_4$.[12]

Experimental Details

Cloning the Gene

As mentioned above, this is the most variable portion of the procedure. In most cases the gene will already have been cloned by some standard technique, and all that will be required is the positioning of the unique restriction site upstream from the gene. Several techniques may be used for this purpose, but perhaps the most general method utilizes so-called synthetic linkers.[13,14] These are short (8 to 10 base pairs long), chemically synthesized DNA duplexes containing the cleavage site for a restriction endonuclease such as *Bam*HI, *Eco*RI, or *Hind*III. For the task at hand, a linker bearing the cutting site for a sticky-end restriction nuclease which does not cut in the gene sequence is chosen. The linker is ligated in considerable molar excess to a flush-ended DNA fragment containing the gene. In a typical reaction, 1–5 pmol of a DNA fragment containing the gene is mixed with a fivefold excess of the chosen linker in a final reaction volume of 25 μl. The ligation reaction is carried out with 1 unit of ligase at room temperature for 2 hr. The reaction mixture is then digested with the restriction enzyme specific for the cutting site carried on the linker. This exposes the sticky ends of the linker molecules joined to the ends of the gene-bearing fragment and digests the excess linkers present in the reaction mixture. The gene-bearing fragment is then cloned into a plasmid vector with a single cleavage site of the type carried on the linker. In the example shown in Fig. 1, for instance, a shear-generated DNA fragment bearing the *cro* gene was cloned into the *Bam*HI site in pBR322 using *Bam*HI linkers.

The cloned gene is now flanked by cutting sites for the restriction nuclease specified by the linker (two *Bam*HI sites for the purposes of this discussion.) All that remains is to remove one of these sites at the carboxyl end of the gene. To do this about 10 μg DNA from the newly

[12] R. Wu, G. Ruben, B. Siegel, E. Jay, P. Spielman, and C. D. Tu, *Biochemistry* **15,** 734 (1976).

[13] C. P. Bahl, K. J. Marians, R. Wu, J. Stawinski, and S. A. Narang, *Gene* **1,** 81 (1976).

[14] R. H. Scheller, R. E. Dickerson, H. W. Boyer, A. D. Riggs, and K. Itakura, *Science* **196,** 177 (1977).

constructed plasmid is partially digested with *Bam*HI, and the resulting mixture of plasmid molecules is separated on an agarose gel. Molecules that have been cut only once are extracted from the gel and treated with DNA polymerase I in the presence of the four deoxynucleoside triphosphates to render the *Bam*HI sticky ends flush. A typical reaction is carried out at 18° for 1 hr in a total volume of 25 μl using 0.25 unit *Pol*I. The *Pol*I-treated molecules are then closed by blunt-end ligation into circles and used to transform *E. coli*. DNA is purified from cultures grown up from several of the resulting transformants. Restriction analysis reveals which plasmids have lost the *Bam*HI site at the carboxyl end and not the site at the amino end of the gene-containing fragment.

It is quite possible that the *Bam*HI site generated above will lie inconveniently far from the amino terminus of the gene of interest. In this case it can be "walked down" the plasmid toward the gene. To do this, plasmid DNA is cut with *Bam*HI, treated with Exo III and S1 as described below to remove a portion of the DNA between the *Bam*HI cut and the gene, and then recircularized in the presence of *Bam*HI linkers.

Using Nucleases to Remove DNA Selectively

Approximately 5 μg of DNA from the plasmid generated above is digested to completion with *Bam*HI. Typically, a DNA concentration of 100 μg/ml is used. Exo III digestion is carried out in the same high-salt restriction buffer used for *Bam*HI digestion. There is no need to inactivate the restriction nuclease. The restricted DNA is cooled to 22° in a water bath, and Exo III is added to a concentration of 6 units of enzyme per microgram of input plasmid. At 22° Exo III digestion proceeds at roughly 8–20 bases per minute per DNA end, depending on the batch of enzyme used. As the reaction proceeds, aliquots are drawn off and pipetted into an equal volume of 2× S1 buffer. Zinc cations in the S1 buffer stop the Exo III reaction.[15] In this manner a collection of DNA molecules can be generated which contain single-stranded regions extending from the *Bam*HI cut for various distances up to and including the beginning of the coding region of the gene. S1 nuclease is added to the reaction mix to a concentration of 1500 Vogt units[10]p/ml. S1 digestion is allowed to continue for 2 hr at 18° and is terminated by a single extraction with phenol. The DNA-containing aqueous layer is then purified by passage over a 1-ml G-50 fine Sephadex column before further use. The net result of the Exo III and S1 digestion is deletion of some or all of the DNA between the *Bam*HI cut and the gene.

The following points concerning the Exo III and S1 digestion process are worth noting: The *Bam*HI site used for the purpose of discussion

[15] C. C. Richardson, I. R. Lehman, and A. Kornberg, *J. Biol. Chem.* **239**, 251 (1964).

could have been almost any other restriction site. Exo III may have a slight preference for sticky ends with a 5′-projection as opposed to blunt ends and does not attack DNA molecules having a sticky end with a 3′-projection. The Exo III concentration was chosen to ensure a slight molar excess of enzyme over DNA ends.[12] Exo III is active over a wide temperature range. The temperature used was chosen to give a conveniently slow rate of digestion. Exo III from one commercial source has proved unsuitable for this use. The S1 concentration and reaction time used were chosen empirically to maximize the number of transformants. High-salt and low-temperature reaction conditions were used in an attempt to minimize the "nibbling" of duplex DNA ends; however, the S1 enzyme used here was capable of removing up to 10 base pairs from the end of a duplex DNA molecule under these conditions. To date, the exact end point of digestion after Exo III and S1 treatment has been determined in about 15 cases. In no case has the reaction left a T as the first undigested base at the 5′-end of the DNA duplex. No attempt has been made to use commercial S1 enzyme.

Inserting the lac Promoter

The sequence of the *lac* promoter-containing fragment used is shown in Fig. 2.[16] *Alu*I may be used to excise the fragment from the bacterial chromosome. Note that the *Alu*I site downstream from the promoter cleaves just two bases before the starting point of translation of the *lacZ* gene. The *Alu*I fragment thus contains, in addition to the promoter, the region coding for most of the leader of the *lacZ* message including its SD sequence. The best source of this fragment is the plasmid pLJ3.[17] In this plasmid the *Alu*I site upstream from the promoter has been converted to an *Eco*RI site, and the *Alu*I site downstream from the promoter has been converted into a *Pvu*II site (*Pvu*II recognizes a six-base-pair sequence which includes the sequence recognized by *Alu*I and leaves the same flush end, as does *Alu*I). Cleavage of pLJ3 DNA with *Eco*RI and *Pvu*II yields five fragments which are easily separated on a preparative 5% acrylamide gel. The smallest fragment is the one desired, the 95-base-pair-long *Eco*RI-*Alu*I fragment.

Insertion of the *Eco*RI-*Alu*I promoter fragment is a simple matter. After the Exo III and S1 treatment described above, plasmid DNA is digested with *Eco*RI, heated to 70° for 10 min to inactivate the *Eco*RI, and ethanol-precipitated. The plasmid DNA is then dissolved in 20 μl of ligation buffer; a fivefold molar excess of the promoter fragment is added, and

[16] W. Gilbert, J. Gralla, J. Majors, and A. Maxam, *in* "Protein-Ligand Interactions" (H. Sund and S. Blauer, eds.), p. 193. de Gruyter, Berlin, 1975.

[17] L. Johnsrude and W. Gilbert, *Proc. Natl. Acad. Sci. U.S.A.* **75,** 5314 (1978).

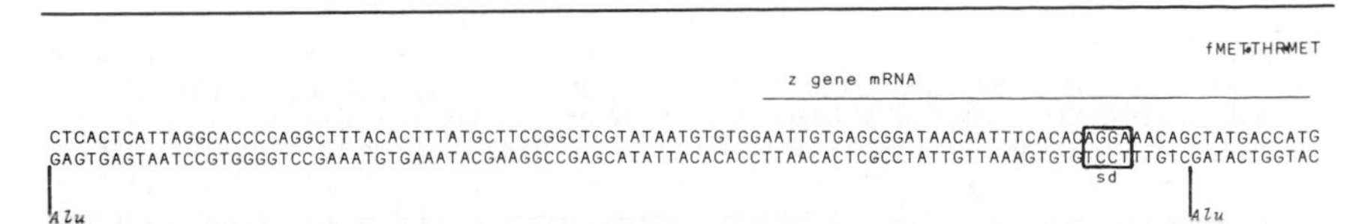

FIG. 2. Diagram of the DNA fragment bearing the promoter of the *lac* operon of *E. coli*. Shown is the DNA sequence of the promoter (*UV5* allele) as it occurs in the *E. coli* chromosome.[16] The solid line indicates the mRNA transcribed from the promoter. The bases coding for the *lac* SD sequence are boxed. Note that the right *Alu*I site occurs just two bases upstream from the start of the *lacZ* gene.

the mixture is ligated for 18 hr at 12°. The resulting DNA is used to transform *E. coli*. In a typical experiment 500 independent *E. coli* transformants might arise from 1 μg of input DNA. Figure 3 shows an example of the product plasmids obtained by the process as outlined in Fig. 1.

There are two further details to note concerning the promoter insertion process. First, the *lac* promoter-containing fragment also bears a *lac* operator. The presence of this operator on the 30–50 plasmid copies in a transformed cell is sufficient to titrate *lac* repressor (which is only present in about 5–10 copies per cell) off the chromosomal copy of the *lac* operator, thus rendering the cell constitutive for β-galactosidase synthesis. For this reason, colonies of cells containing a plasmid-born copy of the *lac* promoter fragment are blue on an agar plate containing the indicator 5-chloro-4-bromo-3-indolyl-β-D-galactoside. Second, it may not always be convenient to use an *Eco*RI site to aide in inserting the promoter-containing fragment. At present this problem can be sidestepped by inserting a blunt-ended promoter-bearing fragment (the 95-base-pair *Alu*I fragment shown in Fig. 2) after the Exo III–S1 digestion step. In the future, the fact that *Pvu*II cuts most plasmid DNAs only a very small number of times should allow construction of plasmids from which suitable *lac* promoter-bearing fragments can be excised with a combination of *Pvu*II and a variety of other restriction nucleases. For instance, the *Hae*III fragment from pLJ3 containing the *Pvu*II-*Eco*RI piece used above could be cloned into the *Eco*RI site of pBR322. This would yield two different plasmids (depending on the orientation of the *Hae*III fragment) from which it would be possible to excise *lac* promoter-bearing fragments by digestion with a combination of *Pvu*II and any of the following enzymes: *Hin*dIII, *Bam*HI, *Sal*I, *Ava*I, or *Pst*I.

The method described above has been used to create *E. coli* strains which overproduce the *cro* protein of λ (see Johnson *et al.*[19] for a discussion of the actual protein isolation procedures used in conjunction with strains carrying these plasmids) and the repressor protein of *Salmonella* phage P22. It has not yet been used in conjunction with an eukaryotic gene. In theory, the ideal placement of the *lac* fragment is two bases up-

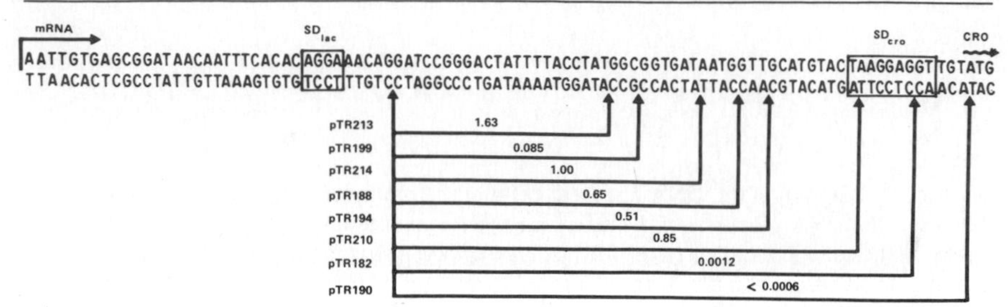

FIG. 3. A summary of the deletion size and *cro* production of selected plasmids constructed by the methods in Fig. 1. Shown is a portion of the sequence of pTR161[5] extending from the starting point of transcription of the *lac* promoter on the left (straight arrow) to the starting point of translation of the *cro* gene on the right (wavy arrow). All plasmids produced by the Exo III–S1 method described in Fig. 1 may be considered deletions of pTR161. The extent of these deletions is indicated by brackets. The numbers on the brackets are the level of *cro* protein as a percentage of total soluble protein in cells transformed with each of the deletion plasmids. For comparison, *cro* protein represents 0.5% of the total soluble protein in cells transformed with pTR161, and a previously reported plasmid[18] in which *cro* transcription is initiated at its own promoter directs synthesis of 0.05% *cro*.

stream from the ATG signaling the starting point of translation of an eukaryotic gene—locating it in the same position relative to the eukaryotic gene it occupies relative to the *lacZ* gene on the *E. coli* chromosome.[19a]

[18] A. Johnson, B. J. Meyer, and M. Ptashne, *Proc. Natl. Acad. Sci. U.S.A.* **75,** 1783 (1978).
[19] A. Johnson *et al.*, this series, Vol. 65, p. 839.
[19a] Recently the techniques described here have been used to produce an unfused eucaryotic protein in *E. coli,* SV40 small t antigen, T. M. Roberts, L. Bikel, R. R. Yocum, D. M. Livingston, and M. Ptashne, *Proc. Natl. Acad. Sci. U.S.A.*, in press.

[35] Use of the λ Phage Promoter P_L to Promote Gene Expression in Hybrid Plasmid Cloning Vehicles

By HANS-ULRICH BERNARD and DONALD R. HELINSKI

The expression of prokaryotic structural genes in bacteria is dependent on DNA sequences upstream from these genes that do not code for a polypeptide chain, e.g., promoters, operators, and ribosome-binding sites. Physical separation *in vitro* of a particular structural gene and its regulatory sequences may be possible by cleavage of the DNA with a suitable restriction endonuclease at a restriction site between these two DNA sequences. A restriction fragment carrying the regulatory sequences of one gene can be ligated *in vitro* to a restriction fragment containing another gene that does not carry its natural regulatory sequences. Cloned on a plasmid or phage vehicle, this new combination of DNA sequences should bring the expression *in vivo* of this latter gene under the control of

ISBN 0-12-181968-X

the regulatory sequences of the first gene. This article describes the construction and use in *Escherichia coli* of plasmid cloning vehicles designed to promote expression of gene inserts. DNA fragments that contain genes deleted for their own promoters can be inserted into any one of four different single restriction sites on these vehicles. Transcription of these genes occurs from the strong phage λ promoter P_L located on the vehicles, from which RNA polymerase can read into the inserted fragments. These vehicles should be useful in promoting high levels of expression of prokaryotic genes that are weakly expressed in their natural genetic environment or in promoting transcription of eukaryotic genes that are poorly expressed in *E. coli*. This construction and properties of the vehicles described in this paper are presented in greater detail elsewhere.[1]

Methods

Growth of Bacteria

All strains used are either grown in L broth or in minimal salts plus 0.2% glucose supplemented with 0.2% casamino acids (Difco certified). Cells carrying a plasmid vehicle containing the phage promoter P_L or P_R under the control of a temperature-sensitive repressor (*cIts*) must be grown at 30°. All other strains are grown at 37°. To select for antibiotic resistance specified by the plasmids, 50 μg/ml kanamycin or 25 μg/ml tetracycline is added.

Transformation

Transformation conditions are a modification of the method of Cohen *et al.*[2] *Escherichia coli* is grown in 20 ml L broth to a $Klett_{540}$ of 70, washed with 10 ml 100 m*M* $MgCl_2$ at 0°, and kept for 20 min in 10 ml 100 m*M* $CaCl_2$ at 0°. The cells are resuspended in 2 ml 100 m*M* $CaCl_2$. To 200-μl fractions of this suspension 1–50 μg DNA in 1–10 μl buffer (66 m*M* Tris-HCl, pH 7.5, 33 m*M* NaCl, 6.6 m*M* $MgCl_2$) is added and left for 30 min at 0°. The mixture is heated for 2 min to 42°, left for 15 min on ice, and then incubated for 2 hr at 30° after the addition of 2 ml L broth. Then 0.2 ml of this suspension is spread per plate containing selective medium.

Purification of Plasmid DNA

Plasmid DNA is purified according to Katz *et al.*,[3] from 250-ml cultures

[1] H.-U. Bernard, E. Remaut, V. M. Hershfield, H. K. Das, D. R. Helinski, C. Yanofsky, and N. Franklin, *Gene* **5**, 59 (1979).

[2] S. N. Cohen, A. C. Y. Chang, and C. L. Hsu, *Proc. Natl. Acad. Sci. U.S.A.* **69**, 2110 (1972).

[3] L. Katz, D. K. Kingsbury, and D. R. Helinski, *J. Bacteriol.* **114**, 557 (1973).

grown overnight in L broth containing kanamycin or tetracycline at 30° or 37° to stationary phase. Cells are pelleted (Sorvall GSA rotor, 15 min, 5000 rpm, 4°) and resuspended in 5 ml TNE buffer (50 m*M* Tris-HCl, pH 8.0, 50 m*M* NaCl, 15 m*M* EDTA). Lysozyme (5 mg) is added, and after a 15-min incubation at room temperature 0.5 ml 0.4% Triton X-100, 50 m*M* Tris-HCl, pH 8.0, and 50 m*M* EDTA. After 10–20 min an increase in viscosity indicates lysis of the cells. In the case of certain strains that are resistant to this procedure, the addition of 0.1% sodium dodecyl sulfate will improve the lysis. The lysate is centrifuged at 20,000 *g* for 20 min at room temperature. CsCl (1 g/ml) is added to the supernatant, followed by the addition of 200 μg ethidium bromide per milliliter previously dissolved in water. This solution is centrifuged for 36 hr at 40,000 rpm in a Beckman Ti50 rotor at 15°. The lower fluorescent band in this gradient is removed by piercing the side of the tube with a syringe needle. This fraction is then extracted three times with isopropanol saturated with water and CsCl and extensively dialyzed against TNE buffer. The DNA is precipitated with 2 volumes of ethanol, dissolved in 100 μl TNE buffer, and stored at 4° or frozen. The yield is approximately 1 μg DNA/ml culture. Note that plasmids carrying P_L should not be amplified by incubating the cells in the presence of chloramphenicol, since runoff replication of the plasmid vehicle by this procedure would result in derepression of the P_L promoter.

Clone Analysis

Analysis of the DNA of individual transformants can be carried out by the following rapid procedure.[4] Twenty-milliliter cultures are grown overnight, centrifuged, and resuspended in 2 ml TNE buffer. After lysis with 1 mg lysozyme and 0.1 ml 0.4% Triton X-100, 50 m*M* Tris-HCl, pH 8.0, and 50 m*M* EDTA, the lysate is centrifuged at 20,000 *g* for 20 min at room temperature. The supernatant is extracted with 1 volume of water-saturated phenol, followed by extraction with 5 volumes of ether, and the DNA is precipitated with 1 volume of isopropanol (1 hr, −20°). The DNA pellet is then dissolved in 100 μl TNE buffer. Crude DNA preparations obtained by this method are cleaved efficiently by all restriction endonucleases used in this study, but must be treated with RNase A prior to loading on gels. With the use of this method, the plasmid DNA of 20–50 individual transformants can easily be obtained in 1 day and studied by restriction analysis. Each clone yields DNA sufficient for 10–20 separate restriction enzyme treatments.

[4] M. Kahn, R. Kolter, C. Thomas, D. Figurski, R. Meyer, E. Remaut, and D. R. Helinski, this volume [17].

Restriction and Ligation

Reactions with *Eco*RI are carried out in 100 m*M* Tris-HCl, pH 7.4, and 50 m*M* NaCl 10 m*M* $MgCl_2$, with *Hpa*I in 10 m*M* Tris-HCl, pH 7.4, 6 m*M* KCl, 10 m*M* $MgCl_2$, 6 m*M* mercaptoethanol and with all other restriction enzymes in 7 m*M* Tris-HCl, pH 7.4, 60 m*M* NaCl, 7 m*M* $MgCl_2$, and 6 m*M* mercaptoethanol. Digestions with several restriction enzymes can either be done at the same time or sequentially after heat-inactivating the first enzyme (20 min, 65°) and adjusting the buffer. For ligation with T4 DNA ligase (0.05 unit/μg DNA), the reaction mixture is adjusted to 66 m*M* Tris, 33 m*M* NaCl, 6.6 m*M* $MgCl_2$, 60 μ*M* ATP, and 1.5 mg/ml dithiothreitol. Blunt-end ligations (between *Hpa*I ends) are carried out in 50 m*M* Tris-HCl, pH 7.5, 5 m*M* $MgCl_2$, 1 m*M* mercapto-ethanol, 10 μ*M* (nucleotides) tRNA, and 60 μ*M* ATP and with 0.5 unit of T4 DNA ligase per microgram of DNA according to Sgaramella *et al.*[5] Cohesive-end ligations are run for 8–16 hr at 14°, and blunt-end ligations for 24–36 hr at 25°. For restriction and ligation reactions, the concentration of plasmid DNA is 1–10 μg/100 μl buffer.

Agarose Gel Electrophoresis

Agarose slab gels (0.8%, Seakem agarose) are run at room temperature, 150 V, using a buffer containing per liter 10.8 g Tris base, 0.93 g EDTA, and 5.5 g boric acid. Approximately 1 μg DNA is loaded into each slot (5 × 5 mm). Size markers, such as *Eco*RI-digested phage λ DNA[6] can be run parallel to cleaved plasmid DNA, and the unknown size of DNA fragments can be determined by interpolation after plotting the migrated distance versus molecular weight of a particular band of semilogarithmic paper.

Expression Experiments

Cells are grown overnight in minimal salts plus 0.2% glucose, 0.2% casamino acids (Difco certified), 50 μg/ml tryptophan, and 50 μg/ml kanamycin or 25 μg/ml tetracycline. These cultures are washed once to remove the antibiotic and inoculated into 150 ml of the same medium not containing the antibiotic to a $Klett_{540}$ of 30. After growing these cultures at 30° to a $Klett_{540}$ of 70–80, they are shifted to 45° for 5 min and kept at 41°

[5] V. Sgaramella, M. Bursztyn-Pettegrew, and S. D. Ehrlich, *in* "Recombinant Molecules: Impact on Science and Society" (R. F. Beers and E. G. Bassett, eds.), p. 57. Raven, New York, 1977.

[6] B. Allet, P. G: N. Jeppensen, K. Y. Karagiri, and H. Delius, *Nature* (*London*) **241**, 120 (1973).

with vigorous shaking. Time points are taken before and at certain time intervals after heat-shocking the cells by removing 50-ml portions. The cells are harvested by centrifugation and kept frozen. After resuspension in 50 m*M* potassium phosphate, pH 7.5, 10% glycerol, 1 m*M* dithiothreitol, 50 m*M* NaCl, and 10 mg pyridoxal phosphate, the cells are sonicated and then centrifuged for 15 min at 10,000 *g*, and the enzymatic activity of the product of the gene on the insert expressed from P_L is determined in the supernatant. The expression of the *trpA* gene that is the test system for the function of the plasmids was determined by the method of Smith and Yanofsky.[7] The reaction between indole and serine to produce tryptophan is measured in the presence of an excess of the *trpB* subunit of tryptophan synthetase. One unit is equivalent to the amount of activity catalyzing the reaction of 0.1 μmol indole/20 min.[7] The amount of total protein in the extract is determined by the method of Lowry.[8]

Rationale for Construction on an Expression Vehicle: Development of Plasmid pHUB12

It was assumed that an effective vehicle for promoting the transcription of a gene on an inserted restriction fragment should have at least the following desirable features: (1) It should carry a strong promoter; (2) this promoter should be under the control of a regulatory gene so that normally the promoter is switched off but can be turned on experimentally; (3) the vehicle should have different restriction sites for the insertion of DNA occurring downstream from the promoter and just once in the vehicle; and (4) the vehicle should be present in a high copy number per cell to provide multiple copies for expression of the DNA insert.

The construction of a vehicle with these properties involved initially the DNA of plasmid pBR322 and phage λ *trp48*. pBR322 contains a single *Eco*RI and a single *Pst*I site (Fig. 1).[9] Cleavage with *Pst*I plus *Eco*RI results in two restriction fragments; the larger fragment has a size of 3.6 kb and contains all functions necessary for replication in addition to the gene(s) that code for resistance to tetracycline. These two functions are not impaired by removal of the smaller *Pst*I-*Eco*RI fragment. This fragment was replaced by a *Pst*I-*Eco*RI fragment of λ *trp48* containing P_L and the *trpDCB* genes. pBR322 and λ *trp48* DNA were separately cleaved with *Pst*I and *Eco*RI, mixed, ligated, and transformed into

[7] O. H. Smith and C. Yanofsky, this series, Vol. 5, p. 794.

[8] O. H. Lowry, N. G. Rosebrough, A. L. Farr, and R. G. Randall, *J. Biol. Chem.* **193,** 265 (1951).

[9] F. Bolivar, R. L. Rodriguez, P. Y. Greens, M. C. Betlach, H. L. Heyneker, H. W. Boyer, J. H. Crosa, and S. Falkow, *Gene* **2,** 95 (1977).

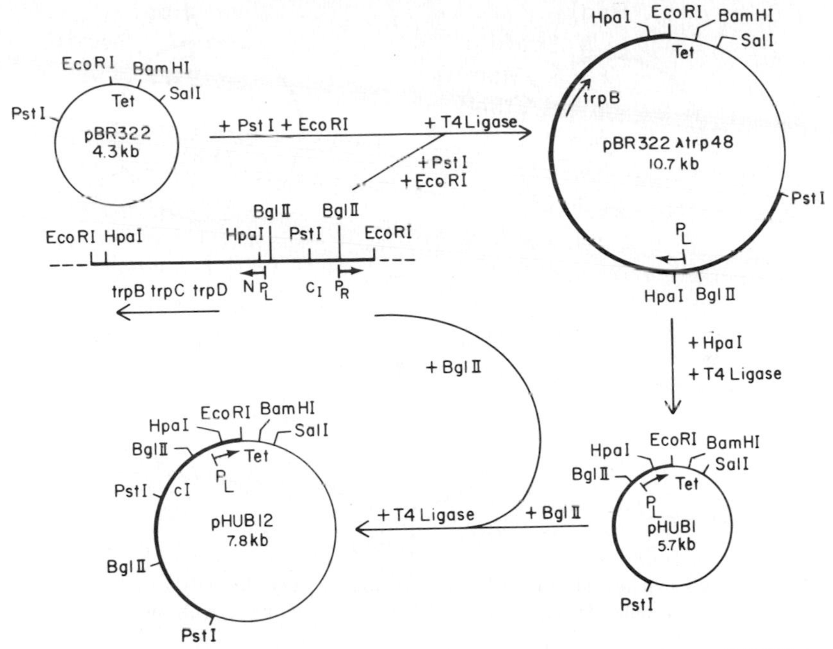

FIG. 1. Construction of plasmid pHUB12. The *Eco*RI fragment of λ *trp48* DNA has a size of 9.3 kb. DNA sequences derived from λ *trp48* DNA are shown by a thick line. Only those genetic loci and restriction sites discussed in the text are indicated.

C600(λ)Δ*trpEB9*. Transformants growing on minimal plates plus glucose supplemented with casamino acids, indole, and tetracycline contained in each case a plasmid of 10.5-kb size that was cleaved by *Pst*I plus *Eco*RI into two fragments. One fragment corresponded in size to the large *Pst*I-*Eco*RI fragment of pBR322. The other one corresponded to the *Pst*I-*Eco*RI fragment of λ *trp48* that contained P_L and provided the *trpB* gene [expressed by the weak *trp* internal promoter p_2[10]] that was selected by growing the transformed $trpB^-$ cells on indole. This plasmid, pBR322λ*trp48*, was cleaved with *Hpa*I to remove a *Hpa*I fragment carrying the *trpDCB* genes. After recyclization with T4 DNA ligase under conditions favoring blunt-end ligation the plasmid DNA was transformed into C600(λ)Δ*trpEB9*. Tet^R $trpB^-$ transformants contained the plasmid designated pHUBl. As shown in Fig. 1, pHUBl possesses four different single restriction sites (*Hpa*I, *Eco*RI, *Bam*HI, and *Sal*I) 0.25–1 kb downstream from P_L that are available for the insertion of genes on DNA fragments that may be transcribed from P_L. pHUBl is stable in phage λ lysogens but cannot be transformed into nonlysogens, presumably because

[10] E. N. Jackson and C. Yanofsky, *J. Mol. Biol.* **69,** 307 (1972).

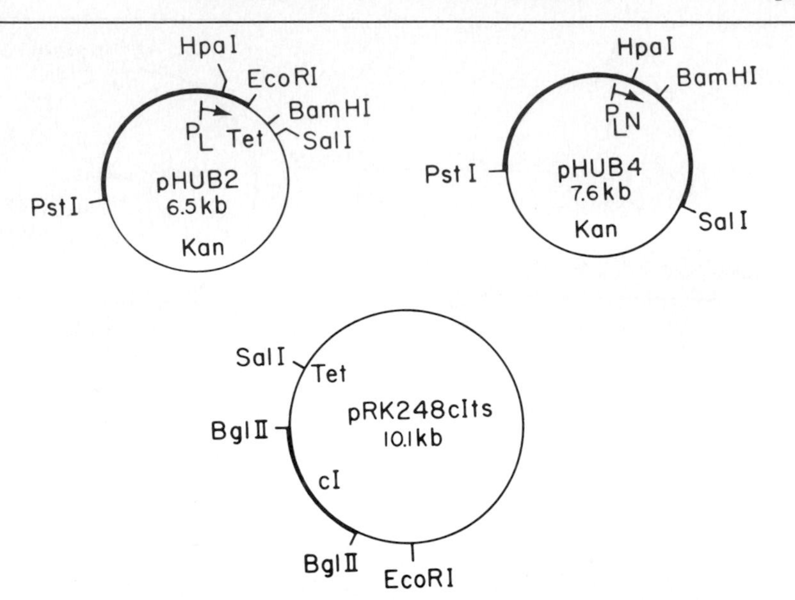

FIG. 2. Maps of plasmids pHUB2, pHUB4 and pRK248*cIts*. DNA sequences derived from λ *trp48* or the related phage λ *trp43*[1] DNA are shown by a thick line.

transcription from the unrepressed P_L interferes with essential plasmid functions including, possibly, replication of the plasmid.

To introduce the expression switch into the vehicle pHUBl, a *Bgl*II fragment of λ *trp48* DNA, coding for a *cI* repressor gene with a mutation to temperature sensitivity,[11] was cloned into the single *Bgl*II site of pHUBl. λ *trp48* DNA and pHUBl DNA were cleaved with *Bgl*II, mixed, ligated and transformed into *E. coli* C600. All Tet^R transformants (grown at 30°) contained a plasmid with two *Bgl*II sites. One *Bgl*II fragment was identical in size to pHUBl, and the other *Bgl*II fragment was derived from λ *trp48* and contained the *cIts* gene. This plasmid was designated pHUB12 (Fig. 1). C600 pHUB12 was immune to phage λ at 30° but lost λ immunity at 42°. In addition, pHUB12 was found to be stably maintained at 30° but was lost during overnight incubation at 42°. These observations indicate that at 30° P_L is essentially switched off by the repressor and turned on at 42°.

Two vehicles similar to pHUB1, designated pHUB2 and pHUB4, have been constructed, and their physical and genetic maps are shown in Fig. 2. Since insertions into the *Bam*HI and *Sal*I site of pHUB1 and pHUB2 inactivate the resistance to tetracycline, pHUB2 offers the ad-

[11] N. C. Franklin, *in* "The Bacteriophage Lambda" (A. D. Hershey, ed.), p. 621. Cold Spring Harbor Lab., Cold Spring Harbor, New York, 1971.

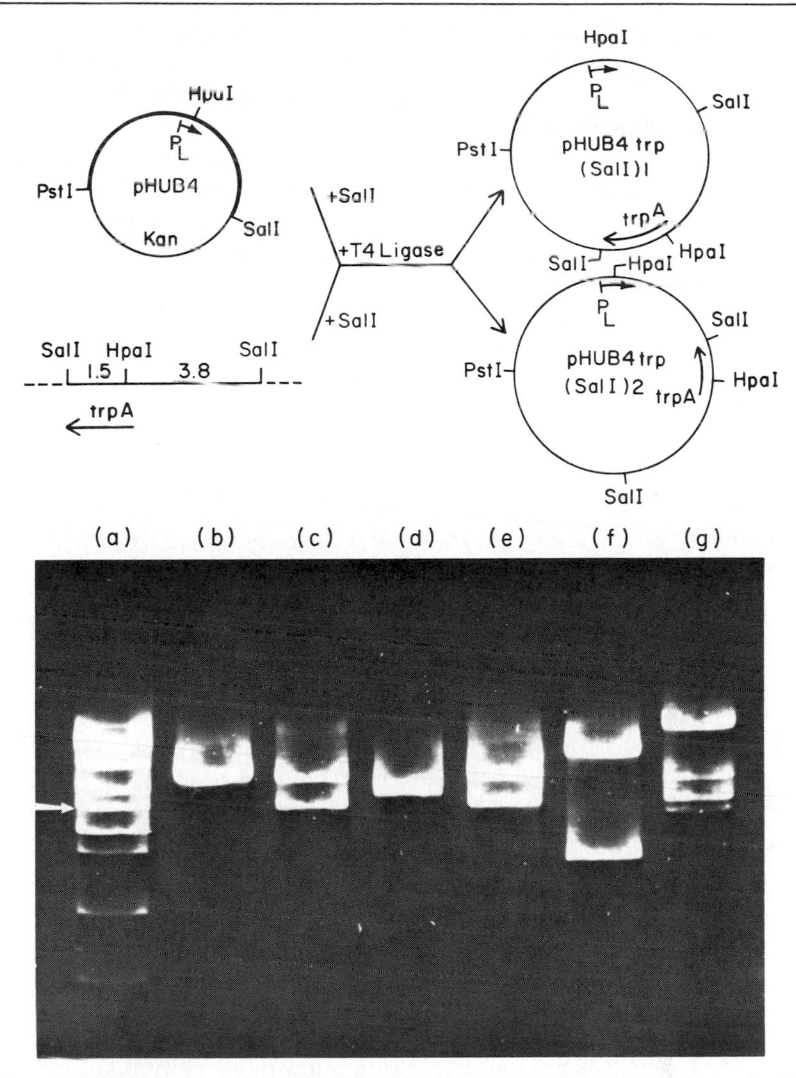

FIG. 3. Schematic representation of the insertion of a *Sal*I fragment containing *trpA* into pHUB4. The gel results distinguish between two orientations of the insert by restriction analysis. (a) *Sal*I digest of ϕ80 *trpE-A* DNA. The band containing *trpD-A*[16] is indicated by an arrow. (b) *Sal*I digest of pHUB4. (c and e) *Sal*I digests of the two hybrid plasmids pHUB4*trpDCBA*(*Sal*I)1 and -2. (These names have been abbreviated in the figure.) The two orientations of the *Sal*I inserts relative to pHUB4 can be distinguished by a *Hpa*I digest (d and f). The plasmid shown in slots c and d contains the *trp* genes in the same orientation as P_L. (g) *Eco*RI digest of phage λ DNA.[6]

vantage that hybrid plasmids can be selected more easily by their remaining resistance to kanamycin. pHUB4, which confers resistance to kanamycin only, has retained the *N* gene of phage λ. The *N* protein has been shown to modify RNA polymerase starting at P_L and may be useful in overcoming transcription termination sequences located before the inserted gene.[12,13] pHUB1, 2, and 4 can be maintained in λ lysogens. Expression experiments with these three vehicles can be carried out either using the host K12ΔHI*cI857* which is a noninducible lysogen and produces a temperature-sensitive repressor from the chromosome, or in a variety of other *E. coli* strains that have been transformed with the plasmid pRK248*cIts*. The hybrid plasmid pRK248*cIts* was constructed from pRK248, a derivative of the P group plasmid RK2[14] by inserting into pRK248 a DNA fragment carrying the *cIts* gene from λ *trp48*. pRK248*cIts* is compatible with the expression vehicles described above and produces a *cIts* repressor that acts in trans on P_L (Fig. 2).

Expression of the Prokaryotic trpA Gene Inserted into Different Sites of pHUB1, pHUB12, pHUB2, and pHUB4

To measure the expression of a prokaryotic gene under the influence of P_L, the *trpA* gene of *Salmonella typhimurium* or *Shigella dysenteriae* was inserted into the *Eco*RI, *Bam*HI, or *Sal*I site of pHUB1, pHUB12, pHUB2, or pHUB4. An *Eco*RI fragment of *S. typhimurium,* a *Bgl*II-*Sal*I fragment, or a Sal I fragment of *S. dysenteriae* was used for these experiments. An example of the construction of a hybrid plasmid for testing expression of a gene insert is shown in Fig. 3. pHUB4 DNA and φ80 phage DNA containing *S. dysenteriae trpE-A* genes were cleaved with *Sal*I, mixed, ligated, and transformed into C600(λ)*trpA33*. All transformants selected for resistance to kanamycin and trp^+ contained pHUB4 with a *Sal*I insert. Since the orientation of *trpA* on the insert was known relative to an asymmetric internal *Hpa*I restriction site, plasmids with P_L and *trpA* having the same orientation could easily be identified.

Table I summarizes the results of the expression experiments. Expression of *trpA* is increased upon heat induction in all combinations of vehicle, host, and restriction site used for insertions when the *trpA* gene is in the correct orientation with respect to transcription starting from P_L. The determined enzymatic activities of the *trpA* protein were reproducibly obtained in different experiments. Quantitative differences seemed to be dependent on the restriction site and on the plasmid that were used. When

[12] C. Epp and M. L. Pearson, *in* "RNA Polymerase" (R. Losick and M. Chamberlin, eds.), p. 667. Cold Spring Harbor Lab., Cold Spring Harbor, New York, 1976.

[13] N. Franklin and C. Yanofsky, *in* "RNA Polymerase" (R. Losick and M. Chamberlin, eds.), p. 693. Cold Spring Harbor Lab., Cold Spring Harbor, New York, 1976.

[14] R. Meyer and D. R. Helinski, in preparation.

TABLE I

P_L-DEPENDENT EXPRESSION OF *trpA* FROM DIFFERENT VEHICLES AT DIFFERENT TIME POINTS AFTER HEAT-INDUCING P_L[a]

		trpA activity [units/mg protein]				
Plasmid	Host	0 min	60 min	120 min	180 min	240 min
pHUB1*trpA*(*Eco*RI)	K12ΔHI	0	9	29	108	
pHUB12*trpA*(*Eco*RI)	C600*trpA33*	1		88		135
pHUB2*trpA*(*Eco*RI)	K12ΔHI	2	20	52	95	114
pHUB2*trpA*(*Eco*RI) plus RK248*cIts*	C600*trpA33*	1		94		86
pHUB2*trpCBA*(*Bam*HI)	K12ΔHI	3	105		100	
pHUB4*trpCBA*(*Bam*HI)	K12ΔHI	29	169		329	
pHUB4*trpDCBA*(*Sal*I)	K12ΔHI	15	51		63	

[a] The designation of the plasmids includes the restriction site that was used for insertion of the fragment containing the *trpA* gene. The relatively high expression from the last two plasmids at $t = 0$ is presumably due to the *trp* internal promoter p_2.[15] The specific activity of the pure *trpA* protein is 5000 units/mg protein.[13]

observed activities are related to the activity of the pure *trpA* protein,[15] it can be estimated that 1.2–6.6% of the total soluble cell protein was expressed in the form of the *trpA* enzyme. When the *trpA* gene is oriented opposite P_L, *trpA* protein was not detected after heat-activating P_L. In cases where the insert contained the *trp* operon internal promoter p_2,[10] p_2-dependent *trpA* expression was reduced when P_L-dependent transcription occurred in the opposite orientation (Figs. 3[6,16] and 4).

Application of the pHUB Plasmids

Though the effectiveness of the described plasmids has been tested largely by measuring the P_L-dependent expression of *trpA*, the results suggest that the application of these plasmids can be extended to other prokaryotic genes. These cloning vehicles may be particularly useful in expressing efficiently genes that in their natural genetic environment are not expressed to a level sufficient for enzymological or protein chemical analysis.

Normally, the availability of restriction sites that encompass a gene to be cloned determine which plasmid and which site may be useful for cloning and expression of the gene from P_L. In the case of cloning an *Eco*RI fragment, pHUB12 may be preferable. Although pHUB1 and pHUB2 in K12ΔHI*cI857* or pHUB2 plus RK248*cIts* directs the expression of comparable levels of the gene product from the insert, pHUB12 specifies the production of its own repressor. For insertions into the

[15] T. E. Creighton and C. Yanofsky, this series, Vol. 17, Part A, p. 365.
[16] C. Yanofsky, personal communication.

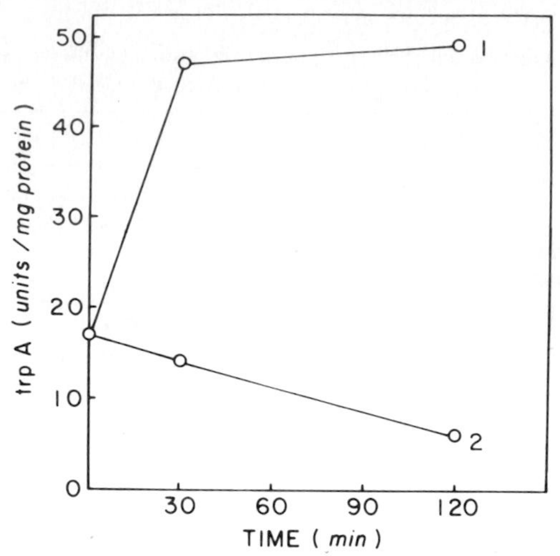

FIG. 4. Expression of *trpA* gene by transcription starting from P_L of pHUB4*trpDCBA*(*Sal*I)1 and 2 in the host K12ΔHI*cI857*. Before P_L is heat-induced ($t = 0$) *trpA* is expressed by the *trp* internal promoter independent of the relative orientation of the *Sal*I insert and the vehicle. Transcription from P_L boosts this expression of *trpA* in the case of pHUB4*trpDCBA*(*Sal*I)1 where P_L and *trpA* have the same direction. Transcription from P_L appears to interfere with the *trp* internal promoter when P_L and *trpA* are facing toward one another (curve 2).

*Bam*HI site (*Bam*HI and *Bgl*II fragments), pHUB4 was found to give a higher level of gene product than pHUB2. For *Sal*I insertions, only pHUB4 has been tested.

When pHUB1, -2, or -4 is used as the cloning vehicle, the host strain that normally contains a deletion mutation of the gene to be inserted into the plasmid should be first transformed with plasmid pRK248*cIts* by selecting for Tet^R transformants at 30°. The selection strain, carrying pRK248*cIts*, can then be used for transformation with the ligated mixture of the vehicle and the insert of choice.

In the expression experiment described in the previous section selection of the insert made use of the *trp* internal promoter p_2.[10] Though this promoter is absent from the *Eco*RI fragment of *S. typhimurium* used in this study, it was found that *trpA* on this fragment could be selected at 30°. This is possibly due to a weak unspecific promoting activity of DNA sequences on this fragment. Transformants selected under these conditions were visible after 48 hr at 30°. It is possible that selection due to unspecific transcription may frequenlty be possible in the case of DNA fragments containing promoterless genes. If the orientation of a gene on a restriction fragment is not known, expression experiments can be done with both orientations of the fragment relative to P_L (as determined by any asymmetric restriction site in the inserted fragment) with the possibility that the expression data may reveal the orientation of the gene (Fig. 4).

[36] Use of Minicells for Bacteriophage-Directed Polypeptide Synthesis

by JOHN REEVE

The major problem in analysis of bacteriophage-directed polypeptide biosynthesis *in vivo* is the concurrent synthesis of polypeptides encoded by the virus and host cell genomes. In some cases the infection results in a rapid inhibition of host cell syntheses, e.g., T5,[1] and in these cases the concurrent synthesis of host and viral polypeptides only occurs very early in the infection. A technique frequently used to obviate this problem is to destroy the host cell DNA by uv light irradiation[2] before infection, so that only the phage DNA is intact and can be expressed. This technique has been successfully and extensively used for the analysis of λ and λ transducing phages. Uv irradiation does not, however, permit an analysis of the *Bacillus subtilis* phage SPP1,[3] as the irradiation destroys the ability of the cell to express both the host and the subsequently infecting viral DNA. Attempts to analyze the expression of Mu phage in uv-irradiated cells have also had only limited success.[4] In addition to the limitation that the uv irradiation technique is not generally applicable, it is clear that the expression of a phage genome in an uv-damaged cell may be different from its expression in a nonirradiated cell.

Infection of anucleate minicells provides a system in which no host DNA is present and uv irradiation is unnecessary.[5] It has been reported that the presence of a plasmid within the minicells greatly stimulates phage expression,[6,7] however, it is not an absolute requirement and successful expression of phages T1,[8] T2,[9] T4,[9] T5,[5] T6,[9] T7,[10] K3,[9] P1,[11] P4,[12] λ,[5,7] Mu,[13] D108,[13] SP01,[14] SP82G,[14] 2C,[15] ϕe,[15] SP50,[15] SPP1,[14,16] SP02,[5]

[1] D. J. McCorquodale and J. M. Buchanan, *J. Biol. Chem.* **243,** 2550 (1968).

[2] M. Ptashne, *Proc. Natl. Acad. Sci. U.S.A.* **57,** 306 (1967).

[3] H. Esche, M. Schweiger, and T. A. Trautner, *Mol. Gen. Genet.* **142,** 45 (1975).

[4] M. Giphart-Gassler and P. van de Putte, *J. Mol. Biol.* **120,** 1 (1978).

[5] J. N. Reeve, *Mol. Gen. Genet.* **158,** 73 (1977).

[6] K. J. Roozen, R. G. Fenwick, Jr., and R. Curtiss, III, *J. Bacteriol.* **107,** 21 (1971).

[7] E. Grzesiuk and K. Taylor, *Virology* **83,** 329 (1977).

[8] J. N. Reeve, unpublished observation (1977).

[9] P. Manning and J. N. Reeve, unpublished observation (1977).

[10] H. Ponta, J. N. Reeve, M. Pfennig-Yeh, M. Hirsch-Kauffmann, M. Schweiger, and P. Herrlich, *Nature (London)* **269,** 440 (1977).

[11] P. Herrlich, E. Lauppe, E. Lanka, H. Schuster, and J. N. Reeve, unpublished observation (1977).

[12] K. Barrett and J. N. Reeve, unpublished observation (1978).

[13] M. Magazin, J. N. Reeve, S. Maynard-Smith, and N. Symonds, *FEMS Microbiol. Lett.* **4,** 5 (1978).

ISBN 0-12-181968-X

ϕ105,[5] ϕ29,[5] and ϕ15[8] in non-plasmid-containing minicells has been demonstrated. The expression of nonviral DNA carried by λ transducing phages[5] or by λ cloning vectors[8] can also be analyzed in infected minicells.

Minicell-Producing Strains

Minicell-producing mutants have been isolated from *Escherichia coli,* several *Salmonella* species, *Erwinia amylovara,* a marine pseudomonad, *B. subtilis,* and *Haemophilus influenzae.*[17] Many genetically constructed derivatives of the originally isolated *Escherichia coli* and *B. subtilis* minicell-producing mutants are available.[17] It is important to note that many minicell-producing strains of *E. coli* are direct derivatives of the original strain P678-54[18] which has the complex genotype: $-F^-$ *thr ara leu* azi^R $tonA^R$ *lacY* $T6^S$ *minA gal* λ^- *minB* str^R *malA xyl mtl thi sup.* A nonsuppressing strain, DS410, was constructed by D. Sherratt by selecting a gal^+ str^R exconjugant after conjugation with the donor Hfr CSH74. This strain is also thr^+ leu^+ and colicin E-sensitive.[19] A spontaneous revertant of this strain capable of growth using maltose as sole carbon source was selected. This strain is permissive for λ development and has been used in the majority of phage infections of *E. coli* minicells.[5,13]

There are two gentically distinct loci (*divIVA* and *divIVB1*) which independently result in minicell production in *B. subtilis.*[20] *divIVB* mutants have been exclusively used in phage infection experiments.[5,14–16] The strain used, CU403 (*thyA thyB metB divIVB1*), has been chosen because (1) it is capable of growth in glucose minimal medium supplemented with only thymine and methionine (no additional casamino acids are required as is usual for *B. subtilis*), (2) it does not readily autolyze during centrifugation through sucrose gradients providing the temperature is kept below 4°, and (3) it is a spontaneously occurring asporogeneous derivative. If a spore-producing strain is used, the spores may contaminate the minicell preparations, as spores and minicells are not separated by the sucrose gradients used to separate minicells from parental cells.

---

[14] J. N. Reeve, G. Mertens, and E. Amann, *J. Mol. Biol.* **120,** 193 (1978).

[15] E. Amann, unpublished observation (1978).

[16] G. Mertens, Ph.D. Thesis, Freie Universität, Berlin (1978).

[17] A. C. Frazer and R. Curtiss, III, *Curr. Top. Microbiol. Immunol.* **69,** 1 (1975).

[18] H. I. Adler, W. D. Fisher, A. Cohen, and A. A. Hardigree, *Proc. Natl. Acad. Sci. U.S.A.* **57,** 321 (1967).

[19] D. Sherratt, personal communication (1978); *Mol. Gen. Genet.* **151,** 151 (1977).

[20] J. N. Reeve, N. H. Mendelson, S. I. Coyne, L. L. Hallock, and R. M. Cole, *J. Bacteriol.* **114,** 860 (1973).

Growth Conditions

Escherichia coli strains produce minicells in all media tested (minimal salts, L-broth, nutrient broth, etc.), and the minicells can be separated from the parental cells without difficulty. In contrast, minicells can only be separated from parental *B. subtilis* cells following growth in minimal medium and preferably at 30° rather than 37°. Only under very slow growth conditions do the individual cells of *B. subtilis* completely separate. Minicells are produced in all media, but in rich media the cells and minicells form chains and cannot be separated by sucrose gradients. A separation can be obtained by exposing the culture to sonication,[21,22] however, the resulting cell wall debris contaminates the minicell preparations, and in subsequent experiments many phages adsorb to this material and not to the minicells.

Minicells are produced at all stages of growth and are usually isolated from cultures just entering the stationary growth phase. In the case of *E. coli* it does not appear to affect significantly the separation of minicells and parental cells or the results of the subsequent minicell infection if a culture is allowed to incubate for several hours at 37° in the stationary phase.[8] The separation of *B. subtilis* cells and minicells is much more difficult if the culture is not harvested immediately upon entering stationary phase, as the parental cell size decreases rapidly and the size difference between minicells and parental cells is therefore much reduced.[8,15]

Minicell Purification

Many techniques have been described and have been recently reviewed.[17] It is important for phage infection experiments that the technique employed does not produce cell wall debris (e.g., sonication,[21,22] penicillin lysis[23]), as this material is difficult to separate from minicells. The technique described below is that which we routinely use; it is not designed for maximum minicell recovery, but it rapidly (~2 hr) separates minicells from large volumes (2 liters) of parental cells with a minimum of handling and produces minicell populations containing one colony-forming unit per 10^4-10^5 minicells. The exposure of cells to high concentrations of sucrose reduces the colony-forming efficiency,[6] and therefore this assay is of limited value. As a simple test for contamination of minicell preparations by parental cells, all preparations are checked by phase-contrast light microscopy at a minicell concentration of $2-5 \times 10^9$ minicells/ml. The observation of 1 cell per microscope field (using 400×

[21] G. G. Khachatourians and C. A. Saunders, *Prep. Biochem.* **3,** 291 (1973).
[22] N. H. Mendelson, J. N. Reeve, and R. M. Cole, *J. Bacteriol.* **117,** 1312 (1974).
[23] S. B. Levy, *J. Bacteriol.* **103,** 836 (1970).

magnification) indicates a cell concentration of approximately 1×10^6 cells/ml. This cell concentration is at least 10-fold higher than is acceptable for the majority of experiments in which radioactive labeling of phage-encoded products is the ultimate objective.

The cells and minicells are removed from a 2-liter suspension by centrifugation at 7000 rpm in a Sorvall GS-3 rotor for 10 min at 4° (8200 *g*). The pellets are resuspended in a total of 20 ml of the supernatant. This resuspension must be very thorough, as any remaining clumps of cells and minicells will travel to the bottom of the subsequent sucrose gradients. We use a magnetic stirring bar and vigorously mix the suspension for 10 min at 4°. The suspension is layered on top of four 30-ml sucrose gradients (approximately 10–30% sucrose) held in cellulose nitrate ultracentrifuge tubes (designed for a Beckman SW27 rotor). The sucrose is dissolved in the appropriate minimal salts medium used for the subsequent phage infections (e.g., RM salts[24] for minicells to be used in λ infection). The gradients are produced by freezing, in a vertical position, 30 ml of a solution of 22% w/v sucrose in each tube and allowing the frozen solution to thaw slowly in the refrigerator (4°) overnight. This freeze–thaw technique allows large numbers of gradients to be made (and stored) without the use of a sucrose gradient-making device. The sucrose gradients are centrifuged for 20 min at 5000 rpm in a Sorvall HB-4 rotor at 4° (4000 *g*). The cellulose nitrate tubes do not exactly fit in the rotor, and an adapter (made from high-strength plastic, not commercially available) is needed to prevent the tubes from collapsing during centrifugation. Other types of tubes can be employed, however, cellulose nitrate tubes have the advantage that they are transparent and that a syringe (10 ml) can be used to remove the minicell band. The needle is pushed through the side of the tube slightly above the bottom of the minicell band, and the minicells are drawn into the syringe. The extreme bottom and top of the minicell band are not collected, as they usually contain parental cells and lysed cells, respectively, in addition to minicells. If nondisposable tubes are used, then the minicell band can be removed from above by use of a syringe or a Pasteur pipette with its tip bent into a J shape.

The minicells are removed from the sucrose solution by centrifugation at 13,000 rpm in a Sorvall SS34 rotor at 4° (20,000 *g*) for 10 min. The pellets are resuspended in a total of 5 ml of the appropriate minimal salts solution, and this suspension is layered onto two 30-ml sucrose gradients as described above. The sucrose gradient centrifugation is repeated, the resulting minicell bands are collected, and the minicells are removed from the suspension by centrifugation and resuspended in a known volume (usually 5 ml) of minimal salts solution. A small sample (0.1 ml) of the

[24] H. Murialdo and L. Siminovitch, *in* "The Bacteriophage Lambda" (A. D. Hershey, ed.), p. 711. Cold Spring Harbor Lab., Cold Spring Harbor, New York, 1971.

minicells is diluted to an A_{660} of 0.2–0.5, and parental cell contamination is checked by phase-contrast microscopy. If the separation of minicells and parental cells is satisfactory, then the optical density of the minicell suspension is accurately measured. The minicells are pelleted by centrifugation at 8000 rpm (7700 *g*) in the SS34 rotor for 10 min at 4°, and the purified minicells are finally resuspended in a known volume of minimal medium containing 30% v/v glycerol. This suspension is divided into small aliquots (usually 1 ml of 2×10^{10} minicells/ml; equivalent to an A_{660} of approximately 2), and these tubes are stored, either at −70° or immersed in liquid nitrogen. Minicells stored in liquid nitrogen have maintained their ability to synthesize RNA and protein for over 1 year.[8] If the level of parental cell contamination is unacceptable after two successive sucrose gradients, a third gradient may be used to improve the separation.

Phage Preparation

Minicell infections are usually carried out in a volume of 0.1 ml containing 2×10^9 minicells. It is therefore necessary to have a sufficiently high-titer phage stock such that addition of phage does not dilute the minicell suspension excessively. The procedure of phage purification and concentration through CsCl density gradients is normally employed.

Minicell Infection

The frozen minicell suspensions are allowed to thaw at room temperature and then diluted 10-fold with the appropriate minimal medium to be used for phage adsorption, e.g., RM medium for λ.[24] [D-Cyclosine (20 μg/ml) is added to all media to prevent cell wall biosynthesis[25,26] and inhibit growth of contaminating parental cells.] The minicells are pelleted and usually resuspended at a concentration of 2×10^{10} minicells ml. Minicell suspensions may be kept on ice for several hours in this state without loss of activity.

Minicells are inactivated by phage ghosts[27] and, as many phage preparations contain particles that cannot be measured as plaque-forming units, it is advisable to determine the optimum ratio of phage to minicells by measuring the stimulation of uridine incorporation by phage infection of minicells. A fixed volume of minicells (50 μl, 2×10^{10} minicells/ml) is mixed with a range of phage volumes, and the incorporation of radioactive uridine into trichloracetic acid (TCA)-precipitable material is measured. The infected minicells are mixed with 2 μl of [5-^{3}H]uridine (at high

[25] G. Mertens and J. N. Reeve, *J. Bacteriol.* **129,** 1198 (1977).
[26] J. N. Reeve, *J. Bacteriol.* **131,** 363 (1977).
[27] M. Vallée, J. B. Cornett, and H. Bernstein, *Virology* **48,** 766 (1972).

TABLE I
T7 RNA POLYMERASE SYNTHESIZED IN MINICELLS OF ICREASING AGE

Length of incubation before T7 infection (hr)	Rifampicin-resistant RNA synthesis[a] (% of zero time)
0	100
1	158
2.5	214
5	159
7	140
12	89
26	21
33	7
48	1

[a] T7 infected minicells were incubated for 40 min at 37°, and the incorporation of [5-^{3}H]uridine into TCA-precipitable material was measured for 5 min between 40 and 45 min after infection.

specific activity, 1–30 Ci/mmol, 1 μCi/μl) plus 0.5 μl of Difco nutrient broth. This mixture is incubated for 30 min at 37°; then 20 μl of 1 mg lysozyme/ml of TES (50 m*M* Tris-HC1, 5 m*M* EDTA, 100 m*M* NaCl, pH 7.6) is then added and incubation continued for 2 min at 37°. The incorporation is stopped, and the minicells lysed by the addition of 20 μl of 2% w/v sarkosyl in TES and 1 ml of 5% ice-cold TCA. The radioactivity in the TCA-precipitable material is measured, and the optimum ratio of phage to minicells is determined. The incorporation of uridine by an uninfected minicell sample is measured as a control. Minicells infected by an excess of phage often incorporate less uridine than the uninfected control minicells. It is still unclear what material is labeled in uninfected minicells, however, the majority of this synthesis is rifampicin-resistant.[8,28]

Minicells and phage are mixed at the optimum ratio and to label radioactively the phage-encoded polypeptides 5 μl of [^{35}S]methionine mixture per 100 μl of infected minicells is added. (Very high specific activity [^{35}S]methionine, up to 1000 Ci/mmol, is diluted 1:10 with Difco methionine assay medium to obtain the [^{35}S]methionine mixture.[29]) Alternatively ^{14}C-labeled protein hydrolysate may be used to label proteins. We normally obtain two- to fivefold more counts incorporated from the [^{35}S]methionine mixture as from the ^{14}C-protein hydrolysate, however, the short half-life of ^{35}S (87.2 days) precludes comparison of samples collected over a long period of experimentation. *Escherichia coli* minicell lysis, preparation of the samples for electrophoresis, polyacrylamide gel electrophoresis, fluorography of the dried gels, etc., are accomplished by

[28] J. N. Reeve and J. B. Cornett, *J. Virol.* **15,** 1308 (1975).
[29] S. B. Levy, *J. Bacteriol.* **120,** 1451 (1974).

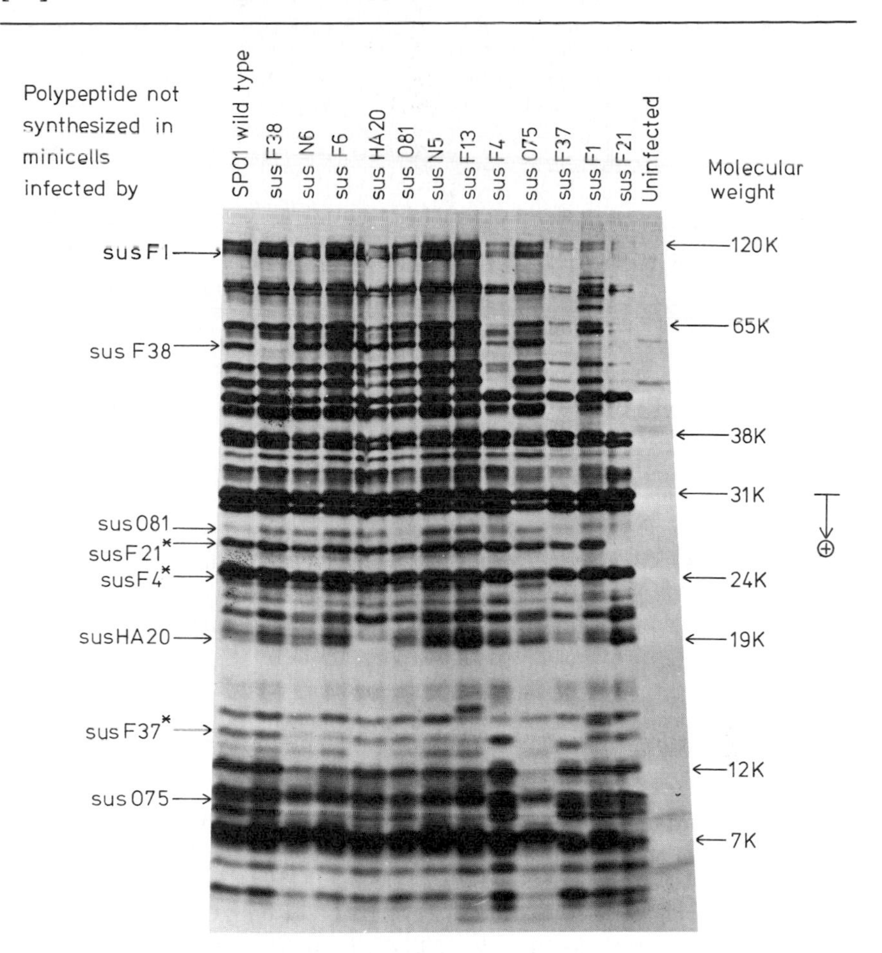

FIG. 1. Autoradiogram showing the polyacrylamide gradient (11–20%) gel electrophoretic separation of polypeptides synthesized in minicells of *B. subtilis* infected by suppressor-sensitive mutants of SP01. The mutants *susF4, susF21,* and possibly *susF37* are pleiotrophic (*), and the arrow to the left of the figure indicates the polypeptide thought to result from the cistron containing the mutation.

the standard techniques used for analysis of radioactively labeled polypeptides synthesized in bacterial cells.[30–32] The minicells produced by *B. subtilis* do not readily lyse in the normal sample lysis buffers, and it is advisable to pellet and resuspend the minicells in 0.5 mg of lysozyme/ml water for 2 min at 37°. Following the brief lysozyme digestion the mini-

[30] U. K. Laemmli, *Nature* (*London*) **227,** 680 (1970).

[31] B. Lugtenberg, J. Meijer, R. Peters, P. van der Hoek, and L. van Alphen, *FEBS Lett.* **58,** 254 (1975).

[32] W. M. Bonner and R. H. Laskey, *Eur. J. Biochem.* **46,** 83 (1974).

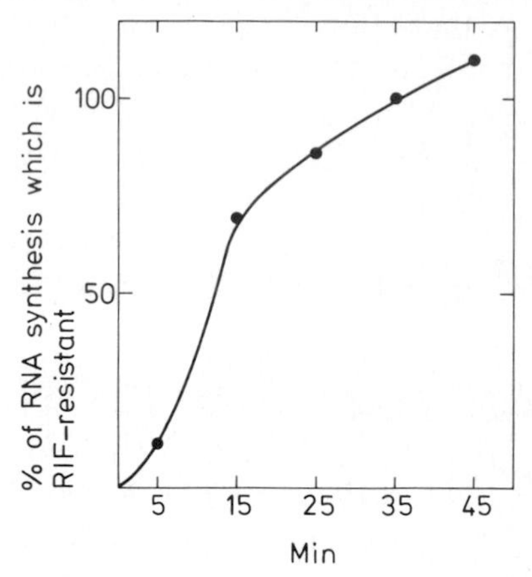

FIG. 2. Synthesis of T7 RNA polymerase in T7-infected minicells. *Escherichia coli* minicells were infected with T7 and incubated at 37°. The rate of RNA synthesis[10] at the times indicated after infection was measured in the presence and absence of rifampicin. Forty minutes after T7 infection of the minicells all RNA synthesis is rifampicin-resistant.

cells can be pelleted (1 min, Eppendorf Model 3200 centrifuge) and resuspended in lysis buffer[3] at 100° to lyse the minicells.

Experimental Applications

Infection of minicells can be used to analyze a number of parameters. The sequential expression of the phage DNA can be observed by infection of the minicells, followed by pulse-labeling of samples over an extended period. Pulse-chase experiments may be used to determine turnover of individual polypeptides. The rate of expression of phage DNA in minicells is much reduced as compared to that of infected cells,[10,14] and in some cases, e.g., λ or P1, little or no synthesis of phage products can be detected during the first 30 min of incubation.[8,11] In contrast to infected cells, infected minicells do not produce a burst of progeny phage and continue to synthesize phage products for many hours. Minicells, themselves, are remarkably stable and, even after extended periods of incubation (24–48 hr) at 37°, are still capable of synthesizing phage products following infection (see Table I). The majority of experiments are designed to label phage-encoded polypeptides, radioactively, and by the use of suppressor-sensitive mutants it is often possible to determine the gene product of a mutated cistron (Fig. 1). It is, however, also possible to detect the activity of phage-encoded products such as T7-encoded RNA polymerase.[8] Following infection of minicells by T7 the RNA synthesis within the minicells becomes 100% rifampicin-resistant, indicating activity of the T7-encoded rifampicin-resistant RNA polymerase and inhibition

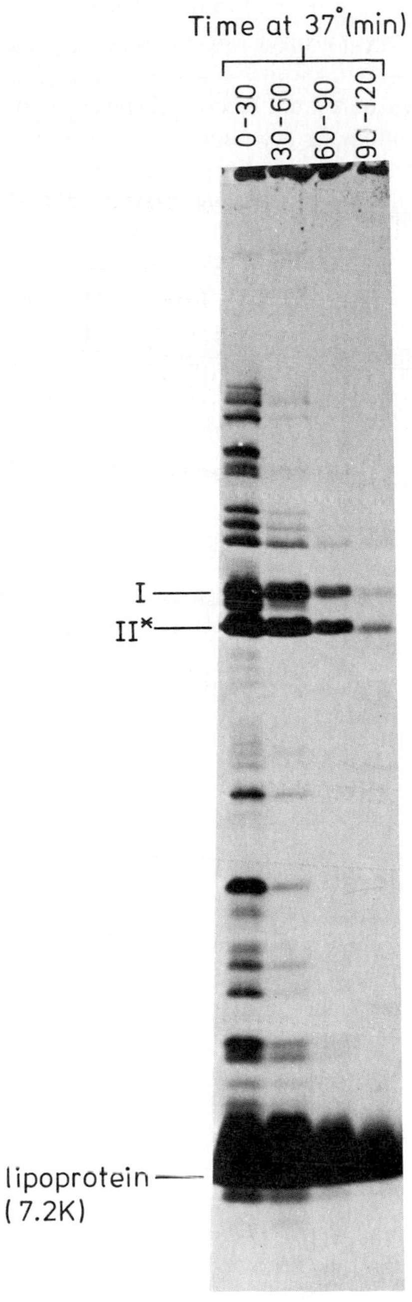

FIG. 3. Decay of endogenous mRNA activity in *E. coli* minicells. Minicells were isolated from an actively growing culture and incubated at 37° in the presence of the [^{35}S]methionine mixture at the times indicated. The radioactively labeled polypeptides synthesized result from the translation of mRNA trapped in the minicells at the time of minicell formation.[33,35] The polypeptides synthesized are mainly those of the outer membrane. The number and rel-

of the rifampicin-sensitive host RNA polymerase (Fig. 2). A simple technique in which the minicells are infected in the presence of chloramphenicol and after a period of incubation are removed from the chloramphenicol and incubated in the presence of rifampicin and [^{35}S]methionine has been described.[5,14] It permits direct identification of the products of cistrons transcribed by the nonmodified host cell RNA polymerase.

The radioactively labeled phage products may be further analyzed as to their location within the infected minicell, e.g., inner or outer cell membrane, cytoplasmic or ribosome-bound. The observations that membrane-bound[33] or ribosomal polypeptides[34] synthesized from λ transducing phages migrate to the correct locations within the minicell indicates that a cellular organization is present within the minicells and that they are not just "bags of enzymes."

Complications

Escherichia coli minicells may contain bacterial mRNA which will result in the synthesis of non-phage-encoded radioactive polypeptides.[33,35] This background can be removed by simply incubating the minicells at 37° before infection to allow degradation of the mRNA (Fig. 3), or the mRNA can be destroyed by a brief exposure to uv irradiation before phage infection.[36]

With the use of unidimensional electrophoretic analysis of the products of λ transducing or λ cloning phages it is often difficult to distinguish between λ polypeptides and non-λ polypeptides. Attempts to inhibit λ phage products by the use of λ lysogenic minicell-producing strains were unsuccessful, indicating that insufficient λ repressor entered the minicells.[37] Recent experiments in which minicells from a strain carrying the plasmid pKB280 (which results in 1% of the *E. coli* cytoplasmic cell protein being λ repressor[38]) were infected have provided a system specifically to permit the synthesis of non-λ polypeptides in the absence of λ polypeptides. The plasmid itself enters the minicells, and it is necessary to uv-irradiate the latter to destroy the synthetic capacity of the plasmid before phage infection.[39]

[33] J. N. Reeve and J. E. Shaw, *Mol. Gen. Genet.* **172,** 243 (1979).
[34] A Subramanian and J. N. Reeve, *FEBS Lett.* **95,** 265 (1978).
[35] S. B. Levy, *Proc. Natl. Acad. Sci. U.S.A.* **72,** 2900 (1975).
[36] P. Herrlich, J. Shaw, and J. N. Reeve, unpublished observation (1977).
[37] J. Shaw, C. Epp, M. Pearson, and J. N. Reeve, *Gene,* in preparation.
[38] K. Backman and M. Ptashne, *Cell* **13,** 65 (1978).
[39] J. N. Reeve, *Nature (London)* **276,** 728 (1978).

FIG. 3 (*continued*). ative amounts of the polypeptides can be varied by growth of the culture in different media.[8] The figure demonstrates that this background synthesis can be avoided by preincubation of the minicells before phage infection.

Summary

Phage infection of minicells provides a simple system to analyze the polypeptide (or RNA) products of a phage genome. In terms of this volume it is particularly important, as it provides a means to express *in vivo* the fragments of DNA cloned in λ vectors.[39] Phage infection of minicells, as opposed to expression of plasmids in minicells, has the advantage of a zero time and that the minicells can be handled in a variety of ways, e.g., inducing or noninducing media, different temperatures, etc., before being required to express the phage DNA. In addition it is not necessary to construct a new minicell strain for each experiment as is the case in the expression of DNA cloned on plasmids.

In terms of analysis of phage development, however, it should be remembered that significant differences exist between an infected cell and an infected minicell. Minicells are generally incapable of supporting significant amounts of phage DNA replication or producing progeny phage.[17,28,37,40,41]

[40] E. Amann and J. N. Reeve, *Biochem. Biophys. Acta* **520,** 82 (1978).
[41] G. Mertens, E. Amann, and J. N. Reeve, *Mol. Gen. Genet.* **172,** 271 (1979).

[37] Measurement of Plasmid Copy Number

By H. MICHAEL SHEPARD and BARRY POLISKY

All known extrachromosomal elements exist in well-defined intracellular copy numbers. For a sizable number of such elements, genetic evidence exists to support the idea that plasmid-specified information is involved in copy number control.[1,2] The ability to determine accurately the intracellular plasmid copy number is essential for analysis of plasmid- and chromosome-specified elements that influence this number.

Several approaches to the problem of determination of plasmid copy number are currently available. The most commonly employed method is centrifugation of total cell lysates[3] or cleared lysates[4] in cesium chloride gradients containing intercalating dye. These methods offer minimal estimates of plasmid copy number, because they require that all plasmid DNA be present as supercoils in the first instance, and that a quantifiable proportion of plasmid be recoverable after high-speed centrifugation to remove chromosomal DNA in the second case. Nicking of supercoils by shearing, the action of nonspecific endonucleases, or disruption of

[1] K. Nordström, L. C. Ingram, and A. Lundback, *J. Bacteriol.* **110,** 562 (1972).
[2] D. H. Gelfand, H. M. Shepard, P. H. O'Farrell, and B. Polisky, *Proc. Natl. Acad. Sci. U.S.A.* **75,** 5869 (1978).
[3] D. D. Womble, D. P. Taylor, and R. H. Rownd, *J. Bacteriol.* **130,** 148 (1977).
[4] D. B. Clewell and D. R. Helinski, *Proc. Natl. Acad. Sci. U.S.A.* **62,** 1159 (1969).

ISBN 01-12-181968-X

plasmid relaxation complexes,[5,6] causes plasmid DNA to band with chromosomal DNA in dye–buoyant density gradients. Similar problems reduce the accuracy of copy number estimates by alkaline sucrose gradient centrifugation.[7] If the plasmid sequences under study are sufficiently different from the chromosome with respect to base composition, cesium chloride isopycnic centrifugation of total cellular DNA will give a reliable estimate of plasmid copy number.[8,9]

An alternative approach to copy number determination that does not require measurement of plasmid DNA directly but rather depends upon gene dosage-dependent production of antibiotic-inactivating enzymes, has recently been demonstrated by Uhlin and Nordström.[10] This method has the limitation that it cannot be employed with plasmids that do not encode a drug-inactivating enzyme.

Determination of plasmid copy number by hybridization analysis of total DNA from plasmid-containing cells avoids the artifacts discussed above. Three methods have been employed: filter saturation hybridization,[11] reassociation kinetic determination of plasmid DNA concentration in solution,[12] and saturation hybridization in solution.[2] Inefficiency of filter hybridization, loss of DNA from filters during hybridization, and other difficulties[13,14] make this method the least desirable hybridization assay. Although the kinetics of reassociation of simple sequence DNA are relatively easy to determine, considerably more work must be done to obtain convincing kinetic data than is necessary for the determination of saturation plateaus. The saturation hybridization method we describe here allows for the determination of plasmid copy number for several different samples simultaneously. It is simple to use, and the results are unambiguous.

In brief, total intracellular DNA from plasmid-containing cells is isolated and labeled to high-specific activity *in vitro* by nick translation with DNA polymerase I. Labeled, single-stranded DNA tracer is then driven into hybrid form with excess purified plasmid DNA. Because of the low sequence complexity of plasmid DNA, the fraction of labeled DNA tracer complementary to plasmid DNA is rapidly driven into hybrid form. Progress of the reaction is monitored by hydroxyapatite chromatography.

[5] M. A. Lovett and D. R. Helinski, *J. Biol. Chem.* **250,** 8790 (1975).
[6] R. J. Rowbury, *Prog. Biophys. Mol. Biol.* **31,** 271 (1977).
[7] B. E. Uhlin and K. Nordström, *Mol. Gen. Genet.* **165,** 167 (1978).
[8] B. Weisblum, M. Y. Graham, T. Gryczan, and D. Dubnau, *J. Bacteriol.* **137,** 635 (1979).
[9] D. Perlman and R. Stickgold, *Proc. Natl. Acad. Sci. U.S.A.* **74,** 2518 (1977).
[10] B. E. Uhlin and K. Nordström, *Plasmid* **1,** 1 (1977).
[11] S. O. Warnaar and J. A. Cohen, *Biochem. Biophys. Res. Commun.* **25,** 554 (1966).
[12] S. Austin, N. Sternberg, and M. Yarmolinsky, *J. Mol. Biol.* **120,** 297 (1978).
[13] D. Gillespie, this series, Vol. 12, Part B, p. 641.
[14] R. J. Britten, D. E. Graham, and B. R. Neufeld, this series, Vol. 29, p. 363.

The fraction of labeled DNA tracer bound to hydroxyapatite at the conclusion of the reaction is equivalent to the fraction of total DNA that is plasmid-specific.

Methods

Isolation of Cellular DNA

The Triton X-100 lysis method (Table I; Clewell and Helinski[4]), without the high-speed centrifugation step to remove chromosomal DNA, gives a good yield of plasmid DNA. Modifications are described below. After Triton X-100 is added and lysis is complete, the chromosomal DNA in the lysate is sheared by trituration with a 10-ml pipette. This step is required to reduce the viscosity of the lysate sufficiently to make it easily extractable with phenol. Following shearing, the lysate is extracted twice with 1 volume of redistilled phenol saturated with 0.1 *M* Tris-HCl, pH 8.0. Most of the phenol is then removed from the aqueous phase by extracting it twice with an equal volume of chloroform. Then, a 0.1 volume of 3 *M* sodium acetate is added to the aqueous phase, and nucleic acid is precipitated with 2 volumes of absolute ethanol for 1 hr or more at $-20°$. Following collection of the precipitate by centrifugation at 7500 rpm for 15 min in a Sorvall GSA rotor, or an equivalent, the pellets are air-dried. The nucleic acid is then resuspended to about 1 mg/ml in TE (10 m*M* Tris-HC1, pH 8, 0.1 m*M* EDTA). Heat-treated (60° for 30 min) RNase A (10 mg/ml in TE) is added to 50 μg/ml. The DNA is then dialyzed overnight at 4° versus 100 volumes of TE. Following dialysis, RNase A is removed by a single phenol and a single chloroform extraction. After ethanol precipitation, pellets are dissolved in a minimal volume of TE and

TABLE I
TRITON X-100 LYSIS PROTOCOL

1. Innoculate 0.1–0.5 liter of medium with a 1:100 dilution of fresh overnight and grow to late exponential phase.
2. Pellet cells at 4° in 250-ml polyallomer centrifuge bottles at 7000 rpm in a Sorvall GSA rotor, or the equivalent.
3. Freeze pellets in a dry ice–acetone bath.
4. Resuspend pellets in 5 ml of 50 m*M* Tris-HCl, pH 8.0, and 25% sucrose. Transfer to 30-ml Corex centrifuge tubes. Incubate on ice for 5 min.
5. Add 1.25 ml of 50 m*M* Tris-HCl, pH 8.0, 25% sucrose, and 6 mg/ml lysozyme. Incubate at 0° for 20 min.
6. Add 2.5 ml of 0.5 *M* EDTA, pH 8.0, mix gently, and leave on ice for 5 min.
7. Add 8.3 ml of Triton lytic mix (25 m*M* Tris-HCl, pH 8.0, 31.3 m*M* EDTA, 0.2% Triton X-100) and gently mix by covering with Parafilm and inverting. Incubate at 0° for 15 min. The remainder of the protocol is described in the text.

dialyzed at room temperature for 4 hr against TE to remove traces of ethanol or salts that may interfere with *in vitro* labeling. A final DNA concentration of 1 mg/ml or greater is convenient at this point. The presence of significant levels of RNA in the preparation can be tested conveniently by electrophoresis of 1–2 μg of the purified DNA in a 0.8% agarose gel for 1 hr at 75 V as described.[2] The presence of contaminating protein or phenol can be determined spectrophotometrically by measuring absorbance at 260, 270, and 280 nm. The ratio of absorbance at 260 nm to that at 280 nm should be about 2.0 for protein-free DNA, and that at 270 nm to that at 260 nm should be about 0.8 for phenol-free preparations. Approximately 12 mg of DNA is usually obtained by these procedures from a 500-ml culture of *E. coli* strain DG75[15] grown in glucose–minimal medium to late exponential phase.

In Vitro Labeling of DNA with Escherichia coli DNA Polymerase I

In vitro labeling with DNA polymerase I is done after potentiating the template by the introduction of single-stranded nicks with DNase I[16,17] or by forming hyperpolymers of sheared DNA.[18] We have employed both methods with equal success. The principal difficulty with the DNase I technique is proper calibration of the enzyme/substrate ratio to give DNA fragments of a size appropriate for hybridization studies [250–600 base pairs (bp)]. Similarly, in employing hyperpolymers for nick translation, proper conditions for shearing DNA to the appropriate size in a Virtis blender must be determined, as well as the additional step of reassociation of the sheared DNA to produce the hyperpolymeric substrate required for efficient labeling.

DNase I Method. The description that follows is similar to that in Rigby *et al.*[17]

1. RNase-free DNase (Worthington or Boehringer-Mannheim) is dissolved to 1 mg/ml in 0.01 *M* HCl. It is then divided into conveniently sized aliquots and stored frozen at −20°. Enzyme stored in this way gives reproducible results for several months. 10× deoxyribonucleoside triphosphates (4×10^{-4} *M* dATP and dCTP, 2×10^{-3} *M* dGTP, P-L Biochemicals) are prepared by dissolving them to 20× in sterile, deionized water and then adding an equal volume of 100% ethanol. The triphosphate

[15] B. Polisky, R. J. Bishop, and D. H. Gelfand, *Proc. Natl. Acad. Sci. U.S.A.* **73,** 3900 (1976).

[16] F. H. Schachat and D. S. Hogness, *Cold Spring Harbor Symp. Quant. Biol.* **38,** 371 (1974).

[17] P. W. J. Rigby, M. Dieckmann, C. Rhodes, and P. Berg, *J. Mol. Biol.* **113,** 237 (1977).

[18] G. A. Galau, W. H. Klein, M. M. Davis, B. J. Wold, R. J. Britten, and E. H. Davidson, *Cell* **7,** 487 (1976).

stocks are stored at −20°. The ^{3}H-dTTP (1×10^{-5} *M*, 60–100 Ci/mmol in 50% ethanol, ICN Chemicals) is used directly.

2. Following thawing to room temperature 25 μl of DNase is diluted into 225 μl of 10 m*M* Tris-HCl, pH 7.5, 5 m*M* $MgCl_2$, and 1 mg/ml bovine serum albumin (BSA) in a 500-μl Eppendorf plastic centrifuge tube. DNase I is then incubated on ice for 2 hr.

3. While DNase activation is occurring, the following mixture is prepared. For each 25 μl of nick translation reaction mixture add 2.5 μl of each 10× deoxyribonucleoside triphosphate and 25 μl of ^{3}H-dTTP to a small (2-ml) sterile shell vial. The vial is then placed, with a loosely fitting cap, in a vacuum dessicator and dried.

4. Following activation, the DNase I is diluted 10^3-fold to a final concentration of 0.1 ng/μl in the same diluent as above.

5. When calibrating the DNase I, 10 to 25-μl reactions can be prepared in 250-μl plastic centrifuge tubes with varying amounts of enzyme. We have found that an enzyme to pBGP120 [17.3 kilobase pairs (kbp)] plasmid DNA ratio of 0.2 ng/μg DNA gives double-stranded fragments, 75% of which are between 250 and 600 bp in length at completion of the labeling reaction. This DNA typically has a specific activity of $1-5 \times 10^6$ cpm/μg.

6. For a final reaction volume of 25 μl and a DNA concentration of 40 μg/ml, 20.5 μl per reaction mix of 0.05 *M* sodium phosphate, pH 7.6, and 2.5 m*M* $MgCl_2$ is used to rinse the shell vial containing the 10× deoxyribonucleoside triphosphates and ^{3}H-dTTP. The mixture is then removed from the shell vial to 250-μl plastic centrifuge tubes. One microliter of DNA (1 mg/ml) and 2 μl of DNase I are added; the contents are mixed at room temperature and then placed on ice. A "zero-time" sample is then taken as described below.

7. Then 2.5 μl of *E. coli* DNA polymerase I (12.5 units, Boehringer-Mannheim) is added and mixed thoroughly with a pipette. The reaction is placed at 14.5°. Incorporation is usually linear for 3–5 hr under these conditions but should be monitored at 0.5 to 1.0-hr intervals. A sample incorporation curve is shown in Fig. 1.

8. The progress of the reaction is monitored by removing 1 μl of the reaction mixture with a calibrated 1- to 5-μl glass capillary and diluting into 1 ml of 0.1 *M* NaCl, 10 m*M* EDTA, pH8, containing 100 μg of BSA on ice. After mixing, 50 μl of the sample is pipetted directly onto a GF/C filter and dried (filter A). To the remainder is added 1 ml of 10% trichloroacetic acid (TCA). After 10 min the precipitate is collected on a GF/C filter, rinsed well with enthanol, and dried (filter B). Percent incorporation is determined from the following formula[19]: [(cpm filter B)/(cpm filter A)]

[19] G. A. Galau, R. J. Britten, and E. H. Davidson, *Proc. Natl. Acad. Sci. U.S.A.* **74,** 1020 (1977).

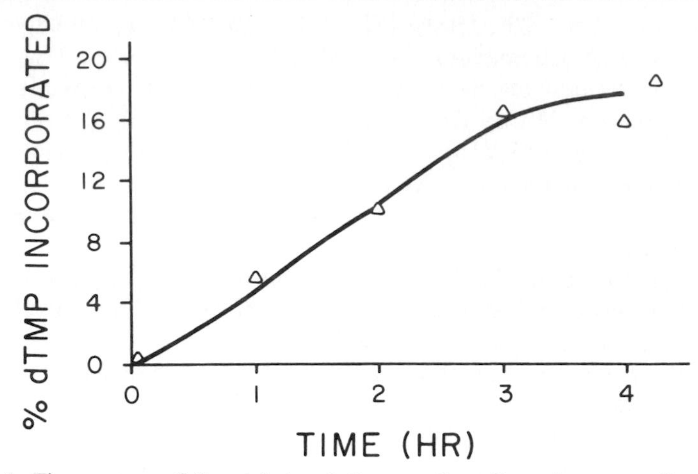

FIG. 1. Time course of the nick translation reaction. One microgram of supercoiled pBGP120 DNA was labeled by the DNase I technique. ^{3}H-dTMP incorporation was monitored as described in the text.

× 100. The percent substitution of dTMP by ^{3}H-dTMP can also be determined from these data.

Hyperpolymer Method. This protocol is essentially that described in Galau *et al.*[18] Fifty micrograms or more of DNA isolated as described above is sheared in 0.03 *M* sodium acetate, pH 8.0, and 67% glycerol in a dry ice–ethanol bath. The shearing is done for 30 min in a total volume of 50 ml in a Virtis 45 blender at a setting of "8." The DNA is then precipitated with ethanol, resuspended in a minimal volume of 0.4 *M* sodium phosphate buffer, pH 6.8 (PB), and dialyzed overnight against the same buffer to remove residual glycerol or other dialyzable contaminants.

Because the bacterial chromosome is 220-fold more complex than the largest plasmid used in these studies (pBGP120, 17.3 kbp) our calculation of incubation time required to form hyperpolymers is usually based upon the complexity of the *E. coli* chromosome (3.75 × 10^6 kbp) which has a $C_0t_{1/2}$ of about 7 mol of nucleotides per liter per second. Therefore, incubation of a DNA mixture to a C_0t of 140 (20× $C_0t_{1/2}$) or greater, would ensure hyperpolymer formation of both pBGP120 plasmid and DG75 chromosomal DNA.[14] This relationship obtains for single-copy plasmids of up to the size of Col V (1.4 × 10^2 kbp, or nearly 4% of the size of the bacterial chromosome; Rowbury[6]). The only mathematical relationship one needs is: (sequence complexity of the chromosome)/(sequence complexity of the plasmid) = ($C_0t_{1/2}$ of the chromosome)/($C_0t_{1/2}$ of the plasmid). A summary of the theory and derivation of the relationship between sequence complexity and reassociation kinetics (and its limitations) is given in Britten *et al.*[14,19]

We have carried out hyperpolymer formation to at least $20\times\ C_0t_{1/2}$ in 0.4 *M* PB. These conditions provide a 4.9-fold acceleration of the reassociation rate over that obtained in 0.12 *M* PB.[14] *Escherichia coli* chromosomal DNA 300–400 bp in length at a concentration of 1 mg/ml ($C_0 = 3.1 \times 10^{-3}$ mol of nucleotides per liter) would require (7 mol liter^{-1} sec)/(3.1×10^{-3} mol liter^{-1}) = 2.2×10^3 sec (37 min) in 0.12 *M* PB or about 8 min in 0.4 *M* PB at 60° to reach $C_0t_{1/2}$ and 2.6 hr to reach $20\times\ C_0t_{1/2}$ in 0.4 *M* PB. For these operations, then, it is clear that the need to form hyperpolymers as a substrate for *in vitro* labeling is not a significant obstacle.

4. In practice, hyperpolymerization is accomplished as follows: (a) 50 μl of sheared DNA at 1 mg/ml in 0.4 *M* PB is drawn into a 100-μl capillary pipette which is then sealed at both ends; (b) the capillary pipette is placed in a boiling water bath for 3 min; (c) the reaction is incubated at 60° for at least 2.6 hr; and (d) the contents of the capillary are expelled and dialyzed against 0.05 *M* sodium phosphate buffer, pH 7.6, and 0.1 m*M* EDTA overnight at 4°.

5. The DNA is then added to the *in vitro* labeling reactions as described above for the DNase I method. The progress of the reaction is monitored as described above.

Purification of Tracer from the Nick Translation Reaction

During *in vitro* labeling using either DNase I or hyperpolymeric DNA as substrate, 20–40% of ^{3}H-dTTP is incorporated into product. Twenty to thirty percent of this incorporation is commonly found to be rapidly renaturing sequences (fold-back DNA) which probably results from strand switching of the DNA polymerase during the reaction. Fold-back DNA and unincorporated radioactivity can be removed using hydroxyapatite chromatography as described below.

The nick translation reaction is diluted to 1 ml with 0.03 *M* sodium phosphate, pH 7.6, and 0.135 *M* NaCl containing 50 μg of purified sheared salmon sperm DNA in a capped vial. The mixture is heated for 5 min in a boiling water bath and then pipetted into a closed, water-jacketed column at 50° containing 0.2 g of DNA-grade hydroxyapatite. The construction of this column is described in Britten *et al.*[14] The reaction mix is incubated for 3 min to allow fold-back DNA to reanneal, the column is opened, and the mixture is forced through the resin with air pressure. Unincorporated radioactivity passes through the column, while the single-stranded tracer and duplex fold-back structures are adsorbed. The column is then thoroughly washed with 2-ml aliquots of loading buffer (at 50°) until background levels of radioactivity are observed in the eluate. The column is heated to 60°, and single-stranded DNA tracer is eluted

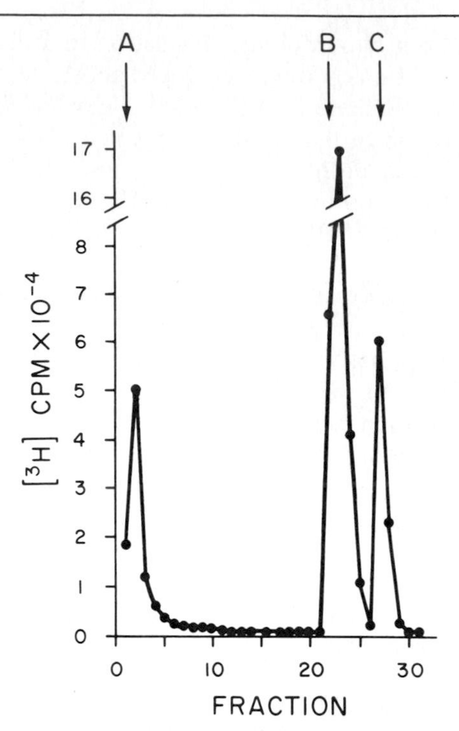

FIG. 2. Purification of single-stranded ^{3}H-labeled tracer DNA. The pBGP120 DNA nick translation mix (Fig. 1) was diluted to 1 ml in 0.03 *M* sodium phosphate, pH 7.6, and 0.135 *M* NaCl in a capped vial, boiled, and applied to hydroxyapatite as described in the text. Ten-microliter aliquots from each 2-ml fraction were added to 1 ml of Instagel, mixed thoroughly, and counted. Unincorporated radioactivity was removed by washing with 0.03 *M* sodium phosphate, pH 7.6, and 0.13 *M* NaCl at 50° (A). Single-stranded (tracer) DNA was eluted by raising the column temperature to 60° and washing with 2-ml aliquots of 0.12 *M* PB (B). Fold-back DNA was recovered by heating the column to 98° and washing with 0.12 *M* PB (C).

with 2-ml aliquots of 0.12 *M* PB at 60°. Fold-back DNA can be recovered by heating the column to 98° and eluting with 0.12 *M* PB at 98°. A typical elution profile is shown in Fig. 2. Fifty micrograms of sheared salmon sperm DNA is added to the pooled tracer fractions, and storage is at 4° over chloroform.

Preparation of Driver DNA

For estimation of tracer reactivity, DNA of the same sort used in the *in vitro* labeling reaction is employed. Because the plasmids investigated in our laboratory are sequence subsets of the pBGP120 plasmid, we use Virtis sheared DNA prepared as described above from *E. coli* containing pBGP120. These reactions are prepared as described above for hyperpolymer formation and are reassociated to 0, 20, 40, and 60 times the $C_0t_{1/2}$ of the *E. coli* chromosomal driver sequences in 0.4 *M* PB with 5–10 × 10^3 cpm of tracer input (i.e., a maximum of 10^{-2} μg). If 10 μg of homologous cellular DNA driver is included with tracer in a total volume of 20 μl, the reaction C_0 (of chromosomal DNA driver) will be 1.6 × 10^{-3} mol of nucleotides per liter and an incubation time of 5 hr will be required to reach 20× $C_0t_{1/2}$. It is convenient to make pools of all the necessary components for each series of time points and then withdraw samples for each time point into capillaries. Reactivities of tracers vary from 65 to 95%.

Hydroxyapatite Assay for Duplex Formation

Capillaries containing hybridization reaction mixtures are broken, and the contents mixed with 4 ml of 0.12 *M* PB containing 50 μg of sheared salmon sperm DNA in a water bath at 60°. Then 0.4 ml is removed into a scintillation vial containing 3.6 ml of 0.12 *M* PB for calculation of recovery. The remaining 3.6 ml is pipetted into a water-jacketed column maintained at 60° and containing 0.25 g of HAP. The column is run exactly as described in Britten *et al.*[14] Single-stranded (unreacted) DNA is eluted with two 4-ml washes of 0.12 *M* PB at 60°, while double-stranded (reacted) DNA is removed by heating the column to 98°. Fractions (4 ml

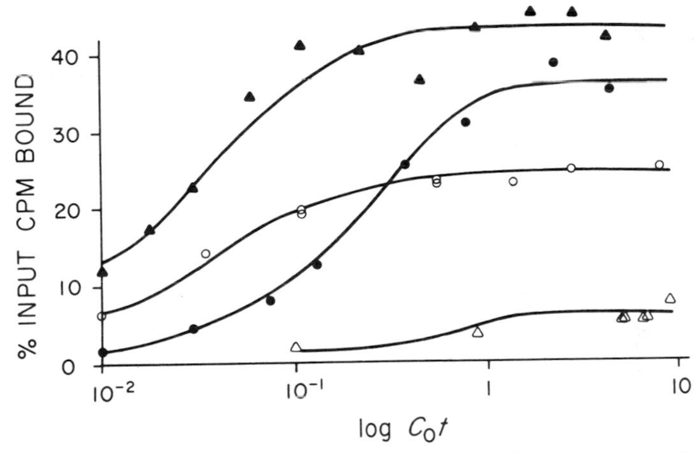

FIG. 3. Saturation hybridization analysis of cellular plasmid DNA content. *In vitro* labeled DG75, DG75/pBGP120, or DG75/p0P1Δ6 DNA was annealed with excess plasmid DNA to increasing driver DNA C_0t values. The percent of input radioactivity bound to hydroxyapatite at the reaction termini are corrected for different tracer reactivities. Tracer self-reaction has been subtracted. △, DG75/pBGP120; ▲, DG75/pBGP120 treated with chloramphenicol; ○, DG75/p0P1Δ6, ●, DG75/p0P1Δ6 treated with chloramphenicol. Sheared pBGP120 DNA was used as driver in all experiments, except that p0P1Δ6 DNA was employed as driver in the case of DG75/p0P1Δ6 (untreated).

each) are collected directly into scintillation vials and then cooled to room temperature. Finally, 10 ml of Instagel (Packard Instruments) is added and the samples are counted. Recovery of radioactivity from the column is 90% or more.

The Saturation Hybridization Assay

This assay works well because plasmid DNA has a much lower sequence complexity than chromosomal DNA. Consequently, very little reaction of nonplasmid sequences in tracer in the presence of sufficient plasmid driver is observed. In our experiments, pBGP120 serves as driver because our other plasmids are subsets of pBGP120 sequences. Because of the ratio of complexity of pBGP120 (17.3 kbp) to that of chromosomal DNA (3.8×10^3 kbp), the plasmid sequences react 2×10^2-fold faster at an equal sequence excess. In tracer with a specific activity of 5×10^6 cpm/μg, containing 15 copies of a 17.3-kbp plasmid per chromosome mass equivalent, about 6% of the input counts per minute is plasmid-specific. If 5×10^4 cpm of tracer is used, this will correspond to $3 \times$

TABLE II
DETERMINATION OF PLASMID COPY NUMBER FROM SATURATION HYBRIDIZATION DATA

Sample[a]	Tracer reactivity[b]	Fraction of DNA plasmid-specific[c]	P/C ratio[d]	Copy number[e]
1. DG75	68	0.015	0.01	
2. DG75/pBGP120	80	0.060	0.06	13
3. DG75/pBGP120 (plus chloramphenicol)	74	0.43	0.75	164
4. DG75/p0P1Δ6	82	0.250	0.33	180
5. DG75/p0P1Δ6 (plus chloramphenicol)	73	0.350	0.54	293

[a] Cells were harvested in late exponential phase, except samples 3 and 5, which were treated with chloramphenicol as described in reference (2). pBGP120 DNA served as driver for samples 1, 2, 3, and 5. p0P1Δ6 DNA served as driver for sample 4. Similar results were obtained for DG75/p0P1Δ6 plasmid copy number when pBGP120 DNA was used as driver.

[b] Determined as described in the text.

[c] After correcting for tracer self-reaction and normalizing reactivities to 100%.

[d] Ratio of the fraction of total DNA that is plasmid divided by the fraction of total DNA that is chromosomal. For example, in DG75/p0P1Δ6, 0.25 of intracellular DNA is plasmid-specific. Thus, the P/C ratio is $0.25/0.75 = 0.33$ in this case.

[e] The copy number is calculated from the obtained P/C ratio divided by the mass ratio if one plasmid copy existed per chromosome mass equivalent. The P/C ratio for pBGP120 (17.3 kbp) at one copy per chromosome (3.75×10^3 kbp) is 4.61×10^{-3}, and for p0P1Δ6 (6.9 kbp) it is 1.84×10^{-3}. With the use of these numbers the copy number of p0P1Δ6 in DG75 is $(0.33)/(1.84 \times 10^{-3}) = 180$.

10^3 cpm or 6×10^{-4} μg of plasmid DNA. One microgram of homologous plasmid driver in a 20-μl reaction mix containing 5×10^4 cpm of the above tracer represents more than a 1600-fold sequence excess. The C_0 (of pBGP120 plasmid DNA driver) would be 1.6×10^{-4} mol of nucleotides per liter, and in 0.12 *M* PB at 60° it would take about 67 min to reach $20\times C_0t_{1/2}$ (6.4×10^{-1} mol of nucleotides per liter), which is equivalent to about twice the amount of time required for the reannealing reaction to go to completion.[19] The tracer background is determined by incubating tracer in the presence of sheared salmon sperm DNA at the concentration used for plasmid drivers. It rarely exceeds 2.0% of the tracer input. Because of this latter observation, we have found that saturation hybridization experiments with satisfactory backgrounds can be obtained by using purified DNA from cleared lysates of chloramphenicol-amplified cells containing ColEl-derived plasmids. The amplification and cleared lysate preparation are done as described in Gelfand *et al.*[2] In general, however, CsCl–ethidium bromide-purified plasmid DNA is used. Sample saturation hybridization curves are shown in Fig. 3. A sample calculation of plasmid copy number from these results is shown in Table II.

[38] Localization of Enzymes in Denaturing Polyacrylamide Gels

By R. P. Dottin, R. E. Manrow, B. R. Fishel, S. L. Aukerman, and J. L. Culleton

Recombinant DNA technology frequently entails the introduction of genes from a given eukaryotic species into bacteria of the cells of another eukaryotic species. In cases where the foreign genes express enzymatic functions, distinguishing the heterospecific enzyme from its endogenous counterpart requires characterization of the gene products by several biochemical criteria.[1–4] We report here a method that is a modification of the procedures of O'Farrell,[5,6] which should be useful in identifying and distinguishing heterospecific enzymes on the basis of subunit molecular weight and isoelectric point differences. This method is based on the observation that sodium dodecyl sulfate (SDS) is effectively re-

[1] V. Struhl and R. W. Davis, *Proc. Natl. Acad. Sci. U.S.A.* **74,** 5255 (1977).

[2] D. Vapnek, J. A. Hautala, J. W. Jacobson, N. H. Giles, and S. R. Kushner, *Proc. Natl. Acad. Sci. U.S.A.* **74,** 3508 (1977).

[3] R. C. Dickson and J. S. Marken, *Cell* **15,** 123 (1978).

[4] M. Wigler, A. Pellicer, S. Silverstein, and R. Axel, *Cell* **14,** 725 (1978).

[5] P. H. O'Farrell, *J. Biol. Chem.* **250,** 4007 (1975).

[6] P. Z. O'Farrell, H. M. Goodman, and P. H. O'Farrell, *Cell* **12,** 1133 (1977).

ISBN 0-12-181968-X

moved from proteins in high concentrations of urea and NP40 during isoelectric focusing (IEF)[7] and that, at least in solution, proteins can be successfully renatured from urea.[8] In this new procedure, proteins are denatured in SDS and mercaptoethanol, resolved on SDS acrylamide gels in one dimension, and then subjected to IEF in urea in a second dimension. If IEF alone provides satisfactory resolution, the first dimension may be omitted. After IEF the enzymes are renatured and localized by the use of specific histochemical stains. The method is sensitive enough to detect small amounts of enzyme in crude extracts and to distinguish polypeptides differing by only a single charged amino acid.

The characterization of colonies containing specific recombinant DNA fragments is often accomplished by translating complementary mRNAs into proteins. This has been limited mainly to genes that synthesize abundant mRNAs whose translation products can be easily identified. Since this method allows localization of cell-free translation products,[9] it should also permit the analysis of genes coding for enzymes that cannot be easily identified on high-resolution gels.

Materials and Reagents

Materials

Enzymes. Uridine diphosphoglucose (UDPG) pyrophosphorylase (EC 2.7.7.9) was obtained from *Dictyostelium discoideum* strain A_3, grown anenically in HL5,[10] and lyophilized. The enzyme is a homomeric octamer with a molecular weight of 400,000 daltons. Both crude extracts (155 units/mg) and partially purified preparations (17,400 units/mg) were used. To purify the enzyme partially, crude extracts were fractionated in streptomycin sulfate (2.5%) and ammonium sulfate (44–60% precipitate) and were further purified by gel filtration and hydroxyapatite chromatography. At this stage, the enzyme was purified approximately 110-fold and was stable if kept frozen in tricine (25 m*M*, pH 7.5) and glycerol (10%). Crude extracts containing UDPG pyrophosphorylase from cells which were allowed to differentiate on Millipore filters for 16 hr[11] were also used in this work. The differentiating sorocarps were washed off the filters and lysed in SDS sample buffer (20 μl/10^7 cells).

Homogeneously pure alcohol dehydrogenase (ADH^F and ADH^D, EC

[7] G. P. Tuszynski, S. R. Baker, J. P. Fuhrer, C. A. Buck, and L. Warren, *J. Biol. Chem.* **253**, 6092 (1978).

[8] K. Weber and D. J. Kuter, *J. Biol. Chem.* **246**, 4504 (1971).

[9] B. Fishel and R. Dottin, in preparation.

[10] W. F. Loomis, Jr., *Exp. Cell Res.* **64**, 494 (1971).

[11] M. Sussman, *Methods Cell Physiol.* **2**, 397 (1966).

1.1.1.1; a gift from W. Sofer) was isolated from *Drosophila melanogaster*. This enzyme is a homomeric dimer of 44,000 daltons. Homogeneously pure lactate dehydrogenase (LDH-B_4^a and LDH-B_4^b, EC 1.1.1.27; a gift from A. Place) was isolated from *Fundulus heteroclitus*. LDH is a homomeric tetramer of 140,000 daltons. Both enzymes were dialyzed in tricine (25 m*M*, pH 7.5) and glycerol (10%) and assayed before use. ADH was 95% inactivated during dialysis.

Rabbit muscle creatine phosphokinase (CPK, 115 units/mg, EC 2.7.3.2) was purchased from Sigma as a lyophilized powder. This enzyme is a homomeric dimer of 81,000 daltons. CPK was resuspended in 20% glycerol before use.

Reagents

Acrylamide and bisacrylamide were obtained from Eastman; the acrylamide was recrystallized from chloroform. *N*,*N*-Diallyltartardiamide (DATD) was purchased from Bio-Rad. Urea solutions were deionized using Bio-Rad AG 50-1-X8(D) mixed-bed resin. Carrier ampholytes were obtained either from Bio-Rad or were gifts from W. Love. Tris was purchased from Schwartz-Mann. Enzymes used in the gel-staining reactions (hexokinase, glucose-6-phosphate dehydrogenase, phosphoglucomutase) were purchased from Boehringer Mannheim. All other reagents used in histochemical staining were obtained from Sigma.

Methods

Assays

UDPG pyrophosphorylase was assayed by modification of the procedure of Dimond *et al.*[12] Phosphoglucomutase and glucose-6-phosphate dehydrogenase were used at 0.08 and 0.35 unit/ml, respectively. One unit is defined as the amount of enzyme necessary to hydrolyze 1 nmol of UDPG at 37°. Staining of gels for UDPG pyrophosphorylase activity was done in reaction solutions containing *p*-iodonitrotetrazolium violet (INT; 400 μg/ml) and phenazine methosulfate (PMS, 40 μg/ml) in addition to the assay components. Staining reactions were done in the dark at room temperature with shaking. After staining, gels were fixed in acetic acid (10%), unless they were to be used for quantitation studies.

ADH was assayed as described by Vigue and Sofer.[13] The staining solution contained potassium phosphate buffer (50 m*M*, pH 7.5), β-NAD^+ (2.44 m*M*), 2-butanol (3.5%), INT (450 μg/ml), and PMS (40 μg/ml).

[12] R. L. Dimond, P. A. Farnsworth, and W. F. Loomis, Jr., *Dev. Biol.* **50,** 169 (1976).
[13] C. Vigue and W. Sofer, *Biochem. Genet.* **11,** 387 (1974).

LDH was assayed and stained as described by Shaw and Prasad,[14] except that the pH of the Tris-HCl buffer was 8.0 and INT was used instead of NBT (300 μg/ml). One unit of ADH or LDH activity is defined as the amount of activity required to reduce 1 nmol of NAD^+ at 37°.

CPK was not assayed prior to use. One unit of activity is defined as the amount of enzyme required to produce 1 μmol of ATP per minute from phosphocreatine at pH 7.4 at 30°. Gel staining was done as described by Harris and Hopkinson[15]; method B, without agar, was used, and INT was substituted for methylthioazolyl tetrazolium (MTT).

Gel Systems

Denaturing IEF. Gels (10 × 0.4 cm) were prepared and run by a modification of the procedure of O'Farrell.[5] They contained urea (7 *M*, 7.5 *M*, 8 *M*, or 9 *M*), NP40 (2%), ampholytes (2%), TEMED (0.15%, ammonium persulfate (0.02%), acrylamide (4.71%), and bisacrylamide (0.27%). The gels used in the quantitation studies contained acrylamide (5%) and DATD (0.83%). Routinely, 2.0-cm plugs containing acrylamide (7.5%) and glycerol (50%) were cast at the bottom of the gel tubes.

The anodic and cathodic reservoirs contained phosphoric acid (0.018 *M*) and degassed sodium hydroxide (0.02 *M*). The pH gradient was established by prefocusing for $\frac{1}{2}$ hr at 200 V, $\frac{1}{2}$ hr at 300 V, and $\frac{1}{2}$ hr at 400 V. The top of each prefocused gel was washed with a sample loading solution containing urea (same concentration as in the gel), NP40 (2%), dithiothreitol (DTT, 1 m*M*), and ampholytes (1%—same ratio as in the gel).

Samples were mixed with an equal volume of SDS sample buffer containing either Tris (6.25 m*M*, pH 6.8) or tricine (25 m*M*, pH 7.5), glycerol (30%), SDS (1 or 2%), and mercaptoethanol (0.5 or 1%). The SDS/protein ratio was usually greater than 40:1. SDS-treated samples were heated to at least 70° for 3 min and diluted in 3 volumes of urea (10 *M*). Prior to loading, the samples were further diluted in sample loading solution such that the final SDS concentration was less than 0.05%. In some experiments, UDPG pyrophosphorylase was not treated with SDS prior to focusing; these samples were denatured in urea by mixing with 100- to 200-μl of sample loading solution. Samples were loaded at either the acidic or the basic end and were focused at 400 V for up to 720 V-h/cm of gel. The gels were cooled during focusing by circulating water (13° for 7 and 7.5 *M* gels, 20° for 8 and 9 *M* gels).

To measure the pH gradients, gels were run in parallel, rinsed in dis-

[14] C. R. Shaw and R. Prasad, *Biochem. Genet.* **4,** 297 (1970).

[15] H. Harris and D. A. Hopkinson, "Handbook of Enzyme Electrophoresis in Human Genetics." North-Holland Publ., Amsterdam, 1977.

tilled water, and sliced into 1-cm segments. Each segment was equilibrated overnight at room temperature in 0.5 ml boiled, distilled H_2O.[16] Similar results were obtained in direct measurements using a contact electrode (Ingold Model 6122).

Native IEF. Gels for the focusing of native enzymes were prepared and run as described above, except that urea was omitted. These gels were loaded at the basic end.

SDS Polyacrylamide Gel Electrophoresis (PAGE). Ten percent polyacrylamide gels (10.5 × 0.4 cm) with 2.5 × 0.4 cm stacking gels were prepared according to the procedure of Laemmli.[17] Bromphenol blue was used as the tracking dye, and the gels were run at a constant voltage of 50 V until 5–6 hr after the tracking dye had migrated off the end of the separating gel. If the gels were to be stained for protein, they were incubated for 12 hr in isopropanol (25%), acetic acid (10%), and Coomassie blue (0.025%); 12 hr in isopropanol (10%), acetic acid (10%), and Coomassie blue (0.0025%); 12 more hr in acetic acid (10%) and Coomassie blue (0.0025%) and then destained in acetic acid (10%).

SDS–Urea PAGE. These gels were prepared and run as described for SDS PAGE, except that the separating gel contained 5.5 *M* urea and the stacking gel contained 2.7 *M* urea.[18]

SDS PAGE IEF Two-Dimensional System. SDS gels were run in the first dimension as described and were equilibrated for the second dimension (4 hr) in urea (7 *M*) and NP40 (2%). The second dimension consisted of a 5% polyacrylamide slab gel (14 × 9.5 × 0.3 cm) containing urea (7 *M*), NP40 (2%), and ampholytes (2%; ratio: 20% pH 3–5, 60% pH 3–10, 20% pH 9–10). This was supported at its base with a 2-cm plug of acrylamide (10%) and glycerol (50%). The tubular first-dimension gels were sealed onto the slab with a small amount of IEF gel solution (same composition as the slab) and then overlayed with agarose (2%). The acidic and basic reservoir solutions were as described above and were changed every 10 hr. The slab gel was focused at 100 V for 10 hr, and then the first-dimensional gel was removed. Focusing was continued at 250 V for an additional 22 hr. The slab gels were rinsed in distilled water, and the pH gradient was measured with a contact electrode. After staining for enzyme activity, the gels were either stained for protein or fluorographed to detect proteins labeled with [^{35}S]methionine.[19]

SDS–Urea PAGE IEF Two-Dimensional System. This system is similar to the one described above, except that the first-dimensional gel contained urea as described by Storti and Rich.[18] After electrophoresis, the

[16] P. G. Rhigetti and J. W. Drysdale, *Lab. Tech. Biochem. Mol. Biol.* **5,** 515 (1976).
[17] U. K. Laemmli, *Nature (London)* **227,** 680 (1970).
[18] R. V. Storti and A. Rich *Proc. Natl. Acad. Sci. U.S.A.* **73,** 2346 (1976).
[19] W. H. Bonner and R. A. Laskey, *Eur. J. Biochem.* **46,** 83 (1974).

first-dimensional gel was equilibrated in urea (7 *M*) and NP40 (2%) for 2 hr.

Renaturation Conditions

After denaturation in SDS and/or urea and IEF in urea and NP40, a number of homomeric enzymes may be renatured *in situ* in polyacrylamide gels in the following manner. The urea is gradually removed by shaking the gels in four changes of renaturation buffer (10 ml for tubular gels, 150 ml for slab gels) at room temperature. The renaturation buffer is changed every hour. The composition of the renaturation buffer depends on the enzyme being studied. All except UDPG pyrophosphorylase require the presence of sulfhydryl-reducing agents during renaturation. The concentration of reducing agent should be kept at a minimum when the staining solutions contain PMS and tetrazolium salts. These compounds are nonenzymatically reduced by sulfhydryl-reducing agents.[20] Therefore, a reducing agent is usually present in the renaturation buffer for only the first two buffer changes.

The compositions of the individual renaturation solutions are:

1. *UDPG pyrophosphorylase:* Tricine (25 m*M*, pH 7.5), NaCl (200 m*M*), glycerol (10%), UTP or uridine (1 m*M*)
2. *CPK:* Tris-HCl (50 m*M*, pH 7.5), NaCl (200 m*M*), glycerol (10%), creatine phosphate or ADP (2 m*M*) with or without DTT (1 m*M*)
3. *ADH:* Potassium phosphate buffer (50 m*M*, pH 7.5), NaCl (200 m*M*), glycerol (10%), β-NAD^+ (1 m*M*) with or without DTT (1 m*M*)
4. *LDH:* Tris-HCl (75 m*M*, pH 8.0), NaCl (200 m*M*), glycerol (10%), β-NAD^+ (1 m*M*) with or without DTT (1 m*M*)

Localization of Denatured Enzymes in Tubular Gels

Once renatured, homomeric enzymes may be localized in polyacrylamide gels using standard histochemical staining procedures. Examples of this are shown in Figs. 1 and 2. Gel a in both Figs. 1 and 2 shows a specific band due to *Dictyostelium* UDPG pyrophosphorylase activity using a partially purified enzyme preparation. Gel b in Fig. 2 demonstrates the specificity of the staining reaction. When the substrate UDPG is omitted from the staining solution, no enzyme stain is detected. Gel b in Fig. 1 shows a gel that has been stained for protein. Numerous bands are seen, demonstrating that this procedure is useful for identifying specific proteins after resolving them in IEF gels. We have tested other enzymes in order to determine the general applicability of this procedure for localizing enzymes *in situ*. Gel c in Fig. 2 shows that CPK can be renatured and localized.

[20] R. Manrow, unpublished observations.

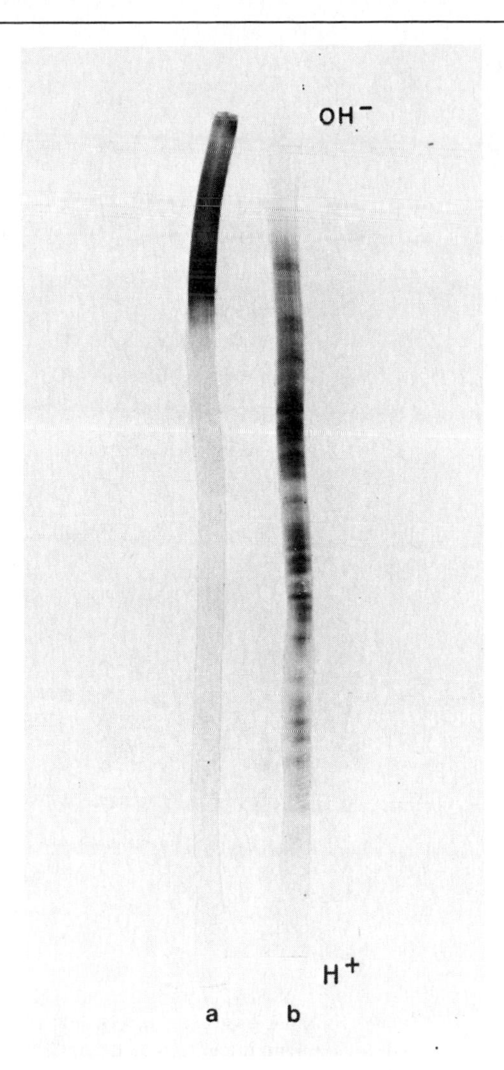

FIG. 1. IEF of urea-treated UDPG pyrophosphorylase. Samples containing 400 units of partially purified UDPG pyrophosphorylase from *D. discoideum* were treated with 7 *M* urea and loaded at the acidic ends of 12.5-cm IEF gels containing urea (7 *M*), NP40 (2%), and pH 3–10 carrier ampholytes (2%). Electrophoresis was done at 500 V for 9000 Vh at 13°. (a) Gel equilibrated for 2 hr in tricine (25 m*M*, pH 7.5) and glycerol (10%) and stained as described under Assays. (b) Gel stained with 0.025% Coomassie blue.

When a higher concentration of enzyme is loaded, an additional band of activity with an isoelectric point 0.2 pH units more basic is observed (data not shown).

This procedure is capable of distinguishing between isoelectric variants of the same enzyme differing by only a single amino acid. This is

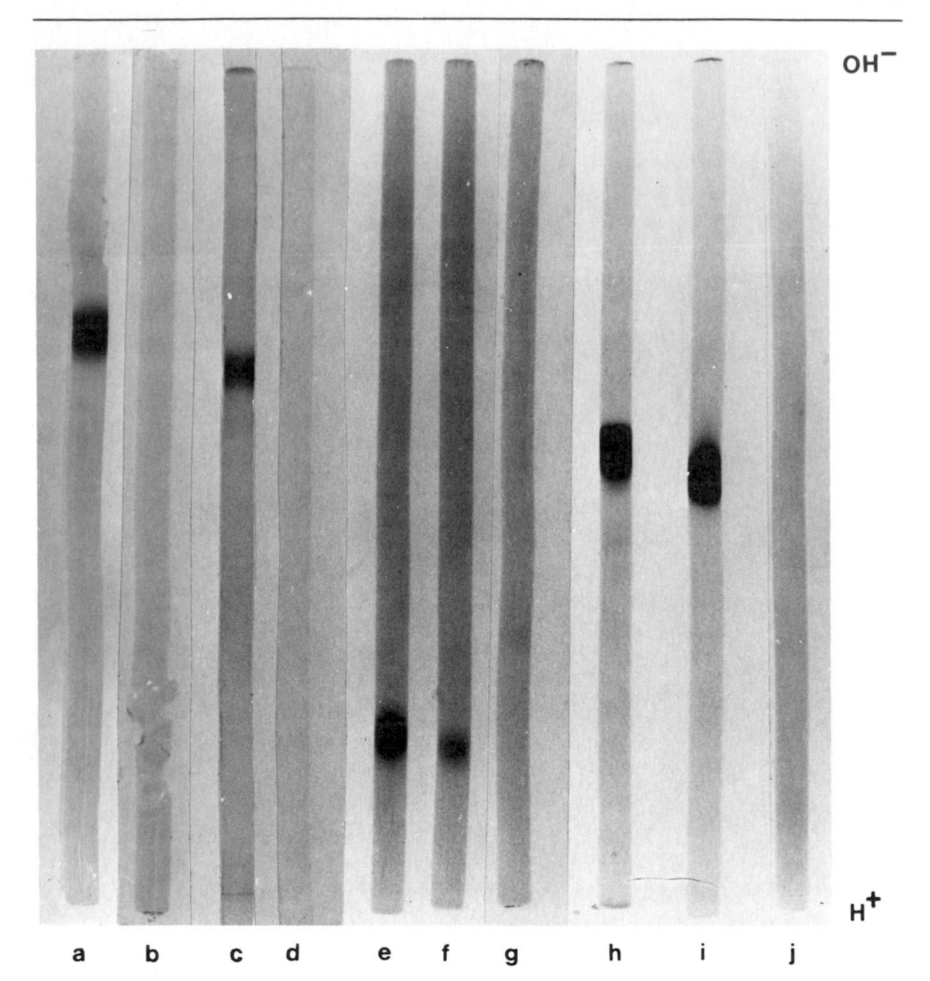

FIG. 2. Renaturation and localization of UDPG pyrophosphorylase, CPK, ADH, and LDH after focusing in urea. Samples containing varying amounts of activity were denatured in SDS and/or urea and focused in gels containing urea as described under Gel Systems. After renaturation, the gels were stained in complete staining solutions or in solutions lacking specific substrates. (a) UDPG pyrophosphorylase, 100 units, urea denatured; gel contained 7 *M* urea and ampholytes in the ratio 60% pH 3–10, 40% pH 9–11. (b) Same as (a); staining solution lacked UDPG. (c) CPK, 2 units, SDS-denatured; gel contained 8 *M* urea and ampholytes in the ratio 80% pH 3–10, 20% pH 3–5. (d) Same as (c); staining solution lacked ADP. (e) ADH^F, 1 unit, SDS-denatured; gel contained 7 *M* urea and ampholytes in the ratio 50% pH 3–10, 50% pH 7–9. (f) ADH^D, 10 units; same conditions as in (e). (g) ADH^F, 10 units; same conditions as in (e); staining solution lacked butanol. (h) LDH-$B_4{}^b$, 100 units, SDS-denatured; gel containing 7 *M* urea and ampholytes in the ratio 90% pH 3–10, 10% pH 9–10. (i) LDH-$B_4{}^a$, 100 units; same conditions as in (h). (j) Same as (i); staining solution lacked lactate.

shown in Fig. 2 (gels e and f). Gel e contains *Drosophila* ADH^F, the wild-type form; gel f contains ADH^D, a mutant form in which one glutamic acid residue has replaced a glycine residue.[21] Consistent with this modification, ADH^D focuses to a pH 0.05–0.10 pH units more acidic than that observed for ADH^F.[22] Similarly, gel h contains LDH-$B_4{}^b$, and gel i contains LDH-$B_4{}^a$, an isoelectric variant (see Fig. 2).

Figure 2 also shows that the staining reactions for CPK (gel d), ADH (gel g), and LDH (gel j) are specific for these enzymes. The substrates ADP, butanol, and lactate, respectively, were omitted from the staining solutions for these gels.

Localization of Enzymes in Two-Dimensional Gels

For purposes requiring resolution greater than that provided by IEF alone, a two-dimensional slab gel technique may be used. This provides SDS electrophoresis in the first dimension and IEF in the second dimension. An example of this procedure for the localization of UDPG pyrophosphorylase from a crude extract of *Dictyostelium* is shown in Fig. 3. Two types of gels were used in the first dimension, one as described by Laemmli[17] (Fig. 3A), and the other as described by Storti and Rich[18] (Fig. 3B). Molecular weight markers and Coomassie blue-stained first-dimensional gels are also shown.

More UDPG pyrophosphorylase activity is recovered when the first-dimensional gel contains urea (Fig. 3B). Since migration in the system is not always according to molecular weight, this higher level of recovered activity may be due to the presence of fewer heterologous, interfering polypeptides in the vicinity of the enzyme subunits at the time of renaturation. Alternatively, this may reflect better equilibration of the first-dimensional gel prior to focusing in the second dimension, thus allowing better subunit focusing.

An example of the resolution of discrete species of abundant cellular proteins in these second-dimensional slab gels is shown in Fig. 3C.

Application of the Two-Dimensional System to Distinguish Isozymes

The high-resolution slab gel technique should be valuable in distinguishing heterospecific enzymes that are isoelectric variants. In addition, it should be useful in characterizing mutant forms having residual activity but different isoelectric points or subunit molecular weights. An example of this is shown in Fig. 4.

[21] M. F. Schwartz and H. Jörnvall, *Eur. J. Biochem.* **68,** 159 (1976).

[22] R. E. Manrow and R. P. Dottin, *Proc. Natl. Sci. U.S.A.*, in press.

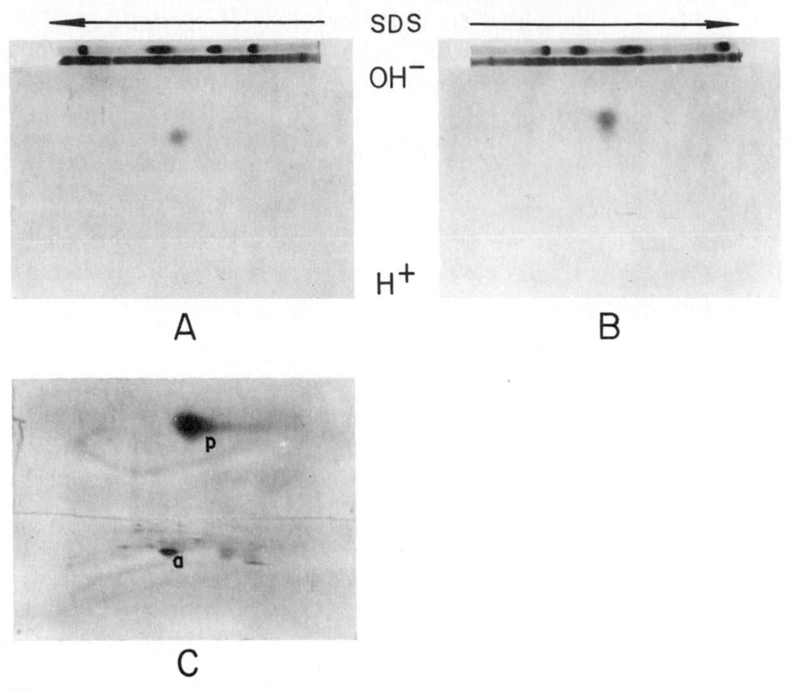

FIG. 3. Localization of UDPG pyrophosphorylase in two-dimensional gels. Crude extracts containing 150 units of activity were run in the first dimension on gels containing either SDS (A) or SDS and urea (B). The second dimensional gels were IEF slab gels containing 7 *M* urea and ampholytes in the ratio 20% pH 3–5, 60% pH 3–10, 20% pH 9–10. After focusing, the slab gels were equilibrated as described and stained for pyrophosphorylase activity for 90 min. Tubular gels stained for protein show the positions of molecular weight markers (phosphorylase b, 94,000; bovine serum albumin, 67,000; ovalbumin, 43,000; carbonic anhydrase, 30,000) and the resolution of *Dictyostelium* proteins in the first dimension. (C) a gel similar to that shown in (A) stained for enzyme activity for 3 hr and subsequently stained for protein with Coomassie blue. Enzyme stain, p; actin, a.

This experiment shows that a new isozyme of *Dictyostelium* UDPG pyrophosphorylase appears in developing cells and is detectable without prior purification of a crude extract. Protein from vegetative cells and from cells isolated from culminating sorocarps was denatured in SDS and mercaptoethanol and electrophoresed on Laemmli first-dimensional gels. The region between 43,000 daltons (ovalbumin) and 67,000 daltons (bovine serum albumin) was cut out of these gels, and the slices were placed on a single slab gel for isoelectric focusing in the second dimension. The focused enzyme was subsequently renatured and stained for activity. Figure 4 (lane b) shows that vegetative extracts contain only a single discrete species of pyrophosphorylase activity. Developing cells, however, contain an additional form which is clearly resolved from the vegetative form even on the steep pH gradient used (Fig. 4, lane a). Lane c in Fig. 4 shows the location of partially purified pyrophosphorylase derived from a vegetative cell crude extract.

It should be noted that these type of data may not be obtainable from native gels stained by histochemical procedures because the resolution is too low. It is the ability of this system to exploit the high resolution of both SDS electrophoresis and IEF that makes it useful. Moreover, in the example cited above, no prior purification of the starting material is required. Therefore, the possibility of artifactual alteration of enzymes during the attempts to separate the isozymes is minimized.

We do not know if the two isozymes are coded for by separate genes, or if the developmental isozyme arises by translation of an altered mRNA molecule or posttranslational modification of the vegetative enzyme. The molecular origin of these two species is currently being investigated.

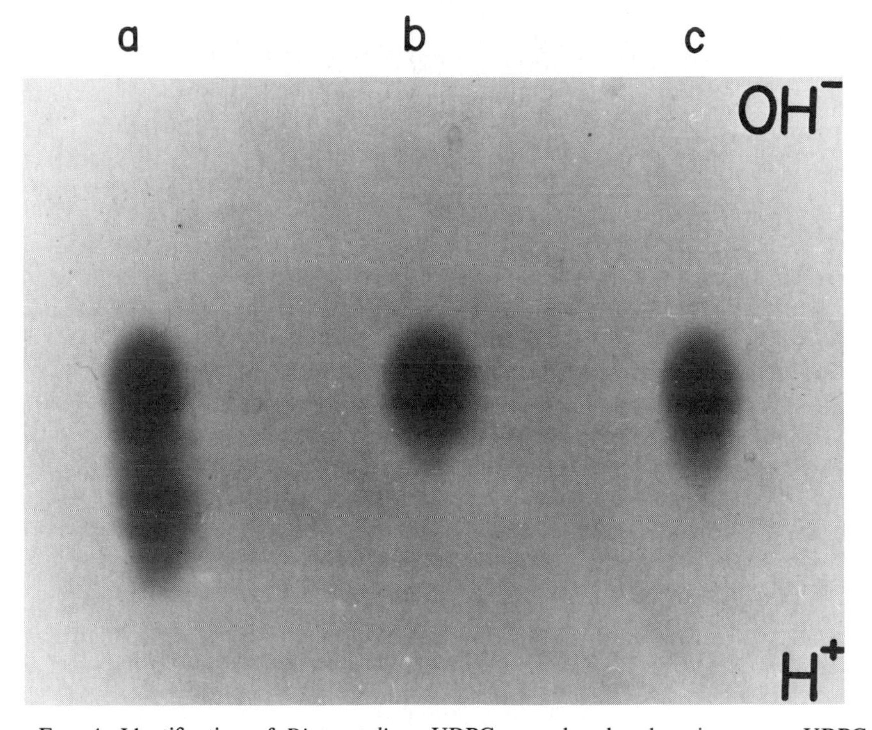

FIG. 4. Identification of *Dictyostelium* UDPG pyrophosphorylase isozymes. UDPG pyrophosphorylase was run on tubular SDS gels as described under Gel Systems. The region corresponding to molecular weights between 43,000 and 67,000 daltons was cut out of these gels, and the slices were applied to an IEF slab gel as described in the text and under Gel Systems. The slab gel contained 8 *M* urea and ampholytes in the ratio 60% pH 3–10, 10% pH 3–5, and 30% pH 9–10. (a) Crude extract, 16-hr developing cells, approximately 12 units. (b) Crude extract, vegetative cells, 4 units. (c) Partially purified enzyme, 4 units.

Quantitation of Enzyme Stain

In some situations it is desirable to quantitate the amount of enzyme stain deposited in these gels. A method for doing so has been developed for enzymes that utilize tetrazolium salts in their staining reactions. To date, this quantitation of enzyme staining has been done for only one enzyme, *Dictyostelium* UDPG pyrophosphorylase, focused in tubular gels. The procedure for this quantitation is:

1. Polyacrylamide IEF gels are prepared and run as described above, except that the periodate-cleavable cross-linking reagent DATD[23] is used instead of bisacrylamide.
2. After staining for a defined period of time, during which the amount of stain deposited is linear with time, the gels are rinsed with cold, deionized water and incubated in water overnight at 4° to remove the glycerol.
3. Segments of gels (2.0–2.6 cm) containing enzyme stain are dissolved in 1 ml sodium *M*-periodate (2%).
4. The released formazan stain is solubilized in *N,N*-dimethylformamide (DMF, final concentration, 75%), and the resulting insoluble material is removed by centrifugation (8000 *g*, 15 min).
5. The supernatant A_{460} is measured, and the total A_{460} is calculated.

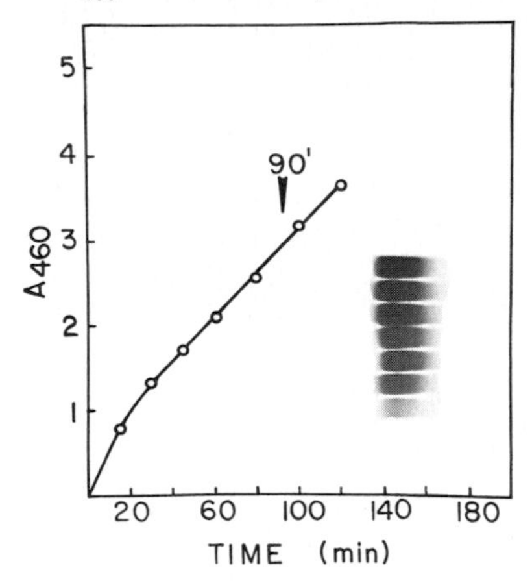

FIG. 5. Kinetics of UDPG pyrophosphorylase enzyme stain deposition. Ten units of partially purified pyrophosphorylase was denatured in urea and focused on tubular DATD-crosslinked gels (see Gel Systems). After renaturation, the gels were stained for definite periods of time in solutions containing fivefold higher concentrations of phosphoglucomutase and glucose-6-phosphate dehydrogenase than those used in the standard staining solution (see text and Fig. 6). Two-centimeter sections containing enzyme stain were dissolved in sodium *m*-periodate, diluted in DMF and centrifuged as described in the text. The supernatant A_{460} for each sample was measured, and total yields of absorbing material were computed. The inset shows the gel section corresponding to each time point.

[23] H. S. Anker, *FEBS Lett.* **7**, 293 (1970).

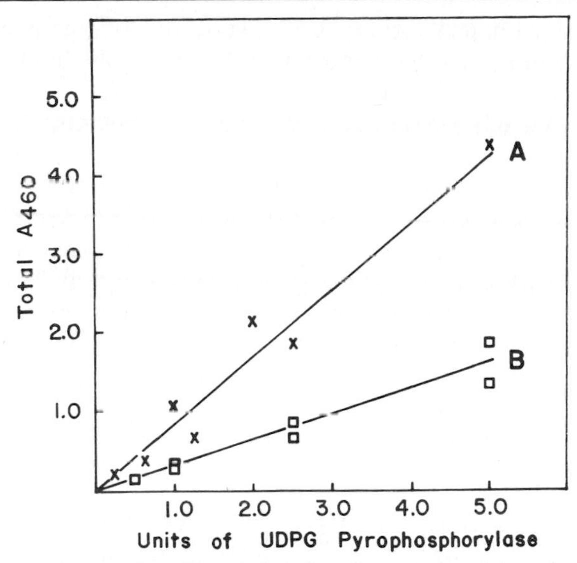

FIG. 6. Quantitation of UDPG pyrophosphorylase renatured in polyacrylamide gels. DATD-crosslinked gel sections (2.0–2.6 cm) containing enzyme stain were dissolved and treated as described in the text and the legend to Fig. 5. Curve A (x): Native enzyme polymerized in gels lacking urea such that a 2.0-cm section contained a defined amount of activity; the ampholyte ratio was 40% pH 3–10, 60% pH 9–10. Curve B (□): Urea denatured enzyme focused in gels containing 7 *M* urea and ampholytes in the ratio 60% pH 3–10, 40% pH 9–10.

We have determined that 460 nm is an absorbance maximum for *p*-iodotetrazolium formazan in 75% DMF (data not shown). An adjacent section of each gel is treated in a similar manner and used as the absorbance blank. An example of the kinetics of enzyme stain deposition is shown in Fig. 5. It can be seen that the amount of stain deposited by 10 units of pyrophosphorylase activity is linear for at least 2 hr.

Quantitation of the recovery of different amounts of applied pyrophosphorylase is shown in Fig. 6. Curve B shows that within the range of activities tested (0.5–5.0 units) the amount of activity recovered in these gels is linearly dependent on the amount of enzyme applied. To achieve this, it was necessary to use concentrations of the coupling enzymes phosphoglucomutase and glucose-6-phosphate dehydrogenase fivefold higher than those used in our standard assay solution. These enzymes couple the formation of glucose 1-phosphate and UTP from UDPG and PP_i by pyrophosphorylase with the reduction of $NADP^+$. When the standard assay concentrations are used, the amount of enzyme stain deposited is nonquantitative with respect to increasing amounts of enzyme activity applied.

Figure 6 also shows the quantitation of stain deposited by native enzyme polymerized in gels lacking urea (curve A). A comparison of curve A with curve B indicates that approximately 40% of the loaded activity is recovered by our procedures.

It is clear that this method of localizing renatured enzymes *in situ* in polyacrylamide gels may be used to detect and quantitate small amounts of enzyme activity. We estimate that approximately 0.05 unit (3.75×10^8 holoenzyme molecules) can be detected in the two-dimensional system using SDS–urea electrophoresis in the first dimension. This amount of activity is within the range synthesized in cell-free translation systems. Although the wheat germ system contains an endogenous pyrophosphorylase activity, it is more acidic than the *Dictyostelium* enzyme (Fishel and Dottin, in preparation). Therefore, it should be possible to detect the synthesis, *in vitro,* of small amounts of activity, since the background activity would be physically separated from the translated activity.

Comments

1. We have observed that the isoelectric point value obtained for a given polypeptide can vary in a complex way with the temperature and the urea concentration used.
2. The equilibrium isoelectric point for SDS-treated UDPG pyrophosphorylase was found to vary depending upon the end of the gel from which it was focused. Enzyme focused from the acidic end gave the same isoelectric point, 8.35, whether it had been SDS-treated or denatured in urea alone. Enzyme focused from the basic end gave variable equilibrium isoelectric points ranging from 8.3 to 8.9.
3. Bisacrylylcystamine (Bio-Rad) was tried as a cleavable gel crosslinking reagent in the quantitation studies; its use was abandoned, however, after we found that the sulfhydryl residues on pyrophosphorylase subunits interacted with the sulfhydryl residues in the gel. The result was the production of artifactual multiple bands of enzyme activity throughout the gel. The nature of these additional species was not determined.

Applications

We suggest that the techniques described above may facilitate the following types of experiments.

1. The identification of heterospecific enzymes in cells receiving foreign genes by somatic cell hybridization, by chromosome or DNA-mediated gene transfer, and by recombinant DNA techniques.
2. The establishment of *in vitro* translation assays for mRNAs that code for enzymes. The translation would be performed with radioactive amino acids and the products resolved on two-dimensional gels. The

radioactive products corresponding to given enzymes could be localized by the inclusion of crude extracts containing these activities. The amount of radioactivity in a given spot would correspond to the relative concentration of a specific mRNA. In cases where the concentration of mRNAs coding for enzyme activities is relatively high, translational products should be active *in situ*.

3. The recognition of restriction fragments that code for enzymes. Once a translation assay is developed, techniques such as hybrid arrested translation (HART) and hybridization selection and translation (HST) of mRNA should be possible.

4. The purification of enzymes for the preparation of antibodies.

5. The identification of mutations affecting the structural genes of enzymes.

6. The characterization, with minimal artifactual alteration, of isozymes in crude extracts.

7. The recognition, in clinical diagnosis, of isozymes and enzyme variants that cannot be adequately resolved on conventional zymograms.

Acknowledgments

We appreciate the valuable suggestions of Stephen Lovell, Peter Knight, and Alan Place. We thank Marc Krauss for assisting in preparing the figures and are grateful to Dorothy Regula for expert typing. This work was supported by a grant from the NSF and a Biomedical Research Support Grant from the NIH. Contribution No. 1033 from the Department of Biology, Johns Hopkins University.

D

Author Index

Numbers in parentheses are reference numbers and indicate that an author's work is referred to although the name is not cited in the text.

E

F

G

I

J

L

N

O

P

Q

R

Y

Z

Subject Index

H

I

K

L

M

N

O

S

T

V

Y